Quality Control for Variable Stiffness Structured Workpiece

Shichang Du · Lifeng Xi

Quality Control for Variable Stiffness Structured Workpiece

Springer

Shichang Du
School of Mechanical and Power Engineering
Shanghai Jiao Tong University
Shanghai, China

Lifeng Xi
School of Mechanical and Power Engineering
Shanghai Jiao Tong University
Shanghai, China

ISBN 978-981-96-5375-1 ISBN 978-981-96-5376-8 (eBook)
https://doi.org/10.1007/978-981-96-5376-8

This Springer imprint is published by the registered company Springer Nature Singapore Pte Ltd.
The registered company address is: 152 Beach Road, #21-01/04 Gateway East, Singapore 189721, Singapore

Foreword

In modern manufacturing, aerospace, automotive, and machinery industries in particular, the demand for precision and reliability has driven the development of increasingly complex workpieces with exacting quality standards. Effective quality control in multistage processes is crucial, especially for variable stiffness workpieces like cylinder blocks and valve shells, where quality deviations can affect performance in subsequent stages. Traditional quality control methods often fall short in addressing these challenges, highlighting the need for advanced techniques to monitor and improve product quality throughout the manufacturing lifecycle.

I am delighted to see that Prof. Shichang Du and Prof. Lifeng Xi have co-authored a book focusing on quality control for variable stiffness structured workpiece. With over 200 peer-reviewed publications in leading journals and conferences, Prof. Du and Prof. Xi have made pioneering contributions to this field, seamlessly blending theoretical advancements with practical industrial applications. Their collaborations with industry leaders, such as the Shanghai Automotive Industry Corporation General Motors Wuling Company (SGMW), have driven significant operational improvements, showcasing the tangible impact of their research. Their unparalleled expertise well positions them to address the critical challenges of quality control for variable stiffness workpieces effectively.

In addressing the challenges associated with variable stiffness workpieces, the book presents innovative solutions for quality control. It explores optimized fixture designs, process parameter tuning, and advanced modeling techniques. Topics under discussion include flatness, waviness, surface monitoring, sealing performance, and state space models. With a comprehensive framework for addressing quality complexities, their work serves as an indispensable resource for advancing quality control in manufacturing. The book offers invaluable insights and practical tools to elevate product quality in complex industrial settings.

Dr. Zhongqin Lin
Academician of Chinese Academy
of Engineering
Chair Professor of School
of Mechanical and Power Engineering
Shanghai Jiao Tong University
Shanghai, China

Preface

Modern manufacturing advancements have facilitated the development of complex variable stiffness structured workpieces, characterized by sophisticated geometries and multistage manufacturing processes. Such workpieces not only present significant challenges in terms of their geometry but also introduce substantial complexity in quality control and evaluation. Surface quality control and assessment are especially critical for these workpieces, as they directly impact the functionality and reliability of the final product.

This book presents a comprehensive analysis of quality control techniques and methodologies relevant to the production of variable stiffness structured workpieces. The book covers a wide range of topics, including geometric error control, process optimization, surface monitoring, and variation management within multistage manufacturing processes. It analyses specific quality challenges encountered in contemporary manufacturing and provides systematic solutions. Particular attention is paid to controlling flatness through optimized fixturing and process parameters, elucidating the mechanisms underlying surface waviness formation via physical modeling, and investigating the application of high-definition measurement technologies to enhance the efficiency of surface quality monitoring.

The structure of the book is based on a series of key quality control issues, including, but not limited to, flatness and waviness, surface monitoring, sealing performance, and variation management in multistage manufacturing processes. Each chapter is dedicated to a specific quality control issue, offering novel solutions that are specifically designed to address the distinctive challenges posed by variable stiffness structured workpieces. The book will be of value to researchers, engineers and students in related fields, offering insights into the application of modern quality control techniques and methods in addressing practical challenges.

In addition to theoretical approaches, the book incorporates numerous real-world case studies, particularly in the area of surface quality control for variable stiffness structured workpieces, offering extensive practical experience. The objective is to facilitate a more profound comprehension of the fundamental aspects of quality control for intricate workpieces and to furnish readers with sophisticated instruments and methodologies to address these challenges.

Furthermore, this work aspires to serve as a point of reference for engineers and technologists engaged in the design and fabrication of variable stiffness structured workpieces. It offers novel perspectives for academic researchers in the domain of quality control and valuable educational resources for students in related disciplines. In addition, for better understanding of the content of this book, readers can refer to *High Definition Metrology Based Surface Quality Control and Applications* (Shichang Du and Lifeng Xi, Springer, ISBN: 978-981-15-0278-1, 2019).

Shanghai, China

Shichang Du
Lifeng Xi

Acknowledgements

We are extremely grateful to the National Natural Science Foundation of China for providing the General Program and Cultivation of Major Research Program (No. 52275499, 92467101), and to the National Key Research and Development Program of China (No. 2022YFF0605700). Many colleagues and students generously volunteered to review sections of the text, offering valuable insights into the techniques discussed and contributing excellent references. We extend our sincere gratitude to the following people: Zhongqin Lin, Jun Ni, Ershun Pan, Sun Jin, Juntong Xi, Zhenqiang Yao, Min Xu, Yongxiang Hu, Yixi Zhao, Lu Chen, Tangbin Xia, Xueping Zhang, Sen Wang, Zhiyuan Yuan, Xiaobo Chen, Bo Zhang, Yanting Li, Guolong Wang, Di Wang, Yongxiang Li, Dong Wang, Xufeng Yao, Delin Huang, Guilong Li, Changping Liu, Lan Fei, Rui Xu, Rui Zhang, Tao Liu, Ke Song, Ziwei Lin, Ziren Zhao, Heng Wang, Tingting Song, Kun Wang, Yaxiang Yin, Yafei Deng, Chen Zhao, Xiaoxiao Shen, Yulu Zhou, Shuo Wang, Xinyu Chen, Xufeng Ji, Peijia Wu, Shiyao Jia, Xuechun Chen, Zhiyuan Xie, Yulun Fang, Yushuo Li, Molin Liu, Fangrui Li, Danyue Shen, Yuheng Wang, Wenhao Yao, and Liang Yan. The above people all come from Shanghai Jiao Tong University. We are grateful for Jun Lv from East China Normal University, Jianjun Shi from Georgia Institute of Technology, Yiping Shao from Zhejiang University of Technology, Shanshan Li and Xianmin Chen from National Key Laboratory of Structural Integrity, Kaibo Wang, Xiaowei Yue, Yanfu Li, and Chen Zhang from Tsinghua University, Hongzhong Huang and Yu Liu from University of Electronic Science and Technology, Jianguo Wu and Xi Zhang from Peking University, Jianbo Yu from Tongji University, Lin Li from University of Illinois Chicago, and Hui Wang from Florida State University, Shanshan Li, and Xianmin Chen from Aircraft Strength Research Institute of China, and Siyang Wang and Chunming Zhang from Weichai Power Co., Ltd. Special acknowledgement goes to three individuals Yaxiang Yin, Chen Zhao and Yulu Zhou who directly helped in editing and reviewing the entire manuscript draft of the book.

We are also grateful for Shanghai Automotive Industry Corporation General Motors Wuling company (SGMW) in Liuzhou, China for offering an extraordinary environment for the engineering practice of surface quality control. The company's skilled engineers, cutting-edge laboratories, and high-precision machinery provide exceptional opportunities for both research and education. This work has greatly benefited from the company's advanced resources, as well as the dedication and enthusiasm of its managers, engineers, and staff. We extend our sincere thanks for their invaluable support.

Contents

Chapter 1
Introduction

1.1 Definitions of Variable Stiffness Structured Workpieces

With the development of modern manufacturing industries, more and more workpieces with complex geometries are designed and manufactured. Three typical workpieces with complex geometries are shown in Fig. 1.1. The definition of variable stiffness structured workpieces includes three key features: multiple geometrical features, multiple geometrical tolerance constraints, and multi-stage manufacturing processes (denoted as 3M features). The variable stiffness structured workpieces have multiple geometrical features (planar surfaces, cylindrical surfaces, cone surfaces, sphere surfaces, and freeform surfaces, etc.) with multiple geometric tolerance constraints (flatness, waviness, roughness, roundness, and cylindrical degree, etc.) and are manufactured by multiple manufacturing processes. The 3M features of variable stiffness structured workpieces bring challenges for products' quality evaluation and quality control. The quality propagation of multiple quality characteristics through multiple manufacturing processes is complicated. The multiscale geometrical specifications influence the final functions of products in different ways. The variable stiffness problem brought by the variable stiffness structure of the workpieces also makes the manufactured surface quality hard to control.

1.2 Scope and Objectives

This book describes the subjects that almost contain all aspects of quality control challenges for variable stiffness structured workpieces. The framework of the book is shown in Fig. 1.2, and the quality control involves the following aspects.

S. Du and L. Xi, *Quality Control for Variable Stiffness Structured Workpiece*,
https://doi.org/10.1007/978-981-96-5376-8_1

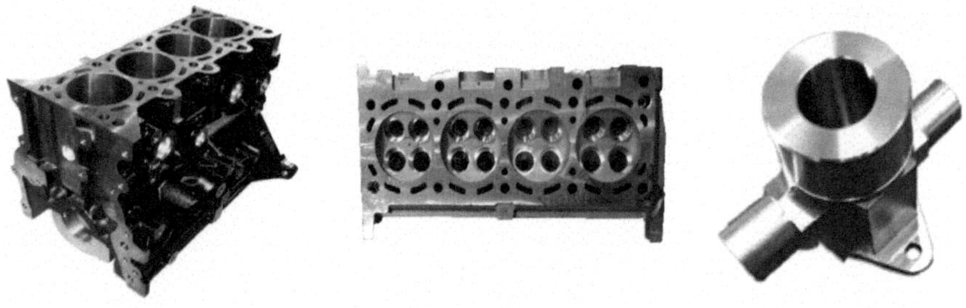

(a) A cylinder block (b) A cylinder head (c) A valve shell

Fig. 1.1 Three typical variable stiffness structured workpieces

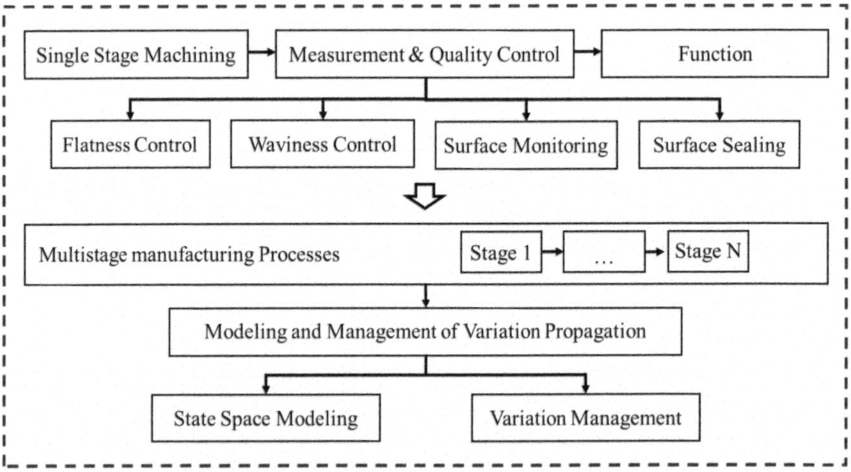

Fig. 1.2 The framework of the book

- **Flatness Control**. The variable stiffness brought by variable geometry makes the face-milled surfaces' flatness hard to control. Based on elastic mechanics and dynamics modeling, the flatness of face-milled surfaces is improved by fixture scheme and process parameters optimization.
- **Waviness Control**. The waviness components of face-milled surfaces are closely related to the trajectories of milling tools. The mechanism of waviness formation is revealed by physical modeling and a segmentation algorithm.
- **Surface Monitoring**. Surfaces' quality such as circularity and cylindricity is vital for workpieces' functions. Based on high definition measurement, surface monitoring and classification methods are proposed to improve the monitoring efficiency of manufacturing processes.

- **Surface Sealing**. The sealing performance of face-milled surfaces is affected by the multiscale surface topography, especially the waviness component. Besides, the complementarity between two rough surfaces is also critical for quality control of mating surfaces' sealing.
- **State Space Modeling**. The variable stiffness structured workpieces are machined in multistage processes. The variation propagation along the production lines is complicated. By improving traditional Stream of Variations theories, a new state space model is built to establish the variation propagation laws of variable stiffness structured workpieces.
- **Variation Management**. The tolerances of variable stiffness structured workpieces are rich and interact with each other, linking with manufacturing processes and costs. To deal with the contradiction between tolerances and costs, a variation management framework is proposed to find the equilibrium point.

1.3 Organization

The whole book is organized into six chapters. In Chap. 2, the flatness of the top surfaces of the cylinder block is focused. Considering the variable stiffness of the geometrical structure, an optimization of the fixture scheme and process parameters is proposed. In Chap. 3, the waviness formed by face-milling trajectories is studied. In Chap. 4, circular and cylindrical profile monitoring and three-dimensional surface classification are discussed. The sealing performance of the top surface of cylinder blocks is detailed in Chap. 5. Chapter 6 is devoted to the modeling and variation management of multistage manufacturing processes.

Chapter 2
Fixturing Scheme and Process Parameters Optimization

2.1 Fixturing Scheme Optimization of Variable Stiffness Structured Workpieces

2.1.1 Introduction

Continuously improving the quality of the surface is a top priority, especially for precision manufacturing. The machined surface quality has always been crucial to meet the requirements of high performance, reliability, and durability [1]. The machined surface deviation is mainly due to the irregular deformation of the workpiece in fixturing and material removing [2]. There is a negative correlation between the machined surface quality and the workpiece deformation caused by different clamping forces and fixture layouts [3]. Thus, the selection of fixturing schemes, i.e., the magnitudes of clamping forces and positions of fixture elements, has a significant influence on the final workpiece quality [4]. Present methodologies in fixturing scheme design depend exclusively on the designer's expertise and traditional guidelines, as opposed to systematic analysis [5]. Therefore, the development of fixturing schemes for precision manufacturing industries is a critical and immediate task. A systematic and scientific approach of optimizing fixturing schemes needs to be implemented to enhance machined surface quality.

Based on the structure stiffness, commonly used workpieces are classified as flexible, rigid, and variable stiffness structure (VSS) workpieces [6, 7]. The stiffness magnitudes of flexible workpieces such as sheet metal, plates, and beams are comparatively low, while the rigidity degree of the workpieces such as non-cavity solid castings is relatively high. Because of the simplicity of the physical construction, the two kinds of flexible and rigid workpieces have the same or slight variation in stiffness. Along with the fast development of the manufacturing industry, workpieces become more and more complicated, and the stiffness magnitude of workpieces has changed greatly. For example, the four-cylinder HT250 grey cast iron engine block is

S. Du and L. Xi, *Quality Control for Variable Stiffness Structured Workpiece*,
https://doi.org/10.1007/978-981-96-5376-8_2

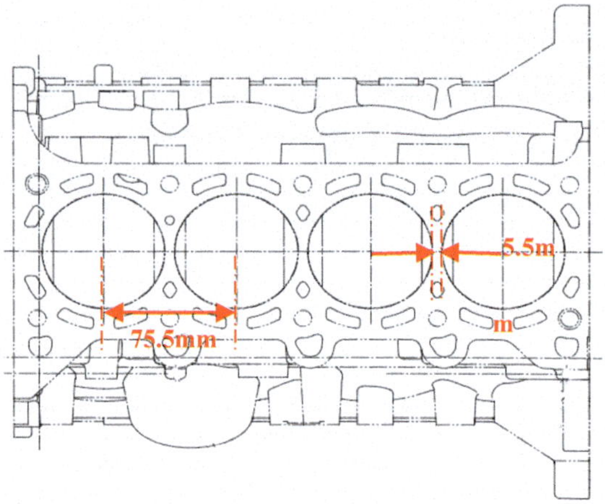

Fig. 2.1 Top view of a four-cylinder engine block

a typical partially thin-walled VSS workpiece [8]. As shown in Fig. 2.1, as bending stiffness is changed between 234.87 N m^2 and 607,537.41 N m^2 [10, 11], the wall thickness varies from 5.5 mm to 75.5 mm [9] Regardless of which class the workpiece falls into, the proper method of deformation analysis is necessary for optimizing the fixturing scheme. In general, the majority of techniques used to analyze deformation fall into two categories: numerical methods and analytical ones [12]. In the past few decades, optimization models for fixturing schemes, particularly those utilizing numerical methods such as the finite element method (FEM), have gained significant attention for their enhanced functionality and adaptability. Nevertheless, the majority of established numerical methods are resource-intensive in terms of computation [13]. Additionally, these methods often operate in a non-transparent manner [14], making it difficult to understand the calculation of deformation and limiting the depth of physical understanding [15]. Conversely, analytical approaches, exemplified by elastic mechanics analysis, can effectively address these limitations.

For sheet metal fixtures, adhering to the "N-2-1" positioning principle, Cai and Hu [16] introduced an approach combining FEM and nonlinear programming to determine the optimal number of positioning points "N", aiming to minimize the deformation of flexible sheet metal. Liu et al. [17] developed a precise cutting force model integrated with a finite element to optimize locator placement during the milling of thin-walled components. Chen et al. [18] formulated a multi-objective model aimed at reducing deformation caused by fixtures for hollow workpieces, utilizing FEM for deformation analysis and a heuristic algorithm for optimization. Rai and Xirouchakis [19] introduced a comprehensive FEM-based verification model for milling processes, predicting deflections and deformations during machining. Bellifa et al. [20] created a new first-order shear deformation theory for the bending and dynamics of functionally graded plates. Ebrahimi and Barati [21] proposed a Navier-type solution for the vibration characteristics of functionally graded nanobeams, based on a third-order shear deformation beam theory. Shan et al. [22] correlated milling force with tool inclination angle in ball-end milling, combining theoretical and experimental methods. They analyzed the impact of tool inclination on milling forces and elastic deformation using FEM and identified the optimal angle for minimal force and deformation.

In the realm of rigid workpieces, Kaya and Ozturk [23] introduced a verification method for workpiece-fixture layouts using FEM, incorporating frictional contact and chip removal dynamics. Carlson [24] applied kinematic analysis to formulate a quadratic sensitivity equation that captures the correlation between locator position errors and the displacement of rigid workpieces within fixtures. Sadd [25] explored plane stress in non-homogeneous isotropic elastic materials, offering straightforward solutions in Cartesian coordinates. Pimenov et al. [26], taking into account milling parameters and tool wear, developed an analytical model to assess flatness deviation and calculate the angular compliance of a face milling system. Wojciechowski [27] examined the surface roughness in hardened steel monolithic torus milling, considering the dynamics of the machining system. They developed a surface roughness model that factors in the impact of cutting forces and tool displacements, highlighting the significant influence of dynamic tool movements on actual surface roughness.

Despite the advancements, research on VSS workpieces remains scarce. Kang et al. [28, 29] utilized computer-aided fixture design verification (CAFDV) to develop a framework that evaluates fixturing stability and locating precision through geometric and kinetic models. Liao and Hu [30, 31] presented an integrated fixture-workpiece system model using FEM, focusing on fixture stability, which was validated through a case study of a V-type engine block. Gong et al. [32] designed an X-ray optics bending mechanism using FEM to minimize surface errors by analyzing the mechanical state and bending resolution.

In conclusion, while numerical approaches to workpiece elastic deformation have been extensively researched, they possess certain limitations that analytical methods could potentially address. However, analytical methods have predominantly been applied to the deformation analysis of flexible and rigid workpieces. The scarcity of analytical methods for deformation calculations has posed a significant challenge to the optimization of fixturing schemes for VSS workpieces, ultimately affecting the quality of machined surfaces.

In this section, a new analytical method grounded in the principles of elastic mechanics has been introduced to enhance the fixturing schemes for VSS workpieces, aiming to enhance surface quality. The core concept involves reducing the clamping forces required for secure fixturing using a static cutting force model, calculating the elastic deformation caused by these forces through elasticity theory, and then employing heuristic algorithms to optimize the model and minimize the deformation affecting the machining surface. The genetic algorithm (GA), inspired by natural selection and reproduction, is a probabilistic, heuristic, and robust optimization technique [33, 34]. As GA focuses solely on the design variables of fixtures and the objective function's fitness value without the need for additional data [35, 36], it is particularly well-suited for addressing the optimization of fixturing schemes.

The subsequent parts of Sect. 2.1 are structured as follows: Sect. 2.1.2 outlines the detailed steps of the proposed method, encompassing force analysis, elasticity analysis, and the optimization model for fixture layout. Section 2.1.3 validates the proposed method's effectiveness and efficiency through a machining experiment on a specific four-cylinder engine block, the B12, comparing the outcomes of the optimized scheme with the original scheme, and includes a FEM simulation of the machining process. Section 2.1.4 summarizes the conclusion of this research.

2.1.2 *The Proposed Approach*

The structure of the proposed method is depicted in Fig. 2.2, with the key steps detailed below:

Step 1: Formulate an exponential model for the static cutting force during the face milling operation to determine the minimum clamping forces required for the mechanical stability of the machining setup.
Step 2: Examine the deformation caused by fixturing on the upper surface of a standard VSS workpiece. Utilizing the superposition principle [37], the VSS elastomer is segmented into three distinct regions: Pier, Span, and Bridge. Following the classical 3-2-1 positioning rule, when a clamping force is applied perpendicularly to the side of a Pier or Span area, the deformation and stress field solutions are calculated using elastic mechanics. The Bridge region's deformation is determined by considering the stress field from the Pier-Bridge interface as the applied load. The displacement fields of each area are computed separately and then combined to represent the total top surface deformation due to fixturing.
Step 3: Construct an optimization model for the fixturing scheme of VSS workpieces, aiming to minimize the surface flatness before machining after fixturing. The primary constraints are the stability of the fixturing and its practical implementation. Given the intricate relationship between the fixturing scheme and the induced surface deformation, which characterizes the problem as a nonlinear programming issue, GA is utilized to address this optimization challenge.

2.1.2.1 Force Analysis

The fundamental role of a fixture is to secure the workpiece in the correct position and orientation throughout the manufacturing process [12]. Force analysis ensures that the fixtures apply adequate forces to keep the workpiece in static equilibrium amidst cutting forces. This research concentrates on the face milling process, a prevalent surface machining technique in the manufacturing sector. Material removal involves a dynamic process influenced by periodic vibrations [38, 39]. However, if the frequency of the cutting forces, specifically the tooth passing frequency [40, 41], significantly differs from the chatter frequencies of the fixture-workpiece system, the effect of these vibrations on the surface quality of the machined part is minimal and can be

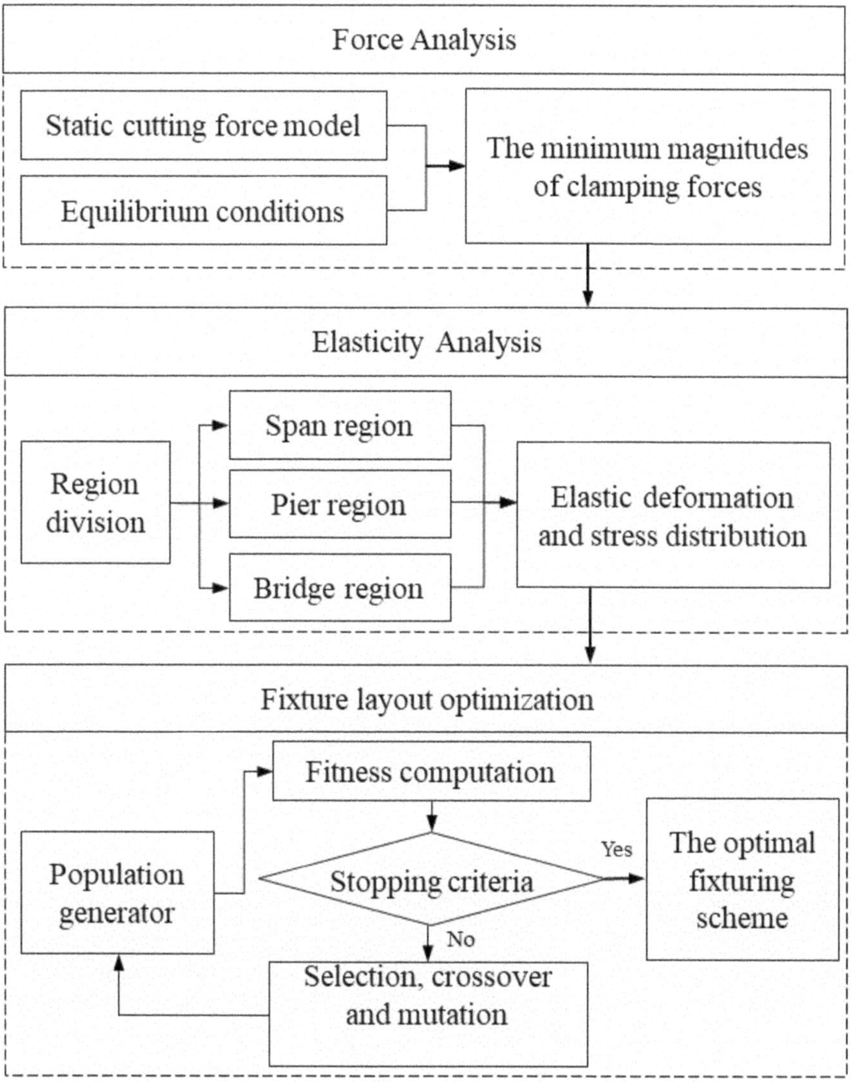

Fig. 2.2 Framework of the proposed approach

disregarded [31]. Consequently, a static model of the cutting force for the face milling operation is established to ascertain the necessary minimum clamping forces.

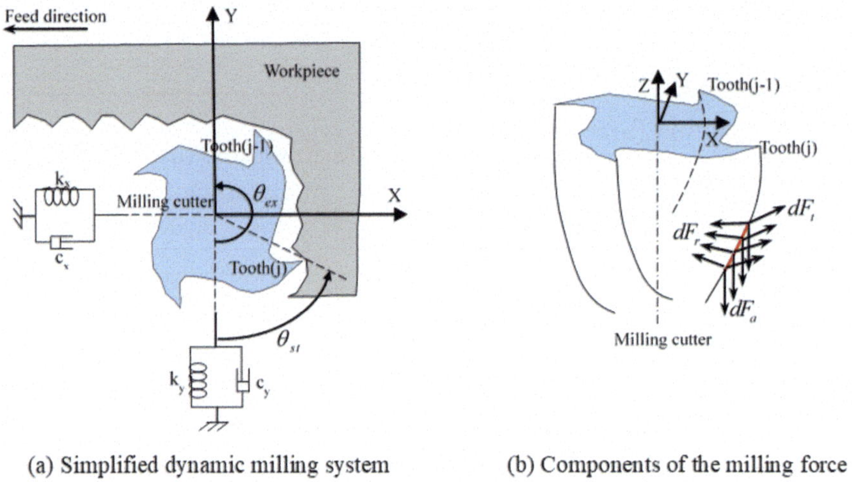

Fig. 2.3 Cutting force modeling for the milling process

Static Cutting Force Model

Typically, the milling system can be approximated as a two-degree-of-freedom vibration system [42] as shown in Fig. 2.3a. Here, θ_{st} represents the initial cut angle and θ_{ex} signifies the exit cut angle. Assuming the milling cutter has N teeth in total, and each tooth's cutting edge is segmented into M microelements, the radial immersion angle for the lth micro element of the cutting edge of jth tooth in contact with the workpiece [42] is indicated as follows:

$$\theta_{j,l} = \theta_{1,0} + \frac{2\pi(j-1)}{N} + \frac{a_{\mathrm{p}} l \tan\beta}{MR_{\mathrm{c}}} \tag{2.1}$$

where $\theta_{1,0}$, a_{p}, β and R_{c} represent the angular position of the first cutting edge's base, axial cutting depth, helix angle of the flank edge, and the ideal cutter radius respectively, for $j = 1, 2, \ldots N$ and $l = 1, 2, \ldots M$.

As shown in Fig. 2.3b, the cutting force acting on the microelement dF can be resolved into three orthogonal components, i.e. the axial force dF_a, the radial force dF_r and the tangential force dF_t.

By summing the force components from all engaged cutting edges and applying a coordinate transformation, we obtain:

$$\begin{cases} F_x = \sum_{j=1}^{N}\sum_{l=1}^{M}[-dF_{t,j,l}\cos\theta_{j,l} - dF_{r,j,l}\sin\theta_{j,l}] \\ F_y = \sum_{j=1}^{N}\sum_{l=1}^{M}[dF_{t,j,l}\sin\theta_{j,l} - dF_{r,j,l}\cos\theta_{j,l}] \\ F_z = \sum_{j=1}^{N}\sum_{l=1}^{M}dF_{a,j,l} \end{cases} \tag{2.2}$$

For the exponential instantaneous model [43, 44], the mean cutting forces along the X, Y, and Z axes can be determined for each rotation of the milling cutter as follows:

$$\begin{cases} F_{A,x} = \frac{a_pN}{2\pi}K_rf_z^{A_x}\int_{\theta_{st}}^{\theta_{ex}}(\sin\theta)^{1-A_x}d\theta \\ F_{A,y} = \frac{a_pN}{2\pi}K_tf_z^{A_y}\int_{\theta_{st}}^{\theta_{ex}}(\sin\theta)^{1-A_y}d\theta \\ F_{A,z} = \frac{a_pN}{2\pi}K_af_z^{A_z}\int_{\theta_{st}}^{\theta_{ex}}(\sin\theta)^{A_z}d\theta \end{cases} \tag{2.3}$$

where K_r,K_t and K_a denote the radial, tangential, and axial cutting parameter coefficients respectively while A_x, A_y and A_z are the corresponding undetermined coefficients, and f_z is the feed per tooth.

By taking the logarithm on both sides of Eq. (2.3), it can be derived that

$$\ln\left(\frac{F_{A,q}}{a_pN}\right) = A_q\ln f_z + B_q \tag{2.4}$$

$$\begin{cases} K_r = \frac{2\pi e^{B_x}}{\int_{\theta_{st}}^{\theta_{ex}}(\sin\theta)^{1+A_x}d\theta} \\ K_t = \frac{2\pi e^{B_y}}{\int_{\theta_{st}}^{\theta_{ex}}(\sin\theta)^{1+A_y}d\theta} \\ K_a = \frac{2\pi e^{B_z}}{\int_{\theta_{st}}^{\theta_{ex}}(\sin\theta)^{A_z}d\theta} \end{cases} \tag{2.5}$$

where B_q, for $q = x, y, z$, is another group of undetermined coefficients substituting K_r,K_t and K_a. Since Eq. (2.4) is obviously a group of linear equations, both A_q and B_q can be obtained by the least square method [45] through multiple sets of experimental tests of cutting forces using different machining parameters. Thus, the average cutting force components along each axis can be calculated by Eq. (2.3) respectively.

Equilibrium Conditions

The primary goal of a fixturing scheme is to precisely position, adequately support, and effectively secure the workpiece throughout the machining process. This is accomplished by strategically positioning fixture components, such as locator pins,

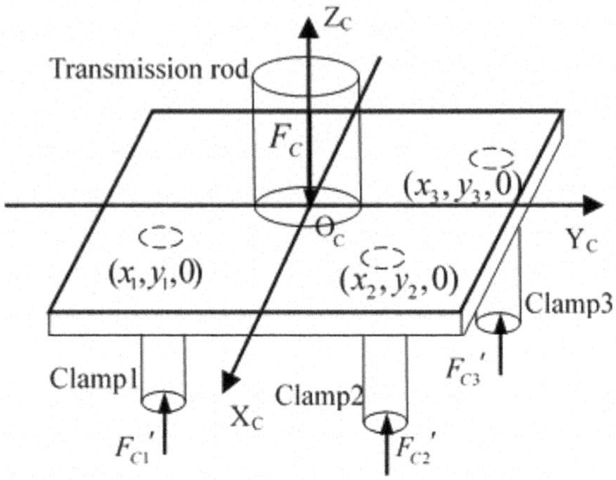

Fig. 2.4 Force distribution of the clamping chuck device

supports, and clamps, around the workpiece and applying clamping forces with suitable magnitude. Overly strong clamping forces can lead to excessive deformation and degraded machining quality, while inadequate forces may compromise the stability of the fixture, potentially causing hazardous incidents. Therefore, it is crucial to determine the clamping forces scientifically. Traditionally, the magnitude of clamping forces in fixture design is largely based on the experience of similar fixtures [46]. In this section, regarding the workpiece-fixture as a unified system, the static equilibrium conditions necessary for fixture stability are analyzed and calculated. To ensure reliable clamping, the theoretical clamping force magnitudes that meet the equilibrium conditions should be adjusted by a safety coefficient ζ_R, typically ranging from 2.5 to 3 for rough machining [47], and from 1.5 to 2 for fine finishing [48, 49]. This can be expressed as follows.

$$F_{C,q} = \zeta_R F_{A,q} \tag{2.6}$$

The clamping force in each direction is delivered by the actuation rod of a clamping chuck device. For example, Fig. 2.4 illustrates a chuck device equipped with three distinct clamps, with their coordinates being $(x_1, y_1, 0)$,$(x_2, y_2, 0)$, and $(x_3, y_3, 0)$ in the three-dimensional Cartesian coordinate system (CCS) $O_C X_C Y_C Z_C$. Consequently, based on the principles of static equilibrium, the distribution of forces for each clamp can be derived by resolving the following system of equations.

$$\begin{cases} F_C = F_{C1} + F_{C2} + F_{C3} \\ F_{C1}|x_1| + F_{C2}|x_2| - F_{C3}|x_3| = 0 \\ F_{C1}|y_1| - F_{C2}|y_2| - F_{C3}|y_3| = 0 \end{cases} \tag{2.7}$$

Similarly, the distribution of clamping forces for other chuck devices equipped with either two or four clamps can also be determined. These optimized clamping forces serve as the loading input for the elasticity analysis model, which is crucial in the proposed methodology.

2.1.2.2 Elasticity Analysis

For VSS workpieces, analyzing the deformation caused by fixturing is both challenging and critical. Taking a simplified four-cylinder engine block as a representative example of VSS workpieces, this research presents a new systematic model grounded in elastic mechanics for deformation analysis, along with the corresponding analytical solutions. Given that the plastic deformation caused by fixturing is considered negligible [12, 29], any deformation mentioned subsequently is assumed to be elastic unless specified otherwise.

Region Division

When a VSS workpiece is subjected to various clamping forces during the fixturing stage before machining, the resulting deformation can be intricate and irregular, making it difficult to analyze and control, which negatively impacts the quality of the machined surface. It is impractical to calculate the displacement field of a VSS workpiece in its entirety analytically. Therefore, a region-based strategy using principles of elastic mechanics is introduced in this section.

Using a four-cylinder engine block as a case study, this section focuses on optimizing the fixturing scheme for face milling, adhering to the 3-2-1 positioning principle. Due to real-world manufacturing constraints, the positions of locator-clamp pairs on the secondary and tertiary positioning planes are typically fixed. In essence, the optimization of fixture layout is primarily focused on the primary positioning plane and its counterpart, which includes three sets of locator-clamp pairs. A key geometric feature of engine blocks is the presence of closely spaced cylinder bores, leading to substantial and continuous variation in wall thickness. Assuming an isotropic elastic modulus, both the flexural and compressive stiffness are directly related to the workpiece's wall thickness, thus creating a variable stiffness structure. For a simplified engine block that retains its critical geometric features (see Fig. 2.5), it is segmented into seven distinct regions by two parallel cuts, labeled Section I and II respectively. Here, d_j (for $j = 1, 2, 3$) represent given distance parameters while a, b, H and R_b donate the length, width, height of the engine block and the radius of the cylinder bore respectively.

Imagine three distinct clamping forces being applied perpendicularly to the sides of region 1, while region 2 interacts with three corresponding locators. Due to their varying stiffness properties, these two areas can be further divided into sub-sections known as Span areas and Pier areas. The other regions, numbered from 3 to 7, are referred to as Bridge areas. The deformation caused by each clamping force in either a Span or Pier area is calculated separately using principles of elasticity. The deformation of Bridge areas is determined by analyzing the elastic behavior of

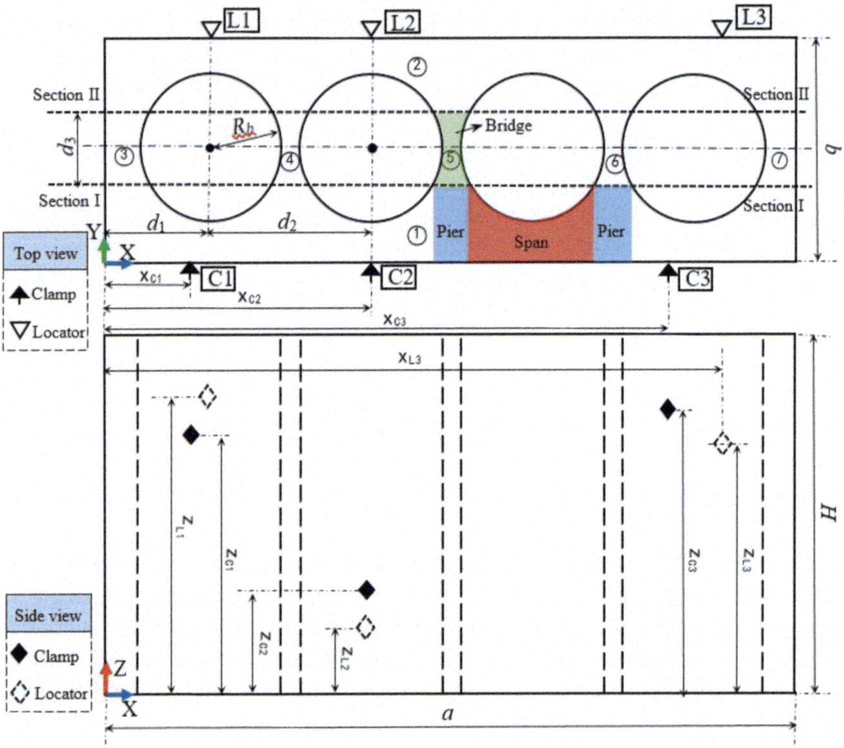

Fig. 2.5 Region division of a simplified four-cylinder engine block

the Span and Pier areas. By applying the superposition principle, the displacement fields of all regions can be combined to yield the overall displacement field after the clamping is in place. It's worth mentioning that the deformation of a VSS workpiece with different geometric characteristics and positioning methods can also be analyzed using a similar elasticity model.

Span Region

The key distinction between a traditional elastic plate of uniform thickness and the Span region lies in the variability of the elastomer's flexural stiffness. Consequently, the Span region is characterized as an elastic plate with fluctuating stiffness. In the scenario where a clamping force is applied perpendicularly to the side of a Span region that adjoins two neighboring Pier regions, as depicted in Fig. 2.6, and referred to as a Pier-Span-Pier configuration, it is necessary to compute the deformation and stress field for this configuration. The system's physical structure is interrupted by the cylinder bore within the Span region, resulting in heightened localized stresses at the interfaces between the Pier and Bridge sections. This occurrence, recognized

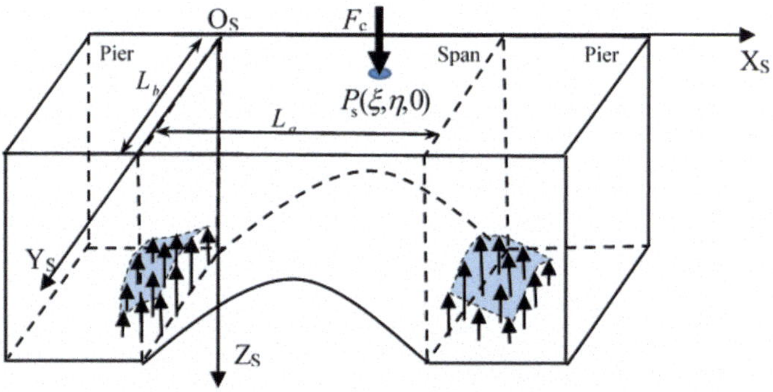

Fig. 2.6 The pier-span-pier system loaded with a perpendicular clamping force

Fig. 2.7 Illustration of the micro element of elastic plate

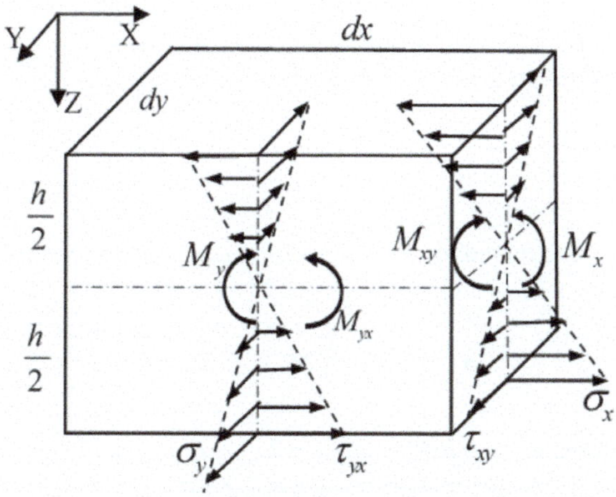

in the academic sphere as stress concentration [50, 51], is typically articulated as:

$$K_{sc} = \frac{\sigma_{max}}{\sigma_{avg}} \tag{2.8}$$

where K_{sc} , σ_{max} and σ_{avg} refer to the stress concentration coefficient, the maximum normal stress, and the average normal stress respectively. For a specific physical structure, K_{sc} is a constant which can be obtained by photo-physical experiments or empirical formulas [52].

An elastic micro element of the elastic plate is shown in Fig. 2.7, where *dx*, *dy*, and *h* denote the length, width, and height of the micro element respectively.

According to [37], the internal moment of this micro element is given by

$$\begin{cases} M_x = \int_{-\frac{h}{2}}^{\frac{h}{2}} \sigma_x z dz = -D\left(\frac{\partial^2 w}{\partial x^2} + \upsilon \frac{\partial^2 w}{\partial y^2}\right) = D\left(\kappa_x + \upsilon \kappa_y\right) \\ M_y = \int_{-\frac{h}{2}}^{\frac{h}{2}} \sigma_y z dz = -D\left(\frac{\partial^2 w}{\partial y^2} + \upsilon \frac{\partial^2 w}{\partial x^2}\right) = D\left(\kappa_y + \upsilon \kappa_x\right) \\ M_{xy} = \int_{-\frac{h}{2}}^{\frac{h}{2}} \tau_{xy} z dz = -D(1-\upsilon)\frac{\partial^2 w}{\partial x \partial y} = D(1-\upsilon)\kappa_{xy} \end{cases} \tag{2.9}$$

where D is the flexural stiffness defined as

$$D = \frac{Eh^3}{12\left(1-\upsilon^2\right)} \tag{2.10}$$

Denoting the external load function as $q(x, y)$, the deflection equation can be represented as

$$\frac{\partial^4 w}{\partial x^4} + 2\frac{\partial^4 w}{\partial x^2 \partial y^2} + \frac{\partial^4 w}{\partial y^4} = \frac{q}{D} \tag{2.11}$$

The boundary conditions for a single Span region can be formulated as

$$\begin{cases} w = 0, \frac{\partial^2 w}{\partial x^2} = 0 \quad \text{for } x = 0, L_a \\ M_y = 0, M_{yx} = 0 \text{ for } y = 0, L_b \end{cases} \tag{2.12}$$

where $L_a = \sqrt{4R_b^2 - d_3^2}$ and $L_b = H$ are the length and width of the Span region respectively and both known constants.

Using the properties of the double trigonometric series [53, 54], under the premise of satisfying the boundary conditions given by Eq. (2.12), the double trigonometric series expansion is applied to the deflection $w(x, y)$ and the external load $q(x, y)$ respectively, and eventually the analytical solution to the deflection is obtained in the form of a function of the given load.

$$w(x, y) = \sum_{m=1}^{\infty}\sum_{n=1}^{\infty} A_{mn} \sin\frac{m\pi x}{L_a} \sin\frac{n\pi y}{L_b} \tag{2.13}$$

$$q(x, y) = \sum_{m=1}^{\infty}\sum_{n=1}^{\infty} C_{mn} \sin\frac{m\pi x}{L_a} \sin\frac{n\pi y}{L_b} \tag{2.14}$$

where A_{mn} and C_{mn} are two undetermined coefficients and both of them can be obtained by the following procedures.

Substituting Eq. (2.13) into Eq. (2.11), yields

$$\pi^4 D \sum_{m=1}^{\infty}\sum_{n=1}^{\infty}\left(\frac{m^2}{L_a^2}+\frac{n^2}{L_b^2}\right)^2 A_{mn}\sin\frac{m\pi x}{L_a}\sin\frac{n\pi y}{L_b}=q(x,y) \tag{2.15}$$

The orthogonality of trigonometric functions is formulated as

$$\int_0^{\frac{L_a}{2}} \sin\frac{m\pi x}{L_a}\sin\frac{m'\pi x}{L_a}dx=\begin{cases}0 & \text{for } m\neq m' \\ \frac{L_a}{2} & \text{for } m=m'\end{cases} \tag{2.16}$$

Multiplying both sides of Eq. (2.14) by $\sin\frac{m\pi x}{L_a}\sin\frac{n\pi y}{L_b}$, and integrating in $[0, L_a]$ and $[0, L_b]$ w.r.t. x and y respectively, yields

$$\int_0^{L_b}\int_0^{L_a} q(x,y)\sin\frac{m\pi x}{L_a}\sin\frac{n\pi y}{L_b}dxdy=\frac{L_a L_b}{4}C_{mn} \tag{2.17}$$

It can be derived that

$$C_{mn}=\frac{4}{L_a L_b}\int_0^{L_b}\int_0^{L_a} q(x,y)\sin\frac{m\pi x}{L_a}\sin\frac{n\pi y}{L_b}dxdy \tag{2.18}$$

Substituting Eq. (2.18) into Eq. (2.14), and then substituting the new equation into Eq. (2.15), the deflection function can be further formulated as

$$\sum_{m=1}^{\infty}\sum_{n=1}^{\infty}\left(\frac{m^2}{L_a^2}+\frac{n^2}{L_b^2}\right)^2 A_{mn}\sin\frac{m\pi x}{L_a}\sin\frac{n\pi y}{L_b}=\sum_{m=1}^{\infty}\sum_{n=1}^{\infty} B_{mn}\sin\frac{m\pi x}{L_a}\sin\frac{n\pi y}{L_b} \tag{2.19}$$

Therein, B_{mn} is another coefficient given by

$$B_{mn}=\frac{4}{\pi^4 D L_a L_b}\int_0^{L_b}\int_0^{L_a} q(x,y)\sin\frac{m\pi x}{L_a}\sin\frac{n\pi y}{L_b}dxdy \tag{2.20}$$

For any $\{(x, y)|0 < x < L_a, 0 < y < L_b\}$, to ensure Eq. (2.19) always holds, coefficients of the trigonometric series on both sides of Eq. (2.19) should be correspondingly equal, thus

$$A_{mn}=\frac{4\int_0^{L_b}\int_0^{L_a} q(x,y)\sin\frac{m\pi x}{L_a}\sin\frac{n\pi y}{L_b}dxdy}{\pi^4 D L_a L_b\left(\frac{m^2}{L_a^2}+\frac{n^2}{L_b^2}\right)^2} \tag{2.21}$$

Substituting Eq. (2.21) into Eq. (2.13), yields

$$w = \frac{4}{\pi^4 L_a L_b D}\sum_{m=1}^{\infty}\sum_{n=1}^{\infty}\frac{\int_0^{L_b}\int_0^{L_a} q(x,y)\sin\frac{m\pi x}{L_a}\sin\frac{n\pi y}{L_b}dxdy}{\left(\frac{m^2}{L_a^2}+\frac{n^2}{L_b^2}\right)^2}\sin\frac{m\pi x}{L_a}\sin\frac{n\pi y}{L_b} \tag{2.22}$$

When a concentrated clamping force F_c, as a special case of external load, is perpendicularly exerted on the point P_s whose coordinates are $(\xi, \eta, 0)$ w.r.t. the CCS $O_S X_S Y_S Z_S$, the deflection can be rewritten as

$$w = \frac{4F_c}{\pi^4 L_a L_b D}\sum_{m=1}^{\infty}\sum_{n=1}^{\infty}\frac{\sin\frac{m\pi\xi}{L_a}\sin\frac{n\pi\eta}{L_b}}{\left(\frac{m^2}{L_a^2}+\frac{n^2}{L_b^2}\right)^2}\sin\frac{m\pi x}{L_a}\sin\frac{n\pi y}{L_b} \tag{2.23}$$

Since the thickness of a Span region $h(x)$ varies continuously along the X-axis direction, it can be formulated as

$$h(x) = \frac{b}{2} - \sqrt{R_b^2 - \left(\frac{L_a}{2} - x\right)^2} \tag{2.24}$$

The flexural stiffness $D(x)$ is consequently derived as

$$D(x) = \frac{E}{12(1-\upsilon^2)}\left[\frac{b}{2} - \sqrt{R_b^2 - \left(\frac{L_a}{2} - x\right)^2}\right]^3 \tag{2.25}$$

Hence, the analytical solution to the deflection function of the Span region is formulated as

$$w = \frac{48F_c(1-\upsilon^2)}{\pi^4 L_a L_b E\left[\frac{b}{2} - \sqrt{R_b^2 - \left(\frac{L_a}{2} - x\right)^2}\right]^3}\sum_{m=1}^{\infty}\sum_{n=1}^{\infty}\frac{\sin\frac{m\pi\xi}{L_a}\sin\frac{n\pi\eta}{L_b}}{\left(\frac{m^2}{L_a^2}+\frac{n^2}{L_b^2}\right)^2}\sin\frac{m\pi x}{L_a}\sin\frac{n\pi y}{L_b} \tag{2.26}$$

Example 1 To demonstrate the proposed method for calculating the deformation of the Span region caused by fixturing before machining, consider a scenario where a concentrated clamping force F_c of 150 N is applied perpendicularly to a point on the flank of the Span region, specifically at point P_s with coordinates (20, 170, 0) relative to the local coordinate system (CCS) $O_S X_S Y_S Z_S$. Assuming the workpiece material is gray cast iron HT250, the dimensions of the Span region and pertinent material properties are summarized in Table 2.1.

According to Eq. (2.26), the deflection w of the Span region is calculated as

Table 2.1 Dimensions of the Span region and material information for Example 1

Symbol	Explanation	Value
L_a	Length of the Span region	64 mm
L_b	Width of the Span region	236 mm
b	Width of the engine block	108 mm
R_b	Radius of the cylinder bore	35 mm
E	Elastic modulus of HT250	140 GPa
υ	Poisson ratio of HT250	0.27

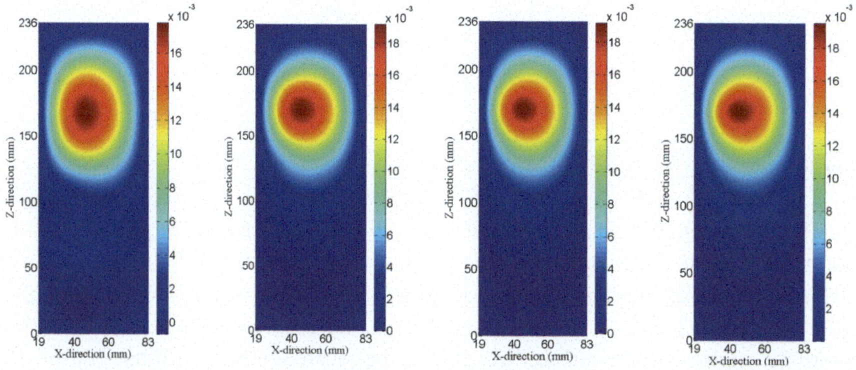

(a) m, n=5 (b) m, n=10 (c) m, n=15 (d) m, n=20

Fig. 2.8 Displacement field of the flank of a Span region loaded with a perpendicular concentrated force calculated by the double trigonometric series method

$$w = \frac{48 \times 150(1 - 0.27^2)}{64 \times 236 \times 1.4 \times 10^5 \pi^4 \left[\frac{108}{2} - \sqrt{35^2 - \left(\frac{64}{2} - x\right)^2}\right]^3}$$

$$\sum_{m=1}^{\infty}\sum_{n=1}^{\infty} \frac{\sin\frac{20m\pi}{64}\sin\frac{170n\pi}{236}}{\left(\frac{m^2}{64^2} + \frac{n^2}{236^2}\right)^2} \sin\frac{m\pi x}{64}\sin\frac{n\pi y}{236}$$

Utilizing MATLAB programming, the displacement field is calculated and depicted in Fig. 2.8. Once the superposition count surpasses 15 instances, that is, when both m and n exceed 15, the displacement field stabilizes, indicating the rapid convergence of the double trigonometric series technique for analyzing the elasticity of the Span region. The clamping force results in a maximum deformation of 0.020 mm in this instance.

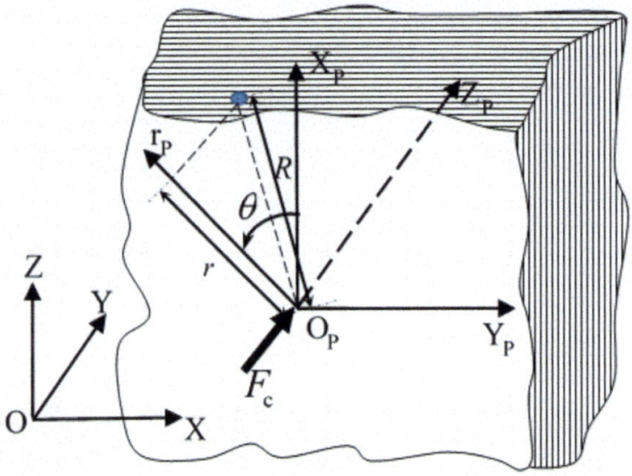

Fig. 2.9 The elastic semi-infinite space and coordinate transformation

Pier Region

The Pier region stands out among the three types of regions for not requiring consideration of variable stiffness effects, and it possesses a relatively high compressive stiffness. However, the elastic analysis of this region remains significant for the following reasons. Firstly, the top surface of a Pier region constitutes an essential part of the engine block's surface, which is the target area for face milling. The deformation caused by fixturing in the Pier region can affect the quality of the surface post-milling. Secondly, the Pier region is in direct contact with the Bridge region. The stress field analysis within the Pier region is fundamental for deriving the analytical solution for stress on the Pier-Bridge interface, which is a crucial input for the elastic analysis of the Bridge region.

In this section, the elastic analysis of the Pier region is approached as an extension of the semi-infinite elastomer problem, incorporating the actual physical boundaries. An elastic semi-infinite space (see Fig. 2.9) is defined as $\{(x, y, z)|-\infty < x, y < +\infty, z \le 0\}$, w.r.t. the local CCS $O_P X_P Y_P Z_P$.

When a clamping force is applied perpendicularly to the side of a Pier region, within a defined range, this scenario can be treated as an elasticity analysis of an elastic semi-infinite space with a boundary subjected to a normal concentrated force, essentially, the classical Boussinesq problem [55, 56]. Using the point of force application as the origin, the analytical solution for the stress field in polar coordinates [56] is provided by Eq. (2.27).

$$\begin{cases} \sigma_r = \frac{F_c}{2\pi}\left[\frac{1-2\nu}{R(R+z)} - \frac{3zr^2}{R^5}\right] \\ \sigma_\theta = \frac{F_c(1-2\nu)}{2\pi}\left[\frac{z}{R^3} - \frac{1}{R(R+z)}\right] \\ \sigma_r = -\frac{3F_c z^3}{2\pi R^5} \\ \tau_{rz} = -\frac{3F_c rz^2}{2\pi R^5} \end{cases} \tag{2.27}$$

Table 2.2 Transformation coefficient matrix of polar and local Cartesian coordinate systems

Coordinate axis	r	θ	Z
XP	$\cos\theta$	$-\sin\theta$	0
YP	$\sin\theta$	$\cos\theta$	0
ZP	0	0	1

The displacement field is given as well.

$$\begin{cases} u = \frac{(1+\nu)F_c}{2\pi E}\left[\frac{rz}{R^3} - (1-2\nu)\frac{r}{R(R+z)}\right] \\ w = \frac{(1+\nu)F_c}{2\pi E}\left[\frac{2(1-\nu)}{R} + \frac{z^2}{R^3}\right] \end{cases} \tag{2.28}$$

where R signifies the radial distance from a point within the elastic semi-infinite space to the point of force application, and r is the distance along the r_p-axis direction.

To streamline subsequent calculations and avoid complications arising from mixed coordinate systems, a coordinate transformation is applied. As detailed in Table 2.2, a transformation matrix is employed to convert the polar coordinate-based elastic solutions into Cartesian coordinates.

By coordinate transformation, the normal stress field can be formulated as

$$\begin{cases} \sigma_x = \sigma_r \cos^2\theta + \sigma_\theta \sin^2\theta \\ \sigma_y = \sigma_r \sin^2\theta + \sigma_\theta \cos^2\theta \\ \sigma_z = \sigma_z \end{cases} \tag{2.29}$$

Substituting Eq. (2.27) into Eq. (2.29), yields

$$\begin{cases} \sigma_x = \frac{F_c(1-2\nu)}{2\pi}\left[\frac{r^2\cos^2\theta - r^2\sin^2\theta}{r^2R(R^2+z)} - \frac{3zr^2\cos^2\theta}{R^5(1-2\nu)} + \frac{zr^2\sin^2\theta}{r^2R^3}\right] \\ \sigma_y = \frac{F_c(1-2\nu)}{2\pi}\left[\frac{r^2\sin^2\theta - r^2\cos^2\theta}{r^2R(R^2+z)} - \frac{3zr^2\sin^2\theta}{R^5(1-2\nu)} + \frac{zr^2\cos^2\theta}{r^2R^3}\right] \\ \sigma_z = -\frac{3F_cz^3}{2\pi R^5} \end{cases} \tag{2.30}$$

One of the basic properties of polar coordinates can be formulated as

$$\begin{cases} x = r\cos\theta \\ y = r\sin\theta \\ R = \sqrt{x^2 + y^2 + z^2} \end{cases} \tag{2.31}$$

Substituting Eq. (2.31) into Eq. (2.30), yields

$$\begin{cases} \sigma_x = \frac{F_c(1-2\nu)}{2\pi}\left[\frac{x^2-y^2}{\sqrt{x^2+y^2}\sqrt{x^2+y^2+z^2}(\sqrt{x^2+y^2+z^2}+z)} \right. \\ \qquad \left. -\frac{3zx^2}{(x^2+y^2+z^2)^{\frac{5}{2}}(1-2\nu)}+\frac{zy^2}{(x^2+y^2+z^2)^{\frac{3}{2}}(x^2+y^2)}\right] \\ \sigma_y = \frac{F_c(1-2\nu)}{2\pi}\left[\frac{y^2-x^2}{\sqrt{x^2+y^2}\sqrt{x^2+y^2+z^2}(\sqrt{x^2+y^2+z^2}+z)} \right. \\ \qquad \left. -\frac{3zy^2}{(x^2+y^2+z^2)^{\frac{5}{2}}(1-2\nu)}+\frac{zx^2}{(x^2+y^2+z^2)^{\frac{3}{2}}(x^2+y^2)}\right] \\ \sigma_z = -\frac{3F_cz^3}{2\pi(x^2+y^2+z^2)^{\frac{5}{2}}} \end{cases} \tag{2.32}$$

Analogously, the displacement field in Cartesian form can be calculated by

$$\begin{cases} u = \frac{(1+\nu)F_c}{2\pi E}\left[\frac{\sqrt{x^2+y^2}z}{(x^2+y^2+z^2)^{\frac{3}{2}}} - (1-2\nu)\frac{\sqrt{x^2+y^2}}{\sqrt{x^2+y^2+z^2}(\sqrt{x^2+y^2+z^2}+z)}\right] \\ v = \frac{(1+\nu)F_c}{2\pi E}\left[\frac{\sqrt{x^2+y^2}z}{(x^2+y^2+z^2)^{\frac{3}{2}}} - (1-2\nu)\frac{\sqrt{x^2+y^2}}{\sqrt{x^2+y^2+z^2}(\sqrt{x^2+y^2+z^2}+z)}\right] \\ w = \frac{(1+\nu)F_c}{2\pi E}\left[\frac{2(1-\nu)}{\sqrt{x^2+y^2+z^2}} + \frac{z^2}{(x^2+y^2+z^2)^{\frac{3}{2}}}\right] \end{cases} \tag{2.33}$$

Denoting the coordinates of an arbitrary mass point of the workpiece w.r.t. the local and global CCSs as $[M] = (x, y, z)^T$ and $[M'] = (x', y', z')^T$ respectively, by homogeneous coordinate transformation, it can be derived that

$$[M] = \begin{bmatrix} 0 & 0 & 1 \\ 1 & 0 & 0 \\ 0 & 1 & 0 \end{bmatrix}[M'] + \begin{bmatrix} -z_c \\ -x_c \\ -y_c \end{bmatrix} \tag{2.34}$$

which can be rewritten as

$$\begin{bmatrix} x \\ y \\ z \end{bmatrix} = \begin{bmatrix} z' - z_c \\ x' - x_c \\ y' - y_c \end{bmatrix} \tag{2.35}$$

where (x_c, y_c, z_c) denotes the coordinates of the force applying point w.r.t. the global CCS.

It should be acknowledged that the premise of an elastic semi-infinite space may not be fully applicable to the Pier region, given the presence of two finite boundaries—the upper and lower surfaces. The analytical solution derived from elastic mechanics for the deformation of the top surface, which is scheduled for machining in subsequent steps, will likely be affected by these finite boundaries. To address this limitation, a virtual plane approach has been introduced. As illustrated in Fig. 2.10, the virtual plane S_λ is delineated by the parameter $\lambda(0 < \lambda < H - z')$, which signifies the gap between the top surface and the aforementioned virtual plane.

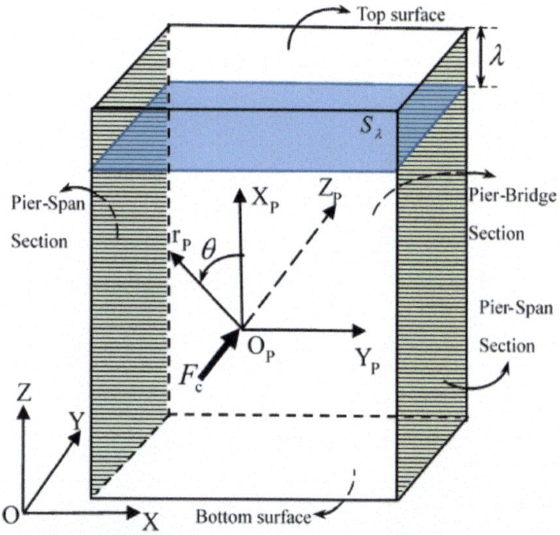

Fig. 2.10 Elasticity analysis of the Pier region using virtual plane

Given that this section prioritizes enhancing the machining quality of the top surface, the precision of the bottom surface's calculation is of lesser importance. Additionally, the other three boundaries are all sectional, comprising two Pier-Span interfaces and one Pier-Bridge interface. Therefore, the part of a Pier region below S_λ can be approximated as an elastic semi-infinite space, while the upper part can be treated as an elastic plate subjected to a load from the virtual plane. Consequently, the stress and displacement field below S_λ can be calculated by Eqs. (2.32) and (2.33) respectively. Using the stress on the virtual plane as the input load function, the deflection at the mid-surface of the upper part S_λ can be calculated using Eq. (2.22).

Example 2 The process for determining the deformation in the Pier region due to clamping before machining is elucidated. A scenario is presented where a 130 N clamping force, designated as F_c, is applied perpendicularly to a specific point on the Pier region's flank, namely point P_P (86, 0, 195) relative to the global CCS OXYZ. A local CCS $O_P X_P Y_P Z_P$ is established with P_P as its origin. The transformation of global coordinates (x', y', z') of any point within the Pier region to local coordinates is facilitated by Eq. (2.35) using $x = z' - 195, y = x' - 86$, and $z = y'$. The Boussinesq condition is met exclusively for points situated beneath the virtual plane S_λ, with the calculation's precision hinging on the parameter λ. Given the VSS workpiece's height of H = 236 mm, the force application point must be such that λ is less than $H - z' = 41$ to ensure S_λ is beyond the force applying position, i.e.λ ranges from 0 to 41 mm. A sensitivity analysis is conducted to find an optimal λ value by comparing analytical displacement solutions with FEM results. Figure 2.11 illustrates the parameter Delta, which represents the absolute percentage deviation between the analytical and FEM numerical solutions. The analysis reveals that Delta initially drops sharply as λ increases from 0, then levels off when λ is approximately

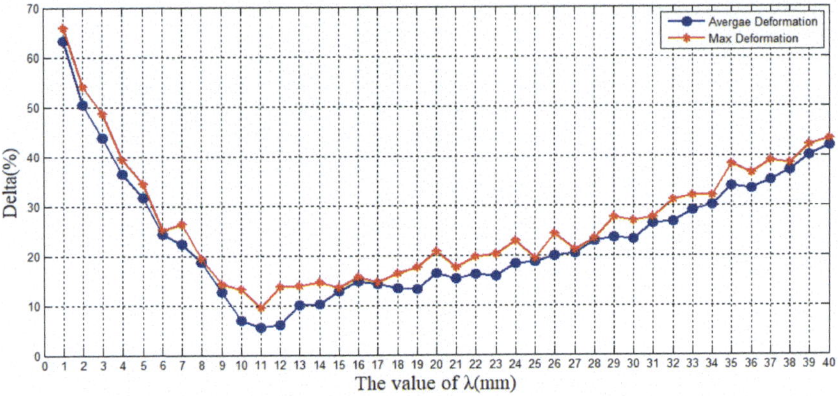

Fig. 2.11 Sensitivity analysis of the value of λ

Fig. 2.12 Displacement field of the top surface of a Pier region with a perpendicular concentrated force loaded on its flank

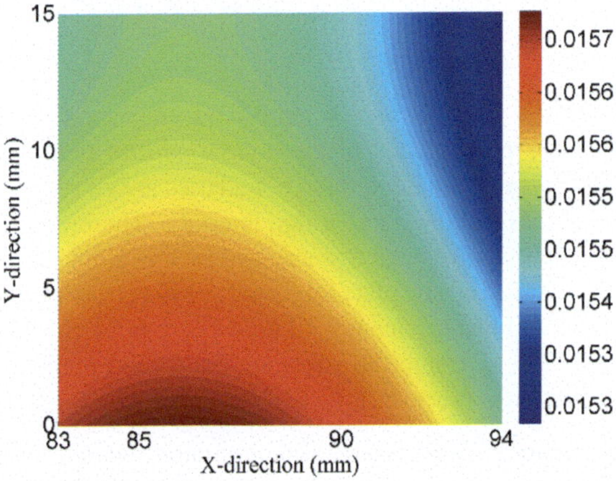

10 mm, and subsequently rises gradually as λ nears its maximum value, for both peak and average deformation.

Therefore,λ is set to 11 mm and the deviation from the conventional FEM-based numerical results is within 10%, which is acceptable in this section. By Eq. (2.32), the stress field of the Pier region below S_λ can be calculated, which becomes the load input of Eq. (2.22) to calculate the deflection of the Pier region above S_λ, and finally, the displacement field of the top surface is obtained, shown in Fig. 2.12. The maximum deformation induced by the clamping force in this example is 0.016 mm.

Bridge Region

As the most representative example of a variable stiffness structure, the Bridge region exhibits several key characteristics. First, it does not experience any external loads

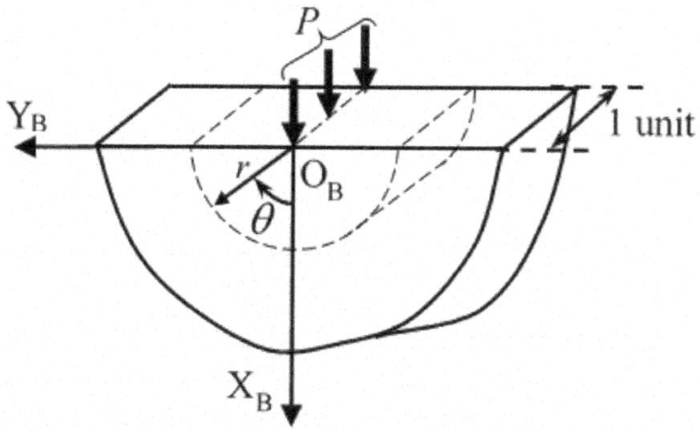

Fig. 2.13 Elasticity analysis of a semi-infinite elastic sheet under the uniformly distributed load exerted on its straight boundary in two-dimensional polar coordinate form

directly applied to it; instead, its deformation is influenced by the loading conditions at the Pier-Bridge interface. Second, the loading conditions at this interface are equal in magnitude but opposite in direction to the stress field derived from the elasticity analyses of the Span and Pier regions. Finally, similar to the treatment of the Span region, the Bridge region can be regarded as a semi-infinite elastic sheet with variable stiffness.

In the context of a semi-infinite elastic sheet of unit width, as depicted in Fig. 2.13 and using two-dimensional polar coordinates, the fundamental equations for elasticity analysis, as referenced in [37], are presented below.

The static equilibrium equations are

$$\begin{cases} \frac{\partial \sigma_r}{r} + \frac{\partial \tau_{r\theta}}{r\partial \theta} + \frac{\sigma_r - \sigma_\theta}{r} = 0 \\ \frac{\partial \sigma_\theta}{r\partial \theta} + \frac{\partial \tau_{r\theta}}{\partial r} + \frac{2\tau_{r\theta}}{r} = 0 \end{cases} \tag{2.36}$$

The strain–displacement relationship can be formulated as

$$\begin{cases} \varepsilon_r = \frac{\partial u}{\partial r} \\ \varepsilon_\theta = \frac{\partial v}{r\partial \theta} + \frac{u}{r} \\ \gamma_{r\theta} = \frac{\partial v}{\partial r} + \frac{\partial u}{r\partial \theta} - \frac{v}{r} \end{cases} \tag{2.37}$$

The elastic constitutive equations are formulated as

$$\begin{cases} \varepsilon_r = \frac{1}{E}(\sigma_r - \upsilon \sigma_\theta) \\ \varepsilon_\theta = \frac{1}{E}(\sigma_\theta - \upsilon \sigma_r) \\ \gamma_{r\theta} = \frac{1}{G}\tau_{r\theta} \end{cases} \tag{2.38}$$

The stress components can be expressed by stress function ϕ as follows.

$$\begin{cases} \sigma_r = \frac{\partial \phi}{r\partial r} + \frac{\partial^2 \phi}{r^2 \partial \theta^2} \\ \sigma_\theta = \frac{\partial^2 \phi}{\partial r^2} \\ \tau_{r\theta} = \frac{\partial \phi}{r^2 \partial \theta} - \frac{\partial^2 \phi}{r\partial r \partial \theta} \end{cases} \tag{2.39}$$

Given that $\phi = -\frac{P}{\pi} r\theta \sin\theta$, based on Eq. (2.39), a basic solution to the stress field can be derived as

$$\begin{cases} \sigma_r = -\frac{2P\cos\theta}{\pi r} \\ \sigma_\theta = \tau_{r\theta} = 0 \end{cases} \tag{2.40}$$

where P is the uniformly distributed load perpendicularly exerted on the straight boundary of an elastic sheet with unit width.

The fundamental solution adheres to Eq. (2.36) and meets the boundary conditions for the straight edge. Specifically, the straight boundary is free from any external loads other than $r = 0$, which is mathematically represented as

$$\begin{cases} \sigma_r|_{\theta=\frac{\pi}{2}, r\neq 0} = 0 \\ \sigma_\theta|_{\theta=\frac{\pi}{2}} = \tau_{r\theta}|_{\theta=\frac{\pi}{2}} = 0 \end{cases} \tag{2.41}$$

With the point of load application at the core, a cylindrical surface of any given radius r experiences a resultant force that is counterbalanced by P. That is,

$$2\int_0^{\frac{\pi}{2}} \sigma_r r \cos\theta d\theta = -P \tag{2.42}$$

It can be proved that the basic stress solution given by Eq. (2.40) satisfies this constraint as well. Hence, Eq. (2.40) is the exact stress solution to elasticity analysis of the semi-infinite elastic sheet under the uniformly distributed load exerted on its straight boundary. And the deformation solution can be calculated by

$$\begin{cases} u = \frac{P}{\pi E}[2(\ln d_B - \ln r)\cos\theta - (1-\upsilon)\theta\sin\theta] \\ v = \frac{P}{\pi E}[2(\upsilon + \ln r - \ln d_B)\sin\theta + (1-\upsilon)(\sin\theta - \theta\cos\theta)] \end{cases} \tag{2.43}$$

where d_B denotes the minimum distance between the origin and the mass point of X-axis whose vertical displacement is equal to 0. The detailed derivation can be found in Appendix 1.

As for the Bridge region (see Fig. 2.14), however, due to stress concentration and different applying positons of clamping forces, there is a strong possibility that the stress load exerted on the Pier-Bridge section is non-coplanar with the middle surface of the Bridge region. According to the Saint–Venant theorem [37], the stress load can be approximate to a biased concentrated force P, and then equivalent to a parallel concentrated force P' and a corresponding balanced torque T composed of P and P''. The magnitude of P' is same as that of P and P' is coplanar with the middle surface of the Bridge region. Denoting the bias of P as d_P, yields $T = Pd_P$.

The analytical solutions provided by Eqs. (2.40) and (2.43) are two-dimensional and are only valid under the premise that the sheet's thickness is 1 unit, an assumption that does not hold for the Bridge region. The thickness of the Bridge region, denoted

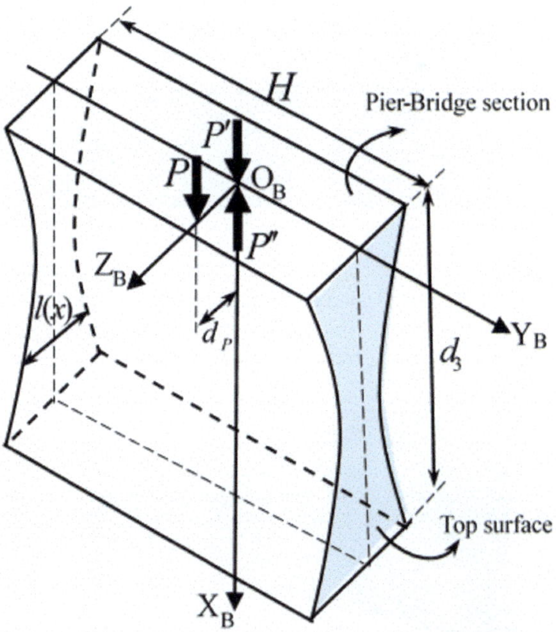

Fig. 2.14 Elasticity analysis of the Bridge region with the Saint–Venant theorem

as $l(x)$, is actually a continuous function of variable x, formulated as

$$l(x) = \frac{d_2}{2} - \sqrt{R_b^2 - \left(\frac{d_3}{2} - x\right)^2} \tag{2.44}$$

Consequently, the impact of variable stiffness must be taken into account, necessitating the adaptation and expansion of the previous solutions.

Induced by the torque T, the deflection of the mid-surface can be calculated by

$$w = \sqrt{R_c^2 - \left(\frac{d_3}{2} - x\right)^2} - \sqrt{R_c^2 - \left(\frac{d_3}{2}\right)^2} \tag{2.45}$$

where R_c denotes the curvature radius and is given by $R_c = \frac{E(l(x))^3}{12Pd_P}$.

It can be seen that the horizontal section area of the Bridge region varies with the variable x, whereas the longitudinal section area remains constant. Therefore, the actual compressive stiffness along the X_B-axis and Y_B-axis can be formulated as $K_x = EHl(x)$ and $K_y = EA_B$ respectively, where A_B is a known constant referring to the area of the top surface of the Bridge region. In order to conform to the assumption of elastic sheet with unit width, i.e., $K_x = E_x H$ and $K_y = E_y d_3$, two modified elastic moduli, $E_x = El(x)$ and $E_y = E\frac{A_B}{d_3}$, are proposed. Thus, by modifying Eq. (2.43) and integrating with Eq. (2.45), the three-dimensional displacement field of the Bridge region can be calculated by

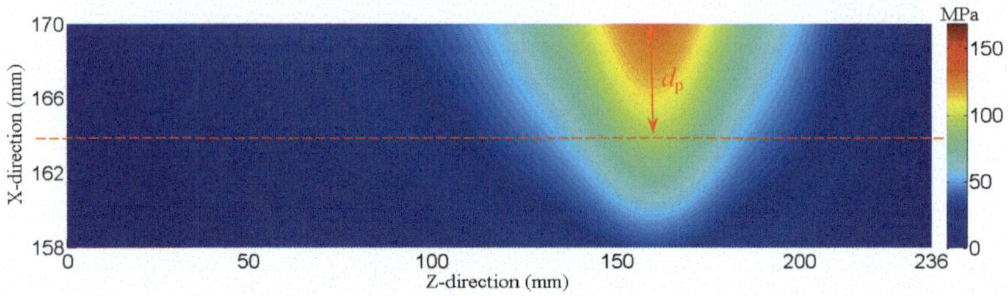

Fig. 2.15 The stress concentration on the edges of the Pier-Bridge section

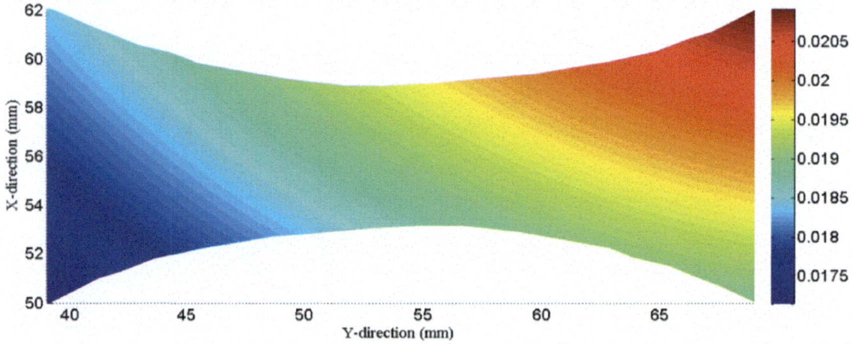

Fig. 2.16 Displacement field of the top surface of a Bridge region with the stress concentration on the edges of the Pier-Bridge section

$$\begin{cases} u = \frac{P}{\pi EI(x)}[2(\ln d_B - \ln r)\cos\theta - (1-\upsilon)\theta\sin\theta] \\ v = \frac{Pd_3}{\pi EA_B}[2(\upsilon + \ln r - \ln d_B)\sin\theta + (1-\upsilon)(\sin\theta - \theta\cos\theta)] \\ w = \sqrt{R_c^2 - (\frac{d_3}{2} - x)^2} - \sqrt{R_c^2 - (\frac{d_3}{2})^2} \end{cases} \tag{2.46}$$

Example 3 To demonstrate the method for calculating the deformation induced by fixturing in the Bridge Span region prior to machining, an example is provided. The analytical stress field solutions from the Span and Pier regions are combined to serve as the loading input for the elasticity analysis. Given the stress concentration at the edges of the Pier-Bridge section (see Fig. 2.15), according to Eq. (2.8), the biased vertical concentrated force P can be calculated by $P = K_{sc}\frac{\iint \sigma_x dydz}{Hl(0)}$. The bias of P is known as $d_P = 0.5l(0)$.

Therefore, given the stress concentration coefficient K_{sc} equals to 3.2, the displacement field of the top surface of the Bridge can be calculated by Eq. (2.46), as shown in Fig. 2.16. The maximum deformation induced by the clamping force in this example is 0.022 mm.

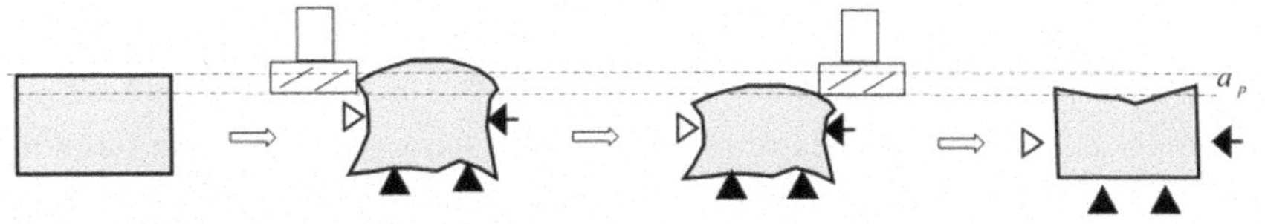

Fig. 2.17 The effect of elastic deformation recovery on final form accuracy

2.1.2.3 Fixture Layout Optimization Model

With the elasticity analysis of the three distinct regions now concluded, the deformation of the VSS workpiece due to fixturing is quantifiable, enabling the potential for effective and efficient clamping force calculations essential for optimizing fixture layout. For VSS workpieces, an inadequate fixture layout selection can cause excessive deformation during the fixturing process, leading to reduced dimensional or geometric precision post-machining. Therefore, it is crucial to develop a suitable model aimed at optimizing the fixture layout.

Objective and Constraints

The primary objective of this research is to enhance the surface quality of the machined workpiece. Because of the elastic deformation that recovers upon fixture removal post-machining (see Fig. 2.17), the deformation caused by fixturing predominantly impacts the form accuracy, particularly the flatness of the machined surface.

It is clear that increased maximum deformation from clamping negatively impacts the quality of the finished machined surface. The optimization model's central goal is to minimize the maximum deformation caused by clamping to improve the flatness of the top surface after face milling. Thus, the objective function is designed to:

$$\min\ S(\boldsymbol{X}) \tag{2.47}$$

where $\boldsymbol{X} = [L_1, L_2, \ldots, L_6, C_1, C_2, \ldots, C_6]^{\mathrm{T}}$ denotes the fixture layout with a 3-2-1 locating principle, composed of six positions of locators and six positions of clamps. With specific machining parameters, including axial depth of cut and feed per tooth, the optimal clamping forces are ascertained through $\boldsymbol{X}$, rendering $S(\boldsymbol{X})$ a function solely dependent on $\boldsymbol{X}$. This function represents the maximum displacement magnitude of the top surface.

The critical considerations for fixture layout design are as follows:

Constraint 1: Fixture elements must not impede the intended tool path; hence, the fixture layout should facilitate machining without obstructing the features under process. Consequently, this section sets the limit to a single locator-clamp pair perpendicular to the tool path to minimize interference.

Constraint 2: The workpiece must possess suitable faces or features that fulfill the necessary quality criteria for interaction with fixture elements. This indicates that not every area on the workpiece is appropriate for support, positioning, or clamping, thereby significantly narrowing the viable options for fixture placement.

Constraint 3: The fixture layout must accurately determine the workpiece's position and orientation relative to the predefined datum planes, ensuring complete constraint of the workpiece's degrees of freedom. Additionally, it is essential to prevent over-locating to avoid excessive constraints.

Computation Rules for GA

Given the nonlinear and intricate nature of the optimization model, GA is utilized for its resolution. In line with the standard protocols of GA, a set of computational rules are formulated as follows:

Rule 1: Fitness computation. Considering that excessive deformation is detrimental and irrelevant to the optimization of fixture layouts, a maximum deformation threshold, denoted as S_0, is established. If the deformation surpasses this threshold, the fitness value is assigned 0. For a given fixture configuration, fitness is determined using a specific calculation method, referred to as:

$$\text{fitness} = \begin{cases} \alpha_S(S_0 - S) & S < S_0 \\ 0 & \text{else} \end{cases} \tag{2.48}$$

where α_S is the fitness weight function of S, providing nonlinear variation for fitness weight if needed, and it can be set to a constant as well.

Rule 2: Selection, crossover, and mutation. To bolster the approach's robustness, a roulette-wheel selection technique is implemented [57]. This technique selects pairs of chromosomes for crossover and mutation based on their fitness profiles, not solely on the most fit individuals. The crossover probability remains fixed, whereas the mutation probability escalates with each successive generation, aiding in the evasion of local optima.

Rule 3: Stopping criteria. The GA employed in this section is deemed to have achieved a global optimum when the highest fitness score is sustained over a predetermined number of generations, signaling the cessation of the iterative process.

2.1.3 Case Study

In this case, a B12 engine's four-cylinder block (see Fig. 2.18) serves as a test case to assess the effectiveness of the proposed method in optimizing the fixturing scheme for VSS workpieces, aiming to enhance surface quality. During assembly, the cylinder block's top surface is in direct contact with one side of the gasket, while the other side of the gasket interfaces with the cylinder head. Consequently, the quality of the top surface, particularly its form accuracy, plays a critical role in the engine's sealing efficiency and overall reliability.

The top surface is machined using the face milling process, adhering to the 3-2-1 positioning principle. Nonetheless, the existing fixturing scheme is primarily based on engineers' experience, indicating a significant potential for enhancement. Thus, this case concentrates on enhancing the form accuracy of the top surface during the face milling process through the optimization of the fixturing scheme.

The parameters predefined for this scenario are detailed in Table 2.3.

2.1.3.1 Milling Force Measurement

To determine the unknown coefficients A_q and B_q in the static cutting force model as per Eq. (2.4), an experimental setup involving nine distinct sets of machining parameters is conducted. The specific machining parameters for this design are detailed in Table 2.4. Given that the spindle speed V_s is not directly included in the static cutting force model, it is set at a constant 400 rounds per minute (rpm) for this particular case.

The milling force measurement experiment is carried out on a DMG-HSC-75 computer numerical control (CNC), as depicted in Fig. 2.19a. Utilizing nine varying sets of machining parameters, a series of face milling operations are conducted on a section of the B12 engine block. The milling forces are captured using a 9027C Kistler three-component dynamometer. The force signals are detected by the acceleration sensors within the clamping device, transmitted via the sensor cable (see Fig. 2.19b),

Fig. 2.18 A four-cylinder block of B12 engine

Table 2.3 The predefined parameters in the case study

Symbol	Explanation	Value
a	Length of the engine block	326.5 mm
b	Width of the engine block	108 mm
H	Height of the engine block	236 mm
d_1	Distance from the first center line to the front face	51 mm
d_2	Distance between two adjacent center line	75.5 mm
d_3	Length of the Bridge region	30 mm
R_b	Radius of the cylinder bore	35 mm
A_B	Top surface area of the Bridge region	47.37 mm^2
ζ_R	Safety coefficient for clamping force	1.7
K_{sc}	Stress concentration coefficient	3.2
λ	Distance between the virtual plane to the top surface	11 mm
S_0	Threshold of the maximum deformation	0.03 mm
α_S	Fitness weight function	10^4
p_c	Crossover probability	0.9
p_m	Mutation probability	0.1–0.3
g_c	Convergence generation	2000

Table 2.4 The experiment design of different machining parameters

Number	f_z (mm)	a_p (mm)	N
1	0.1	0.1	6
2	0.12	0.15	8
3	0.14	0.2	10
4	0.16	0.12	8
5	0.18	0.17	10
6	0.2	0.22	6
7	0.22	0.08	10
8	0.24	0.12	6
9	0.26	0.18	8

and then amplified and recorded by the signal acquisition system (see Fig. 2.19c). The data is sampled over a period of 30 s.

As illustrated in Fig. 2.20, the raw signals are filtered using a second-order low pass filter, with a stable milling phase of 8 s extracted. The filter's lower cutoff frequency is set at 50 Hz, and the upper at 100 Hz.

The waveform of each signal segment exhibits two peaks and one valley, signifying an increase in milling force when the cutter engages with the Pier and Bridge regions and a decrease when it encounters the Span regions. The milling force variations align with changes in the radial cutting depth, influenced by the geometric characteristics

(a) CNC machining center (b) Vise-workpiece system (c) Signal acquisition device

Fig. 2.19 Experimental setup of milling force measurement

of the VSS workpiece. Given the quarter-section of the B12 engine block used in this experiment, the predominant feature is the cylinder bore, making the results logical.

By averaging the milling forces along each axis for the different machining parameters and applying least squares fitting to Eq. (2.4), the coefficients for the static cutting force model are determined and presented in Table 2.5.

The milling force measurement experiment has enabled the calculation of the unknown coefficients in the static cutting force model, facilitating the determination of the optimal clamping forces. These forces will serve as the loading input for the elasticity analysis. The B12 engine block, constructed from gray cast iron HT250, has material properties listed in Table 2.6, which will be utilized in both the elasticity analysis and the FEM simulation.

2.1.3.2 Optimized Fixturing Scheme Verification

Initial Fixturing Scheme

Due to the specific physical layout of the B12 engine block, the placement of locators and clamps is not arbitrary, as indicated by the second constraint in the fixture layout design. For the main locating datum plane and its counterpart, there are seven potential positions available for both locating (from $L_{3\text{-}1}$ to $L_{3\text{-}7}$) and clamping (from $C_{3\text{-}1}$ to $C_{3\text{-}7}$) purposes, as shown in Fig. 2.21. The fixture layout requires the selection of three potential locating positions and either three or four potential clamping positions.

The existing fixturing scheme employed in production, known as the initial scheme, is primarily engineered based on practical experience. Locators at positions $L_{3\text{-}1}$, $L_{3\text{-}3}$, $L_{3\text{-}6}$, $C_{3\text{-}1}$, $C_{3\text{-}3}$, and $C_{3\text{-}5}$, are used for positioning and clamping purposes, respectively. The initial clamping forces are set at 130N, 210N, and 235N for clamps $C_{3\text{-}1}$, $C_{3\text{-}3}$, and $C_{3\text{-}5}$, respectively.

A custom set of combined L-shaped plates, as shown in Fig. 2.22, is engineered to facilitate the positioning and clamping of the B12 engine block. The setup includes

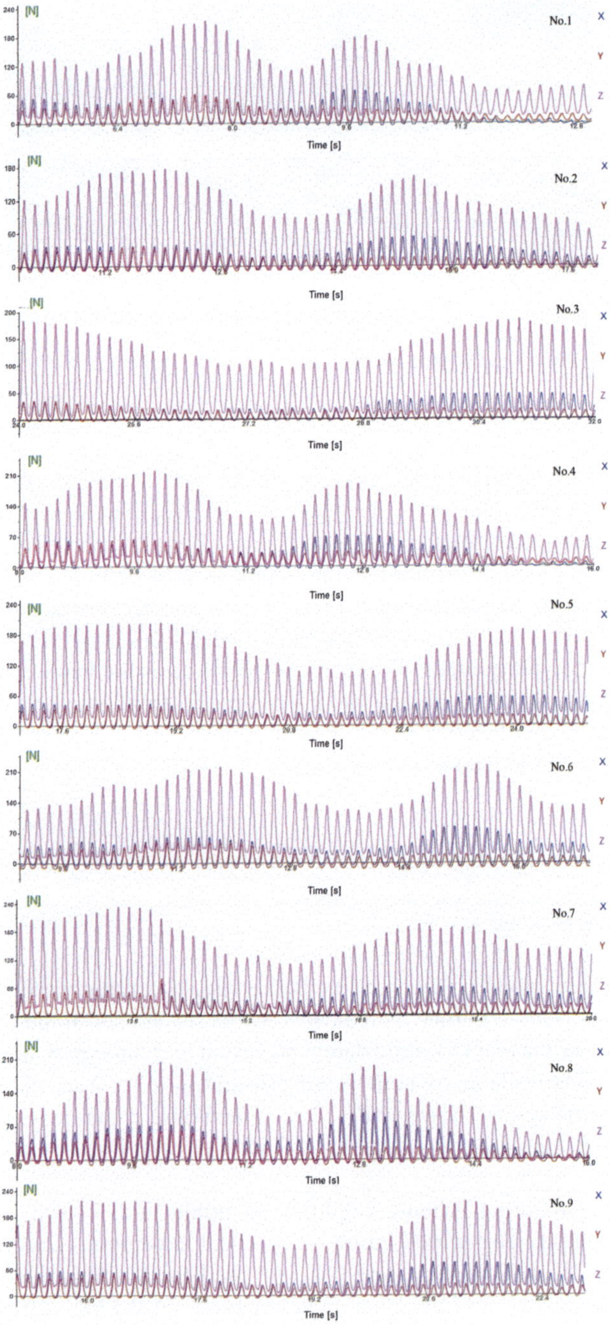

Fig. 2.20 The filtered milling force signal in the stable stage

Table 2.5 Undetermined coefficients of static cutting force model with 400 rpm spindle speed

Coefficients	A_x	A_y	A_z	B_x	B_y	B_z
Value	0.2413	0.3341	0.5113	6.5408	6.8744	7.1703

Table 2.6 Material properties of gray cast iron HT250

Density	Elastic modulus	Poisson ratio
7200 kg/m^3	140 GPa	0.27

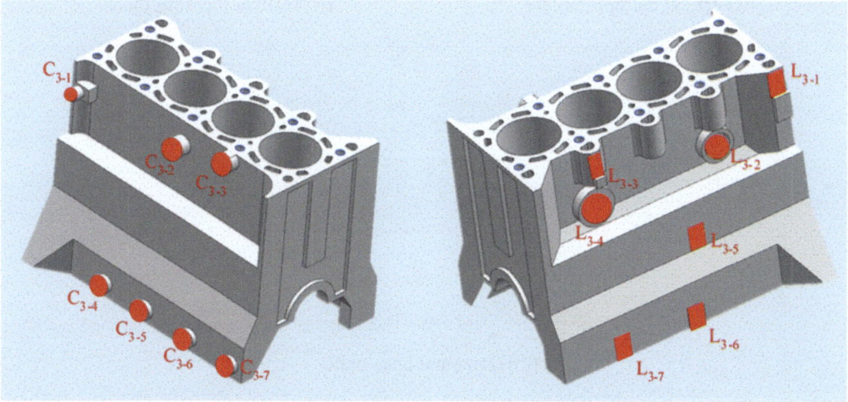

Fig. 2.21 Candidate positions for locating and clamping on the primary locating datum plane and its opposite plane

two vertical plates, one designated for clamping and the other for locating. These plates are drilled with holes at specific locations and sizes to allow bolts to position and apply clamping forces at the correct spots on the engine block. The vertical plates are then secured to horizontal base plates using screws. These horizontal base plates are anchored to the worktable to ensure the stability of the combined L-shaped plates during both fixturing and machining processes.

The assembly drawing generated by Unigraphics NX 10.0 and the experimental setup of the initial fixturing scheme are shown in Fig. 2.23

Optimized Fixturing Scheme

For a given fixture layout, the optimal clamping forces can be determined using Eq. (2.4), informed by the milling force measurement data presented in Table 2.5. With these optimized clamping forces as the loading input, the deformation caused by fixturing in the Span, Pier, and Bridge regions can be calculated using Eqs. (2.26), (2.33), and (2.46), respectively. These deformations are then combined to derive the

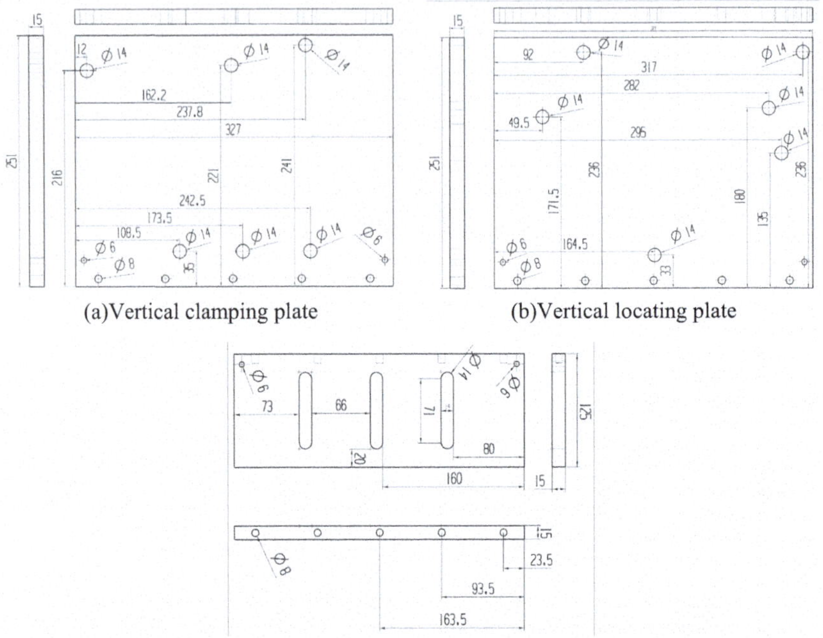

(a)Vertical clamping plate

(b)Vertical locating plate

(c)Horizontal base plate

Fig. 2.22 Three views of the combined L-plates

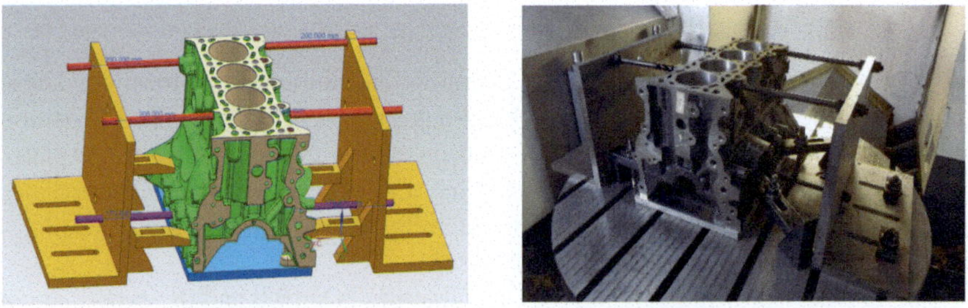

(a)Assembly simulation of the initial fixturing scheme

(b)Experimental setup of the initial fixturing scheme

Fig. 2.23 The initial fixturing scheme

analytical solution for the total fixturing-induced deformation. By solving the optimization model, which aims to minimize the maximum deformation due to fixturing, using the Genetic Algorithm (GA), the optimal fixture layout is determined. The convergence of the GA is depicted in Fig. 2.24.

It can be seen that there is a staged increase in maximum fitness within the generations from 1 to 1600. Ultimately, the termination criteria are met as the maximum

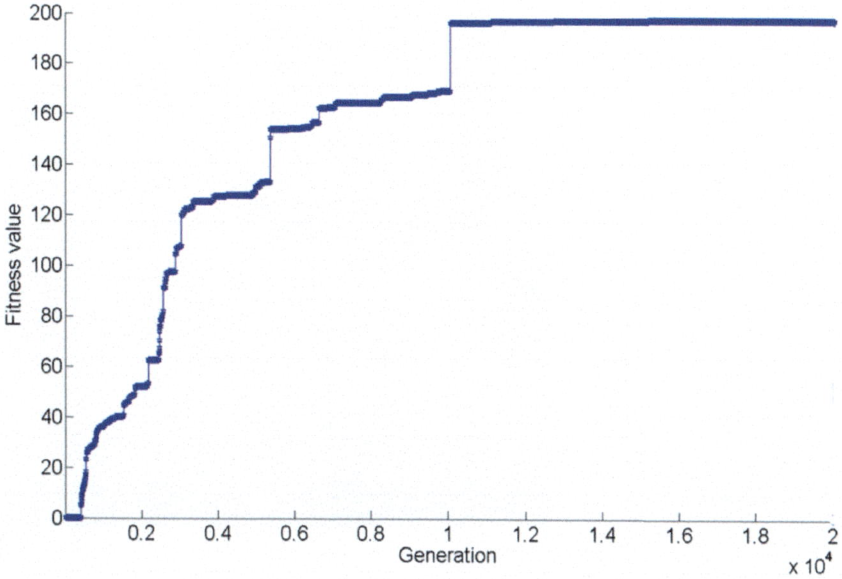

Fig. 2.24 The trend of the maximum fitness value with increasing iterations

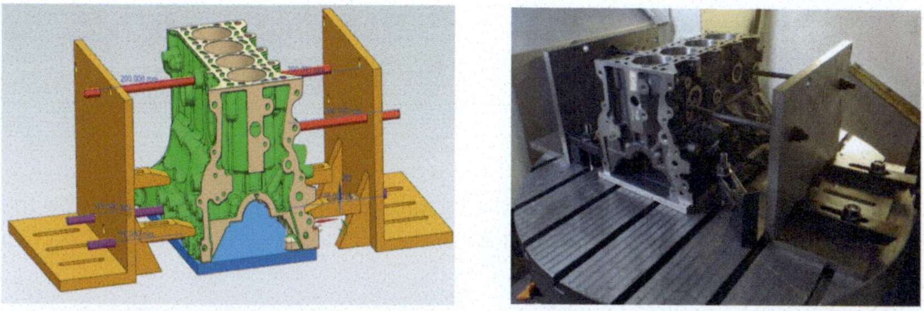

(a) Assembly simulation of the optimized fixturing scheme

(b) Experimental setup of the optimized fixturing scheme

Fig. 2.25 The optimized fixturing scheme

fitness score remains consistently at 196 for more than 2000 generations, signifying that the iterative process has reached the global optimum.

The GA-derived theoretical optimal solution specifies positioning at $L_{3\text{-}2}$, $L_{3\text{-}4}$, and $L_{3\text{-}6}$, and clamping at $C_{3\text{-}2}$, $C_{3\text{-}4}$, and $C_{3\text{-}6}$ with forces of 145 N, 95 N, and 120 N, respectively. The corresponding maximum deformation of the top surface due to fixturing is 0.0172 mm. Figure 2.25 illustrates both the assembly drawing and the experimental setup for the optimized fixturing scheme.

To apply the clamping force with precise magnitude, a torque wrench, as shown in Fig. 2.26, with a capacity of up to 200 Nm, is utilized in the fixturing process.

Fig. 2.26 Clamping force exertion by the torque wrench

Table 2.7 Pre-tightening coefficient for different contact conditions of friction surfaces

K_t	Lubrication	No lubrication
Finishing surface	0.01	0.012
Roughing surface	0.013–0.015	0.018–0.026
Oxidized surface	0.02	0.024
Galvanized surface	0.018	0.022

The relationship between the pre-tightening torque M_t applied to the bolts by the wrench and the clamping force F_c applied to the engine block by the bolts [58] can be expressed as follows:

$$M_t = K_t F_c D_b \tag{2.49}$$

where K_t denotes the pre-tightening coefficient, varying with different contact conditions of friction surfaces given in Table 2.7, while D_b denotes the diameter of the bolt.

In this scenario, the bolt diameter D_b is consistently 14 mm, and K_t is assigned a value of 0.02. Consequently, the pre-tightening torques required for both the initial and optimized fixturing arrangements are detailed in Table 2.8. It should be noted that the rounding error is within an acceptable range, attributed to the torque wrench's minimum precision threshold.

Table 2.8 Pre-tightening torque needed for different fixturing scheme (N·m)

Clamping position	Initial fixturing scheme	Optimized fixturing scheme
$C_{3\text{-}1}$	36	0
$C_{3\text{-}2}$	0	41
$C_{3\text{-}3}$	59	0
$C_{3\text{-}4}$	0	27
$C_{3\text{-}5}$	66	0
$C_{3\text{-}6}$	0	34
$C_{3\text{-}6}$	0	0

Fig. 2.27 Machining process within the CNC machining center

Verification

(1) Experimental verification

A validation experiment of the proposed methodology is carried out through machining. Face milling is executed on a pair of engine blocks within the DMG-HSC-75 CNC machining center, as depicted in Fig. 2.27. A disc milling cutter, measuring 160 mm in diameter and equipped with 8 cubic boron nitride (CBN) teeth, is utilized. The cutting parameters are configured with a cutting speed of 64,000 mm/min, a depth of cut at 0.15 mm, and a feed rate of 320 mm/min. Quaker 370 KLG cutting fluid is applied throughout the process. The machining is meticulously monitored to preclude any unexpected issues.

The LEITZ-PMM-XI coordinate measuring machine (CMM) system, as depicted in Fig. 2.28, is employed to assess the flatness of the top surface post-milling. For each engine block, a CMM probe captures the dimensions of 50 randomly selected

Fig. 2.28 Machined surface quality measurement by the CMM system

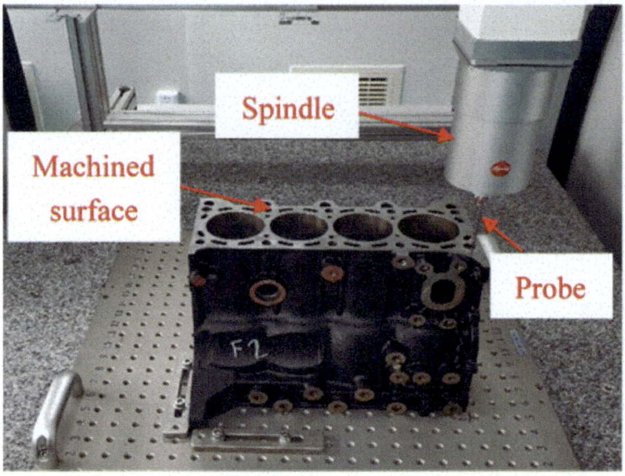

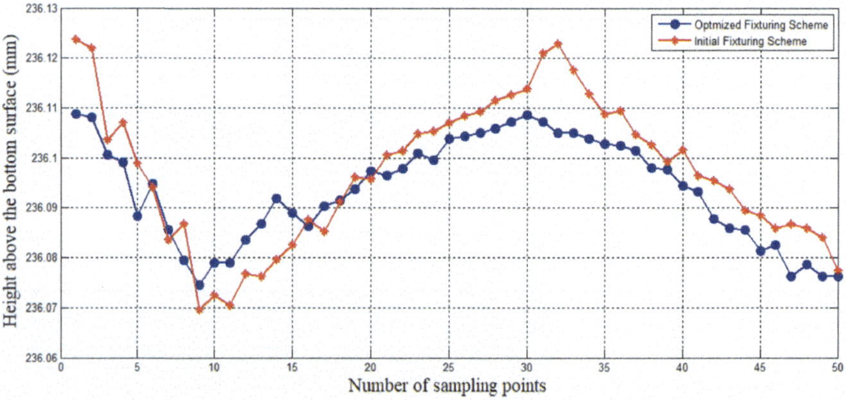

Fig. 2.29 The measured height of the sampling points

points on the top surface. The recorded heights of these sampling points are presented in Fig. 2.29.

Following the coordinate transformation, the flatness of the milled top surface is determined using the least squares method. The flatness measure for the top surface milled with the initial fixturing scheme is 0.0254 mm, and with the optimized scheme, it is 0.0231 mm. The results indicate that the proposed method has enhanced the flatness by approximately 9.1%.

Table 2.9 provides a detailed comparison between the initial and optimized schemes, focusing on three key performance indicators: maximum deformation due to fixturing, machined surface flatness, and computational expense. The results demonstrate that the proposed method effectively enhances both the maximum deformation from fixturing and the quality of the machined surface. Additionally, the efficiency of the proposed method is confirmed by its reasonable computational cost.

Table 2.9 Comparison between the conventional method and the proposed approach

Evaluating indicators	Initial scheme	Optimized scheme	Improved
Max. fixturing induced deformation	0.0201 mm	0.0172 mm	14.4%
Flatness of the machined surface	0.0254 mm	0.0231 mm	9.1%
Computation cost	0.1 h	0.5 h	Acceptable

(a)Mesh module (b)Load module (c)Interaction module

Fig. 2.30 Main modules of the dynamic explicit FEM

(2) Simulation verification

To further substantiate the effectiveness of the proposed approach, dynamic explicit FEM [59] is utilized to simulate the face milling process on the top surface of the B12 engine block using ABAQUS/CAE 6.14-1. The main components of the simulation are depicted in Fig. 2.30.

In the meshing stage, three-dimensional solid elements are employed for discretization, with a total of 1,887,415 nodes and 4,264,337 elements, including 1,168,234 linear hexahedral elements of type C3D8 and 3,096,103 linear tetrahedral elements of type C3D4. The high-quality mesh ensures the simulation's convergence and accuracy. In the loading stage, boundary conditions, and clamping forces are applied to the FEM model, with all input parameters matching those of the actual machining experiment. In the interaction stage, interaction properties, particularly the mechanical characteristics between the tool and the engine block's top surface, are specified. For the tangential behavior, the friction model is set to Penalty, while for the normal behavior, the pressure overclosure is set to Hard Contact.

The visualization results of the milled top surface of the engine block are shown in Fig. 2.31.

The sole variation between the two simulation scenarios lies in the fixturing scheme applied. By extracting the displacement field program, the flatness of the milled top surface is calculated, revealing an approximate 11.2% enhancement in the machined surface quality when employing the proposed fixturing scheme. The simulation outcomes align with those from the actual machining experiment, validating the proposed approach's efficacy in enhancing surface quality.

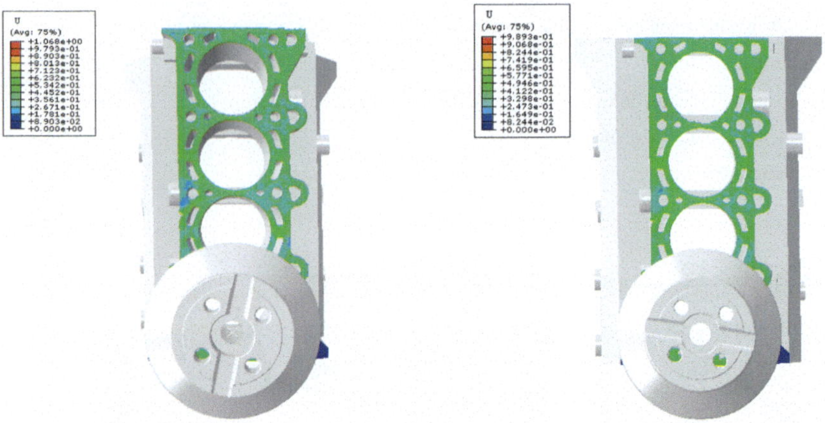

(a) The initial fixturing scheme (b) The optimized fixturing scheme

Fig. 2.31 Dynamic explicit FEM simulation of face milling with different fixturing schemes

2.1.4 Conclusion

This section presents a comprehensive methodology for optimizing the fixturing schemes of VSS workpieces to enhance surface quality. A static cutting force model is formulated to determine the minimum clamping forces required for the static equilibrium of the fixture-workpiece system during machining. The core of this approach is an elasticity-based model, offering an alternative to the traditional FEM. By employing an innovative regional division strategy, a standard VSS workpiece is categorized into three distinct regions based on geometric characteristics. The approach provides analytical solutions for deformation and stress fields in each region, which are more straightforward and insightful compared to numerical solutions. Moreover, these analytical solutions facilitate the efficient resolution of the optimization model using iterative algorithms. A case study of a four-cylinder B12 engine block is conducted to validate the proposed methodology, with the theoretical optimal fixturing scheme showing a 9.1% improvement in machined surface flatness over the initial scheme.

The practical significance of the models introduced in this section is the provision of an analytical framework for optimizing fixturing schemes for face milling and other machining processes that adhere to a 3-2-1 locating principle. This optimization process is significantly more efficient than the FEM-based numerical methods, offering a substantial advantage for practical manufacturing applications. Furthermore, the optimized fixturing scheme has proven effective in improving the quality of the machined surface.

For future research, three primary areas are identified for further exploration. First, the elasticity analysis model could be expanded to accommodate VSS workpieces with more intricate geometric features, necessitating advanced mechanics knowledge and mathematical proficiency. It is important to weigh the costs of algorithm development against computational time savings. Second, to gain a deeper understanding of the fixture-workpiece system during the machining process, an analytical relationship between clamping and cutting forces could be established based on kinematic and dynamic principles. Third, the dynamic cutting force model and the stability of the machining system against chatter could be investigated to further enhance surface quality. Since cutting forces directly affect the machined surface, the frequency and amplitude of vibrations can impact the surface's micro-texture, including waviness and roughness.

2.2 Process Parameters Optimization of Variable Stiffness Structured Workpieces

2.2.1 Introduction

Owing to its efficiency and the quality of the machined surfaces it produces, face milling is widely used in industries like automotive, shipbuilding, and aerospace[11, 60, 61]. However, regenerative chatter, which can be triggered by inappropriate process parameters, remains a significant challenge that hinders the enhancement of machining efficiency [40, 62], the quality of the machined surfaces [41, 63], and the extension of tool life [39, 64].

Researchers have long been dedicated to the dynamics modeling and prediction of the chatter phenomenon to enhance our grasp of its underlying physics. Building upon the seminal work of Tobias [65] and Tlusty [66] on classical chatter theories, a variety of methods have emerged for modeling and predicting chatter, spanning analytical, numerical, and experimental approaches. Sridhar et al. [67] crafted an analytical framework to elucidate the dynamics of general milling, which was numerically resolved. Zheng et al. [68] introduced an analytical technique to pinpoint the detrimental spindle speeds and the maximum permissible axial depth of cut within a 2-DOF planar isotropic milling system. Minis et al. [69, 70] put forth theoretical models for the anticipation of both linear and nonlinear chatter in milling. Movahhedy and Mosaddegh [71], leveraging Timoshenko beam theory for frequency response analysis, constructed a finite element model that accounts for gyroscopic effects to foresee chatter in high-speed milling scenarios. Ding et al. [72] advanced a full-discretization approach for milling stability prediction, stemming from a direct integration method. Li et al. [73] presented a Runge–Kutta-based complete discretization method (RKCDM) for assessing milling chatter stability. Expanding on the findings of Ding and Li's studies [72, 73], Dai et al. [74] implemented the explicit precise integration method (PIM) to enhance convergence velocity and predictive precision. Ismail and Soliman [75] devised an experimental strategy for swift stability lobe identification in milling by tracking a chatter indicator's response while incrementally adjusting spindle speed. Lee and Park [76] engineered a capsule-type modular machine tool capable of handling diverse processes in a single setup, with its dynamic performance confirmed through experimental FRF testing and simulation. Tuysuz and Altintas [77], employing substructuring and perturbation techniques, proposed an innovative reduced-order time-domain model that offers a fourfold increase in computational efficiency over their earlier frequency-domain model [78].

The stability lobe diagram (SLD) remains a prominent tool in chatter prediction, garnering ongoing interest from researchers. Smith and Tlusty [79] devised an algorithm for generating stability lobes through time-domain simulations in milling processes. Cao et al. [80] introduced a novel approach to creating SLDs for high-speed milling, incorporating variable spindle dynamics. Quintana et al. [81] employed sound mapping to ascertain the SLD in milling operations. Altintas and Budak [82] applied the zero-order analytical (ZOA) method to forecast chatter stability lobes. Medrol and Altintas [83] advanced the ZOA method, introducing a multifrequency approach to refine the transfer function by incorporating tooth passing frequency harmonics. Jensen and Shin [84, 85], expanding on Budak and Altintas' dynamics modeling to three dimensions, developed an algorithm for predicting stability lobes in face milling based on frequency domain analysis. Niu et al. [86], in more recent studies, investigated regenerative chatter patterns in general milling, factoring in cutting parameters, tool geometries, and runout effects. Caliskan et al. [87] proposed an innovative online chatter detection technique through vibration energy monitoring. Tang et al. [88] introduced a new stability analysis method to evaluate the impact of tool orientation on machining efficiency in five-axis bull-nose end milling.

Following the exploration of dynamics modeling and chatter prediction, scholars have turned their attention to the optimization of process parameters. Baek et al. [89] examined how insert runout errors and feed rate variations impact surface roughness and dimensional precision in face-milling operations, utilizing a surface roughness model to develop an optimization model for feed rate that seeks to maximize material removal rate while adhering to surface roughness specifications. Song et al. [90] introduced an optimization technique for cutting parameters that accounts for both self-excited and forced vibrations in peripheral milling. Zhang and Ding [91] established optimization protocols for machining parameters to achieve chatter-free milling. Additionally, Zhang et al. taking into account the dynamic uncertainties of the milling system [92] and ensuring no interference between the tool and the flexible workpiece [93], have proposed optimization models for process parameters that address these considerations.

The studies discussed earlier offer viable models for dynamic behavior, chatter anticipation, and parameter tuning in the majority of standard milling operations involving workpieces with uninterrupted surfaces. However, the literature seldom addresses scenarios with workpieces featuring discontinuous surfaces. Given that dynamic cutting forces are inherently linked to the tool's path and the workpiece's material layout [49, 94, 95], the precise interaction between the disc milling tool and the workpiece must be considered for enhanced accuracy in chatter prediction during face milling. To achieve a more precise assessment of dynamic cutting forces, it is essential to incorporate the workpiece's attributes, including its physical structure and the continuity of the surface being machined. Traditional models, however, predominantly concentrate on workpieces with uninterrupted surfaces, which can result in inaccurate chatter stability predictions when applied to workpieces with discontinuous surfaces. As the manufacturing sector advances rapidly, the physical structures of workpieces are becoming more intricate [14, 96, 97]. For example, the four-cylinder engine block shown in Fig. 2.1, composed of gray cast iron HT250, exemplifies a workpiece with discontinuous surfaces. This engine block features four cylinder bores evenly spaced along the midline of its top surface, along with several smaller holes for assembly, bolt installation, and fluid circulation. These features significantly increase the complexity of the workpiece's geometry and its stiffness distribution. Accurately modeling the dynamics during face milling on the top surface of such an engine block is challenging due to the drastic changes in workpiece geometry [1]. Without an appropriate and dependable dynamic cutting force model for chatter prediction, it becomes difficult to optimize process parameters for face milling operations on workpieces with discontinuous surfaces.

Nevertheless, a comprehensive and readily implementable optimization technique that incorporates the intricacies of face milling has not yet been established. Zhang et al. [98] employed a Taguchi design approach to enhance surface finish in CNC face milling. Drawing on the Taguchi methodology, Asiltürk and Neseli [99] applied response surface analysis (RSA) to identify optimal cutting parameters for multiple objectives in a CNC turning operation. Mhapsekar et al. [100] suggested integrating manufacturing constraints with a density-based topology optimization algorithm for additive manufacturing applications. Lu et al. [101] developed an optimization framework for cutting parameters in high-speed end milling of SKD61 tool steel, leveraging grey relational analysis in conjunction with principal component analysis to address the model. Pang and Kishawy [102] utilized a genetic algorithm to perform an inverse analysis on the primary shear zone, aiming to ascertain the five constants within Johnson–Cook's material constitutive model under metal cutting conditions. The majority of studies on process parameter optimization concentrate on the heuristic or statistical algorithms themselves, often neglecting the physical aspects of the machining process. There is a particular scarcity of suitable physical models for the face milling of workpieces with discontinuous surfaces, which results in imprecise chatter predictions and suboptimal selection of process parameters. In response, this section introduces a dynamics-based analytical method aimed at maximizing the efficiency of chatter-free machining by optimizing the parameters for face milling operations on workpieces with discontinuous surfaces. The influence of cutter-workpiece interaction is captured through the calculation of the dynamic number of cutting inserts in contact with the workpiece, a concept known as cutting insert engagement.

The primary innovation of this section is the introduction of an analytical method for optimizing process parameters, which accounts for the impact of cutting insert engagement in the dynamics modeling of the face milling system. This method is mathematically formulated based on the geometric properties of the disc milling cutter and workpieces with discontinuous surfaces.

The structure of the rest of Sect. 2.2 is as follows: Section 2.2.2 outlines the detailed steps of the proposed method, encompassing dynamic cutting force modeling, calculation of cutting insert engagement, prediction of chatter stability, sensitivity analysis, and optimization of process parameters. Section 2.2.3 presents the validation of the proposed method through a machining experiment on a four-cylinder engine block, along with a comparison of the experimental outcomes with those of the traditional approach [84]. Section 2.2.4 summarizes the findings of this section.

2.2.2 *The Proposed Approach*

2.2.2.1 Dynamic Cutting Force Modeling

Without loss of generality, the face milling system can be simplified to a 2-DOF vibration system [66, 103] as shown in Fig. 2.32. Therein, two spring damping subsystems are employed to generate vibrations along the X-axis and Y-axis respectively, and θ_{st} is the starting angle while θ_{ex} is the exiting angle. Supposing that there are a total of N_c inserts on the disc milling cutter and the cutting edge of each insert is divided into M micro elements, for the lth micro element of the cutting edge of jth insert engaged with the workpiece, the radial immersion angle is denoted as

$$\theta_{j,l} = \theta_{1,0} + \frac{2\pi(j-1)}{N_c} + \frac{a_p l \tan\beta}{MR_c} \tag{2.50}$$

where $\theta_{1,0}$, a_p, β and R_c denote angular displacement of the bottom of the first cutting edge, axial cutting depth, angle of the flank edge, and ideal radius of the disc milling cutter respectively, for $j = 1, 2, \ldots N_c$ and $l = 1, 2, \ldots M$.

In this simplified 2-DOF face milling model, the cutting forces are projected in two orthogonal directions, denoted by F_x and F_y respectively, which can be formulated as

$$\begin{cases} m_x\ddot{x}(t) + c_x\dot{x}(t) + k_x x(t) = F_x(t) \\ m_y\ddot{y}(t) + c_y\dot{y}(t) + k_y y(t) = F_y(t) \end{cases} \tag{2.51}$$

where m_x, m_y, c_x, c_y, k_x and k_y are the mass, damping, and stiffness of the cutter-workpiece system along the X-axis and Y-axis respectively.

As shown in Fig. 2.33, the cutting force on one of the micro elements of a cutting edge dF can be decomposed into three mutually perpendicular components, i.e. the axial, radial, and tangential force, denoted by dF_a, dF_r and dF_t respectively. These components are transformed into the machine coordinate system and represented as

$$\begin{cases} dF_{x,j,l} = -dF_{t,j,l}\cos\theta_{j,l} - dF_{r,j,l}\sin\theta_{j,l} \\ dF_{y,j,l} = -dF_{t,j,l}\sin\theta_{j,l} - dF_{r,j,l}\cos\theta_{j,l} \\ dF_{z,j,l} = dF_{a,j,l} \end{cases} \tag{2.52}$$

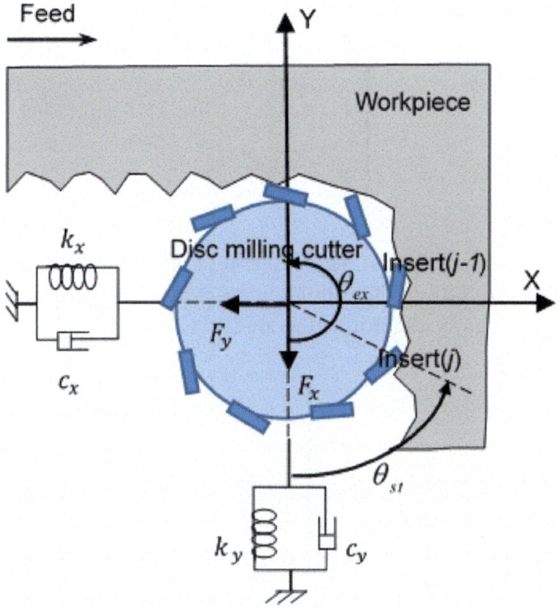

Fig. 2.32 The simplified 2-DOF face milling model

Accounting for the actual contact between the disc milling cutter and the workpiece, cutting forces arise at the cutting edges of the inserts that are in contact with the workpiece, with forces on the remaining inserts being considered negligible. It is important to recognize that the number of inserts in contact with the workpiece, represented by N, is a dynamic parameter that will be ascertained in the subsequent section. Summing the force components on all cutting edges actively involved in material removal at a given instant, yields

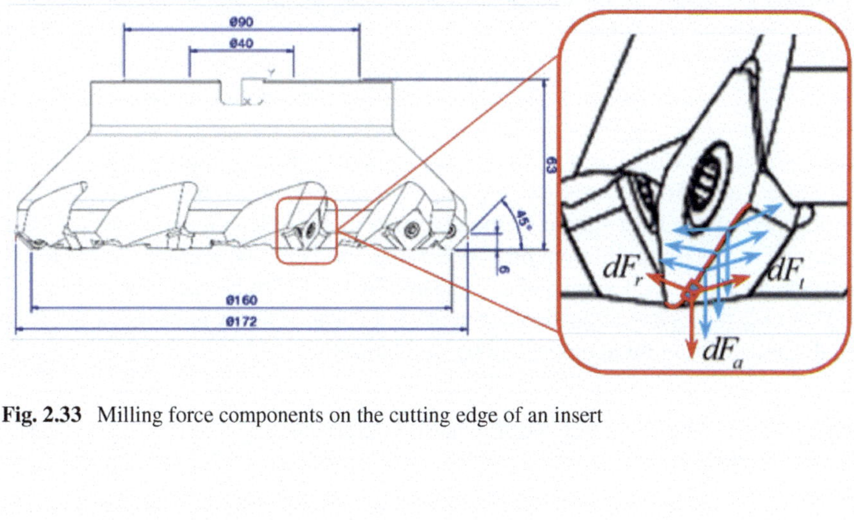

Fig. 2.33 Milling force components on the cutting edge of an insert

$$\begin{cases} F_x = \sum_{j=1}^{N}\sum_{l=1}^{M}[-dF_{t,j,l}\cos\theta_{j,l} - dF_{r,j,l}\sin\theta_{j,l}] \\ F_y = \sum_{j=1}^{N}\sum_{l=1}^{M}[dF_{t,j,l}\sin\theta_{j,l} - dF_{r,j,l}\cos\theta_{j,l}] \\ F_z = \sum_{j=1}^{N}\sum_{l=1}^{M} dF_{a,j,l} \end{cases} \tag{2.53}$$

In the context of the exponential instantaneous model referenced in [44, 45], the average cutting force components along each axis for each rotation of the disc milling cutter can be determined using Eq. (2.54).

$$\begin{cases} F_{A,x} = \frac{a_p N}{2\pi} K_r f_z^{A_x} \int\limits_{\theta_{st}}^{\theta_{ex}} (\sin\theta)^{1-A_x} d\theta \\ F_{A,y} = \frac{a_p N}{2\pi} K_t f_z^{A_y} \int\limits_{\theta_{st}}^{\theta_{ex}} (\sin\theta)^{1-A_y} d\theta \\ F_{A,z} = \frac{a_p N}{2\pi} K_a f_z^{A_z} \int\limits_{\theta_{st}}^{\theta_{ex}} (\sin\theta)^{A_z} d\theta \end{cases} \tag{2.54}$$

Therein,K_r, K_t and K_a denote the radial, tangential, and axial cutting coefficients respectively, given by

$$\begin{cases} K_r = \frac{2\pi e^{B_x}}{\int_{\theta_{st}}^{\theta_{ex}} (\sin\theta)^{1+A_x} d\theta} \\ K_t = \frac{2\pi e^{B_y}}{\int_{\theta_{st}}^{\theta_{ex}} (\sin\theta)^{1+A_y} d\theta} \\ K_a = \frac{2\pi e^{B_z}}{\int_{\theta_{st}}^{\theta_{ex}} (\sin\theta)^{A_z} d\theta} \end{cases} \tag{2.55}$$

And f_z refers to the feed per insert.

Taking the logarithm on both sides of Eq. (2.54), yields

$$\ln\left(\frac{F_{A,q}}{a_p N}\right) = A_q \ln f_z + B_q \tag{2.56}$$

where A_q and B_q (for $q = x, y, z$) are pending coefficients substituting K_r, K_t and K_a. Since Eq. (2.56) is obviously a group of linear equations, both A_q and B_q can be obtained by the least square method through multiple sets of experimental tests of cutting forces using different process parameters. Thus, the average cutting force components along each axis can be calculated by Eq. (2.56) respectively.

Denoting the regeneration displacement vector at an arbitrary moment t as $\{\Delta(t)\}$, the dynamic cutting force model can be formulated as

$$\{F(t)\} = \frac{1}{2} a_p K_t [A(t)]\{\Delta(t)\} \tag{2.57}$$

where $A(t)$ is the directional coefficient matrix which is a periodic function varying with time and the radial immersion angle of the milling cutter. The angular frequency of $A(t)$ is $\omega = \frac{\Omega N_c}{60}$.

Using the matrix form, Eq. (2.57) can be further expressed as

$$\begin{Bmatrix} F_x \\ F_y \end{Bmatrix} = \frac{1}{2} a_p K_t \begin{bmatrix} a_{xx} & a_{xy} \\ a_{yx} & a_{yy} \end{bmatrix} \begin{Bmatrix} \Delta x \\ \Delta y \end{Bmatrix} \tag{2.58}$$

where a_{xx}, a_{xy}, a_{yx} and a_{yy} are the directional coefficients.

To eliminate the time dependency of $A(t)$, the directional coefficients are decomposed into Fourier series components and averaged individually. Building on Budak and Altintas's [104,105] demonstration that higher-order harmonics of periodic functions have negligible impact on prediction accuracy, retaining only the zero-order term of the Fourier series yields a time-invariant directional coefficient matrix denoted as $[A_0]$.

$$[A_0] = \frac{N}{2\pi} \begin{bmatrix} \alpha_{xx} & \alpha_{xy} \\ \alpha_{yx} & \alpha_{yy} \end{bmatrix} \begin{Bmatrix} \Delta x \\ \Delta y \end{Bmatrix} \tag{2.59}$$

where a_{xx}, a_{xy}, a_{yx} and a_{yy} are termed as the average directional coefficients and given by

$$\begin{cases} \alpha_{xx} = 2 \int_{\theta_{st}}^{\theta_{ex}} (\sin 2\theta + K_r(1 - \cos 2\theta)) d\theta \\ \alpha_{xy} = 2 \int_{\theta_{st}}^{\theta_{ex}} (1 + \cos 2\theta + K_r \sin 2\theta) d\theta \\ \alpha_{yx} = 2 \int_{\theta_{st}}^{\theta_{ex}} (-1 + \cos 2\theta + K_r \sin 2\theta) d\theta \\ \alpha_{yy} = 2 \int_{\theta_{st}}^{\theta_{ex}} (-\sin 2\theta + K_r(1 + \cos 2\theta)) d\theta \end{cases} \tag{2.60}$$

Therefore, the dynamic cutting force model for the face milling process can be rewritten as

$$\{F(t)\} = \frac{N a_p}{4\pi} K_t \boldsymbol{\alpha} \{\Delta(t)\} \tag{2.61}$$

Therein, $\boldsymbol{\alpha}$ is independent of time and only varies with the radial immersion angle.

2.2.2.2 Determination of Cutting Insert Engagement

The primary cause of chatter in face milling is typically the regenerative effect, which is intimately linked to dynamic cutting forces. When the cutting inserts interact with the workpiece, periodic forces, and surface irregularities are produced through the cyclical compression and friction between the inserts and the workpiece with each spindle rotation. The vibrations between the cutter and workpiece are perpetuated by the surface irregularities created in the previous rotation of the disc milling cutter, known as the regenerative effect. For workpieces with straightforward physical structures and uniform material distribution, the milling process is consistent, and the engagement of cutting inserts is stable, often leading to its omission in prior studies. However, with workpieces featuring discontinuous surfaces and intricate physical structures, such as numerous cavities and chambers, the engagement of cutting inserts is influenced by the cutting path and workpiece attributes, resulting in a more erratic regenerative process. The dynamic cutting force model described by Eq. (2.61) is dependent on N. Therefore, overlooking the impact of cutting insert engagement, especially for workpieces with discontinuous surfaces, can result in inaccurate chatter stability predictions.

This section focuses on establishing an appropriate dynamic cutting force model for workpieces with discontinuous surfaces, using the face milling of a representative workpiece featuring multiple cylindrical bores as a case study. Figure 2.34 illustrates the variation of cutting insert engagement, denoted as N, along the feed direction. Given that the disc milling cutter's diameter D_c exceeds the radial depth of cut a_e, the milling operation is characterized by intermittent engagement. Additionally, the material distribution causes N to decrease when the cutter encounters the cylindrical bores and increases otherwise. In conclusion, while the pattern of N's variation is intricate, it is calculable.

Before the determination of cutting insert engagement, several definitions of parameters used in the derivation process are introduced in Table 2.10.

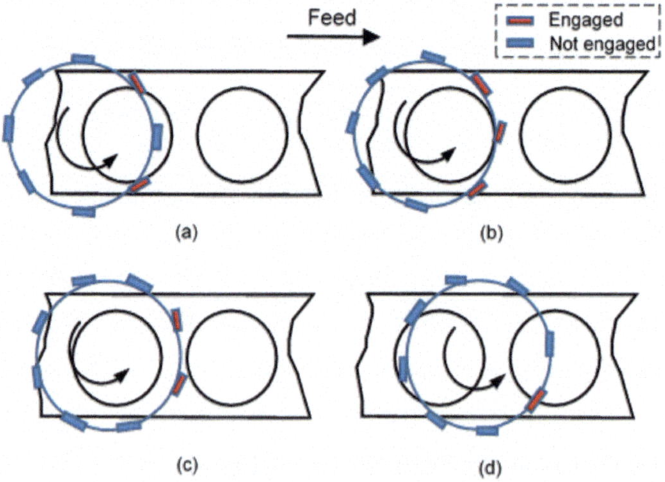

Fig. 2.34 Variation of cutting insert engagement along the feed direction **a** $N = 2$ **b** N switches from 2 to 3 **c** N switches from 3 to 2 **d** N switches from 2 to 1

Table 2.10 Definitions of parameters used in the derivation process

Parameter	Definition	Function
Cutting duration [s]	$t_c = t_{now} - t_0$	Duration from the start of cutting (t_0) to observation (t_{now})
Cutting distance [m]	$L_c = V_f t_c$	Maximum distance between edges of cutter and workpiece along feed direction
Cutting speed [m/s]	$V_c = \frac{\pi D_c \Omega}{60}$	Linear velocity of each insert
Micro element of cutting distance [m]	$\Delta L_c = V_f \frac{\pi D_c}{V_c}$	Feed per revolution of the disc milling cutter

Based on these definitions, there are two important and reasonable hypotheses proposed in this section.

Hypothesis 1 In the face milling of a standard workpiece featuring non-continuous surfaces, the operation is completed by a disc-shaped milling tool in a single stroke, starting from the front face and moving to the back face, with the cutting trajectory traversing the centers of all cylindrical holes. In other words, the cutting path aligns with the median line of the upper surface. This assumption closely mirrors real-world manufacturing practices and facilitates subsequent mathematical analysis.

Hypothesis 2 The cutting speed V_c significantly exceeds the feed rate V_f, and the feed per revolution (denoted by ΔL_c) is minimal, treating it as a minor segment of L_c. The variation in N during each rotation is solely dependent on the cutting arc length, as depicted in Fig. 2.35. L_k and L_{k+1} (for $k = 1, 3, 5, 7$) represent the horizontal distances from the front face to the closest and farthest generatrix of the $(k/2 + 1/2)^{th}$ cylindrical bore. The cutting arc length is a function of L_c and is intricately linked to the geometric properties of the workpiece. By considering the cutting arc length, the entire face milling operation can be segmented into multiple stages. For each stage, the engagement of the cutting insert can be ascertained by defining a piecewise function based on the cutting arc length.

Denoting the radius of a cylindrical bore, the length, and width of a typical workpiece by R_b, L_a and L_b respectively, the cutting insert engagement can be determined phase by phase, and the illustrations for each phase are summarized in Fig. 2.36.

Phase I The onset of *Phase I* in the face milling process occurs as the disc milling cutter's edge makes contact with the front face, initiating the generation of cutting forces through the interaction with the workpiece. This phase is characterized by $0 \leq L_c \leq L_1$, with the radial cutting depth a_e specified by

$$a_e = \sqrt{D_c^2 - (D_c - 2L_c)^2} \tag{2.62}$$

It should be observed that two distinct scenarios may arise and require separate consideration. Depending on the dimensions of the disc milling cutter and the engine

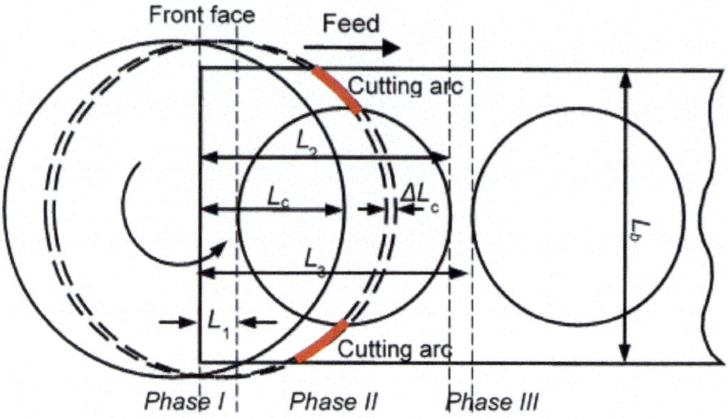

Fig. 2.35 Phase division based on the cutting arc

block being machined, a unique phase, referred to as *Phase I**, may occur prior to the transition into *Phase II*.

(i) In the context of $\sqrt{D_c^2 - (D_c - 2L_1)^2} \le L_b$, as depicted in Fig. 2.36a, the cutting arc length increases progressively during *Phase I*. This scenario is regarded as the norm due to the continuous change in the cutting arc length function. With each rotation of the disc-shaped milling tool, the number of inserts, N, that come into contact with the workpiece is directly proportional to the arc length. Therefore, the interaction of cutting inserts during *Phase I* can be formulated as follows:

$$N = \frac{N_c}{\pi} \arccos \frac{D_c - 2L_c}{D_c} \tag{2.63}$$

(ii) In the scenario of $\sqrt{D_c^2 - (D_c - 2L_1)^2} > L_b$, as shown in Fig. 2.36b, *Phase I* transitions directly into *Phase I**. During this transition, the effective cutting edge a_e increases from 0 to L_b before the disc milling cutter reaches the first cylindrical bore. This occurs due to the large diameter D_c and the short length L_1. Throughout *Phase I**, the cutting arc length remains constant before the transition to *Phase II*. The engagement of cutting inserts at the beginning of *Phase I** can be ascertained using formula

$$N = \begin{cases} \frac{N_c}{\pi} \arccos \frac{D_c - 2L_c}{D_c} & 0 \le L_c \le \frac{D_c - \sqrt{D_c^2 - L_b^2}}{2} \\ \frac{N_c}{\pi} \arcsin \frac{L_b}{D_c} & \frac{D_c - \sqrt{D_c^2 - L_b^2}}{2} < L_c \le L_1 \end{cases} \tag{2.64}$$

Phase II Upon the disc milling cutter encountering the first cylindrical bore, the face milling operation transitions into *Phase II*, characterized by $L_1 \le L_c \le L_2$. The cutting arc is divided into two symmetrical segments due to the absence of material in the cylindrical bore. The combined length of the cutting arcs initially decreases

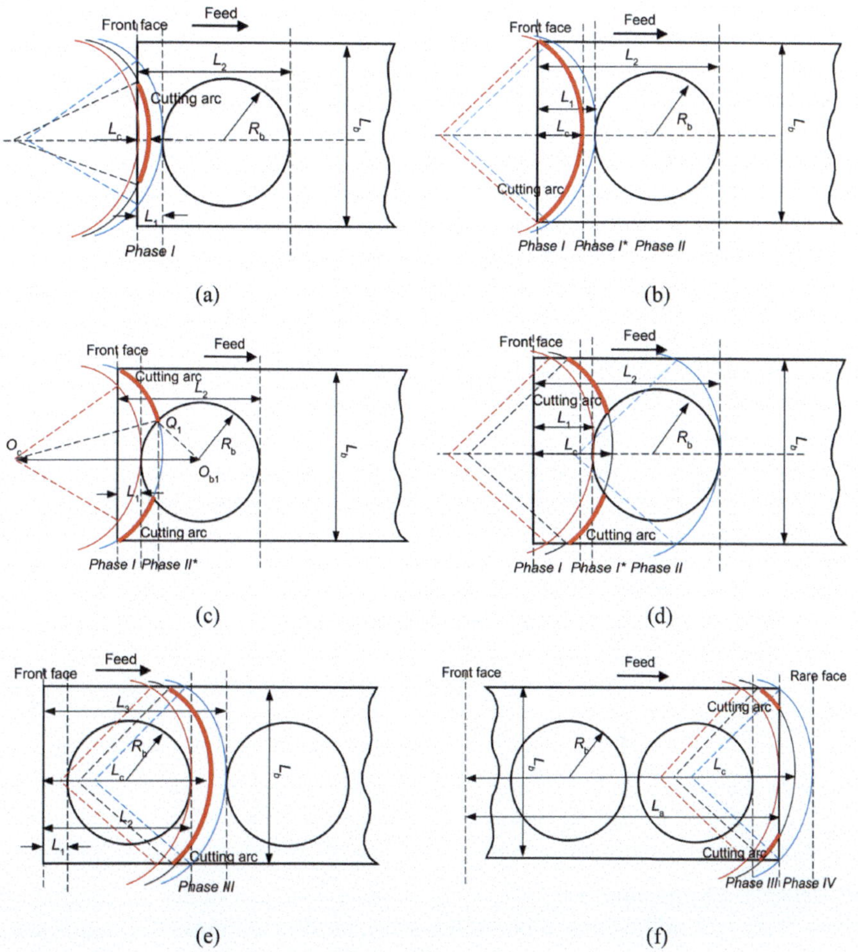

Fig. 2.36 Cutting insert engagement for different phases **a** *Phase I* without the emergence of *Phase I** **b** *Phase I* with the emergence of *Phase I** **c** *Phase II* with the emergence of *Phase II** **d** *Phase II* without the emergence of *Phase II** **e** *Phase III* **f** *Phase IV*

and then subsequently increases. As with *Phase I*, there are two distinct scenarios for cutting insert engagement in *Phase II* that require separate analysis.

(i) In the context of $\sqrt{D_c^2 - (D_c - 2L_1)^2} < L_b$, *Phase II** may occur, as indicated in Fig. 2.36c. It is important to recognize that the conditions defining *Phase II** differ from those of *Phase I*, as both phases would be irrelevant if $\sqrt{D_c^2 - (D_c - 2L_1)^2}$ were simply equal to L_b. Moreover, *Phase I** and *Phase II** are mutually exclusive within a single face milling operation, and it is possible that neither of these special phases occurs throughout the entire process. The engagement of cutting inserts in the event of *Phase II** can be expressed as

$$N = \begin{cases} \frac{N_c}{\pi}\left(\arccos\frac{D_c-2L_c}{D_c} - \arccos\frac{4S_{t1}}{D_cL_{y1}}\right) & L_1 \le L_c \le \frac{D_c-\sqrt{D_c^2-L_b^2}}{2} \\ \frac{N_c}{\pi}\left(\arcsin\frac{L_b}{D_c} - \arccos\frac{4S_{t1}}{D_cL_{y1}}\right) & \frac{D_c-\sqrt{D_c^2-L_b^2}}{2} < L_c \le L_2 \end{cases} \tag{2.65}$$

In this context, L_{y1} represents the distance between the centers of the disc milling cutter and the first cylindrical bore, while S_{t1} denotes the area of the triangular region $Q_1O_cO_b$, which is calculable using Heron's Formula as referenced in [106]. The detailed derivation of this calculation is presented in the Appendix 2.

(ii) In scenario $\sqrt{D_c^2-(D_c-2L_1)^2} \ge L_b$, the radial cutting depth a_e reaches the width of the engine block, L_b, prior to the disc milling cutter making contact with the first cylindrical bore, as depicted in Fig. 2.36d. The engagement of cutting inserts during *Phase II* can be expressed by

$$N = \frac{N_c}{\pi}\left(\arcsin\frac{L_b}{D_c} - \arccos\frac{4S_{t1}}{D_cL_{y1}}\right) \tag{2.66}$$

Phase III This phase is defined by $L_2 \le L_c \le L_3$, the two individual cutting arcs merging to form a unified cutting arc, as shown in Fig. 2.36e. Throughout *Phase III*, the length of the cutting arc is stable and is denoted by

$$N = \frac{N_c}{\pi}\arcsin\frac{L_b}{D_c} \tag{2.67}$$

Phase IV Following multiple iterations of *Phase II* and *Phase III*, which are not detailed here, the face milling process advances until it fully traverses the last cylindrical bore, subsequently transitioning into *Phase IV* as shown in Fig. 2.36f. *Phase IV*, delineated by parameter $L_a \le L_c \le L_a + \frac{D_c-\sqrt{D_c^2-L_b^2}}{2}$, constitutes the concluding phase of the face milling operation. At this stage, the disc milling cutter commences its departure from the workpiece's top surface, with the cutting arc length dividing into two symmetrical segments that gradually diminish to zero. The engagement of cutting inserts in Phase IV is described by

$$N = \frac{N_c}{\pi}\left(\arcsin\frac{L_b}{D_c} - \arccos\frac{D_c-2L_c+2L_a}{D_c}\right) \tag{2.68}$$

For ease of comparison, Table 2.11 provides a compilation of the mathematical formulas for cutting insert engagement N across various phases.

As shown in Fig. 2.37, there are three possible cases for the face milling process, dependent on the diameter of the disc milling cutter as well as the geometric characteristic of the workpiece.

To encapsulate the findings, the phase-based calculation of cutting insert engagement plays a crucial role in enhancing the precision of dynamic cutting force estimation and milling chatter forecasting. By analytically defining the interaction between

Table 2.11 Formulas of cutting insert engagement N for different phases

Phase	Formula of cutting insert engagement	In case of
$I\ (0 \le L_c \le L_1)$	$N = \frac{N_c}{\pi} \arccos \frac{D_c - 2L_c}{D_c}$	$\sqrt{D_c^2 - (D_c - 2L_1)^2} \le L_b$
	$\begin{cases} \frac{N_c}{\pi} \arccos \frac{D_c - 2L_c}{D_c} & \left(0 \le L_c \le \frac{D_c - \sqrt{D_c^2 - L_b^2}}{2}\right) \\ \frac{N_c}{\pi} \arcsin \frac{L_b}{D_c} & \left(\frac{D_c - \sqrt{D_c^2 - L_b^2}}{2} < L_c \le L_1\right) \end{cases}$	$\sqrt{D_c^2 - (D_c - 2L_1)^2} > L_b$
$II\ (L_1 \le L_c \le L_2)$	$\begin{cases} \frac{N_c}{\pi}\left(\arccos \frac{D_c - 2L_c}{D_c} - \arccos \frac{4S_{t1}}{D_c L_{y1}}\right) & \left(L_1 \le L_c \le \frac{D_c - \sqrt{D_c^2 - L_b^2}}{2}\right) \\ \frac{N_c}{\pi}\left(\arcsin \frac{L_b}{D_c} - \arccos \frac{4S_{t1}}{D_c L_{y1}}\right) & \left(\frac{D_c - \sqrt{D_c^2 - L_b^2}}{2} < L_c \le L_2\right) \end{cases}$	$\sqrt{D_c^2 - (D_c - 2L_1)^2} < L_b$
	$\frac{N_c}{\pi}\left(\arcsin \frac{L_b}{D_c} - \arccos \frac{4S_{t1}}{D_c L_{y1}}\right)$	$\sqrt{D_c^2 - (D_c - 2L_1)^2} \ge L_b$
$III\ (L_2 \le L_c \le L_3)$	$\frac{N_c}{\pi} \arcsin \frac{L_b}{D_c}$	–
$IV\ \left(L_a \le L_c \le L_a + \frac{D_c - \sqrt{D_c^2 - L_b^2}}{2}\right)$	$N = \frac{N_c}{\pi}\left(\arcsin \frac{L_b}{D_c} - \arccos \frac{D_c - 2L_c + 2L_a}{D_c}\right)$	–

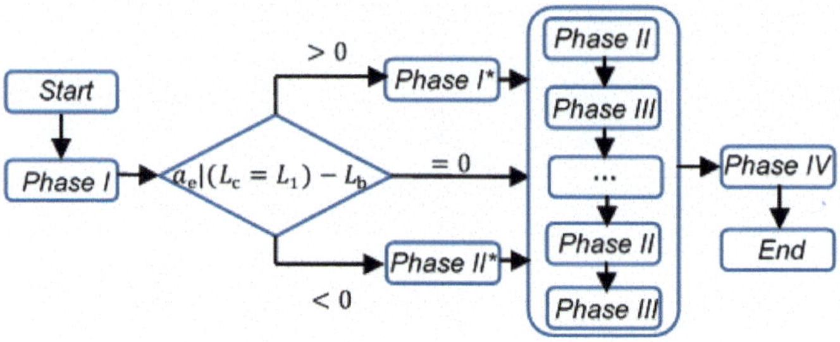

Fig. 2.37 Three possible cases for the face milling process

the disc milling cutter and the workpiece, this method effectively captures the chatter stability for face milling operations on workpieces with discontinuous surfaces.

2.2.2.3 Chatter Stability Prediction

Given the effectiveness and efficiency of the stability lobe diagram (SLD) method in predicting and mitigating chatter in engineering applications, this section aims to derive the SLD for the face milling process. Chatter stability hinges on the dynamic response of the machining system, typically represented by the frequency response function (FRF). For a two-degree-of-freedom (2-DOF) cutter-workpiece vibration system, the FRF at the chatter frequency ω_c can be formulated as follows:

$$[G(i\omega_c)] = \begin{bmatrix} G_{xx}(i\omega_c) & G_{xy}(i\omega_c) \\ G_{yx}(i\omega_c) & G_{yy}(i\omega_c) \end{bmatrix} \tag{2.69}$$

where i is the unit of imaginary number and the chatter frequencies $\omega_c = \omega_{c1}, \omega_{c2}, \ldots, \omega_{cn}$ can be obtained by simulation or experiments, which are usually around the dominant modes of the vibration system.

Considering the regenerative effect, the dynamic cutting force model can be further derived as

$$\{F(t)\}e^{i\omega_c t} = \frac{1}{2}a_p K_t(1 - e^{-i\omega_c \tau})[A_0][G(i\omega_c)]\{F(t)\}e^{i\omega_c t} \tag{2.70}$$

where τ is the time delay, equivalent to the insert passing period, namely $\tau = \frac{60}{\Omega N_c}$.

The necessary and sufficient condition of nonsingular solutions to Eq. (2.70) is termed the chatter characteristic equation, which is given by

$$\det\{[I] - \Lambda[G_0(i\omega_c)]\} = 0 \tag{2.71}$$

Therein, $[I]$ is the identity matrix while $G_0(i\omega_c)$ is the directional coefficient transfer matrix defined by

$$G_0(i\omega_c) = \frac{2\pi}{N_c}[A_0]G(i\omega_c) \tag{2.72}$$

And Λ is the eigenvalue, which is given by

$$\Lambda = \frac{N_c}{4\pi}a_p K_t(1 - e^{-i\omega_c\tau}) \tag{2.73}$$

Note that Λ is a complex number. Denoting the imaginary part and the real part of Λ as Λ_I and Λ_R respectively, Λ can be rewritten as $\Lambda = \Lambda_R + i\Lambda_I$. Substituting the Euler's formula [107] $e^{-i\omega_c\tau} = \cos(\omega_c\tau) - i\sin(\omega_c\tau)$ into Eq. (2.73), yields

$$\begin{cases} \Lambda_R = \frac{N_c}{4\pi}a_p K_t(1 - \cos(\omega_c\tau)) \\ \Lambda_I = \frac{N_c}{4\pi}a_p K_t \sin(\omega_c\tau) \end{cases} \tag{2.74}$$

And Eq. (2.74) can be further derived as

$$\frac{\Lambda_I}{\Lambda_R} = \frac{\sin(\omega_c\tau)}{1 - \cos(\omega_c\tau)} \tag{2.75}$$

The critical value of the axial cutting depth can be calculated by

$$a_{p\_\lim} = \frac{2\pi(\Lambda_R^2 + \Lambda_I^2)}{N_c K_t \Lambda_R} \tag{2.76}$$

The corresponding spindle speed can be obtained by

$$\Omega = \frac{60}{N_c((2k_L + 1)\pi - 2\arctan(\Lambda_I/\Lambda_R))} \tag{2.77}$$

where $k_L = 0, 1, 2\ldots$, which gives successive stability lobes.

The SLD can be generated by the following main steps, and the pseudo code for programming is exhibited in Fig. 2.38.

Step 1: Establish the dynamic cutting force model considering cutting insert engagement and regenerative effect

Step 2: Obtain the FRF $[G(i\omega_c)]$ at each chatter frequency based on the dynamic cutting force model

Step 3: Select chatter frequencies $\omega_c = \omega_{c1}, \omega_{c2}, \ldots, \omega_{cn}$ around the dominant modes and scan each of them

Step 4: Set the initial axial cutting depth $a_p = a_{p\_0}$

Step 5: Calculate the critical axial cutting depth $a_{p\_\lim}$ and the corresponding spindle speed Ω by solving the chatter characteristic equation Eq. (2.71)

Step 6: Increase the axial cutting depth gradually, namely $a_p = a_{p\_0}, a_{p\_1}, \ldots, a_{p\_n}$, and repeat Step 5

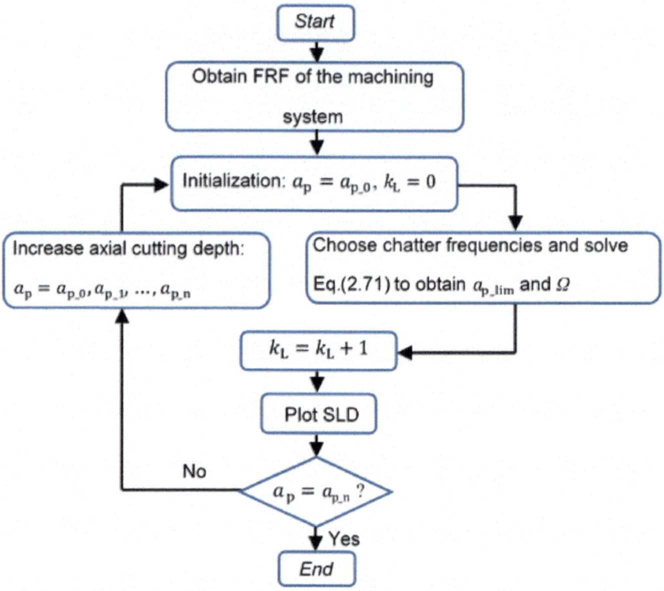

Fig. 2.38 Pseudo code for generating the stability lobe diagram for face milling process

Step 7: Assemble the critical axial cutting depth a_{p_lim} at each corresponding spindle speed Ω and then plot the chatter stability lobes

Utilizing the SLD obtained from the aforementioned procedures, it is possible to forecast the chatter stability for the face milling of workpieces with discontinuous surfaces. Selecting appropriate process parameters before the machining operation is essential to minimize the likelihood of chatter occurrence.

2.2.2.4 Sensitivity Analysis

The face milling process stability is influenced by a multitude of process parameters, such as the feed per insert f_z, axial cutting depth a_p, spindle speed Ω, total number of inserts N_c, and the diameter of the disc milling cutter D_c. Consequently, a sensitivity analysis of each pertinent parameter is essential before the optimization of process parameters. The axial cutting depth in relation to the absolute stability region, denoted as a_{p_abs}, characterizes a critical aspect of the SLD. As shown in Fig. 2.39, a_{p_abs} is employed to assess the absolute "height" of the stability lobe curve.

Spanning the range of spindle speeds from the initial Ω_{st} to the final Ω_{fi}, a qualitative assessment of the impact of various process parameters on the SLD can be derived using the single-variable approach. Drawing on typical engineering practices, the default parameters are outlined in Table 2.12.

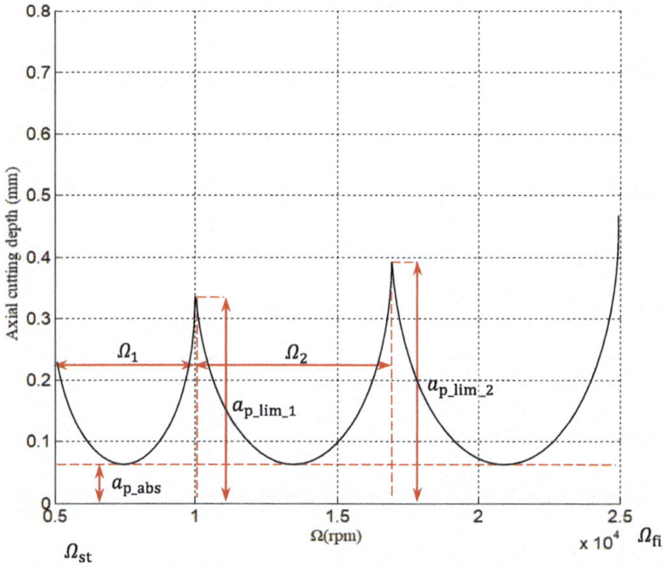

Fig. 2.39 Illustration of evaluation indexes used in sensitivity analysis

Table 2.12 Default set of parameters given by engineering experience

Type	Parameter	Default level
Process	f_z (mm)	0.02
	D_c (mm)	160
	N_c	12
Geometric	R_b (mm)	35
	L_b (mm)	108
Material	E (GPa)	140
	υ	0.27

It should be noted that this section focuses solely on the optimization of process parameters, with all geometric and material properties of the workpiece being held at constant values. By varying one parameter from its baseline level, while maintaining all others constant, a set of numerical experiments are performed to produce stability lobe diagrams, and the outcomes are presented below.

As depicted in Fig. 2.40, there is a downward trend in the stability lobes with the growth of f_z, leading to a nonlinear decline in a_{p_abs}.

As Eq. (2.54) indicates a direct relationship between the average cutting force and f_z, the dynamic interactions between the disc milling cutter and the workpiece lead to increased forced vibrations with a higher f_z. A higher feed rate, while enhancing machining efficiency, also compromises processing stability, highlighting a crucial trade-off to be addressed in the optimization of process parameters.

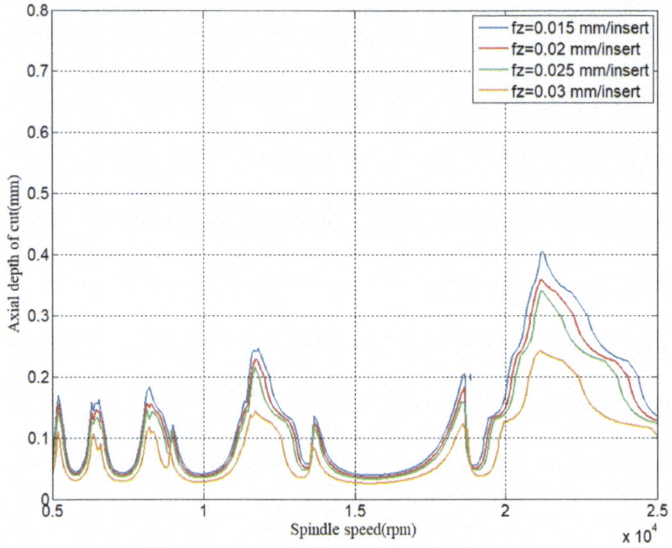

Fig. 2.40 Trend chart of stability lobe curves with varying feed per insert

Figure 2.41 demonstrates the stability lobes' downward trajectory as N_c increases. This pattern is primarily due to the rising excitation frequency, aligning well with findings from established literature [71, 73]. Nevertheless, because processing efficiency is directly related to N_c, this represents a significant trade-off to consider in the optimization of process parameters.

Figure 2.42 illustrates the upward movement of the stability lobes as D_c increases. Based on Eqs. (2.63)–(2.68), it is established that the number of inserts in contact with the workpiece, N, decreases monotonically with the diameter of the disc milling cutter, D_c, within a specific range. Consequently, by judiciously increasing D_c, the amplitude of forced vibrations can be diminished, enhancing the dynamic stability of the machining system. This insight will significantly facilitate the simplification of the process parameter optimization model in the subsequent section.

Drawing from the outcomes of the sensitivity analysis, a model for optimization can be formulated and resolved to enhance machining efficiency while taking into account stability and quality in the process.

2.2.2.5 Optimization of Process Parameters

The primary objective of this section is to enhance the efficiency of chatter-free machining. The material removal rate, a standard measure of machining efficiency, is characterized by the following definition:

$$Q_{\mathrm{MRR}} = a_p a_e f_z \Omega N_c \tag{2.78}$$

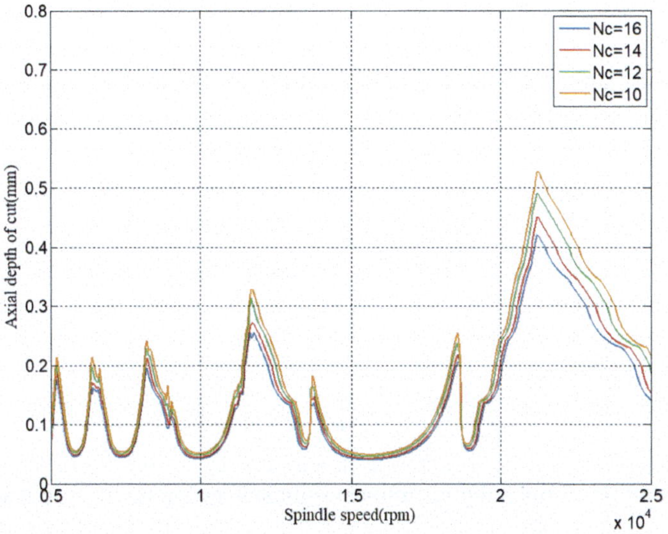

Fig. 2.41 Trend chart of stability lobe curves with varying number of inserts

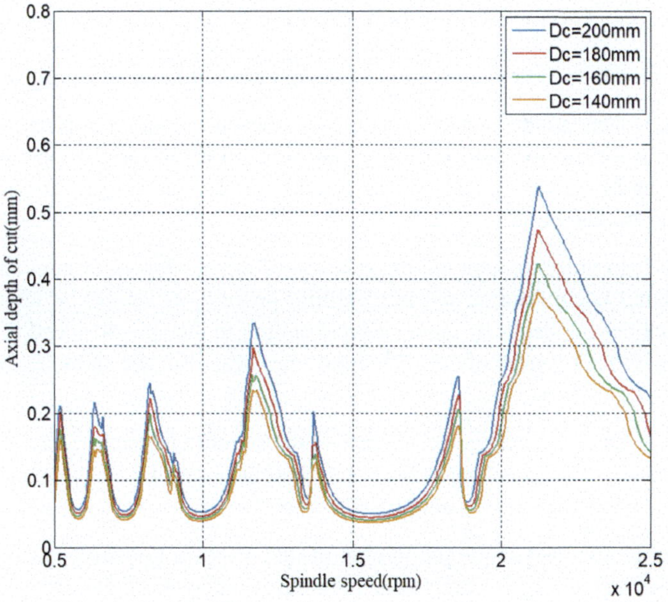

Fig. 2.42 Trend chart of stability lobe curves with varying diameter of disc milling cutter

It is evident that increasing the values of process parameters results in a higher material removal rate. All factors on the right side of Eq. (2.78), with the exception of the radial cutting depth a_e, can be set before machining, are independent of each other, and remain constant once established. However, due to the changing geometry of the workpiece, a_e changes with the cutting distance L_c throughout the face milling operation. Consequently, Q_{MRR} is time-dependent, and its average value $\overline{Q}_{MRR}$ is required for assessing machining efficiency, which complicates optimization modeling. To address this issue, this section introduces a simplified objective function. According to Eq. (2.62), by considering L_c as a constant, a_e becomes a monotonically increasing function of D_c in *Phase I*. It is also demonstrated that in subsequent phases of the face milling process, the average value of a_e is independent of D_c due to the symmetry of the cylindrical bores, as per Hypothesis 1. Furthermore, based on the positive assessment of D_c in Sect. 2.2.2.4, moderately increasing D_c enhances both the system's dynamic stability and the level of a_e. In essence, the optimization model's objective can be streamlined by setting D_c to its maximum feasible value D_c^*, eliminating a_e from the right side of Eq. (2.78), and maximizing the product of the remaining four factors, thereby effectively optimizing machining efficiency. Thus, the objective function is formulated as:

$$\max a_p f_z \Omega N_c \tag{2.79}$$

Main constraints in process parameter design are summarized as follows.

Constraint 1: The spindle speed Ω should be chosen within an appropriate range. The lower and upper bounds are jointly determined by material properties of the workpiece and the load capacity of the machine tool, which are set to Ω_{st} and Ω_{fi} respectively in this section.

Constraint 2: In order to maintain the stability of the machining system and ensure that no chatter occurs during the face milling process, a_p should be less than the critical axial cutting depth $a_{p\_lim}(\Omega)$ corresponding to the spindle speed Ω. That is, the point with coordinates of (a_p, Ω) should be located in the stable region of the SLD.

Constraint 3: Since the evaluation indexes of the stability lobe curve worsen rapidly with the growth of f_z according to Sect. 2.4, there is no need to set an explicit upper bound for f_z while the lower bound is set to zero.

Constraint 4: Obviously, the total number of inserts N_c is a positive integer, and its highest possible level N_c^* is equal to the maximum number of inserts on a disc milling cutter with a diameter of D_c^*. Note that N_c is usually set to an even number to improve the symmetry of the cutting forces in engineering practice, which has significantly reduced the feasible domain of this problem.

Therefore, the optimization model is subjected to

$$\begin{cases} \Omega_{st} < \Omega < \Omega_{fi} \\ 0 < a_p < a_{p\_lim}(\Omega) \\ f_z > 0 \\ N_c \in \{2, 4, \ldots, N_c^*\} \end{cases} \tag{2.80}$$

Given the nonlinear and intricate nature of the optimization model as defined by Eqs. (2.79) and (2.80), a genetic algorithm (GA) is utilized to address the model. Following the standard protocols of GA, the principal computational rules are detailed below.

Rule for variable discretization: Eq. (2.80) shows that all process parameters, except for N_c, are continuous, posing challenges for GA in solving the optimization model. Thus, a discretization rule is essential. The main challenge is to set the search step size for each continuous variable. An overly large step size may lead GA to converge on local optima rather than the global optimum, while an excessively small step size can result in unnecessary computations, diminishing the efficiency of the proposed method. Moreover, parameters with excessive decimal places (beyond the machine tool's adjustment precision) are impractical for engineering applications. Therefore, balancing solution effectiveness, computational efficiency, and practical applicability, the search step sizes for a_p, Ω and f_z are determined to be 0.01 mm, 100 rpm, and 0.001 mm, respectively.

Rule for selection, crossover, and mutation: To bolster the robustness of the optimization process, a roulette-wheel selection method [108] is employed. This method involves the random selection of two chromosomes based on their fitness's cumulative distribution function, rather than relying solely on the highest fitness for selection, to undergo crossover and mutation. The probability of crossover remains constant, while the mutation probability increases incrementally with each generation, enhancing the algorithm's capacity to evade local optima.

Rule for stopping criteria: The optimization process will cease when the highest fitness level remains unchanged for a specified number of generations, which is designated as 200 in this context, indicating that the GA has reached the global optimum.

2.2.3 *Case Study*

2.2.3.1 Dynamic Cutting Force Measurement

To confirm the accuracy of the proposed dynamic cutting force model that incorporates the impact of cutting insert engagement, six distinct sets of face milling tests are conducted, each with varying process parameters, for the purpose of measuring dynamic cutting forces. The specific process parameters used for these force measurements are outlined in Table 2.13.

Table 2.13 Experiment design for cutting force measurement

Case No.	f_z (mm)	a_p (mm)	Ω (rpm)	N_c
(1-1)	0.01	0.05	5000	6
(1-2)	0.02	0.2	28,000	8
(1-3)	0.016	0.13	15,000	10
(1-4)	0.012	0.28	20,000	8
(1-5)	0.024	0.15	17,000	10
(1-6)	0.026	0.22	23,000	6

The experimental setup for dynamic cutting force measurement is illustrated in Fig. 2.43. The machining operations are conducted in a computer numerical control (CNC) machining center of type DMG-HSC-75 (see Fig. 2.43a). According to the type of the CNC, the highest possible level of diameter of the disc milling cutter is 200 mm. Thus D_c^* is set to 200 mm in this section. Cutting fluid of type Quaker-370-KLG is used during the face milling process. The components of dynamic cutting force are measured by a three-component dynamometer of type Kistler-9027C (see Fig. 2.43b). The acceleration sensors of the dynamometer are embedded in the vise to detect the signals of dynamic cutting forces. The sampling period for each single operation is set to 30 s. The whole experimental process is under closely supervised conditions to ensure that no anomalous problems emerge.

Impact testing can be used to determine the frequency response functions (FRFs) of the machine tool system. The disc milling cutter has a cantilever length of 70 mm. To minimize uncertainty in the impact testing, it is performed three times in each direction, and the average FRF values are considered as the final measurements. The "Peak-picking" method [109] is applied to extract the modal parameters of the machine tool system. Figure 2.44 displays the real and imaginary parts of the FRFs measured. It is assumed in this section that the cross FRFs of the machine tool system are equivalent.

Limited by the size of the dynamometer vise (with a maximum opening distance of about 100 mm), the face milling operations are performed on the top surface of a quarter of the four-cylinder engine block made of gray cast iron (HT250). Since a complete cylinder bore is contained in this quarter, the most representative geometric characteristics of the engine block are reserved. The analysis of cutting insert engagement proposed in Sect. 2.2 is still applicable to this case. Based on the geometry dimensions of the disc milling cutter and the engine block given in Table 2.14, it can be obtained that $\sqrt{D_c^2 - (D_c - 2L_1)^2}$ is around 111.6 mm, which is larger than L_b. Therefore, the cutting insert engagement in this case can be calculated with the emergence of *Phase I**.

In the course of the experiments, it was observed that chatter manifested in cases (1-2), while other cases remained more stable. Using case (1-4) as a stable cutting scenario in comparison to the unstable case (1-2), the dynamic cutting force components along the X-axis and their corresponding frequency spectra are depicted in Fig. 2.45. In both stable and unstable cutting scenarios, the cutting force waveform

Fig. 2.43 Experimental setup for dynamic cutting force measurement **a** CNC machining center **b** fixturing by the dynamometer vise

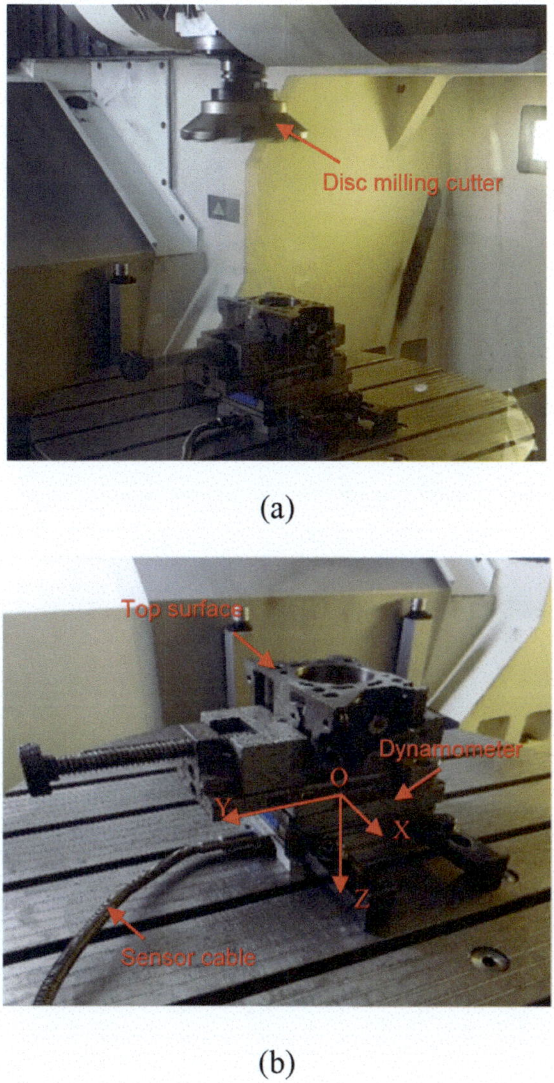

exhibits two peaks and one valley (see Fig. 2.45a, c). This indicates that the cutting force first increases then diminishes, and subsequently increases again as the face milling operation progresses. When the disc milling cutter approaches the closest generatrix of the cylindrical bore, the cutting arc length begins to shorten, and the number of inserts in contact with the workpiece decreases. However, as the edge of the disc milling cutter passes the center of the cylindrical bore, both values increase once more. The cutting force magnitude follows a pattern similar to that of the cutting insert engagement. Furthermore, in the stable cutting case (1-4), the predominant frequency of the cutting forces matches the tool passing frequency of 330 Hz (see

Fig. 2.44 Frequency response functions of the machine tool system **a** G_{xx} **b** G_{xy} **c** G_{yy}

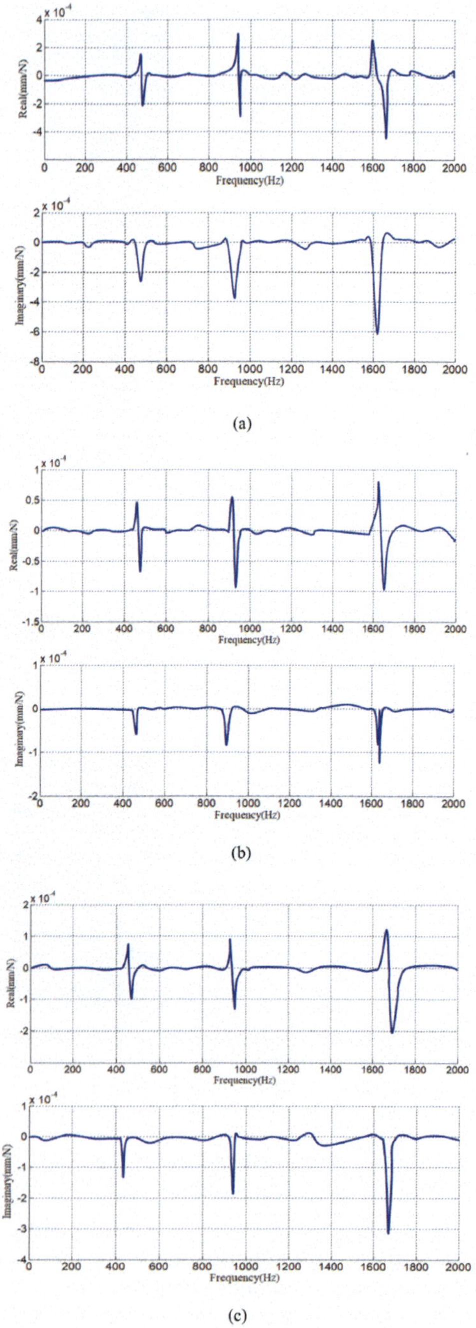

Table 2.14 Geometry dimensions (mm) for cutting insert engagement

D_c	L_1	L_2	L_3	L_4	L_5	L_6	L_7	L_8	L_a	L_b	R_b
200	17	87	92.5	162.5	168	238	243.5	313.5	326	108	35

Fig. 2.45b). In contrast, due to chatter vibrations, the cutting force magnitude in the unstable case (1–2) rises considerably, and the chatter frequency of 1650 Hz becomes dominant (see Fig. 2.45d), which corresponds to one of the natural frequencies of the machining system. Thus, the experimental outcomes are logical and align with the proposed theoretical framework.

2.2.3.2 Validation of Optimized Process Parameters

To confirm the benefits of the proposed method, a traditional approach [84] is applied to establish a baseline design of process parameters for comparative analysis. Utilizing GA to solve the optimization model, which aims to maximize the equivalent processing efficiency, the optimal set of process parameters is derived. Table 2.15 presents a comparison between the baseline and optimized process parameter designs.

Figure 2.46 displays the fitness convergence of the GA. It is observed that the maximum fitness experiences a phased increase as the generations progress from 1 to 1000. The termination condition is ultimately met when the maximum fitness remains consistently at 3380.3 for more than 200 generations, indicating that the iterative process has reached the global optimum. It should be noted that the default process parameter design is bypassed during optimization, as indicated by the red dashed line in Fig. 2.46, which is a common occurrence for GA [110, 111].

The SLDs for both the optimized and default process parameter sets are depicted using MATLAB R2013b, as illustrated in Fig. 2.47. It is observed that the optimized SLD shows a significant enhancement over the default one. To assess the accuracy of the chatter predictions made by the SLDs from the proposed method, 30 face milling experiments are performed on the top surface of a full-scale four-cylinder engine block. In the case of the optimized SLD, a disc milling cutter with a diameter of 200 mm and 10 cubic boron nitride (CBN) inserts are utilized, with a feed per insert f_z of 0.023 mm. For the default SLD, a disc milling cutter with a diameter of 160 mm and 12 CBN inserts are used, with f_z set to 0.02 mm. Table 2.16 provides the specific details for the axial cutting depth a_p and spindle speed Ω.

These 30 experimental cases are indicated in Fig. 2.47 with red diamonds and blue circles. For the default SLD, 10 cases are labeled from (2-1) to (2-15), where empty red diamonds denote stable cuts and filled ones indicate chatter. For the optimized SLD, the remaining 10 cases are labeled from (3-1) to (3-15), with empty blue circles representing stable cuts and filled ones representing chatter. Notably, case (2-8) and case (3-8) are executed using the default and optimal process parameter designs,

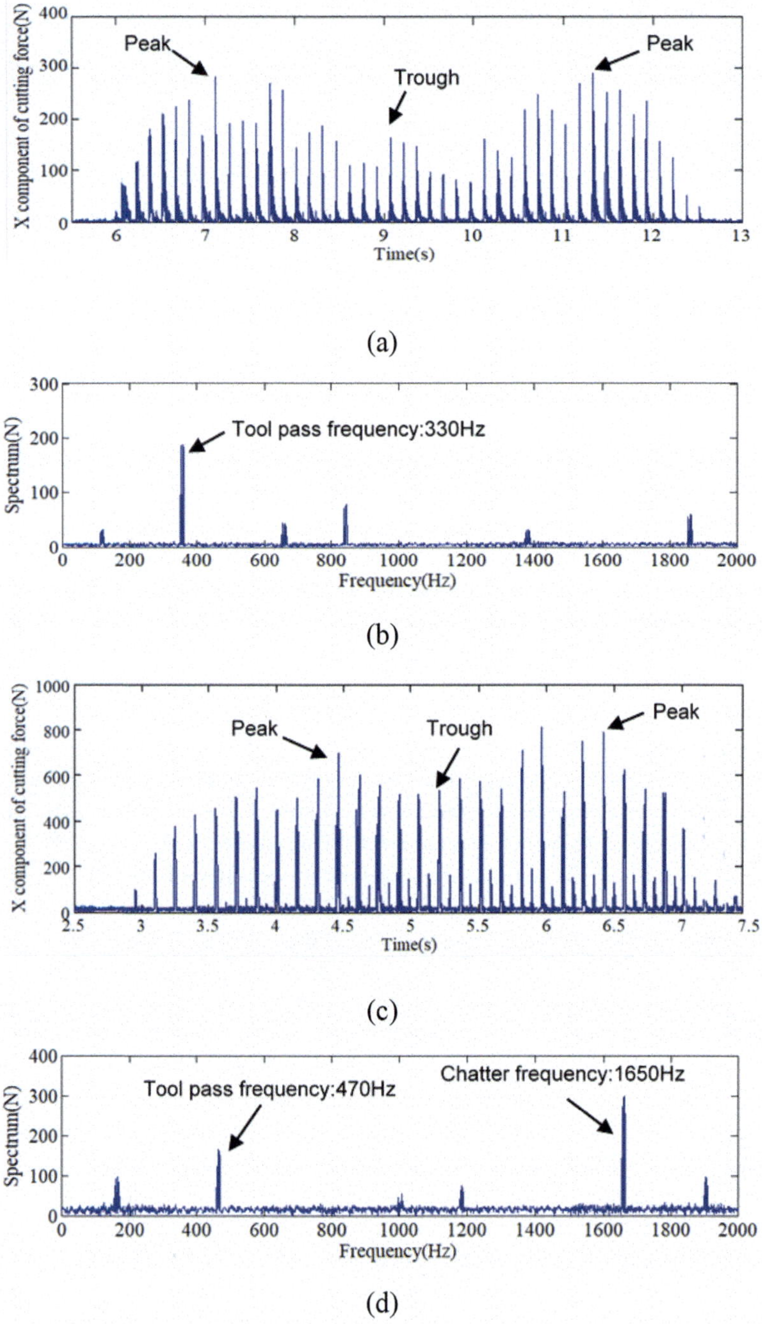

Fig. 2.45 Dynamic cutting forces and spectra for stable cutting case and chatter case **a** dynamic cutting forces of stable cutting case (1-4) **b** cutting force spectrum of stable cutting case (1-4) **c** dynamic cutting forces of chatter case (1-2) **d** cutting force spectrum of chatter case (1-2)

Table 2.15 Comparison between the default design and optimized design

Process parameter	Default design	Optimized design
f_z (mm)	0.02	0.023
D_c (mm)	160	200
N_c	12	10
a_p (mm)	0.5	0.69
Ω (rpm)	21,200	21,300

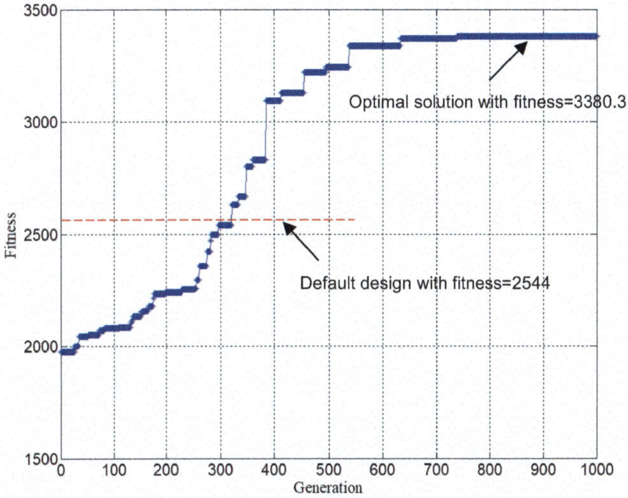

Fig. 2.46 The trend of the maximum fitness value as generation grows

respectively, and both exhibit stable cutting, aligning with the SLDs predicted by the proposed approach.

The observation indicates that case (3-3) was expected to be unstable as per the optimized SLD, yet no chatter was actually detected during its machining process. Beyond any potential experimental inaccuracies, two plausible explanations exist for this discrepancy. Firstly, the axial cutting depth and spindle speed for case (3-3) are close to one of the sub-lobes of the optimized SLD, potentially due to computational variances stemming from the geometric properties of the engine block. In other words, while simplifying the engine block facilitates computation, it also introduces computational discrepancies, resulting in sub-lobes near the critical axial cutting depth. Secondly, chatter prediction complexity involves various factors, including the dynamic behavior of the machine spindle and the stability of the fixture system, which are treated as controlled variables and assumed constant in this section. These factors could impact the prediction's accuracy to some extent. Nonetheless, the chatter prediction accuracy based on the proposed method is sufficiently effective

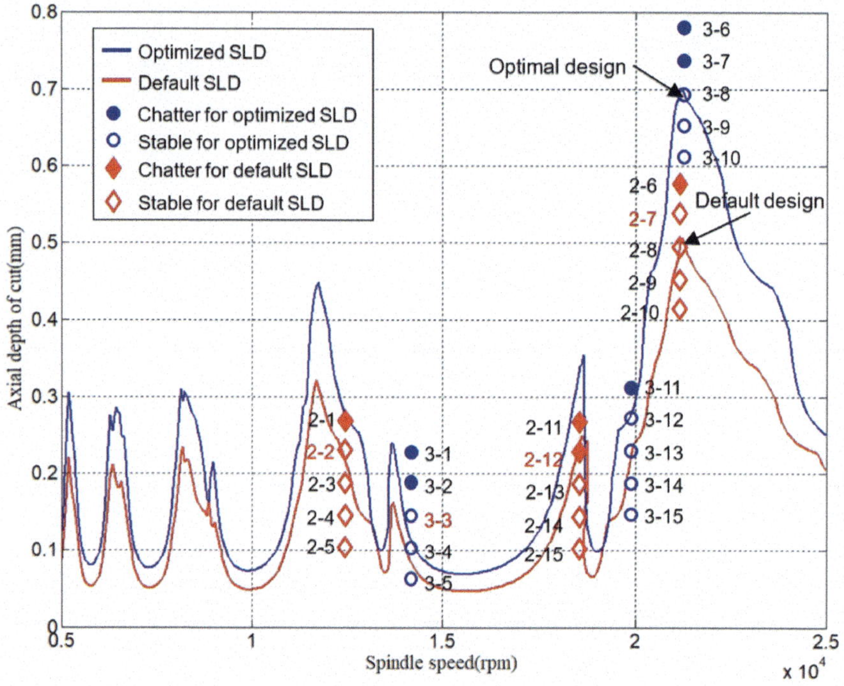

Fig. 2.47 Predicted and measured chatter stability results

and within acceptable limits, confirming that the optimization of process parameters is founded on a reliable dynamic model.

The experimental arrangement is depicted in Fig. 2.48. A 3-2-1 alignment principle is applied for securing the workpiece, with the engine block being positioned and secured by a custom fixture as seen in Fig. 2.48a. The flatness of the machined surface is assessed using a LEITZ-PMM-XI type coordinate measurement machine (CMM) system, as illustrated in Fig. 2.48b. For each measurement, a CMM probe randomly samples and measures 50 points across the top surface. The measurement outcomes for cases (2-3), (3-2), (2-8), and (3-8) are presented in Fig. 2.49.

It is evident that the surface finish of the stable cutting cases, namely cases (2-3), (2-8), and (3-8), is significantly superior to that of the chatter case, case (3-2). The flatness of the machined surfaces for each scenario is determined using the method of least squares. Additionally, the average material removal rate (MRR) for each scenario is derived using Eqs. (2.62) and (2.78). The computed results are summarized in Table 2.17.

It is found that chatter not only exacerbates the flatness deviation but also results in an elevation bias of the machined surface mean height (approximately 0.02 mm above the anticipated level), further highlighting the importance of chatter prediction in achieving surface finish quality. The results also show that the proposed method

Table 2.16 Axial cutting depth and spindle speed for validation of chatter prediction

Default SLD			Optimized SLD		
Case No.	a_p (mm)	Ω (rpm)	Case No.	a_p (mm)	Ω (rpm)
(2-1)	0.27	12,500	(3-1)	0.23	14,200
(2-2)	0.23	12,500	(3-2)	0.19	14,200
(2-3)	0.19	12,500	(3-3)	0.15	14,200
(2-4)	0.15	12,500	(3-4)	0.11	14,200
(2-5)	0.11	12,500	(3-5)	0.07	14,200
(2-6)	0.58	21,200	(3-6)	0.77	21,300
(2-7)	0.54	21,200	(3-7)	0.73	21,300
(2-8)	0.5	21,200	(3-8)	0.69	21,300
(2-9)	0.46	21,200	(3-9)	0.65	21,300
(2-10)	0.42	21,200	(3-10)	0.61	21,300
(2-11)	0.27	18,300	(3-11)	0.31	19,900
(2-12)	0.23	18,300	(3-12)	0.27	19,900
(2-13)	0.19	18,300	(3-13)	0.23	19,900
(2-14)	0.15	18,300	(3-14)	0.19	19,900
(2-15)	0.11	18,300	(3-15)	0.15	19,900

has nearly increased the average MRR by 53.3%, from 178,589 to 273,805 mm^3/min. Furthermore, by significantly enhancing the average MRR under chatter-free conditions through process parameter optimization, the quality of the machined surface is also maintained.

In this section, a case study on engine block faces is detailed to demonstrate the assessment of 3D surface form error using high-definition measurement (HDM) data. The derived 3D form error characteristics are contrasted with flatness, and the efficacy of the proposed approach is examined.

2.2.4 Conclusions

This section introduces an analytical method for optimizing process parameters, grounded in dynamic modeling for face milling operations on workpieces with discontinuous surfaces. A new dynamic cutting force model is developed, fully accounting for the influence of cutting insert engagement. The section offers a practical solution for enhancing process efficiency without chatter, using a four-cylinder engine block as a case study.

In this section, we present an analytical technique to optimize process parameters, which is rooted in the dynamic modeling of face milling for workpieces featuring discontinuous surfaces. We have formulated a novel dynamic cutting force model that incorporates the impact of cutting insert engagement. Through the application of

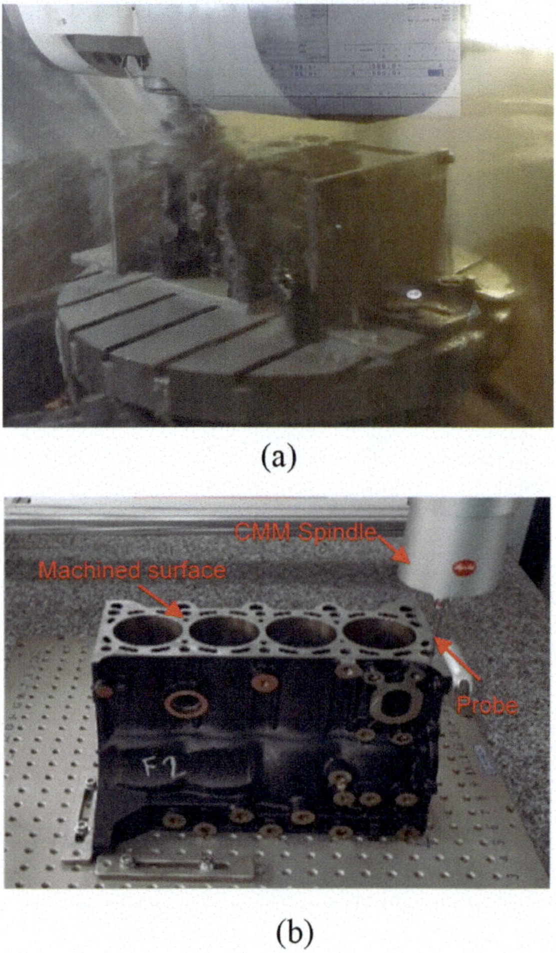

Fig. 2.48 Experimental setup for validation of optimized process parameters **a** face milling process on the four-cylinder engine block **b** flatness measurement by the CMM system

a four-cylinder engine block, this research provides a viable method to boost process efficiency in the absence of chatter.

The face milling operation is segmented into multiple stages, with dynamic responses for each stage determined by the engagement of cutting inserts. Furthermore, SLDs with sufficient precision are established for predicting chatter in the face milling of workpieces with discontinuous surfaces. A sensitivity analysis is conducted to measure the impact of varying process parameters on the SLDs. Drawing on the findings from this analysis, an optimization model is formulated to maximize the equivalent MRR, and the GA is utilized to address this model. To validate the proposed method, a suite of machining tests is carried out to assess dynamic cutting forces and to confirm the optimized process parameters. The experimental outcomes align well with the theoretical predictions, confirming that the

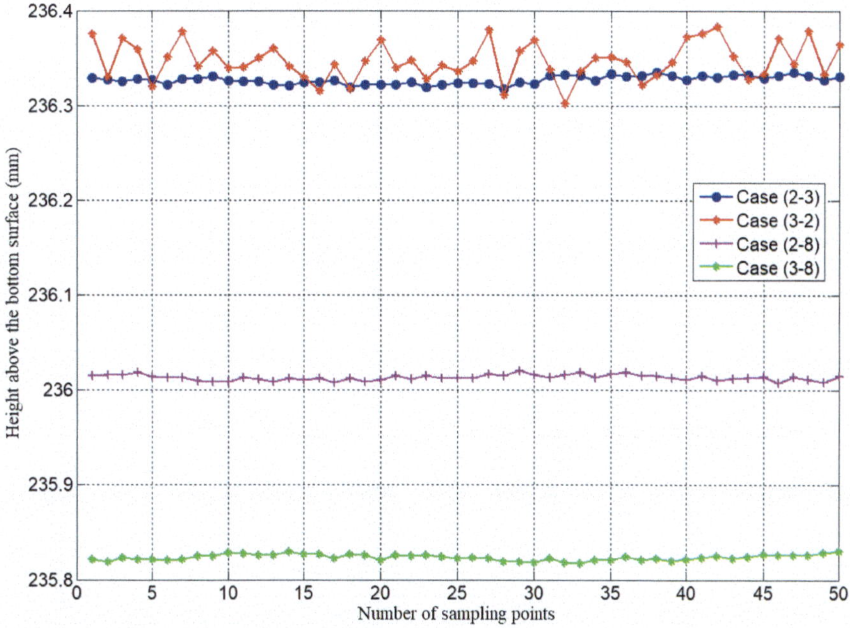

Fig. 2.49 Measured heights of the sampling points

Table 2.17 Experimental results for each case

Case No.	(2-3)	(3-2)	(2-8)	(3-8)
Mean height of machined surface (mm)	236.33	236.35	236.02	235.83
Flatness of machined surface (mm)	0.0226	0.0721	0.0230	0.0232
Average MRR (mm^3/min)	67,864	75,396	178,589	273,805

proposed approach can significantly enhance processing efficiency while maintaining the quality of the machined surface.

Given that the top surface of a four-cylinder engine block represents a common discontinuous surface, the proposed approach's success with the engine block suggests its applicability to optimizing process parameters for other discontinuous surface workpieces. This method can serve as a reference for industrial engineers to select the best process parameter settings at the design stage, thereby maximizing processing efficiency without chatter.

For future research, additional internal factors influencing chatter stability, such as the spindle's gyroscopic effect and fixture rigidity, should be considered. Additionally, determining the cutter-workpiece interaction for peripheral and end milling processes will be crucial to enhancing the versatility of the proposed dynamic model.

Appendix 1

Derivation of Eq. (2.43):

Substituting Eq. (2.40) into Eq. (2.38), yields

$$\varepsilon_r = -\frac{2P\cos\theta}{\pi Er} \tag{2.81}$$

$$\varepsilon_\theta = \frac{2\upsilon P\cos\theta}{\pi Er} \tag{2.82}$$

$$\gamma_{r\theta} = 0 \tag{2.83}$$

Substituting the above equations into Eq. (2.37), yields

$$\frac{\partial u}{\partial r} = -\frac{2P\cos\theta}{\pi Er} \tag{2.84}$$

$$\frac{\partial v}{r\partial\theta} + \frac{u}{r} = \frac{2\upsilon P\cos\theta}{\pi Er} \tag{2.85}$$

$$\frac{\partial v}{\partial r} + \frac{\partial u}{r\partial\theta} - \frac{v}{r} = 0 \tag{2.86}$$

Equation (2.84) can be rewritten as

$$u = \int \varepsilon_r dr = -\frac{2P}{\pi E}\ln r\cos\theta + g(\theta) \tag{2.87}$$

where $g(\theta)$ is a function of variable θ.

Substituting Eq. (2.87) into Eq. (2.85), yields

$$v = \int (r\varepsilon_\theta - u)d\theta = \frac{2\upsilon P}{\pi E}\sin\theta + \frac{2P}{\pi E}\sin\theta\ln r - G(\theta) + f(r) \tag{2.88}$$

where $G(\theta)$ is the primitive of $g(\theta)$,while $f(r)$ is a function of variable r.

Substituting Eqs. (2.87) and (2.88) into Eq. (2.86), it can be further derived that

$$\frac{2(1-\upsilon)P}{\pi E}\sin\theta + g'(\theta) + G(\theta) = f(r) - rf'(r) \tag{2.89}$$

To keep Eq. (2.89) always valid, both side of this equation shall be 0.

By solving two ordinary differential equations, the general solutions to $f(r)$ and $g(\theta)$ are obtained respectively:

$$f(r) = C_1 r \tag{2.90}$$

$$g(\theta) = -\frac{(1-\upsilon)P}{\pi E}\sin\theta + C_2\sin\theta + C_3\cos\theta \tag{2.91}$$

where C_1,C_2 and C_3 are undetermined constant coefficients.

Therefore, the displacement field can be represented as

$$u = -\frac{2P}{\pi E}\ln r\cos\theta - \frac{(1-\upsilon)P}{\pi E}\sin\theta + C_2\sin\theta + C_3\cos\theta \tag{2.92}$$

$$v = \frac{2P}{\pi E}(\ln r + \upsilon)\sin\theta + \frac{(1-\upsilon)P}{\pi E}(\sin\theta - \theta\cos\theta) + C_2\cos\theta - C_3\sin\theta + C_1 r \tag{2.93}$$

To calculate above three constant coefficients, considering the boundary conditions shown in Fig. 2.13, the horizontal displacement remains 0 for all mass points on the X-axis.

$$v|_{\theta=0} = C_2 + C_1 r = 0 \tag{2.94}$$

Thus $C_1 = C_2 = 0$.

The vertical displacement of the mass points on the X-axis can be expressed as

$$u|_{\theta=0} = -\frac{2P}{\pi E}\ln r + C_3 \tag{2.95}$$

For an elastic sheet loaded by a vertical concentrated force on its straight boundary, if the vertical extent is sufficiently long, the influence range of the load will be limited. Since d_B denotes the minimum distance between the local coordinate origin and the point on X-axis whose vertical displacement is 0, it can be derived that

$$u|_{\theta=0,r=d_B} = -\frac{2P}{\pi E}\ln d_B + C_3 = 0 \tag{2.96}$$

Thus $C_3 = \frac{2P}{\pi E}\ln d_B$.

Appendix 2

The distance between the centers of the disc milling cutter and the first cylindrical bore can be formulated as

$$L_{y1} = \frac{D_c}{2} + R_b - L_c + L_1 \tag{2.97}$$

According to the Heron's Formula, he area of the triangular $Q_1 O_c O_b$ can be given by

$$S_{t1} = \sqrt{p\left(p - \frac{D_c}{2}\right)(p - R_b)(p - L_{y1})} \tag{2.98}$$

where p is an intermediate parameter, given by

$$p = \frac{1}{2}\left(\frac{D_c}{2} + R_b + L_{y1}\right) \tag{2.99}$$

References

1. Du SC, Liu T, Huang DL (2018) A fast and adaptive bi-dimensional empirical mode decomposition approach for filtering of workpiece surfaces using high definition metrology. J Manuf Syst 46:247–263
2. Zeng SS, Wan XJ, Li WL, Yin ZP, Xiong YL (2012) A novel approach to fixture design on suppressing machining vibration of flexible workpiece. Int J Mach Tools Manuf 58:29–43
3. Wan XJ, Zhang Y, Huang XD (2013) Investigation of influence of fixture layout on dynamic response of thin-wall multi-framed work-piece in machining. Int J Mach Tools Manuf 75:87–99
4. Hashemi H, Shaharoun AM, Sudin I (2014) A case-based reasoning approach for design of machining fixture. Int J Adv Manuf Technol 74:113–124
5. Meshreki M, Kövecses J, Attia H (2008) Dynamics modeling and analysis of thin-walled aerospace structures for fixture design in multi-axis milling. J Manuf Sci Eng 130:031011-1–12
6. Tonietti G, Schiavi R, Bicchi A (2005) Design and control of a variable stiffness actuator for safe and fast physical human/robot interaction. In: International conference on robotics and automation, proceedings of the 2005 IEEE, pp 526–531
7. Alblalaihid K, Lawes S, Kinnell P (2016) Variable stiffness probing systems for micro-coordinate measuring machines. Precis Eng 43:262–269
8. Varadarajan N, Nagarajaiah S (2004) Wind response control of building with variable stiffness tuned mass damper using empirical mode decomposition/hilbert transform. J Eng Mech 130(4):451–458
9. Du SC, Fei L (2016) Co-Kriging method for form error estimation incorporating condition variable measurements. J Manuf Sci Eng 138:041003-1–16
10. Wang M, Xi LF, Du SC (2014) 3D surface form error evaluation using high definition metrology. Precis Eng 38(1):230–236
11. Du SC, Liu CP, Huang DL (2015) A shearlet-based separation method of 3D engineering surface using high definition metrology. Precis Eng 40:55–73
12. Vasundara M, Padmanaban KP (2014) Recent developments on machining fixture layout design, analysis, and optimization using finite element method and evolutionary techniques. Int J Adv Manuf Technol 70:79–96
13. Yi SN, Xu L, Cheng GD (2015) FEM formulation of homogenization method for effective properties of periodic heterogeneous beam and size effect of basic cell in thickness direction. Comput Struct 156:1–11

14. Huang DL, Du SC, Li GL (2017) A systematic approach for on-line minimizing volume difference of multiple chambers in machining processes based on high definition metrology. J Manuf Sci Eng 139(8):081003-1-17
15. Yi SN, Cheng GD, Xu L (2016) Stiffness design of heterogeneous periodic beam by topology optimization with integration of commercial software. Comput Struct 172:71–80
16. Cai W, Hu SJ (1996) Deformable sheet metal fixturing: principles, algorithms, and simulations. J Manuf Sci Eng 118:318–324
17. Liu SG, Zheng L, Zhang ZH (2006) Optimal fixture design in peripheral milling of thin-walled workpiece. Int J Adv Manuf Technol 28:653–658
18. Chen WF, Ni LJ, Xue JB (2008) Deformation control through fixture layout design. Int J Adv Manuf Technol 38:860–867
19. Rai JK, Xirouchakis P (2008) Finite element method based machining simulation environment for analyzing part errors induced during milling of thin-walled components. Int J Mach Tools Manuf 48:629–643
20. Bellifa H, Benrahou KH, Hadji L (2016) Bending and free vibration analysis of functionally graded plates using a simple shear deformation theory and the concept the neutral surface position. J Braz Soc Mech Sci Eng 38:265–275
21. Ebrahimi F, Barati MR (2016) A nonlocal higher-order shear deformation beam theory for vibration analysis of size-dependent functionally graded nanobeams. Arab J Sci Eng 41:1679–1690
22. Shan C, Lv X, Duan W (2016) Effect of tool inclination angle on the elastic deformation of thin-walled parts in multi-axis ball-end milling. Procedia CIRP 56:311–315
23. Kaya N, Ozturk F (2003) The application of chip removal and frictional contact analysis for workpiece-fixture layout verification. Int J Adv Manuf Technol 21(6):411–419
24. Carlson JS (2001) Quadratic sensitivity analysis of fixtures and locating schemes for rigid parts. J Manuf Sci Eng 123:462–472
25. Sadd MH (2010) Some simple cartesian solutions to plane non-homogeneous elasticity problems. Mech Res Commun 37:22–27
26. Pimenov DY, Guzeev VI, Krolczyk G, Mia M, Wojciechowski S (2018) Modeling flatness deviation in face milling considering angular movement of the machine tool system components and tool flank wear. Precis Eng 54:327–337
27. Wojciechowski S (2011) Machined surface roughness including cutter displacements in milling of hardened teel. Metrol Meas Syst 18(3):429–440
28. Kang Y, Rong Y, Yang JA (2003) Geometric and kinetic model based computer-aided fixture design verification. J Comput Inf Sci Eng 3:187–199
29. Kang Y, Rong Y, Yang JC (2003) Computer-Aided fixture design verification. Part 1. The framework and modelling. Int J Adv Manuf Technol 21:827–835
30. Liao YG, Hu SJ (2000) Flexible multibody dynamics based fixture-workpiece analysis model for fixturing stability. Int J Mach Tools Manuf 40:343–362
31. Liao YG, Hu SJ (2001) An integrated model of a fixture-workpiece system for surface quality prediction. Int J Adv Manuf Technol 17:810–818
32. Gong X, Lu Q, Wang Y, Song Y (2018) Mechanical design and finite element analyses of surface bending mechanism for X-ray optics. Precis Eng 53:107–119
33. Doostie S, Hoshiar AK, Nazarahari M, Lee S, Choi H (2018) Optimal path planning of multiple nanoparticles in continuous environment using a novel adaptive genetic algorithm. Precis Eng 53:65–78
34. Lai HY, Jywe WY, Chen CK, Liu CH (2000) Precision modeling of form errors for cylindricity evaluation using genetic algorithms. Precis Eng 24:310–319
35. Qiu MK, Ming Z, Li JY, Gai KK, Zong ZL (2015) Phase-Change memory optimization for green cloud with genetic algorithm. IEEE Trans Comput 64(12):3528–3540
36. Krishnakumar K, Melkote SN (2000) Machining fixture layout optimization using the genetic algorithm. Int J Mach Tools Manuf 40:579–598
37. Timoshenko SP, Goodier JN (1934) Theory of elasticity. McGraw-Hill Book Co., New York

38. Du SC, Liu T, Huang DL (2017) An optimal ensemble empirical mode decomposition method for vibration signal decomposition. J Vib Acoust 139:031003-1–18
39. Suriano S, Wang H, Shao CH, Hu SJ, Sekhar P (2015) Progressive measurement and monitoring for multi-resolution data in surface manufacturing considering spatial and cross correlations. IIE Trans 47:1033–1052
40. Du SC, Liu CP, Xi LF (2015) A selective multiclass support vector machine ensemble classifier for engineering surface classification using high definition metrology. J Manuf Sci Eng 137(1):011003
41. Du SC, Huang DL, Wang H (2015) An adaptive support vector machine-based workpiece surface classification system using high-definition metrology. IEEE Trans Instrum Meas 64(10):2590–2604
42. Shirase K, Altintas Y (1996) Cutting force and dimensional surface error generation in peripheral milling with variable pitch helical end mills. Int J Mach Tools Manuf 36(5):567–584
43. Ko JH, Cho DW (2005) 3D ball-end milling force model using instantaneous cutting force coefficients. J Manuf Sci Eng 127:1–12
44. Wan M, Zhang WH, Dang JW, Yang Y (2009) New procedures for calibration of instantaneous cutting force coefficients and cutter runout parameters in peripheral milling. Int J Mach Tools Manuf 49:1144–1151
45. Wan M, Pan WJ, Zhang WH, Ma YC, Yang Y (2014) A unified instantaneous cutting force model for flat end mills with variable geometries. J Mater Process Technol 214:641–650
46. Li B, Melkote SN (2001) Fixture clamping force optimization and its impact on workpiece location accuracy. Int J Adv Manuf Technol 17:104–113
47. Roy U, Liao J (2002) Fixturing analysis for stability consideration in an automated fixture design system. J Manuf Sci Eng 124:98–104
48. Varadarajan KM, Culpepper ML (2007) A dual-purpose positioner-fixture for precision six-axis positioning and precision fixturing part II. Characterization and calibration. Precis Eng 31:287–292
49. Wang M, Shao YP, Du SC, Xi LF (2015) A diffusion filter for discontinuous surface measured by high definition metrology. Int J Precis Eng Manuf 16(10):2057–2062
50. Wang J, Yu T, Ding W, Fu Y, Bastawros AF (2018) Wear evolution and stress distribution of single CBN superabrasive grain in high-speed grinding. Precis Eng 54:70–80
51. Arola D, Williams CL (2002) Estimating the fatigue stress concentration factor of machined surfaces. Int J Fatigue 24:923–930
52. Sburlati R (2013) Stress concentration factor due to a functionally graded ring around a hole in an isotropic plate. Int J Solids Struct 50:3649–3658
53. Tang AJ, Liu ZQ (2008) Deformations of thin-walled plate due to static end milling force. J Mater Process Technol 206:345–351
54. Zhang HY, Wang MY, Li J, Zhu J (2016) A soft compressive sensor using dielectric elastomers. Smart Mater Struct 25(3):035045-1-7
55. Field FA (1964) A comment on the uniqueness of the solution to boussinesq's problem. J Appl Mech 31(3):547
56. Gao X, Hao F, Fang DN, Huang ZP (2013) Boussinesq problem with the surface effect and its application to contact mechanics at the nanoscale. Int J Solids Struct 50:2620–2630
57. Kaya N (2006) Machining fixture locating and clamping position optimization using genetic algorithms. Comput Ind 57:112–120
58. Wang S, Rong W, Wang L, Pei Z, Sun L (2018) A novel inchworm type piezoelectric rotary actuator with large output torque: design, analysis and experimental performance. Precis Eng 51:545–551
59. Hughes TJR (1987) The finite element method: linear static and dynamic finite element analysis. Prentice-Hall Inc., New Jersey
60. Wang M, Ken T, Du S, Xi L (2015) Tool wear monitoring of wiper inserts in multi-insert face milling using three-dimensional surface form indicators. J Manuf Sci Eng 137(3):031006
61. Shao Y, Yin Y, Du S, Xia T, Xi L (2018) Leakage monitoring in static sealing interface based on three dimensional surface topography indicator. J Manuf Sci Eng 140(10):101003

62. Abele E, Fiedler U (2004) Creating stability lobe diagrams during milling. CIRP Ann 53(1):309–312
63. Eksioglu C, Kilic ZM, Altintas Y (2012) Discrete-time prediction of chatter stability, cutting forces, and surface location errors in flexible milling systems. J Manuf Sci Eng 134(6):061006
64. Suriano S, Wang H, Hu SJ (2012) Sequential monitoring of surface spatial variation in automotive machining processes based on high definition metrology. J Manuf Syst 31:8–14
65. Tobias SA (1965) Machine tool vibration. Wiley, New York
66. Tlusty J (1999) Manufacturing process and equipment. Prentice Hall, New Jersey
67. Sridhar R, Hohn RE, Long GW (1968) A stability algorithm for the general milling process. ASME J Eng Ind 90(2):330–334
68. Zheng CM, Wang JJJ, Sung CF (2014) Analytical prediction of the critical depth of cut and worst spindle speeds for chatter in end milling. J Manuf Sci Eng 136(1):011003
69. Minis I, Yanushevsky R, Tembo A, Hocken R (1990) Analysis of linear and nonlinear chatter in milling. CIRP Ann 39(1):459–462
70. Minis I, Yanushevsky R (1993) A new theoretical approach for the prediction of machine tool chatter in milling. J Eng Ind 115(1):1–8
71. Movahhedy MR, Mosaddegh P (2006) Prediction of chatter in high speed milling including gyroscopic effects. Int J Mach Tool Manuf 46:996–1001
72. Ding Y, Zhu L, Zhang X, Ding H (2010) A full-discretization method for prediction of milling stability. Int J Mach Tool Manuf 50:502–509
73. Li Z, Yang Z, Peng Y, Zhu F, Ming X (2016) Prediction of chatter stability for milling process using runge-kutta-based complete discretization method. Int J Adv Manuf Technol 86:943–952
74. Dai Y, Li H, Xing X, Hao B (2018) Prediction of chatter stability for milling process using precise integration method. Precis Eng 52:152–157
75. Ismail F, Soliman E (1997) A new method for the identification of stability lobes in machining. Int J Mach Tools Manuf 37(6):763–774
76. Lee S, Park JK (2018) Experimental verification of dynamic behavior of a capsule-type modular machine tool for multifunctional processes. J Manuf Sci Eng 140(1):014501
77. Tuysuz O, Altintas Y (2018) Time-domain modeling of varying dynamic characteristics in thin-wall machining using perturbation and reduced-order substructuring methods. J Manuf Sci Eng 140(1):011015
78. Tuysuz O, Altintas Y (2017) Frequency domain prediction of varying thin-walled workpiece dynamics in machining. J Manuf Sci Eng 139(7):071013
79. Smith S, Tlusty J (1993) Efficient simulation programs for chatter in milling. CIRP Ann 42(1):463–466
80. Cao H, Li B, He Z (2012) Chatter stability of milling with speed-varying dynamics of spindles. Int J Mach Tool Manuf 52:50–58
81. Quintana G, Ciurana J, Ferrer I, Rodriguez CA (2009) Sound mapping for identification of stability lobe diagrams in milling processes. Int J Mach Tools Manuf 49(3):203–211
82. Altintas Y, Budak E (1995) Analytical prediction of stability lobes in milling. CIRP Ann Manuf Technol 44(1):357–362
83. Merdol SD, Altintaş Y (2014) Multi frequency solution of chatter stability for low immersion milling. J Manuf Sci Eng 126:459–466
84. Jensen SA, Shin YC (1999) Stability analysis in face milling operations, part 1: theory of stability lobe prediction. J Manuf Sci Eng 121(4):600–605
85. Jensen SA, Shin YC (1999) Stability analysis in face milling operations, part 2: experimental validation and influencing factors. J Manuf Sci Eng 121(4):606–614
86. Niu J, Ding Y, Geng Z, Zhu L, Ding H (2018) Patterns of regenerative milling chatter under joint influences of cutting parameters, tool geometries, and runout. J Manuf Sci Eng 140(12):121004
87. Caliskan H, Kilic ZM, Altintas Y (2018) On-Line energy-based milling chatter detection. J Manuf Sci Eng 140(11):111012

88. Tang X, Zhu Z, Yan R, Chen C, Peng F, Zhang M, Li Y (2018) Stability prediction based effect analysis of tool orientation on machining efficiency for five-axis bull-nose end milling. J Manuf Sci Eng 140(12):121015
89. Baek DK, Ko TJ, Kim HS (2001) Optimization of feedrate in a face milling operation using a surface roughness model. Int J Mach Tools Manuf 41(3):451–462
90. Song Q, Ju G, Liu Z, Ai X (2014) Subdivision of chatter-free regions and optimal cutting parameters based on vibration frequencies for peripheral milling process. Int J Mech Sci 83:172–183
91. Zhang X, Ding H (2013) Note on a novel method for machining parameters optimization in a chatter-free milling process. Int J Mach Tools Manuf 72:11–15
92. Zhang X, Zhu L, Zhang D, Ding H, Xiong Y (2012) Numerical robust optimization of spindle speed for milling process with uncertainties. Int J Mach Tools Manuf 61:9–19
93. Zhang X, Zhang D, Cao L, Huang T, Leopold J, Ding H (2017) Minimax optimization strategy for process parameters planning: toward interference-free between tool and flexible workpiece in milling process. J Manuf Sci Eng 139(5):051010
94. Nguyen H, Wang H, Hu S (2012) Characterization of cutting force induced surface shape variation using high-definition metrology. In: ASME 2012 international manufacturing science and engineering conference, pp 641–650
95. Altintas Y, Lee P (1996) A general mechanics and dynamics model for helical end mills. Annals CIRP 45(1):59–64
96. Huang D, Du S, Li G, Zhao C, Deng Y (2018) Detection and monitoring of defects on three-dimensional curved surfaces based on high-density point cloud data. Precis Eng 53:79–95
97. Li G, Du S, Huang D, Zhao C, Deng Y (2019) Elastic mechanics-based fixturing scheme optimization of variable stiffness structure workpieces for surface quality improvement. Precis Eng 56:343–363
98. Zhang JZ, Chen JC, Kirby ED (2007) Surface roughness optimization in an end-milling operation using the taguchi design method. J Mater Process Technol 184:233–239
99. Asiltürk I, Neşeli S (2012) Multi response optimisation of CNC turning parameters via taguchi method-based response surface analysis. Measurement 45(4):785–794
100. Mhapsekar K, McConaha M, Anand S (2018) Additive manufacturing constraints in topology optimization for improved manufacturability. J Manuf Sci Eng 140(5):051017
101. Lu HS, Chang CK, Hwang NC, Chung CT (2009) Grey relational analysis coupled with principal component analysis for optimization design of the cutting parameters in high-speed end milling. J Mater Process Technol 209(8):3808–3817
102. Pang L, Kishawy HA (2012) Modified primary shear zone analysis for identification of material mechanical behavior during machining process using genetic algorithm. 134(4):041003
103. Kim HS, Ehmann KF (1993) A cutting force model for face milling operations. Int J Mach Tools Manuf 33(5):651–673
104. Budak E, Altintas Y (1998) Analytical prediction of chatter stability in milling-part I: general formulation. J Dyn Syst Meas Contr 120(1):22–30
105. Budak E, Altintas Y (1998) Analytical prediction of chatter stability in milling-part II: application of the general formulation to common milling systems. J Dyn Syst Meas Contr 120(1):31–36
106. Dunham W (1990) Journey through genius: the great theorems of mathematics. Wiley, New York
107. Svetlik M, Radojicic M, Radovic S, Simic-Muller K (2018) Justifying Euler's formula through motion in a plane. Math Enthusiast 15(3):397–406
108. Aderiani AR, Wärmefjord K, Söderberg R (2018) A multistage approach to the selective assembly of components without dimensional distribution assumptions. J Manuf Sci Eng 140(7):071015
109. Schmitz TL, Smith KS (2008) Machining dynamics: frequency response to improved productivity. Springer Science & Business Media, Berlin
110. Soori M, Arezoo B, Habibi M (2016) Tool deflection error of three-Axis computer numerical control milling machines, monitoring and minimizing by a virtual machining system. J Manuf Sci Eng 138(8):081005

111. Bosetti P, Bort CMG, Bruschi S (2013) Identification of Johnson-Cook and Tresca's parameters for numerical modeling of AISI-304 machining processes. J Manuf Sci Eng 135(5):051021

Chapter 3
Evaluation and Modeling of Tool Marks in Face Milling Processes

3.1 Segmentation of Workpiece Surfaces with Tool Marks

3.1.1 Introduction

In recent years, the quanlity of surface contours has become more crucial within the mechanical manufacturing industry. The textures of surfaces at varying scales have distinct effects on their functionality. Notably, the mid-scale surface textures play a pivotal role in determining the sealing capabilities of a surface. The surface patterns generated by milling are predominantly characterized by their periodic tool marks.

Utilized frequently for achieving smooth surfaces that ensure effective sealing, face milling is a standard machining process. Figure 3.1 provides a schematic depiction of the surface resulting from face milling. Within this context, a tool mark is defined as the peak of the surface profile, as illustrated in the sectional view of the milled surface presented in Fig. 3.1. The segment between two successive tool marks represents the portion of the surface machined by a single tool tooth during its complete rotation.

In the assembly of automotive engines, the sealing efficacy of the mating surfaces, namely, the top surfaces of the cylinder blocks and the undersides of the cylinder heads, is crucial and is achieved through face milling. The sealing effectiveness is closely linked to the surface's topographical features. Empirical studies have shown that mid-scale surface elements, specifically waviness, significantly influence sealing capabilities [1]. Malburg pioneered a technique to correlate waviness parameters with sealing functions [2], which Shao et al. later advanced to three-dimensional scenarios, introducing 3D waviness-based leakage metrics [3]. Liao et al. utilized wavelet decomposition to isolate the waviness component from the face-milled surfaces measured by high-definition metrology (HDM), revealing that this component is predominantly composed of tool marks [4]. These studies collectively suggest a correlation between the presence of tool marks and the sealing

S. Du and L. Xi, *Quality Control for Variable Stiffness Structured Workpiece*,
https://doi.org/10.1007/978-981-96-5376-8_3

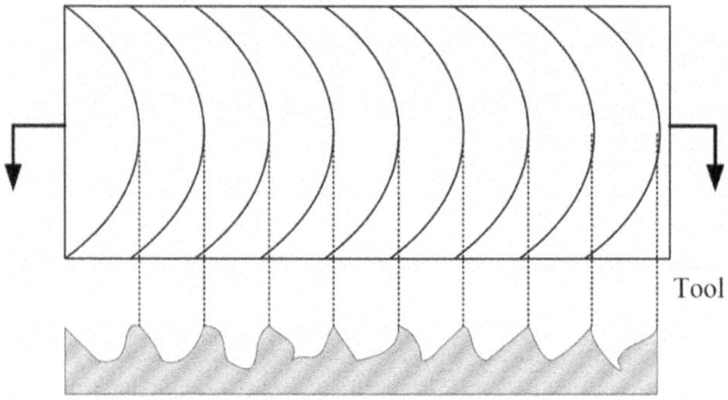

Fig. 3.1 Surface appearance of face milling, single tooth

efficiency of surfaces. Given that tool marks are shaped by the milling parameters, optimizing these parameters could enhance the surface's sealing properties.

Investigations into the correlation between tool marks and milling parameters are primarily conducted through two methodologies: surface simulation and surface monitoring. Surface simulation involves incorporating milling parameters into models to replicate surface topography, while surface monitoring uses measured surface topography as input to infer milling parameters for process monitoring. Surface simulation often constructs dynamic models of milling operations to produce simulated surface topographies. For instance, Shun et al. developed a numerical model that integrates data from *ABAQUS* and *MATLAB* to forecast the surface topography generated by face milling [5]. Sun et al. introduced a comprehensive simulation model to predict 3D face-milled surface topography, taking into account milling parameters and various initial setup errors [6]. Li et al. established a kinetic model for tool cutting to simulate turning and milling surface topographies and examined the impact of machined surface texture on contact performance [7].

Surface monitoring of milled surfaces necessitates the use of HDM technology for tool mark assessment. The Shapix, a newly developed HDM device, has been effectively implemented in automotive engine manufacturing facilities [8]. This instrument is capable of capturing a large surface area and producing a 3D height map in approximately 40 s. The device boasts a measurement resolution of 150 μm in the x–y plane and 0.05 μm in the z-axis. Utilizing this innovative measurement platform, a variety of studies have been conducted, encompassing surface quality assessment [9], monitoring and diagnostics [3], surface filtering [10–13], classification [14, 15], and forecasting [16, 17]. A recently published volume offers more comprehensive insights into HDM-based surface quality control and its applications [18]. Figure 3.2 illustrates the HDM instrument alongside an engine block face as measured by HDM.

Utilizing sophisticated HDM technology, a variety of methods for monitoring milled surfaces have been introduced. Nguyen et al. conducted modeling and experimental studies on face milling processes, establishing a correlation between surface patterns derived from straightened tool marks and variations in cutting force [19].

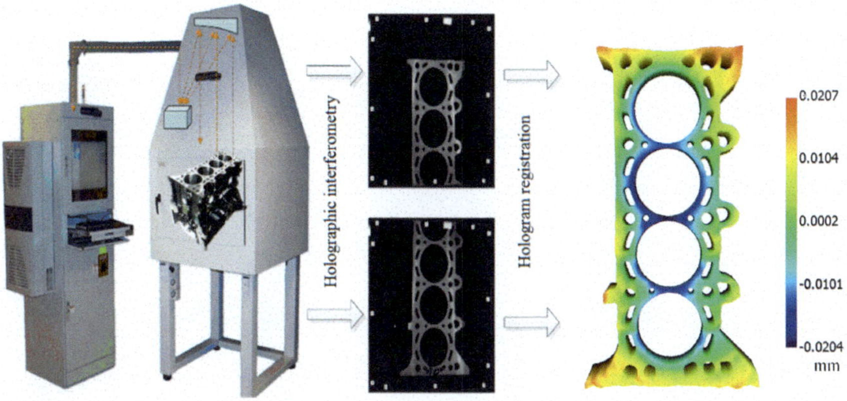

Fig. 3.2 Measurement by HDM

Subsequently, Nguyen et al. developed a technique to monitor spindle alignment tilt and deflection at individual tool marks, leveraging HDM-acquired surface data [20]. Wang et al. employed 3D surface form indicators extracted from tool mark images, which were converted from HDM measurements, to track the wear of wiper inserts [21].

Figure 3.3 displays a surface from an automobile engine's cylinder block. The grayscale image, derived from the HDM-measured surface's point cloud [9], has pixel values that correspond to surface height measurements. The image reveals periodic tool marks, which are visually discernible and encode detailed information about the milling process, including tool geometry, spindle inclination, and milling parameters. Existing research has primarily concentrated on general statistical characteristics of surface variations; however, a more in-depth analysis of individual tool marks could uncover additional insights. For example, the cross-sectional view of tool marks (perpendicular to the tool tooth path) can reveal the tool geometry and its changes throughout the machining process, an aspect often neglected by traditional statistical analysis of tool marks. To undertake such a nuanced examination of each tool mark, it is essential to initially segment the surface topography into separate regions, with each region corresponding to a specific tool tooth path and demarcated by tool marks. Consequently, an appropriate surface segmentation technique is necessary to achieve the task of partitioning the surface as described.

Nevertheless, the sole existing standard for surface segmentation is ISO 16610-85, which is part of the morphological areal filters [22]. The standard employs watershed segmentation with wolf pruning to divide a surface into separate regions, characterized by features such as peaks, pits, and saddle points and delineated by course lines and ridge lines (hills and dales). However, the direct application of the watershed algorithm to images often results in over-segmentation, producing numerous small, irregular catchment basins that lack significance. To address this issue, various region merging algorithms have been suggested [23, 24]. For instance, Barré and Lopez have proposed merging regions with areas below a specified threshold with their adjacent

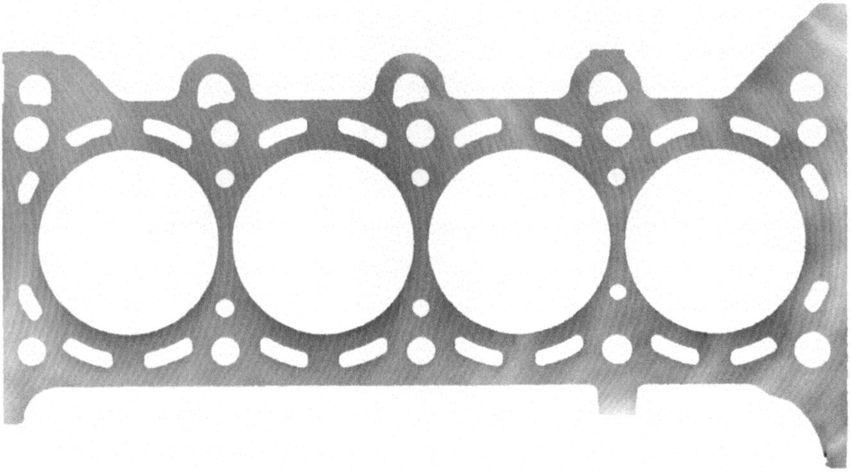

Fig. 3.3 The top surface from a cylinder block of automobile engines

regions, continuing this process until all segmented regions exceed the threshold [24].

The watershed segmentation technique and its variations have achieved success in segmenting certain types of machined surfaces, particularly those with particulate characteristics. As demonstrated in ISO 16610-85, surfaces such as a grinding wheel and a car body panel have achieved satisfactory segmentation results [22]. The watershed segmentation concept originates from geomorphology, offering a novel perspective on machined surface topography that is closely tied to surface functionality. However, it exhibits a significant limitation: for surfaces lacking particulate features, such as those produced by turning and milling processes, the segmentation outcomes are frequently inadequate and bear no relevance to the machining operations. This section addresses this limitation by introducing an enhanced segmentation approach. The methodology presented herein serves as a significant addition and extension to ISO 16610-85, marking the first attempt to connect surface topography with machining processes through surface segmentation.

The surface segmentation approach is composed of four key stages: automatic selection of periodic subsurface, enhancement of tool marks through local thresholding, identification, and repair of broken tool marks, and application of watershed segmentation. This approach is designated as "STRW," an acronym derived from the principal operations at each stage: Subsurface selection, local Thresholding, Repair of tool marks, and Watershed segmentation.

The subsequent sections of this section are structured as follows: Section 3.1.2 provides an in-depth exposition of the "STRW" methodology. Section 3.1.3 presents three case studies that validate the efficacy of the methodology and contrast its segmentation outcomes with those of multiscale watershed segmentation. Finally, Sect. 3.1.4 deliberates on the constraints of the "STRW" methodology and suggests avenues for future enhancement.

3.1.2 The Proposed Methodology

The "STRW" segmentation methodology, as depicted in Fig. 3.4, encompasses four principal stages:

Step 1: The initial measured surface implicitly reveals the tool marks pattern. To facilitate the extraction of tool marks in subsequent stages, the surface must undergo filtering to yield a subsurface that more distinctly exhibits tool marks. Given the unknown spatial frequency of tool marks, the optimal subsurface for extraction is traditionally selected based on engineers' expertise. This step introduces the concept of "periodic degree" to automate the subsurface selection process, with the subsurface exhibiting the highest periodic degree identified as the most suitable for tool marks extraction.
Step 2: To distinctly separate tool marks from their background, a thresholding technique is applied to the chosen subsurface. The subsurface may exhibit form errors that fluctuate, adversely impacting the extraction of tool marks. To mitigate this, a local thresholding method is proposed, which determines the threshold based on the characteristics of a small surrounding area of a point, thus accommodating surface variations. This results in the creation of a binary image of tool marks by applying the local thresholding method to the selected subsurface.

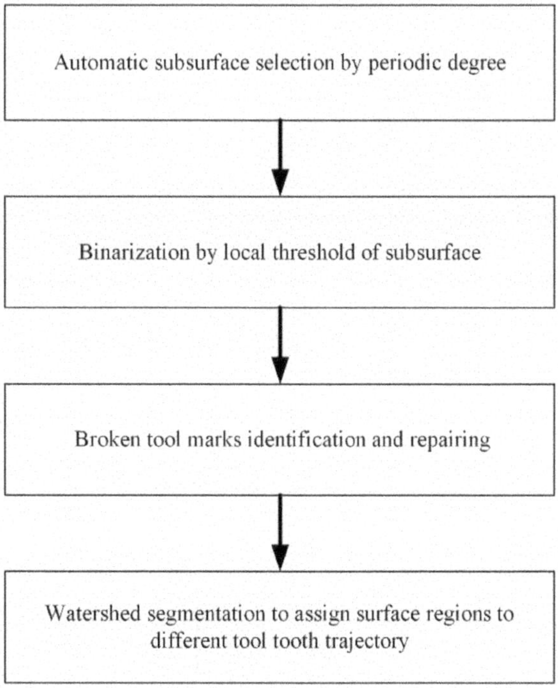

Fig. 3.4 Framework of the proposed STRW surface segmentation methodology

Step 3: Post-thresholding, tool marks are isolated from their background. However, due to the irregularities of actual surfaces, tool marks may be fragmented. Fragmented tool marks are identified using an adaptive distance threshold and subsequently repaired employing a convex-hull-based tool marks repair algorithm.
Step 4: The ultimate step involves dividing the surface into regions corresponding to different tool tooth trajectories. Building on the previous steps, the direct application of watershed segmentation to the binary image post-tool mark repair is expected to yield a satisfactory segmentation outcome.

3.1.2.1 Automatic Subsurface Selection by Periodic Degree

The raw measured surface implicitly captures the pattern of tool marks, necessitating initial filtration to identify the subsurface that best highlights these marks. Given that tool marks possess a consistent frequency, the subsurface displaying them should exhibit a pronounced periodic characteristic. ISO 25178-2 outlines various surface texture parameters, including the texture aspect ratio (S_{tr}) and the auto-correlation length (S_{al}) [25]. S_{al} is associated with the wavelength of the predominant frequency components of a surface, while Str quantifies surface anisotropy. Nonetheless, no existing parameter is designed to gauge the periodic degree of a surface.

In this section, a new concept termed "periodic degree" (PD) is introduced to measure the periodicity of a profile or surface. Figure 3.5a, c, and e illustrate three functions with varying degrees of periodicity. It is evident that function (3.5) (a) exhibits the highest periodicity, whereas function (3.5)e shows the least. The 1D auto-covariance functions (ACV) (defined in Eq. (3.1)) depicted in Fig. 3.5b, d, and f clearly delineate the differences in periodicity, allowing PD to be defined as the ratio of the height of the secondary peak (H_{sp}) to that of the highest peak (H_{hp}), as per Eq. (3.2). The calculated periodic degrees are $PD_a = 125.6/157.1 = 0.80$, $PD_c = 2.44/15.05 = 0.16$, $PD_e = 0$, aligning with visual assessment. It should be noted that in this context, a "peak" is defined as a local maximum surrounded by an ascending and descending path, each with a minimum length of three points. In accordance with this definition, Fig. 3.5f lacks peaks, and in such cases, the periodic degree is designated as zero.

$$ACV(i') = \frac{1}{nx}\sum_{i=1}^{nx-i'} y(i)y(i+i'), 0 \le i' \le nx - 1 \tag{3.1}$$

$$PD = \frac{H_{sp}}{H_{hp}} \tag{3.2}$$

$$y_i \in Peak \text{ if} \begin{cases} y_i > y_{i-1} > y_{i-2} > y_{i-3} \\ y_i > y_{i+1} > y_{i+2} > y_{i+3} \end{cases} \tag{3.3}$$

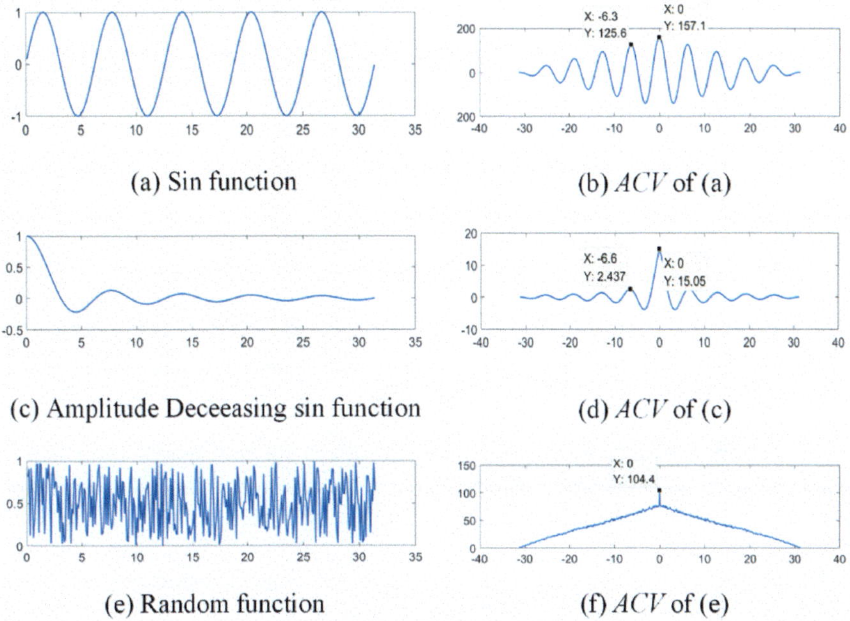

Fig. 3.5 Three types of function with different PD and their ACV

The concept of the auto-covariance function, along with the definition of a "peak" and the periodic degree, can be expanded to two-dimensional scenarios. The areal auto-covariance function for three-dimensional surfaces is articulated by Eq. (3.4).

$$AACV(j', i') = \frac{1}{nx \cdot ny} \sum_{i=1}^{nx-i'} \sum_{j=1}^{ny-j'} z(j, i)z(j + j', i + i'), 0 \leq j' \leq ny - 1, 0 \leq i' \leq nx - 1 \quad (3.4)$$

The definition of 2D periodicity degree mirrors that of the 1D scenario, as specified by Eq. (3.2). A "peak" in two dimensions is identified as a point that exceeds the values of its forty-eight closest neighbors, as depicted in Fig. 3.6. The distance between any two points in 2D is calculated using the chessboard distance metric, detailed in Eq. (3.5). Consequently, the distances from the yellow, orange, and green points to the reference point are 1, 2, and 3 respectively. Thus, in 2D, a peak is formally characterized by Eq. (3.6), aligning with its 1D counterpart.

$$d(P_{mn}, P_{ij}) = \max(|m - i|, |n - j|) \quad (3.5)$$

$$z_{ij} \in Peaks \text{ if } z_{ij} > z_{mn} > z_{pq} > z_{st}, d(z_{ij}, z_{mn}) = 1, d(z_{ij}, z_{pq}) = 2, d(z_{ij}, z_{st}) = 3 \quad (3.6)$$

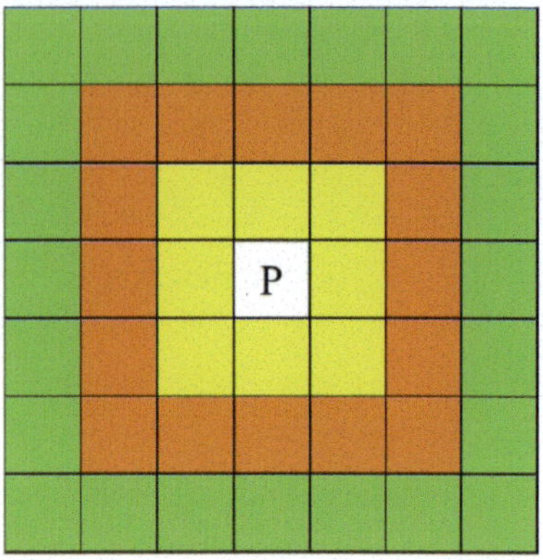

Fig. 3.6 A 2D peak and its neighbors

The selection of the most appropriate subsurface for tool marks extraction is based on the calculation of the periodic degree of the auto-covariance function across various subsurfaces, with the subsurface exhibiting the highest periodic degree being chosen for the task.

To illustrate the efficacy of this subsurface selection algorithm, a simulated surface, as defined by equation, is utilized for testing

$$z(x, y) = -0.5\cos(2\pi x/6.4) + 0.1\cos(2\pi x/3) + 0.05x + 0.05y + 0.05 randn(256) \tag{3.7}$$

where the cosine term of wavelength 6.4 is regarded as periodic tool marks, the cosine term of wavelength 3 is the simulated periodic noises, and the last three terms are regarded as form error and random noises.

The dimensions of the simulated surface are 51.2 mm by 51.2 mm, with a sampling interval of 0.2 mm in both horizontal directions. A biorthogonal wavelet decomposition using bior4.4 is performed on this surface, achieving a five-level decomposition. Figure 3.7a presents the simulated surface in three-dimensional representation, while Fig. 3.7b–f display the decomposed subsurfaces D1 through D5 (denoting $D_i = D_i^H + D_i^V + D_i^D$, with wavelet decomposition being a recursive process of $A_i = A_{i+1} + D_{i+1}$). Visually inspecting these figures, Fig. 3.7e appears to possess the most pronounced periodicity. The auto-covariance functions of the subsurfaces are depicted in Fig. 3.8, with the last three graphs exhibiting similar characteristics, making it challenging to discern the most periodic subsurface through visual assessment alone. The periodic degrees, as defined in this section, are computed and tabulated in Table 3.1, revealing that D4 possesses the highest value, which corroborates the visual observation from Fig. 3.7. Consequently, subsurface D4 is chosen in this step.

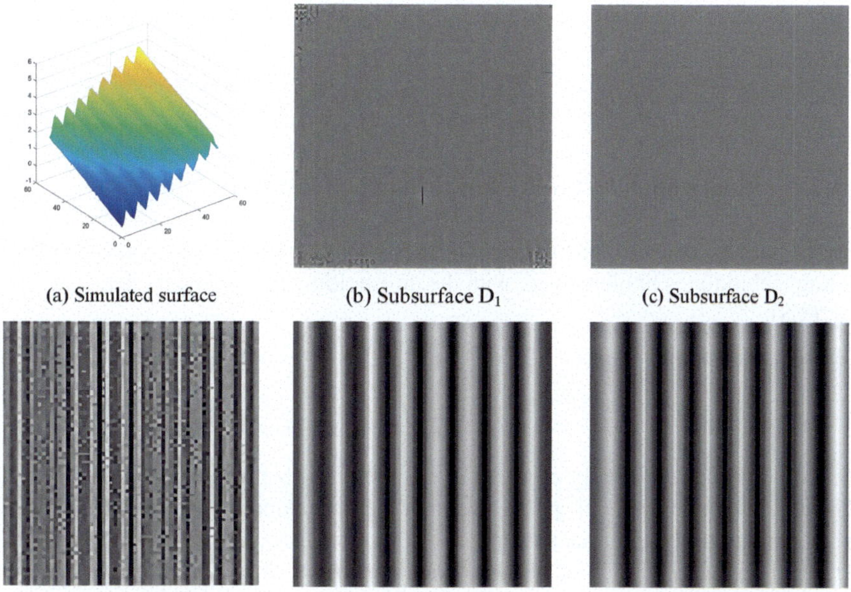

(a) Simulated surface (b) Subsurface D_1 (c) Subsurface D_2 (d) Subsurface D_3 (e) Subsurface D_4 (f) Subsurface D_5

Fig. 3.7 Simulated surface and decomposed subsurfaces

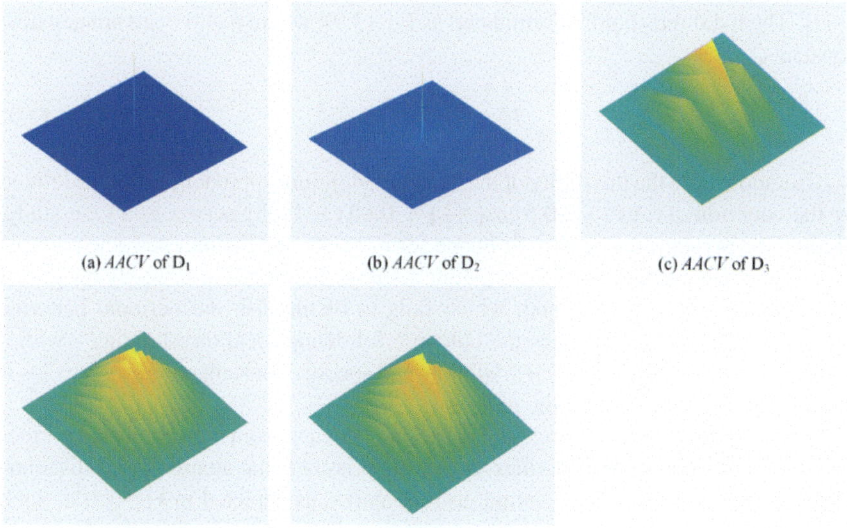

(a) $AACV$ of D_1 (b) $AACV$ of D_2 (c) $AACV$ of D_3 (d) $AACV$ of D_4 (e) $AACV$ of D_5

Fig. 3.8 AACV plots of subsurfaces

Table 3.1 Periodic degree of decomposed subsurface D1 ~ D5

Subsurface	D_1	D_2	D_3	D_4	D_5
Periodic degree	0.013	0.086	0.539	0.742	0.693

3.1.2.2 Binarization by Local Thresh-Hold of Subsurface

For the purpose of distinguishing tool marks from their background, an image thresholding technique is implemented, with the thresholding process being delineated by Eq. (3.8).

$$B(x, y) = \begin{cases} 1 \text{ if } I(x, y) > T \\ 0 \text{ if } I(x, y) \leq T \end{cases} \tag{3.8}$$

Thresholding is typically executed using two primary methods: global and local thresholding. The classical global thresholding technique is Ostu's method [26], which aims to find an optimal threshold T^* to maximize the variance between classes. An advanced adaptive method involves calculating a threshold for each point (x, y) in the image, based on local properties within a defined neighborhood of (x, y). Within this approach, the local threshold is determined by the standard deviation and mean height of the points surrounding a given pixel. Denote the standard deviation and mean height of the neighboring points centered at (x, y) by σ_{xy} and m_{xy}, respectively. The local threshold is formulated as Eq. (3.9), where a and b are nonnegative constants.

$$T_{xy} = a\sigma_{xy} + bm_{xy} \tag{3.9}$$

To demonstrate the flexibility of the local thresholding approach, a surface defined by the equation $z(x, y) = -0.5\cos\left(\frac{2\pi x}{6.4}\right) + 0.05x + 0.05y$ serves as a case study. Figure 3.9a presents the surface in image form, exhibiting distinct periodic patterns over a gradient background. Figure 3.9b displays the result after applying global thresholding via Ostu's method, which fails to distinguish the periodic patterns. In contrast, Fig. 3.9c illustrates the outcome following local thresholding—with a neighborhood region set to a 7×7 window and specific constant $a = 1.5,\ b = 1$—it effectively captures the periodic patterns.

Following local thresholding, tool marks may deviate from the anticipated perfection due to machining irregularities. A prevalent issue is the unanticipated discontinuity within what should be continuous tool marks, as depicted in Fig. 3.10a. Such discontinuities can impede accurate segmentation. Consequently, there is a clear need for an algorithm capable of automatically mending these breaks, as exemplified in Fig. 3.10b.

For the purpose of mending the discontinuities in tool marks, it is essential to initially pinpoint the fractured segments of the marks. Given that the gap size of tool mark breaks is typically shorter than the distance between nearby tool marks, the

(a) Simulated surface (b) Glocal thresh-holding (c) Local thresh-holding

Fig. 3.9 Simulated surface and its binary form by different thresh-holding methods

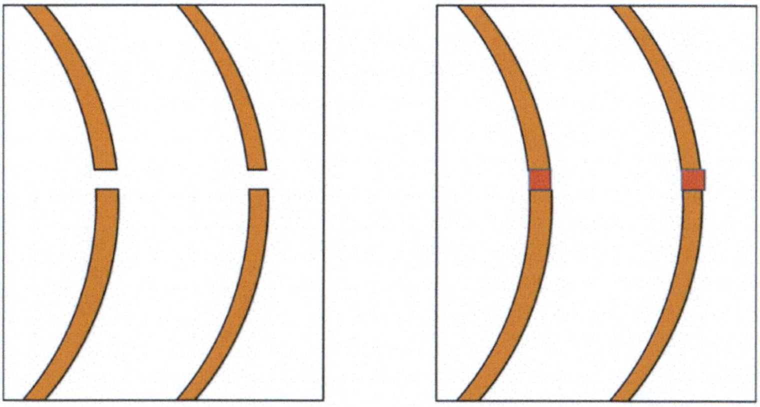

(a) Breakage of continuous tool marks (b) Expected repair of broken tool marks

Fig. 3.10 Unexpected breakage of tool marks and their repairing scheme

Inter-Region Distance (*IRD*) matrix can be employed to differentiate broken marks from those that are adjacent. The matrix element *IRD(i,j)* denotes the minimum distance between a region labeled i and another region labeled j, as articulated by Eq. (3.10).

$$IRD(i,j) = \min\sqrt{(x_p^i - x_q^j)^2 + (y_p^i - y_q^j)^2}\ (p = 1, 2, \cdots, N_i, q = 1, 2, \cdots, N_j, i \neq j) \tag{3.10}$$

where N_i is the number of points in region X_i and N_j is the number of points in another region X_j. The vector min(*IRD*) contains the minimum value from each column of the *IRD* matrix. The maximum value within min(*IRD*) approximates the distance between adjacent tool marks. Therefore, the adaptive distance threshold

DT, used to distinguish broken marks from adjacent ones, can be set as a fraction of max(min(IRD)) based on engineering experience (with a recommended proportion coefficient C between 0.5 and 0.9), and DT is defined by Eq. (3.11).

$$DT = C \times \max(\min(IRD)) \tag{3.11}$$

If the minimum value in the IRD vector for a region i, denoted as min($IRD(i)$), is less than the adaptive distance threshold DT, then the region i is classified as part of the broken tool marks. Conversely, the region i is categorized as part of the continuous tool marks. By repairing the broken tool marks and combining them with the continuous tool marks, a fully restored image of the tool marks can be achieved. This approach of segmentation and repair is schematically represented in Fig. 3.11.

The algorithm for "tool mark repairing" is specifically crafted to address sub-images that encompass all fragmented tool marks. It employs a "convex hull" generation technique to facilitate the repair process.

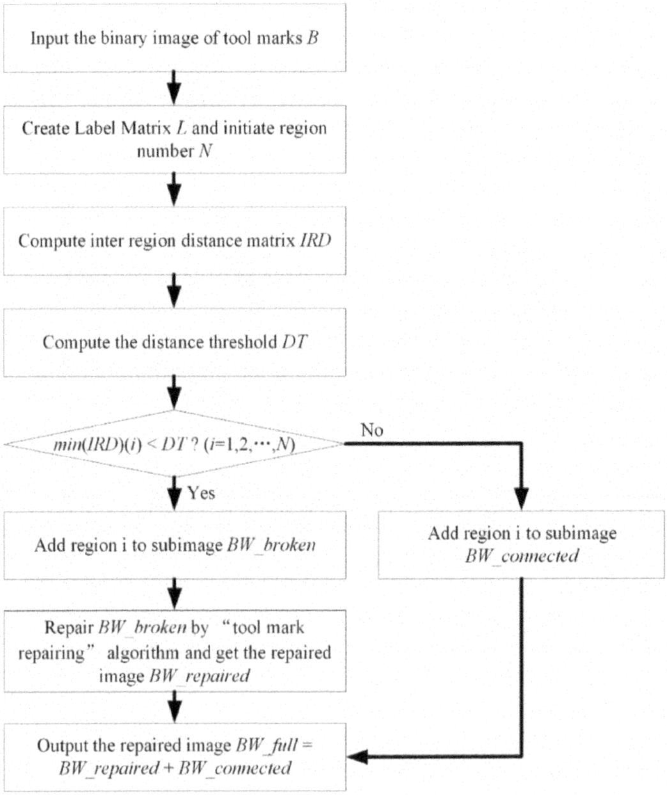

Fig. 3.11 The work-flow of the broken tool marks identification and repairing

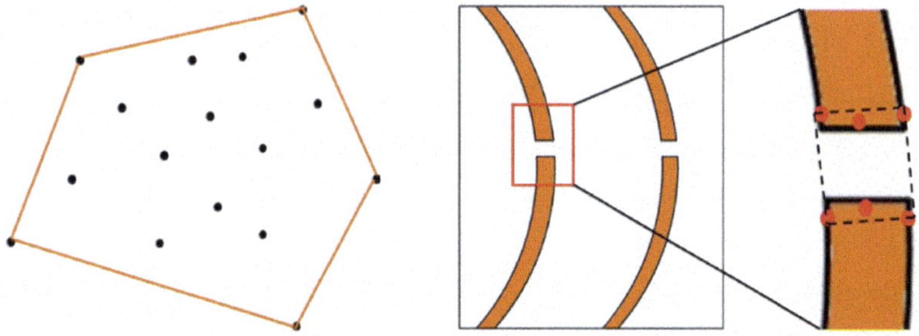

(a) Convex hull example (b) The idea repairing breakage by convex hull

Fig. 3.12 The concept of convex hull and its application to repair tool marks' breakage

In mathematical terms, the convex hull of a set of points S in the Euclidean plane is the smallest convex set that encompasses S, and it comprises all convex combinations of the points within S. The convex hull can be formally defined by formula (3.12),

$$Convhull(X) = \left\{ \sum_{i=1}^{|X|} \alpha_i x_i \middle| \forall i : \alpha_i \geq 0 \wedge \sum_{i=1}^{|X|} \alpha_i = 1, x_i \in X \right\} \tag{3.12}$$

where $|X|$ represents the number of points in set X. Figure 3.12a provides an example of a point set along with its convex hull. The methodology for repairing breaks using the convex hull is depicted in Fig. 3.12b. The process begins by identifying edge points near the breakage gap, proceeds with the construction of a convex hull from the set of these points, and concludes with filling the region within the convex hull. The quickhull algorithm, a well-known method for generating convex hulls [27], is utilized to connect the disjoint segments. The comprehensive steps of the tool mark repairing algorithm are illustrated in Fig. 3.13.

To demonstrate the effectiveness of the proposed repair algorithm, a straightforward example of fragmented tool marks is presented. Figure 3.14a displays the simulated tool marks with visible breaks, while the sub-images *BW_broken* and *BW_connected* are depicted in Fig. 3.14b and c, respectively. The convex hulls employed in the repair of the breaks are indicated with red outlines in Fig. 3.14d. The sub-image of the repaired tool marks, *BW_repaired*, along with the complete image, *BW_full*, are showcased in Fig. 3.14e and f, respectively.

Once the broken tool marks have been repaired, the periodic pattern of the tool marks becomes evident. For the purpose of surface segmentation, it is preferable to have dividing lines of one-pixel width to enhance the precision of the segmentation. The objective of this concluding step is to reduce the width of the tool marks to a single pixel while ensuring that these thinned marks closely follow the centerline of the original tool marks.

Classical watershed segmentation is instrumental in accomplishing this task. Figure 3.15a presents a simulated diagonal tool mark, and the dividing line obtained

Algorithm 1 : Tool mark repairing algorithm

Input: a binary image of broken tool marks *BW_broken* and distance threshold *DT*

Output: a binary image of repaired tool marks *BW_repaired*

```
s=regionprops(BW_broken,'PixelList');len=length(s); BW_repaired=BW_broken;
For i=1:len-1 do
  For j=i+1:len do
    M1=length(s(i).PixelList); M2=length(s(j).PixelList)
     For h=1:M1 do
       For k=1:M2 do
          If dist(s(i).PixelList(h,:) ;s(j).PixelList(k,:))< 1.5*DT do
                        cr = [cr;s(i).PixelList(h,:);s(j).PixelList(k,:)];
              End if
          End for
       End for
    cr = unique(cr,'rows'); t = convhull(cr(:,1),cr(:,2));
     BW_R =roiploy(BW_broken, cr(t:,1),cr(t:,2));
    BW_repaired=BW_repaired+BW_R;
  End for
End for
```

Fig. 3.13 Tool mark repair algorithm

through watershed segmentation is nearly central to the diagonal, with the exception of the two ends, as depicted in Fig. 3.15b.

3.1.3 Case Study

3.1.3.1 Case Study I

STRW Algorithm Experiment

Surfaces generated by milling and turning operations exhibit distinct periodic tool marks. To evaluate the proposed surface segmentation methodology, the top surface of a B12 cylinder block from an automotive manufacturing plant is selected for analysis. The milling was performed on an EX-CELL-O Machining Center, utilizing a CBN milling cutter with a diameter of 200 mm. The operational parameters included

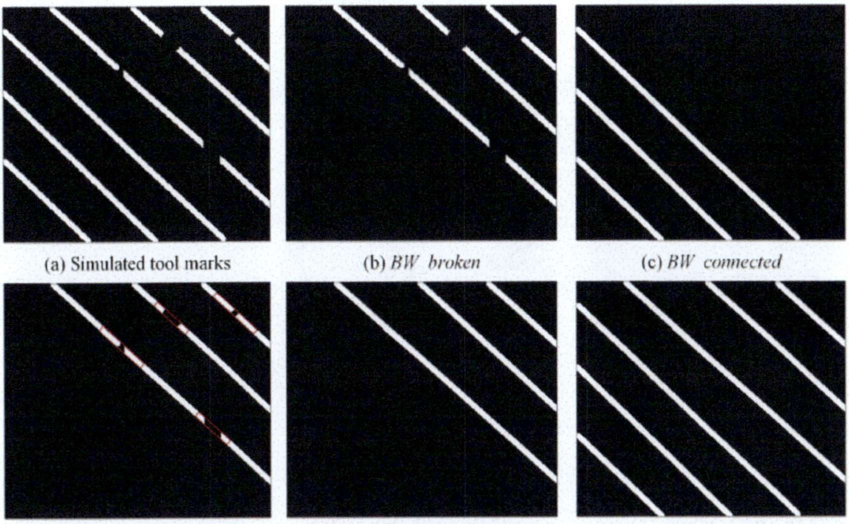

(a) Simulated tool marks (b) *BW_broken* (c) *BW_connected* (d) convex hulls for repair (e) *BW_repaired* (f) *BW_full*

Fig. 3.14 Simulated tool marks with breakage and its repairing procedures

(a) A tool mark example (b) Watershed segmentation of (a)

Fig. 3.15 A diagonal tool mark and its watershed segmentation

a cutting speed of 1300 rpm, a depth of cut at 0.5 mm, and a feed rate of 3360 mm/min. The ShaPix3D® 3000 series, an HDM instrument, was employed for measuring the cylinder block. It boasts a vertical resolution of 0.05 μm and a lateral resolution of 150 μm, with a height measurement range of ± 5 mm and vertical accuracy reaching 1 μm. The instrument's field of view spans 280 mm × 280 mm, and it can capture up to 4 million sampling points per view. For workpieces that extend beyond the field of

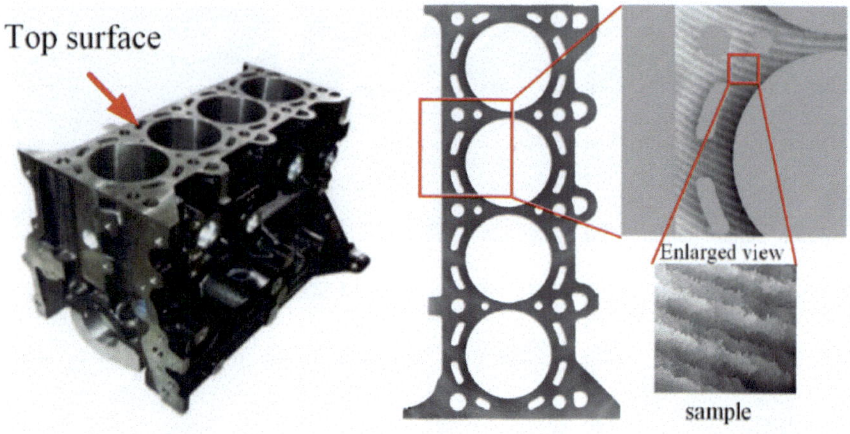

(a) A cylinder block (b) A sample from the top surface

Fig. 3.16 B12 cylinder block and a sample from its top surface

view, the HDM is capable of merging multiple point clouds to create a comprehensive surface representation.

The B12 cylinder block is depicted in Fig. 3.16a, with a sample of its top surface displayed in Fig. 3.16b. An enlarged portion of Fig. 3.16b highlights the clear periodic tool mark characteristics. To facilitate the demonstration of the segmentation process, a smaller section of the surface is selected for detailed study, as illustrated in Fig. 3.16b.

The sampled surface data is decomposed into various sub-bands using a biorthogonal wavelet transform with the bior4.4 wavelet, as illustrated in Fig. 3.17.

For the purpose of identifying the optimal subsurface for segmentation, the areal auto-covariance function of each subsurface is determined and depicted in Fig. 3.18. It is evident that the central peak, corresponding to a zero offset, is the most prominent in every *AACV* plot. Visual inspection of Fig. 3.17 suggests that subsurface D_4 exhibits the most pronounced periodicity. The periodic degree values presented in Table 3.2 corroborate this observation. Consequently, subsurface D_4 is chosen for subsequent processing.

The local thresholding outcome for subsurface D_4 is depicted in Fig. 3.19a, with the neighborhood size set to a 7×7 square window. The threshold is denoted as $T = 1.4\sigma + m$, where a and b in Eq. (3.9) represent empirical constants that can vary with different surface production techniques. Noise appears as white blocks in Fig. 3.19a, and to address this, white blocks with areas smaller than a predefined threshold are eliminated, with the threshold set to 10 in this instance. The image following noise removal is presented in Fig. 3.19b, and the region labels are displayed in Fig. 3.19c.

Observations from Fig. 3.19b indicate a breakage in the right-top tool mark. To identify such breakages, the *IRD* matrix is computed and presented in Table 3.3. The

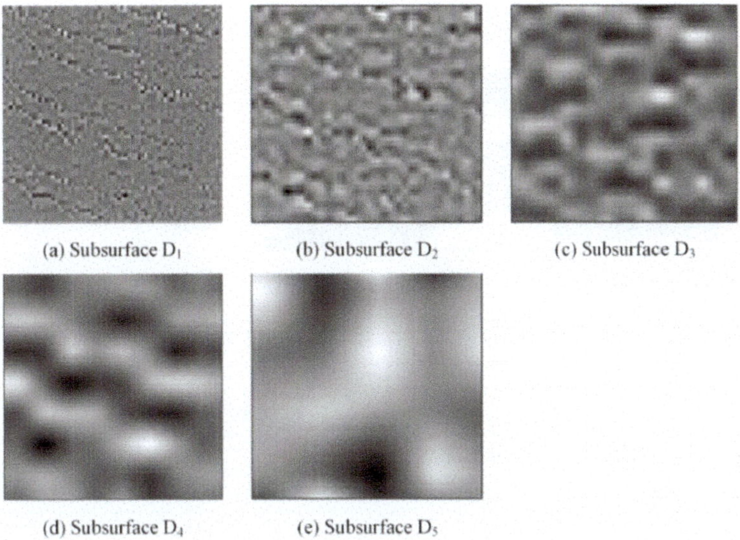

Fig. 3.17 Decomposed subsurfaces of sampled surface in Case Study I

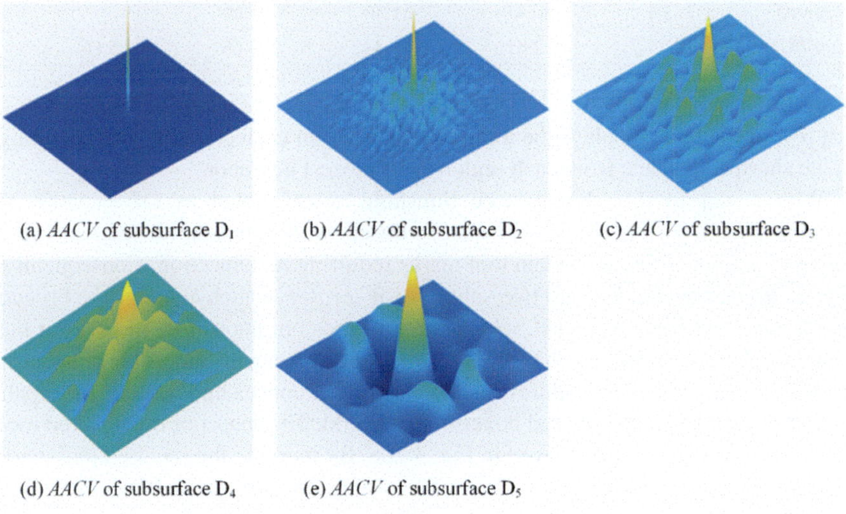

Fig. 3.18 AACV of subsurface D1–D5 in Case Study I

Table 3.2 Periodic degree of subsurface D1–D5 in Case Study I

Subsurface	D_1	D_2	D_3	D_4	D_5
Periodic degree	0.038	0.181	0.300	0.398	0.045

(a) Local thresh-holding (b) Noise removal (c) Label numbers

Fig. 3.19 Local thresh-holding of D4 and its noise removal in Case Study I

Table 3.3 The inter region distance matrix IRD in Case Study I

Distance	Region1	Region2	Region3	Region4	Region5
Region1		12.81	31.06	10.77	12.04
Region2	12.81		13.34	30.41	30.41
Region3	31.06	13.34		50.01	51.86
Region4	10.77	30.41	50.01		5.66
Region5	12.04	30.41	51.86	5.66	
min(*IRD*)	10.77	12.81	13.34	5.66	5.66

last row of the table displays the minimum value from each column, corresponding to the shortest distance from each region to its closest neighbor.

With C set to 0.5, the distance threshold is calculated as $DT = 50\% \times \max(\min(IRD)) = 50\% \times 13.34 = 6.67$. Based on this threshold, Region 4 and Region 5 are identified as broken tool marks requiring reconnection. Consequently, Fig. 3.19b can be divided into two subsets: *BW_broken*, which contains the broken tool marks, and *BW_connected*, which contains the continuous tool marks. For the broken tool marks illustrated in Fig. 3.20b, the "tool mark repair" algorithm detailed in Sect. 2.3 is implemented to address the breakage. The convex hulls used in the repair process are highlighted with red boxes in Fig. 3.20d. The subset of the repaired tool marks, *BW_repaired*, is displayed in Fig. 3.20e. By merging the repaired tool mark image *BW_repaired* with the image of the connected tool marks BW_connected, the complete repaired tool mark image, *BW_full*, is assembled and presented in Fig. 3.20f.

Utilizing the classical watershed algorithm on the fully repaired tool mark image, the segmentation outcomes overlaid on the original surface topography and the selected subsurface are depicted in Fig. 3.21a and b, respectively. Figure 3.21c assigns a unique color to each segmented region for clearer visualization. The watershed lines are distinctly marked in red to improve their visibility.

A cross section perpendicular to tool marks is plotted in. The points on the tool marks are marked with A, B, C and D. It is clear that A, B, C and D are local

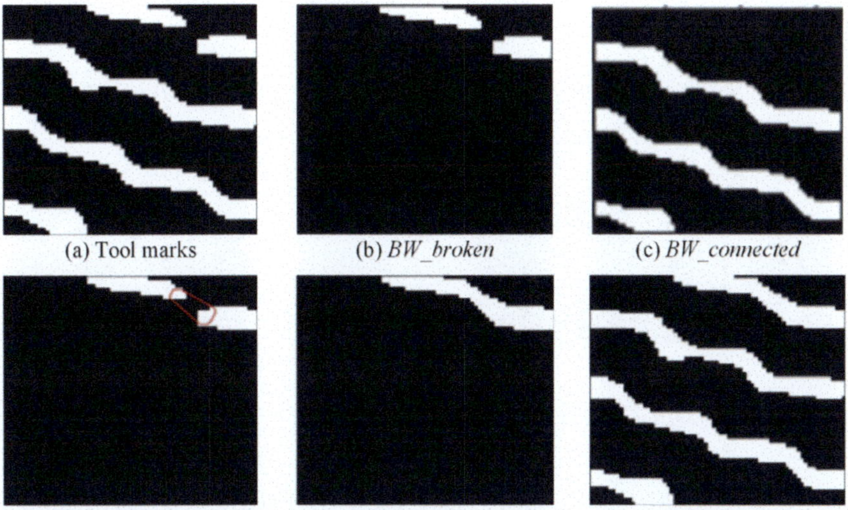

(a) Tool marks (b) *BW_broken* (c) *BW_connected*

(d) Concex hull for repair (e) *BW_repaired* (f) *BW_full*

Fig. 3.20 Tool marks repairing procedures in Case Study I

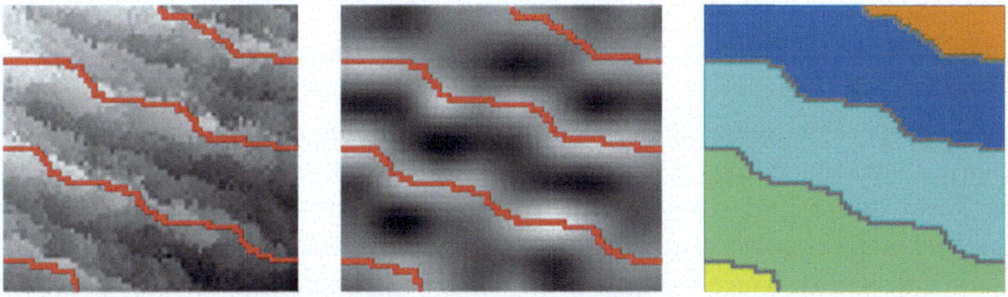

(a) Segmentation shown on original surface (b) Segmentation shown on selected subsurface (c) Segmentation result labelled with color

Fig. 3.21 Final surface segmentation result in Case Study I

peak points, which is consistent with the concept of tool marks in. Generally, the tool shape and scar will be negative replicated in surface and repeated at the feed spacing. This phenomenon also holds in this case, the shapes of profile AB, BC and CD are similar to each other. The average feed spacing calculated from Fig. 3.22b is $f = \frac{AB+BC+CD}{3} = \frac{2.7+3+3}{3} = 2.9$ mm. The cutting speed was 1300 rpm, and feed rate was 3360 mm/min. The theoretical feed spacing is $f = \frac{3360}{1300} = 2.58$ mm. The relative error is $r = \frac{2.9-2.58}{2.58} = 12.4\%$.

A cross-sectional view orthogonal to the tool marks is presented in Fig. 3.22, with points A, B, C, and D identified on the tool marks. It is evident that A, B, C, and D are local peak points, aligning with the concept of tool marks as shown in Fig. 3.1. Typically, the tool shape and its engagement leave a negative impression on the surface, recurring at the feed intervals. This pattern is observed here, with profiles

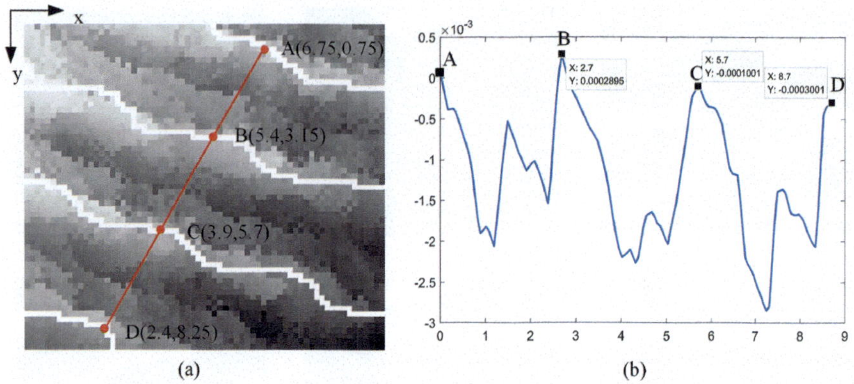

Fig. 3.22 A cross-section view of tool marks in Case Study I

AB, BC, and CD exhibiting similar shapes. The average feed spacing, derived from Fig. 3.22b, is calculated as $f = \frac{AB+BC+CD}{3} = \frac{2.7+3+3}{3} = 2.9$ mm. Given the cutting speed of 1300 rpm and a feed rate of 3360 mm/min, the theoretical feed spacing is $f = \frac{3360}{1300} = 2.58$ mm. The relative error is then calculated as $r = \frac{2.9-2.58}{2.58} = 12.4\%$.

Comparison with Multiscale Watershed Segmentation Algorithm

For the purpose of evaluating the segmentation algorithm introduced in this section against conventional approaches, the multiscale watershed algorithm put forth by Barré and Lopez [24] is implemented on the same surface dataset. This multi-scale technique employs a sequence of area thresholds to execute multiscale watershed segmentation, which has been observed to yield improved outcomes in certain scenarios over the standard watershed algorithm. The underlying concept involves filling in catchment basins that are deemed insignificant, specifically those with areas below a given threshold, and subsequently identifying the watershed lines on the modified surface to achieve a refined segmentation. The pseudocode illustrating the steps of this algorithm is depicted in Fig. 3.23.

Figure 3.24 presents the outcomes of the multiscale watershed segmentation, where the area threshold A ranges from 1 to 57.7 in increments of 8.1. When juxtaposed with Fig. 3.21, it becomes evident that Fig. 3.24 fails to achieve effective segmentation; the watershed lines are disordered and lack any tangible physical significance. Moreover, the multiscale segmentation does not exhibit a tendency to converge towards the segmented result depicted in Fig. 3.21. Consequently, it can be inferred that traditional watershed segmentation techniques are not adept at effectively partitioning surfaces that exhibit periodic tool marks.

Given that the top surface of the B12 engine block represents just one variety of milled surfaces, this section extends the investigation to include two additional engineering surfaces to further substantiate the efficacy of the proposed methodology.

Algorithm 2: Multiscale watershed segmentation algorithm

Input: a gray image of surface topography *I*, area threshold *Amin*, *Amax* and step *Astep*

Output: a cell array of segmentation results *S*

```
For A=Amin:Astep:Amax do
  L=watershed(I); area = regionprops(L,'area'); min_area = min(area);
  While min_area<A do
    area_small=area<A;index=find(area_small==1)
    For i=1:length(index) do
      region=(L==(index(i)));I(region)=min(I(edge(region)));
    End for
    L=watershed(I);area = regionprops(L,'area'); min_area = min(area);
  End while
  S((A-Amin)/Astep+1)=L;
End for
```

Fig. 3.23 Multiscale watershed segmentation algorithm

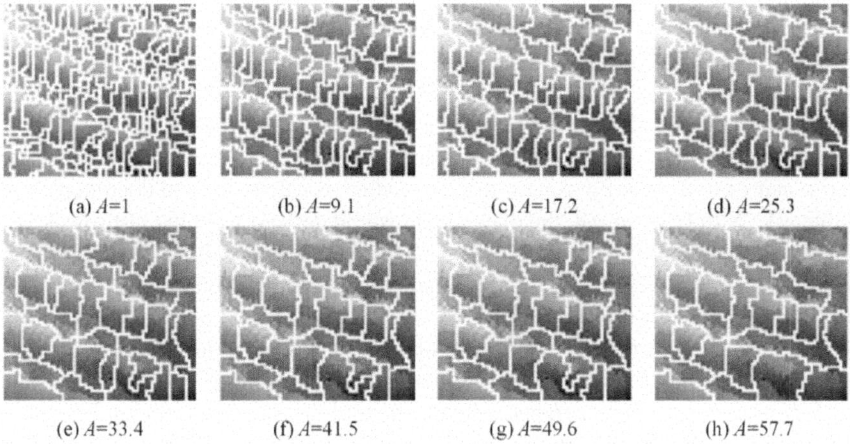

Fig. 3.24 Results of multiscale watershed segmentation in Case Study I

3.1.3.2 Case Study II

In this scenario, a sample from the top surface of an N12 cylinder block is examined. The N12 block underwent milling at a cutting speed of 1200 rpm, with a depth of cut set to 0.6 mm and a feed rate of 3000 mm/min. Figure 3.25 depicts the N12 cylinder block along with the height map of its top surface as measured by Shapix.

(a) N12 cylinder block

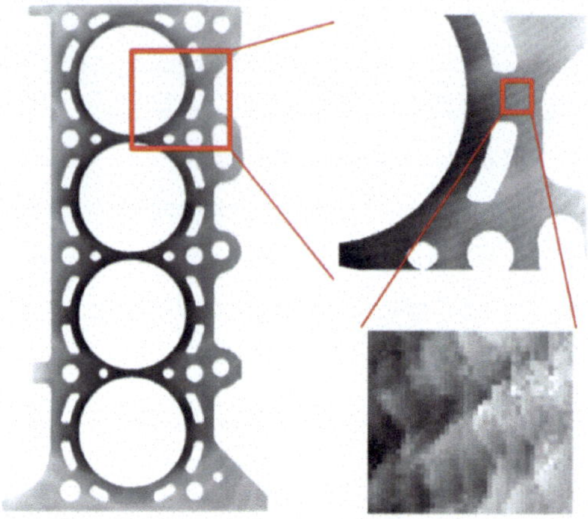

(b) A sample from the top surface

Fig. 3.25 N12 cylinder block and a sample from its top surface

Following a similar approach to case 1, a segment of the surface was extracted for segmentation purposes. The wavelet transform using bior4.4 is applied to decompose the sample into subsurfaces, as illustrated in Fig. 3.26, and their auto-covariance plots are displayed in Fig. 3.27. The periodic degrees for each subsurface are detailed in Table 3.4. Given that D_4 exhibits the highest periodic degree, it is chosen for subsequent analysis.

D_4 undergoes local thresholding with a specific value $T = 1.2\sigma + m$ to create a binary image, with each region labeled as seen in Fig. 3.28a. To address the detection and repair of tool mark breakages, the *IRD* matrix is calculated and documented in Table 3.5. With C set to 0.8, the distance threshold derived from the last row is DT $= 0.8 \times \max(\min(IRD)) = 0.8 \times 10.63 = 8.50$. Based on this threshold, Regions 1, 3, 4, and 5 are identified as broken tool marks that require reconnection. The repair procedures are detailed in Fig. 3.28.

Utilizing the classical watershed algorithm on the image of the fully repaired tool marks, the resulting segmentation outcomes are depicted in Fig. 3.29.

A cross-sectional view perpendicular to the tool marks is presented in Fig. 3.30, with points A, B, and C indicated on the tool marks. While A, B, and C are not precisely local peak points, their positions are in close proximity to local peaks, and the shapes of the profiles AB and BC are similar to each other. The average feed spacing, derived from Fig. 3.30b, is calculated as $f = \frac{AB+BC}{2} = \frac{5.04}{2} = 2.52$ mm. The theoretical feed spacing, based on the feed rate and cutting speed, is $f = \frac{3000}{1200} = 2.5$ mm. The relative error is then calculated as $r = \frac{2.52-2.5}{2.5} = 0.8\%$.

Figure 3.31 illustrates the results of the multiscale watershed segmentation with the area threshold A ranging from 1 to 44.4 in increments of 6.2. It is evident that the multiscale watershed segmentation approach fails to effectively segment surfaces with periodic tool marks into distinct tool tooth trajectories.

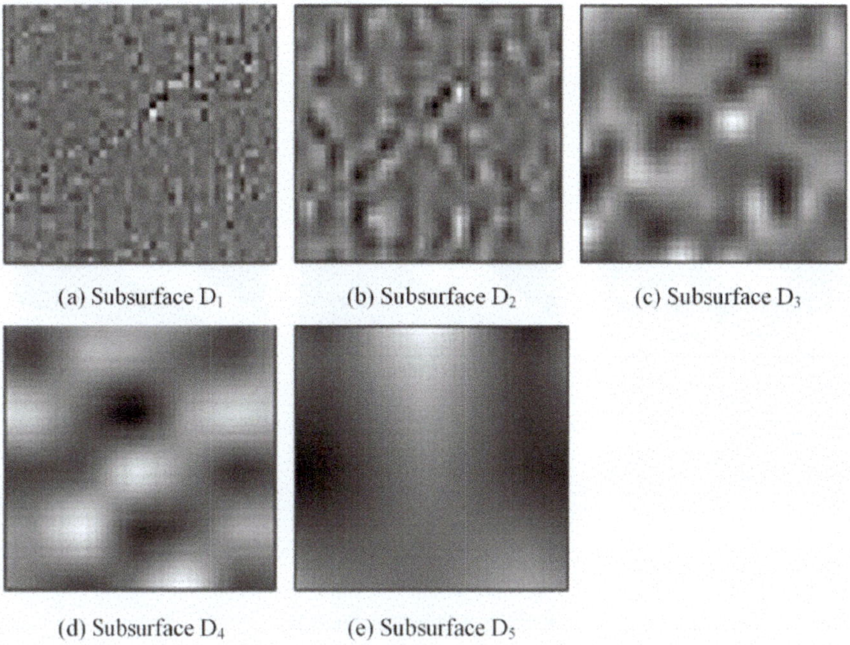

Fig. 3.26 Decomposed subsurfaces of sampled surface in Case Study II

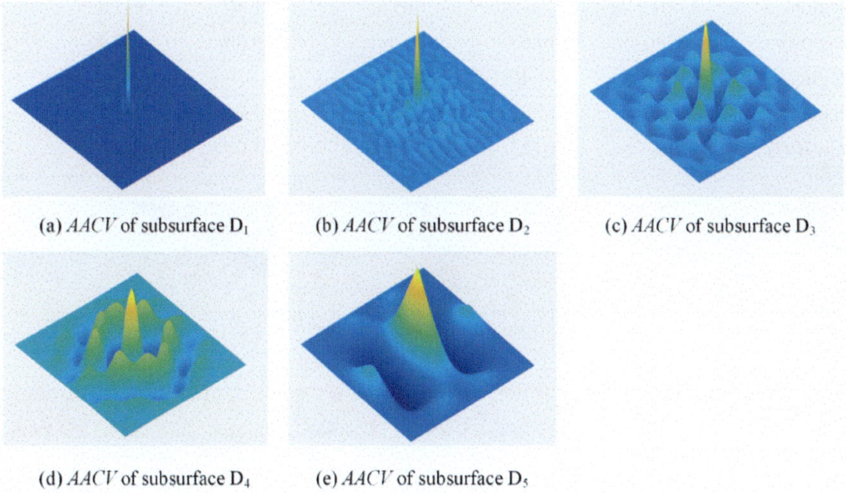

Fig. 3.27 AACV of subsurface D1–D5 in Case Study II

Table 3.4 Periodic degree of subsurface D1–D5 in Case Study II

Subsurface	D_1	D_2	D_3	D_4	D_5
Periodic degree	0.111	0.173	0.273	0.447	0.221

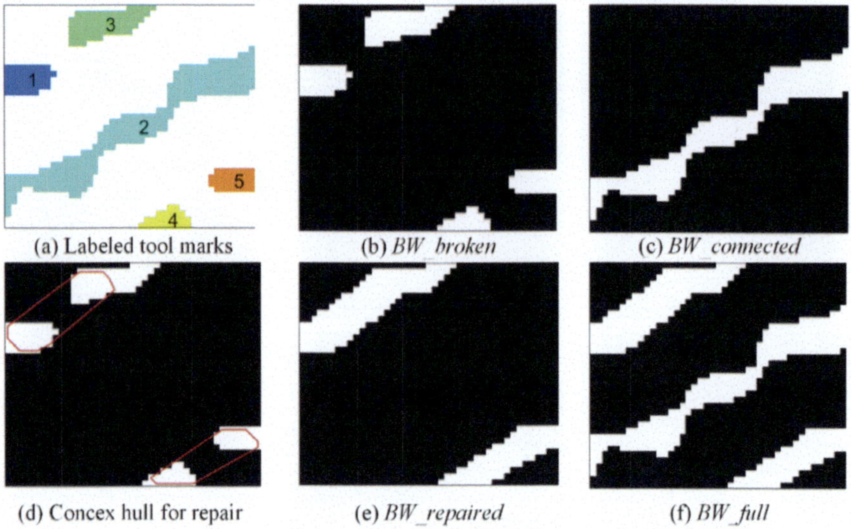

Fig. 3.28 Tool marks labelling and repairing procedures in Case Study II

Table 3.5 The inter region distance matrix IRD in Case Study II

Distance	Region1	Region2	Region3	Region4	Region5
Region1		12.04	5.66	27.80	31.76
Region2	12.04		10.63	11.40	11.31
Region3	5.66	10.63		29.83	27.46
Region4	27.80	11.40	29.83		7.07
Region5	31.76	11.31	27.46	7.07	
min(*IRD*)	5.66	10.63	5.66	7.07	7.07

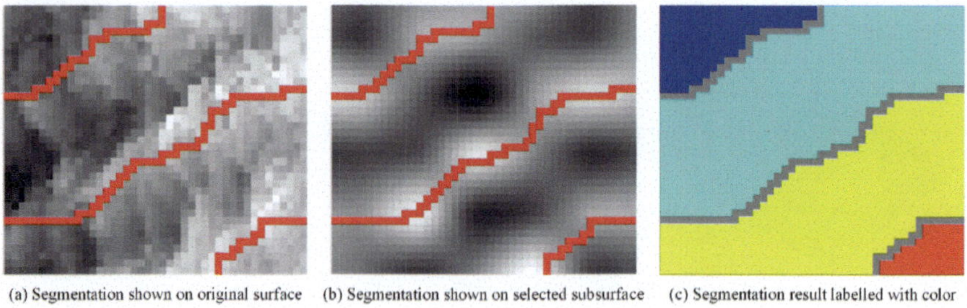

Fig. 3.29 Final surface segmentation result in Case Study II

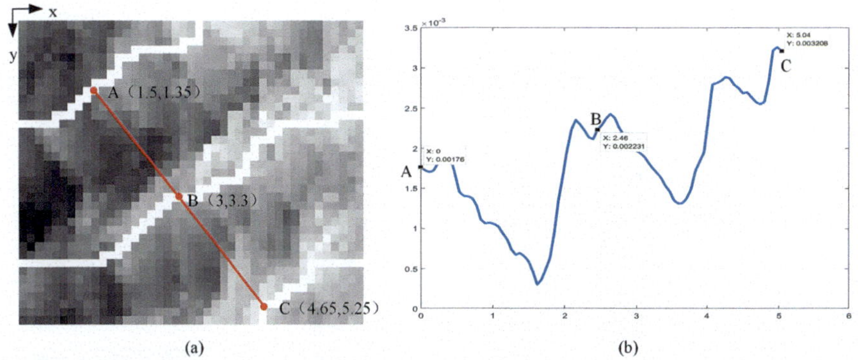

Fig. 3.30 A cross-section view of tool marks in Case Study II

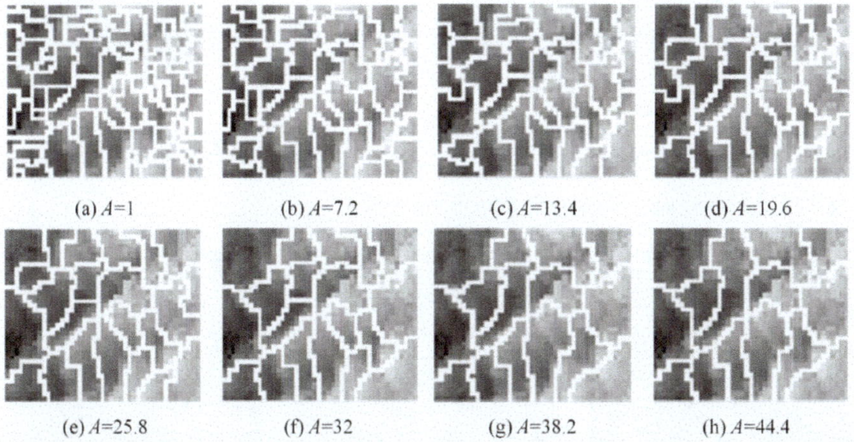

Fig. 3.31 Results of multiscale watershed segmentation in Case Study II

3.1.3.3 Case Study III

To further validate the efficacy of the proposed algorithm, a sample from the top surface of a bearing cap has been selected for analysis. The top surface of the bearing cap was milled at a cutting speed of 1400 rpm, with a depth of cut at 0.5 mm and a feed rate of 3800 mm/min. Figure 3.32a displays the bearing cap, and the height map of the left top surface, as measured by HDM, is presented in Fig. 3.32b. A segment from this measurement is extracted to evaluate the proposed methodology.

The subsurfaces decomposed using the wavelet transform with bior4.4 and their *AACV* plots are depicted in Figs. 3.33and 3.34, respectively. The periodic degrees for each of these subsurfaces are detailed in Table 3.6. Subsurface D_4 is chosen once more for further analysis, as it exhibits the highest periodic degree.

(a) Bearing cap (b) A sample from the top surface

Fig. 3.32 A bearing cap and a sample from its top surface

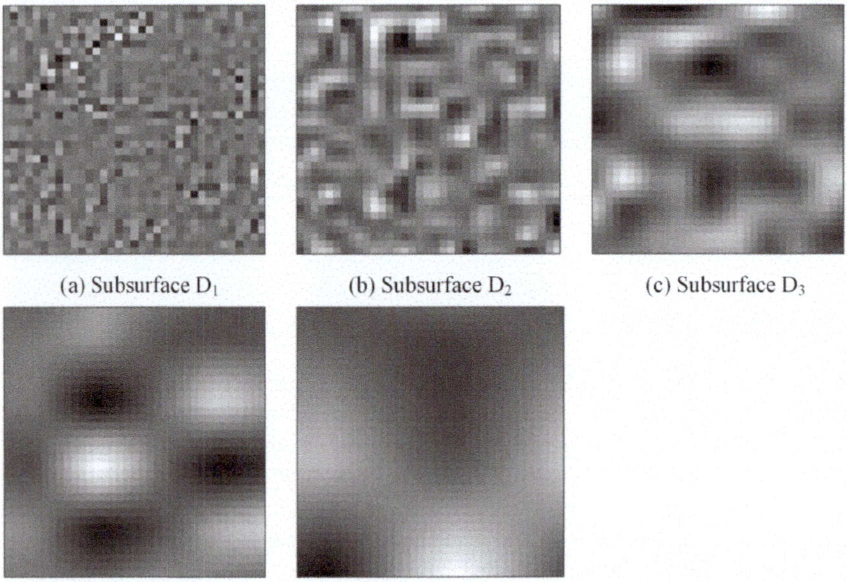

(a) Subsurface D_1 (b) Subsurface D_2 (c) Subsurface D_3

(d) Subsurface D_4 (e) Subsurface D_5

Fig. 3.33 Decomposed subsurfaces of sampled surface in Case Study III

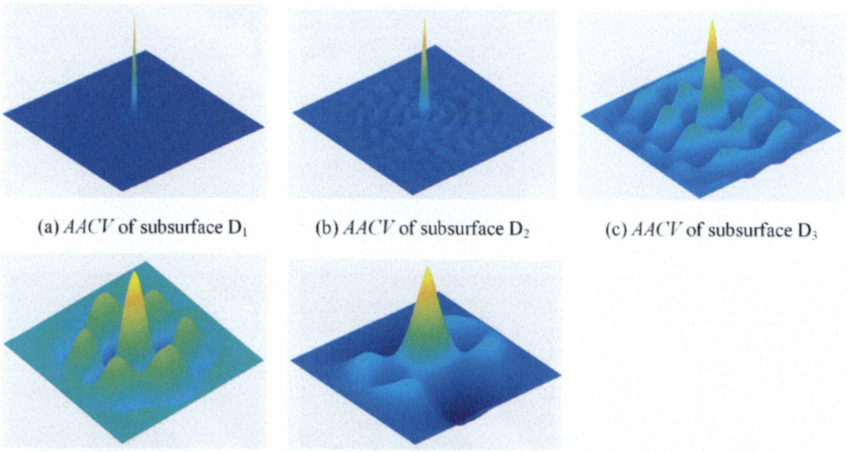

(a) *AACV* of subsurface D_1 (b) *AACV* of subsurface D_2 (c) *AACV* of subsurface D_3

(d) *AACV* of subsurface D_4 (e) *AACV* of subsurface D_5

Fig. 3.34 AACV of subsurface D1–D5 in Case Study III

Table 3.6 Periodic degree of subsurface D1–D5 in Case Study III

Subsurface	D_1	D_2	D_3	D_4	D_5
Periodic degree	0.081	0.109	0.237	0.392	0.189

Employing the same local thresholding parameters as used in case 2, the labels for each mark are depicted in Fig. 3.35a. To identify and mend any breaks in the tool marks, the *IRD* matrix is computed and documented in Table 3.7. With C set to 0.8, the distance threshold is calculated as $\mathrm{DT} = 0.8 \times \max(\min(IRD)) = 0.8 \times 10.63 = 8.50$. Based on this threshold, Regions 2, 3, and 5 are identified as broken tool marks requiring reconnection. The repair procedures are detailed in Fig. 3.35. The ultimate results of the watershed segmentation are presented in Fig. 3.36.

A cross-sectional view perpendicular to the tool marks is presented in Fig. 3.37, with points A, B, and C indicated on the tool marks. Although point B is not precisely a local peak, its position is in close proximity to a local peak, and the shapes of the profiles AB and BC are similar to each other. The average feed spacing, derived from Fig. 3.37b, is calculated as $f = \frac{AB+BC}{2} = \frac{5.1}{2} = 2.55$ mm. The theoretical feed spacing, based on the feed rate and cutting speed, is $f = \frac{3800}{1400} = 2.71$ mm. The relative error is then calculated as $r = \frac{2.71-2.55}{2.71} = 5.9\%$.

Figure 3.38 displays the outcomes of the multiscale watershed segmentation with the area threshold *A* ranging from 1 to 50 in increments of 7. Consistent with the findings from cases 1 and 2, the multiscale watershed segmentation is unsuccessful in effectively segmenting surfaces that feature periodic tool marks.

Across the three cases, it has been observed that the points on the tool marks are either local peaks or very close to them; each profile segment, which is a negative

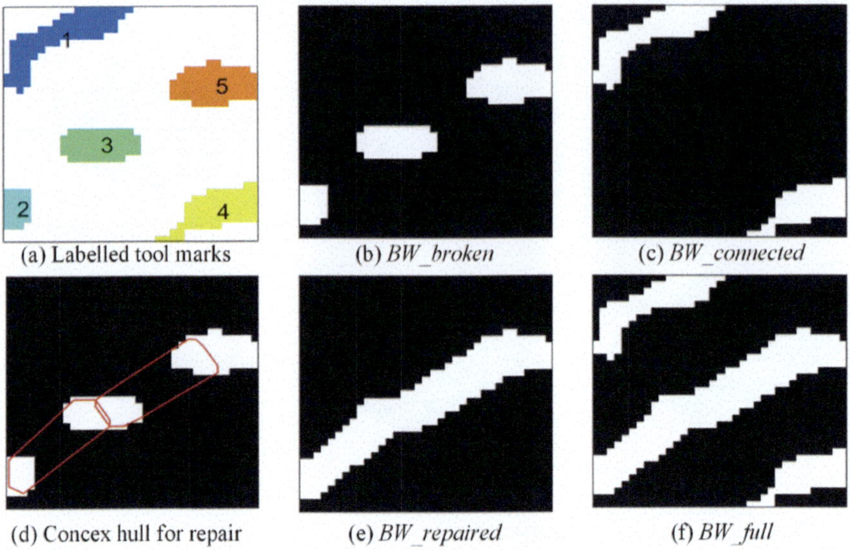

Fig. 3.35 Tool marks labelling and repairing procedures in Case Study III

Table 3.7 The inter region distance matrix *IRD* in Case Study III

Distance	Region1	Region2	Region3	Region4	Region5
Region1		15	9.22	27.29	11.31
Region2	15		7.81	18.25	24.41
Region3	9.22	7.81		10.63	7.81
Region4	27.29	18.25	10.63		12.37
Region5	11.31	24.41	7.81	12.37	
min(IRD)	9.22	7.81	7.81	10.63	7.81

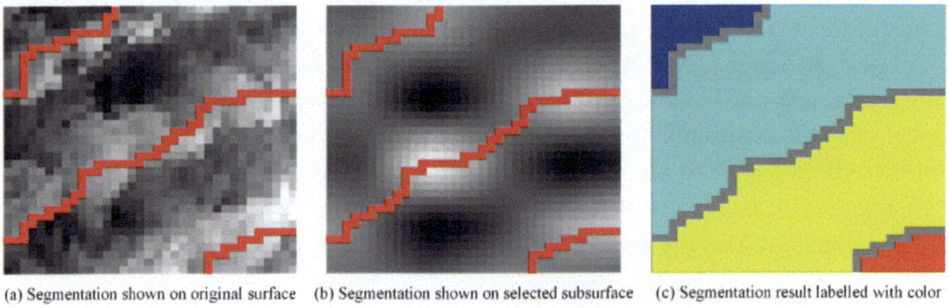

Fig. 3.36 Final surface segmentation result in Case Study III

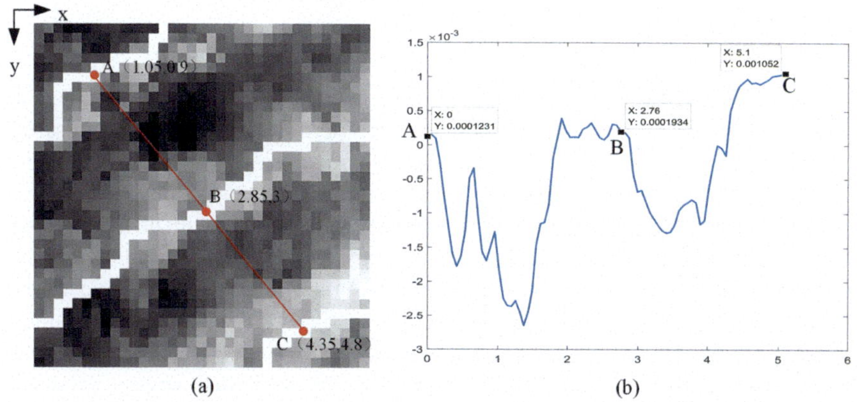

Fig. 3.37 A cross section view of tool marks in Case Study III

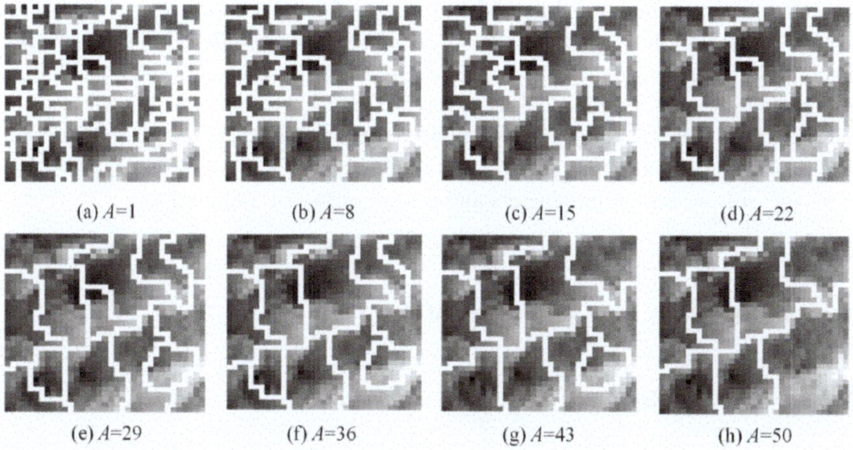

Fig. 3.38 Results of multiscale watershed segmentation in Case Study III

replica of the tool shape, is similar to one another; and the feed spacing calculated from the segmentation images is near to the theoretical values derived from the cutting parameters, with an average relative error of approximately 6.4%. Moreover, the multiscale watershed segmentation has failed to correctly partition the surfaces in all three cases. Consequently, it can be concluded that the "STRW" segmentation methodology is capable of effectively segmenting milling surfaces that possess periodic tool marks.

3.1.4 Conclusion

This section introduces an enhanced surface segmentation approach known as "STRW," designed to segment surfaces featuring periodic tool marks into discrete regions, each corresponding to a specific tool tooth trajectory. The process begins with decomposing the measured surface into various subsurfaces, employing a novel "periodic degree" concept as the basis for subsurface selection. Following this, the chosen subsurface undergoes local thresholding to distinguish tool marks from their background. Surface irregularities may result in broken tool marks, which are detected using an adaptive distance threshold and subsequently repaired using a convex-hull-based algorithm. Ultimately, classical watershed segmentation is utilized to divide the surface into distinct, meaningful regions. The "STRW" methodology serves as a significant addition to ISO 16610-85, offering a novel method to connect surface topography with machining processes.

The "STRW" methodology is particularly effective for continuous surfaces characterized by periodic tool marks. For machined surfaces lacking periodic tool marks, segmentation may necessitate specialized treatment tailored to their specific machining processes. In the case of milled surfaces with significant holes and bores, as depicted in Fig. 3.3, additional measures are required to connect tool marks separated by these features. Future research will focus on classifying and parameterizing tool marks across the entire surface of cylinder blocks to explore the relationship between milling process parameters and the properties of tool marks.

3.2 Evaluation and Modeling of Tool Marks

3.2.1 Introduction

Surface texture, also known as surface finish or surface topography, refers to the micro characteristics of a machined surface, which include the minor undulations of the actual surface from an ideal flat plane [28, 29]. This texture plays a direct role in influencing the mechanical properties of workpieces, including contact stiffness [30], fatigue strength [31], vibration resistance [32], and sealing performance [33]. The assessment of surface texture quality can be conducted at different observation scales, with surface roughness and waviness being the respective indicators for these evaluations [34–36].

Extensive research has been conducted to understand the formation and assessment of surface texture under diverse machining conditions. Villa et al. [37], in the context of machining processes with defined tool geometry, created a mathematical model that represents the machined surface profile as an ordered sequence of tool marks. From this model, a pattern identification procedure was developed to identify suitable reference points on the profile [38]. Hadad and Ramezani [39] examined the cutting tool geometry and developed a simulation model for the surface texture pattern generated in face milling processes. Salisbury et al. [40] proposed a geometric-kinematic model to study the surface topography generation in a single-pass surface grinding process, and further developed a wheel surface model integrated with a surface grinding process model for surface texture simulation and evaluation [41]. Baek et al. [42] proposed a physical model to analyze the impact of insert runout errors and feed rate variations on surface texture in face milling operations. Considering the effects of varying dynamics, Kiss et al. [43] presented a finite element method (FEM)-based model for surface error calculations and stability conditions in milling operations involving flexible workpieces. Shen et al. [44] established a neural network prediction model for the dynamic performance of machine tool joint surfaces, investigating the influence of surface texture on joint surface dynamics. Batsch [45] developed a mathematical model for the honing process of helical gears with external teeth, aiming to determine the honing tool profile for gears with profile modifications. The majority of published research has been validated through experimental cases using traditional measurement equipment such as the coordinate measurement machine (CMM). However, due to the sampling density limitations of CMM, some valuable surface texture information is inevitably overlooked.

In recent years, high definition metrology (HDM), a non-contact measurement technology employing laser holographic interferometry, has garnered significant scholarly attention [14, 46, 47]. HDM's primary benefits over CMM are its efficiency and sampling density [13, 48]. This technology enables the rapid generation of a 3D height map for large and geometrically complex surfaces, comprising millions of data points, within seconds [12]. In surface texture measurement, HDM outperforms traditional methods in terms of effectiveness and efficiency. Analysis of the high-resolution point cloud data from HDM can uncover more insightful information about the machining process, which is crucial for controlling and enhancing surface quality. Consequently, this study employs HDM for surface texture measurement and feature assessment due to its superior measurement precision and speed compared to conventional methodologies.

Periodic tool marks, a key element of surface texture, are prevalent in milling and honing operations, offering substantial insights into the manufacturing processes. In face milling, these marks are influenced by the interaction between workpieces and cutting inserts, tool conditions such as wear and insert runout, and the relative vibration between the tool and workpiece. During the computer numerical control (CNC) machine-based face milling, the spindle may exhibit a slight tilt, typically ranging from 0.1 to 0.3 mm over a 1000 mm length in the feed direction, causing a lack of perfect parallelism between the disc milling cutter and the workpiece surface. This spindle tilt mitigates the back-cutting effect and reduces additional flank wear on the inserts, thereby extending tool life. The spindle's condition also impacts the dynamic performance of the machining system [49, 50], necessitating consideration of chatter stability in the presence of spindle tilt. The engagement mode between the workpiece and cutting tool alters due to spindle tilt, leading to distinct tool mark patterns on the surface texture. HDM is essential for assessing and enhancing the quality of a face milled surface by measuring tool mark characteristics. Depending on the spindle's inclination during face milling, three distinct surface textures with tool marks are left on the workpiece's machined surface, as depicted in Fig. 3.39.

The benefits and drawbacks of the three spindle inclination types are detailed in Table 3.8. This section, due to its scope, concentrates primarily on surface textures marked by unidirectional tool marks resulting from the positive inclination of the spindle in face milling processes. It is important to recognize that the concave shape of the machined surface is a result of not just spindle tilt, but also cutting forces, fixture stability, and various other elements. In this context, given the small degree of inclination used, spindle tilt is not considered the primary factor. Furthermore, the impact of form accuracy on surface quality assessment can be largely eliminated through appropriate filtering techniques.

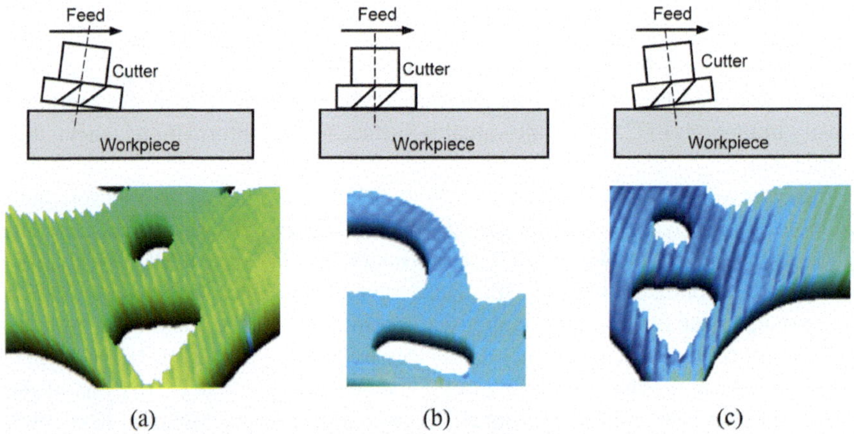

Fig. 3.39 Typical surface texture with different kinds of tool marks: **a** unidirectional tool marks caused by positive inclination of spindle, **b** crisscross tool marks without spindle inclination, and **c** unidirectional tool marks caused by negative inclination of spindle

Table 3.8 Advantages and disadvantages of three kinds of spindle inclination

Inclination	Positive	Zero	Negative
Advantages	(1) No back-cutting and longer tool life (2) Relatively simple and regular texture pattern	(1) Better flatness of the machined surface (2) Easier to be applied for conventional machine	(1) Relatively simple and regular texture pattern
Disadvantages	(1) Spindle tilt induced slightly concave form	(1) Back-cutting induced additional tool wear (2) Hard to control the texture pattern	(1) Severe tool wear (2) Poor surface quality

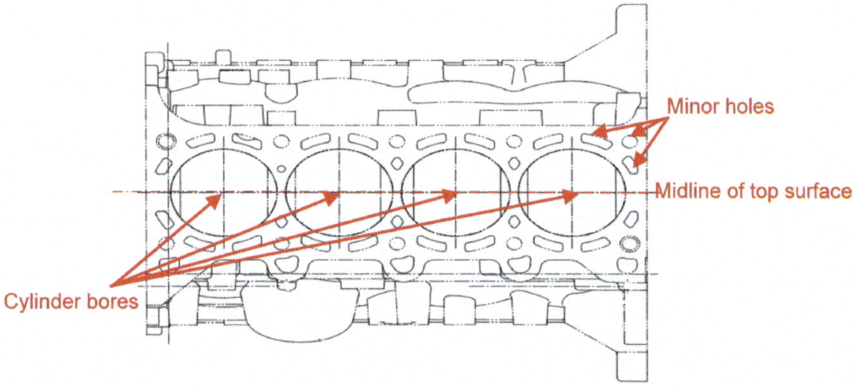

Fig. 3.40 Top view of a four-cylinder engine block

Surface texture characteristics, notably the waviness of machined surfaces, are heavily influenced by the pattern of tool marks, impacting the operational performance of workpieces significantly. HDM data informed an enhanced method by Yin et al. [51] to prevent over-segmentation in the segmentation of surfaces exhibiting tool marks. This method's ability to accurately segment sub-surfaces marked by tool marks is crucial for the precise estimation of sealing performance. Shao et al. [52] leveraged the 3D surface topography obtained from HDM to create a model for detecting leakage in static sealing interfaces. Ren and Wang [53] introduced a transfer learning (TL) framework that integrates high-resolution HDM data with low-resolution CMM data for surface variation modeling. While these studies offer viable models for predicting and monitoring based on HDM-derived surface textures, there is a scarcity of physical models focused on assessing and producing surface textures with tool marks for workpieces featuring discontinuous surfaces.

The complexity of workpiece geometry has grown with the advancement of the manufacturing sector [54]. Instances such as the four-cylinder engine block, which features discontinuous surfaces, are representative of this trend, as shown in Fig. 3.40.

The top surface of the engine block is characterized by four cylinder bores and numerous minor holes, which exacerbate the complexity of the workpiece geometry and result in a discontinuous cutter-workpiece interface. Given that the surface texture, marked by tool marks, is substantially influenced by the vibrational behavior of the machining system—behavior that is contingent upon the mode of cutter-workpiece interaction [19, 55, 56]—it becomes imperative to consider the face milling dynamics for the optimization of processing parameters to achieve greater efficiency.

Incorporating workpiece geometry and pivotal aspects of the face milling process, scholarly investigation into HDM-based solutions for associated challenges has progressed in recent years. Wang et al. [57] crafted a framework employing principal curves and quality control charts for the efficient surveillance of spatial variations within HDM data. Expanding on a sequential approach for global and localized detection of shape deviations in HDM data, Suriano et al. [58] advanced a framework tailored for HDM data from automotive engine head machining, effectively pinpointing defective areas on non-conforming components. Tai et al. [59] introduced a feed rate optimization technique aimed at enhancing surface flatness in face milling, aligning the tool feed rate with the cutter's axial force and the workpiece's local flexibility, utilizing HDM insights. Zhou et al. [60] devised a functional morphing approach, leveraging pre-milling HDM surface data and process variables to forecast post-milling surface quality in a multistage manufacturing context. Nguyen et al. [20] presented a physical model for spindle setup tilt and deflection surveillance, utilizing HDM-measured surface data. While data-driven and statistical methods may encounter constraints with intricate physical processes, current physical models, as per a review of the state-of-the-art, fall short in accurately depicting the actual face milling dynamics. Moreover, HDM data application predominantly targets flatness, with waviness information, particularly tool mark characteristics, being underutilized due to the absence of robust waviness characteristic assessments. the International Organization for Standardization (ISO)-defined waviness parameters, serving as broad surface texture evaluation metrics, struggle to link with processing parameters through physical modeling, as they overlook the distinctive morphological attributes of tool marks. This section is propelled by these constraints, recognizing the significance of waviness as a key element of surface texture for enhancing surface quality.

In summary, the assessment of surface texture marked by tool traces using HDM data has been inadequately addressed, posing a distinct challenge to the physical simulation of tool marks on workpieces with interrupted surfaces during face milling. To address this challenge, a novel and comprehensive HDM-based methodology for enhancing surface texture quality is introduced. Initially, the tool mark evaluation metrics are delineated through HDM data preprocessing, which are intended to supersede the ISO waviness parameters for predicting surface texture quality through physical simulation. Subsequently, the physical simulation of tool marks is detailed, with a thorough examination of the influence of insert runout, tool wear, and chatter stability on surface texture formation within the model. Following this, a physical-informed optimization model is formulated and resolved utilizing the non-dominated sorting genetic algorithm-II (NSGA-II) [61], aiming to identify the optimal machining parameters for surface quality enhancement. Ultimately, the efficacy of the proposed methodology is validated through an experimental case study on four-cylinder engine blocks. The primary contribution of this section is the elucidation of the physical correlation between the face milling mechanism, which is contingent upon machining parameters and workpiece geometry, and the evaluation metrics of surface texture marked by tool traces, with an inclusive consideration of spindle tilt, runout, tool wear, and chatter stability—a relationship that has not been thoroughly investigated in the existing literature.

The subsequent sections of this research are structured as follows: Section 2.2 outlines the specific steps of the proposed methodology, encompassing HDM data preprocessing, physical simulation, and physics-informed optimization modeling. Section 2.3 showcases the efficacy of the proposed methodology through a case study on the face milling process of a representative workpiece with interrupted surfaces, with a detailed discussion and analysis of the experimental findings. Section 3.2.4 concludes the research with a summary of the findings.

3.2.2 The Proposed Approach

The schematic of the proposed methodology is depicted in Fig. 3.41, with the principal steps detailed below.

Step 1: Raw HDM-measured data undergoes preprocessing, which includes the removal of outliers and filtering. From the filtered point cloud of the surface texture, tool marks are extracted. To assess the characteristics of these marks, three evaluation indices are established.

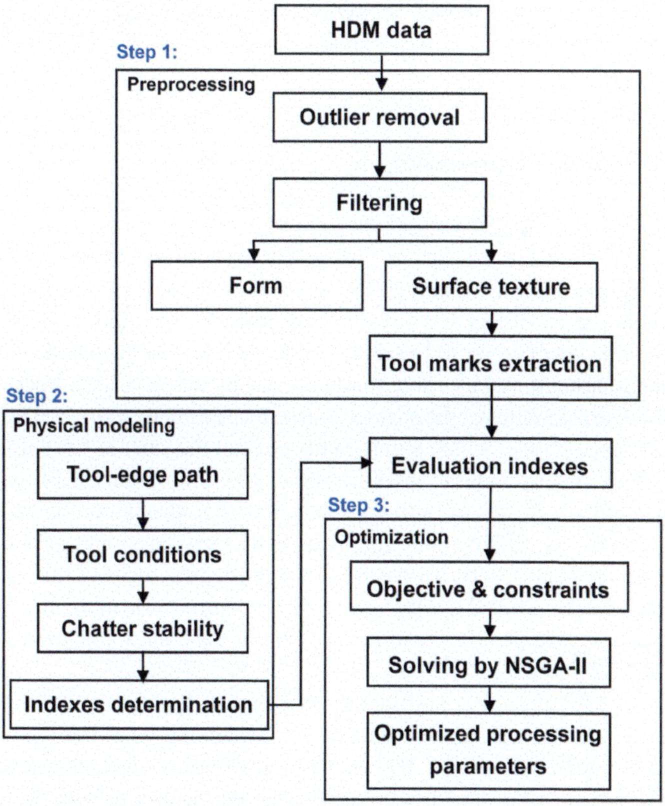

Fig. 3.41 Framework of the proposed approach

Step 2: Using the face milling process of a typical workpiece featuring a discontinuous surface as a case study, a physical model for surface texture with tool marks is developed. The generation of the tool-edge path, a key element of surface texture marked by tool traces, is examined. Additionally, considering the intermittent nature of cutter-workpiece engagement, the impacts of tool conditions and chatter stability on surface texture are analyzed. Consequently, the evaluation indices are ascertained based on this physical model.
Step 3: A physical-informed optimization model is constructed with the goal of enhancing surface texture quality. The variables encompass standard processing parameters used in face milling. The aim is to optimize the evaluation indices of tool marks simultaneously to reduce the waviness of the surface resulting from face milling. The primary constraints are cutting stability and the practicality of operation. Given the intricate mathematical interplay between processing parameters and evaluation indices, this problem is characterized as a multi-objective, nonlinear programming challenge. The NSGA-II, recognized as an advanced algorithm for multi-objective optimization [62, 63], is engaged to address this optimization model. Subsequently, the optimized processing parameters are implemented in actual experiments to validate the efficacy of the proposed methodology.

3.2.2.1 HDM Data Preprocessing

Outlier Removal and Filtering

In the realm of practical engineering, HDM-gathered raw data may encompass redundant elements such as outliers and low-pass components. Typically, outliers are categorized into two classes (refer to Fig. 3.42). Type A: These are isolated and recessed points on the measured surface, potentially caused by ambient light interference or the reflective characteristics of the surface under measurement. Type B: These are the extremely elevated points, which significantly exceed the surface's nominal height, often a result of dust accumulation or carbon deposition on the measured surface. The presence of both classes of outliers is detrimental to the extraction and analysis of features. Moreover, HDM data comprises form (low-pass) and surface texture (high-pass) components. Since form error is not the central focus of this section, the surface texture can be derived through a suitable decomposition of the HDM-derived point cloud data. Consequently, the removal of outliers and the application of filtering are essential steps.

For a given workpiece, the nominal distance from the measured surface to the datum plane is established, and the deviation between the measured value and the nominal height can be ascertained. By establishing a tolerance level for the maximum allowable deviation as the threshold, outliers can be identified. Specifically, if the Z-axis coordinate value of a data point P exceeds this threshold, point P is classified as an outlier. For further insights on threshold determination for outlier removal, reference to literature [48] is advised.

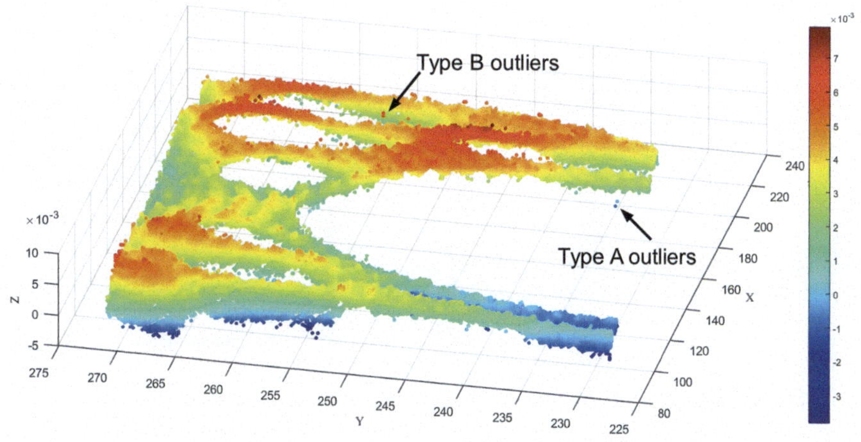

Fig. 3.42 Outliers in raw HDM data

Following the removal of outliers, the HDM data can be separated into form and surface texture components based on frequency coefficients. As per the guidelines in ISO 16610-21 [64], the areal Gaussian filter is the prevailing standard for 3D surface filtering. However, its application to discontinuous surfaces may result in significant edge distortions in the filtered outcomes, compromising data precision. To mitigate the edge effect, this section employs an extended tetrolet transform method, as proposed in literature [12], for filtering purposes.

Tool Marks Extraction

In the face milling operation utilizing a multi-insert disc milling cutter, the tool-edge orbits around the spindle's centerline, while the cutter head progresses along the feed direction. Consequently, the typical form of a single tool mark is cycloidal. As an example, the cycloidal tool marks on the face-milled top surface of an engine block are depicted in Fig. 3.43.

Due to the continuous change in the tool-edge path direction with processing time, the shape characteristics of tool marks across various regions exhibit disparities. For instance, moving along the feed direction, the spacing between consecutive tool marks in the outer zone is significantly greater than in the inner zone. Typically, shape features aligned perpendicular to the feed direction offer the most accurate representation of tool mark attributes [65]. Consequently, the extraction of tool marks necessitates the conversion of cycloidal marks into a straightened form, ensuring they are consistently oriented perpendicular to the feed direction.

The transformation to straighten the original cycloidal tool marks is appropriate for face milling scenarios that do not involve back-cutting. As illustrated in Fig. 3.44, the arc $\widehat{AC}$ represents a cycloidal tool mark generated by the path of a single insert

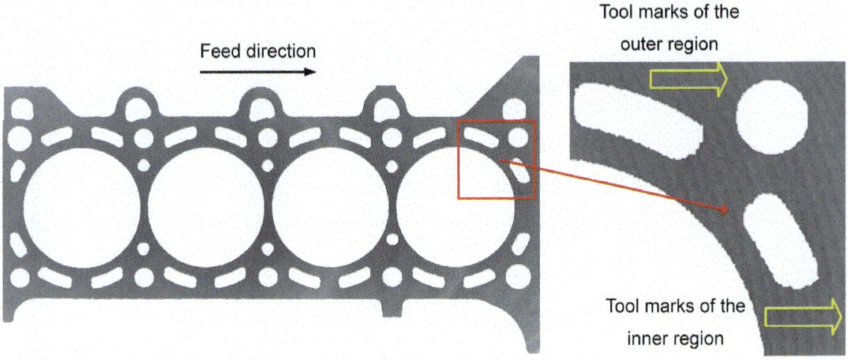

Fig. 3.43 Original cycloidal tool marks on the face milled top surface of an engine block

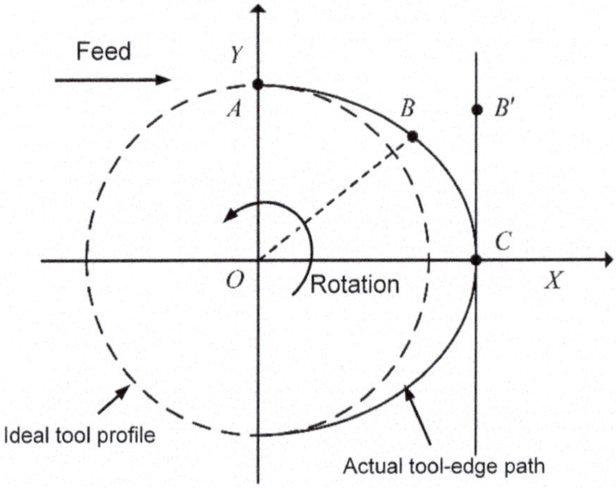

Fig. 3.44 Straightening transformation for face milling induced tool marks

during one complete revolution. With the disc milling cutter advancing along the X-axis at a rate of f and the spindle rotating at a speed of Ω, and by defining the radius of the disc milling cutter as R, the initial coordinates of the cutter center as $[x_O, y_O]$, and the machining time when the insert reaches a specific point B on arc $\widehat{AC}$ as t_B, the coordinates of point B are determined by the following formula:

$$\begin{cases} x_B = x_O + R\sin\Omega t_B + f t_B \\ y_B = R\sin\Omega t_B \end{cases} \tag{3.13}$$

By straightening transformation, the coordinates of point B′ on the corresponding straightened tool mark can be formulated by

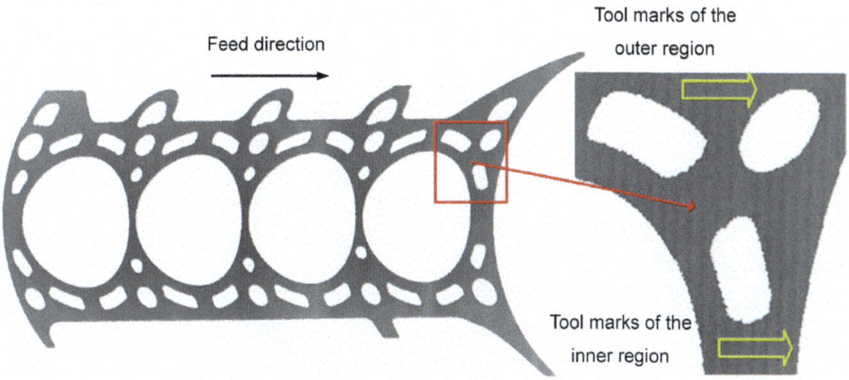

Fig. 3.45 Straightened tool marks on the face milled top surface of an engine block

$$\begin{cases} x_{B'} = x_C = x_O + R + f\frac{\pi}{2\Omega} \\ y_{B'} = -\int_{\pi/2\Omega}^{t_B} \sqrt{\left(\frac{dx_B}{dt}\right)^2 + \left(\frac{dy_B}{dt}\right)^2} dt \end{cases} \tag{3.14}$$

Figure 3.45 illustrates the conversion of cycloidal tool marks into their straightened equivalents. The shape characteristics of tool marks extracted from both the inner and outer regions are now consistent, which has substantially facilitated the extraction of tool mark features.

Evaluation Indexes Definition

In accordance with the most recent ISO standard [25], a suite of areal surface texture parameters has been clearly delineated and broadly implemented across the industry. These include height parameters (S_q, S_p, S_v, S_z, S_a), shape parameters (S_{sk}, S_{ku}), and spatial parameters (S_{al}, S_{tr}), among others. However, these ISO parameters serve as general assessment indices for areal surface texture and are predominantly data-driven and outcome-focused. Establishing a clear physical connection between these indices and the machining process is challenging, indicating a deficiency in the correlation between the machining mechanism and the resulting surface texture, particularly for tool marks generated by face milling.

To provide a quantitative description of the surface texture characteristics with tool marks based on the physical model presented in this section, three morphological evaluation indices have been established. As depicted in Fig. 3.46, these indices include the average peak-valley difference h_d, the wavelength spacing s_λ, and the height fluctuation h_f. The nominal peaks are indicated by solid lines, while the nominal valleys are represented by dotted lines. It is important to note that even after the HDM data has been thoroughly cleansed through outlier removal, filtering, and tool marks extraction, the morphology of the surface texture with tool marks may not resemble an approximate sine wave. Instead, a concave phenomenon might

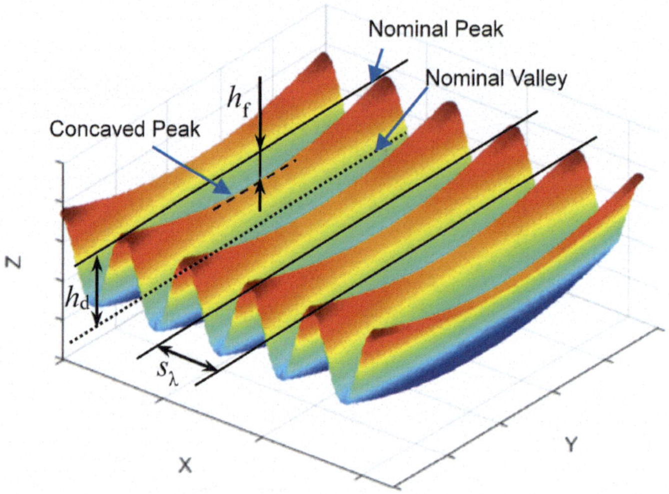

Fig. 3.46 Evaluation indexes for surface texture with tool marks

appear, and the lowest height of the peak of a single tool mark is referred to as the concaved peak, which is delineated by dashed lines in Fig. 3.46. The causes for this can be twofold. Firstly, along the X-axis, the amplitudes of each tool mark may vary due to tool wear and insert runout. Secondly, along the Y-axis, the height of a single tool mark is not constant, often a result of spindle tilt. Unlike form error, height fluctuation should not be removed by filtering, as it encompasses the micro characteristics of the surface texture and should be preserved.

For a series of consecutive adjacent tool marks within the sampling region, labeled sequentially as $\{1, 2, \ldots, s\}$, the average peak-valley difference is calculated as the mean of the height disparities between the nominal peaks and their corresponding nominal valleys for each individual tool mark. This can be mathematically expressed as:

$$h_{\mathrm{d}} = \frac{1}{s}\sum_{i=1}^{s}(\max\{z_i\} - \min\{z_i\}) \tag{3.15}$$

where z_i represents the Z-axis coordinates of the peak and valley, respectively, on the i-th tool mark. Essentially, the average peak-valley difference h_{d} exerts the most substantial influence on surface waviness and is advised to be minimized for enhancing surface quality.

Following the straightening transformation, all tool marks align perpendicular to the feed direction (X-axis), with the abscissas of the peaks of the tool marks denoted by the set $\{x_1, x_2, \ldots, x_s\}$. Given that the sampled tool marks are evenly spaced, the wavelength spacing s_λ remains constant. To mitigate the impact of measurement error, s_λ is determined by calculating the average of the intervals between the tool marks as follows:

$$s_\lambda = \frac{1}{s-1}\sum_{i=1}^{s-1}(x_{i+1} - x_i) \tag{3.16}$$

The wavelength spacing s_λ significantly influences the determination of the cutoff wavelength λ_f, which is essential for differentiating waviness components from form components [66]. For a precise assessment of waviness, it is crucial to maximize the difference between s_λ and λ_f, that is, to maximize the value of $\lambda_f - s_\lambda$. In this section, λ_f is set at 0.8 mm, aligning with one of the recommended values for λ_f as per the ISO standard [22].

The third evaluation index, h_f, captures the height fluctuation within a single straightened tool mark, independent of the machined surface's form error. Within the sampled region, height fluctuation is quantified by averaging the height differences between the nominal peak and its corresponding concaved peak for each individual tool mark. This can be mathematically represented as:

$$h_f = \frac{1}{s}\sum_{i=1}^{s}(\max\{z_i|x = x_i\} - \min\{z_i|x = x_i\}) \tag{3.17}$$

where $\{z_i|x = x_i\}$ represents the set of Z-axis coordinates of the data points at the peak of the i-th tool mark. An increased value of h_f correlates with a decrease in surface texture quality.

3.2.2.2 Physical Modeling

The surface texture of a machined surface is intricately linked to the material removal process during face milling and is significantly affected by various factors, including tool conditions, dynamic characteristics, and processing parameters. Consequently, a holistic physical model of the face milling process is essential to establish the mapping relationship between quality evaluation indices and the most pertinent contributing factors. In this section, the physical model incorporates considerations of spindle tilt, runout, tool wear, and chatter stability.

The general schematic diagram for the face milling process is depicted in Fig. 3.47. In practice, the face milling process typically encompasses both roughing and finishing operations. Material removal can be accomplished through multiple passes, consisting of n_a axially parallel passes and n_r radially parallel passes. The total radial cutting depth, S_{ae}, is equal to n_r times the actual radial cutting depth a_e. Similarly, the total axial cutting depth, S_{ap}, is determined by $S_{ap} = n_a a_p$. When employing a large disc milling cutter, a single radial pass may suffice for radial material removal, implying $n_r = 1$ and thus $S_{ae} = a_e$. Additionally, there may be an offset, denoted by Δe, from the geometric central line (GCL) of the disc milling cutter to that of the workpiece, which complicates mathematical deduction. Therefore, the proposed physical model concentrates on a single pass of the disc milling cutter with a tool

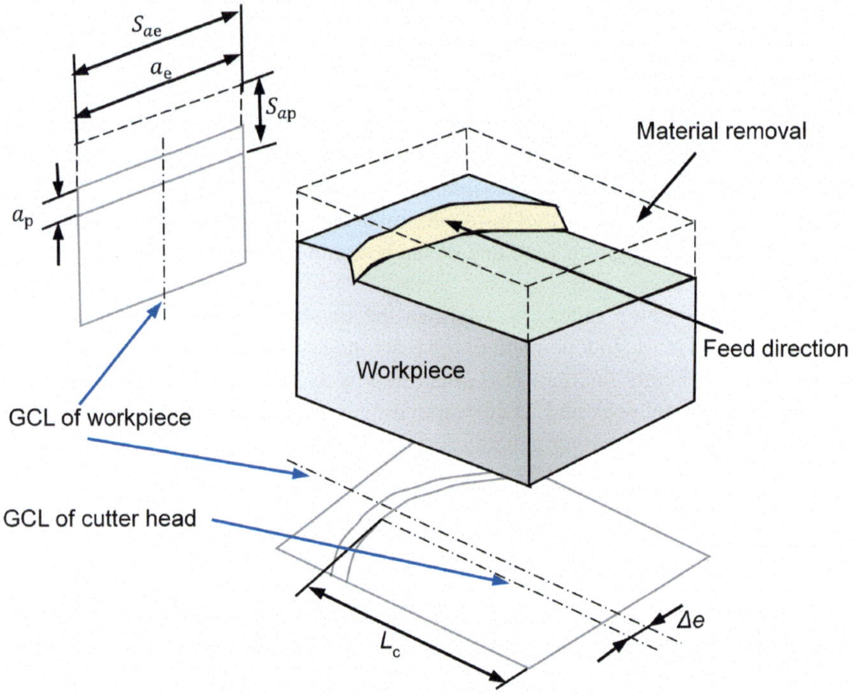

Fig. 3.47 General schematic diagram for face milling process

path aligned with the GCL of the workpiece. In essence, for mathematical simplicity, n_a is set to 1 and Δe is set to 0 in this research.

Tool-Edge Path Generation

The path traced by the tool-edge is a principal element of tool marks, particularly in scenarios that involve the use of a large disc milling cutter and a spindle with tilt. As illustrated in Fig. 3.48, the cutter head is attached to a spindle with a slight inclination ϕ_0 in the feed direction, resulting in the bottom plane of the cutter head and the workpiece surface not being parallel. Consequently, the actual lead angle of the cutting insert ϕ_a is influenced by the spindle's tilt.

The top view of a typical face milling setup is displayed in Fig. 3.49. The insert-workpiece engagement point rotates counter-clockwise around the center of the disc milling cutter at a speed of Ω, while the cutter head is fed along the X-axis at a rate of f. A total of N serially numbered inserts are mounted on the cutter head. Timing begins from the moment the inserts come into actual contact with the workpiece. For a specified cutting duration t_c, the tool-edge path generated by the first engaged insert can be described as:

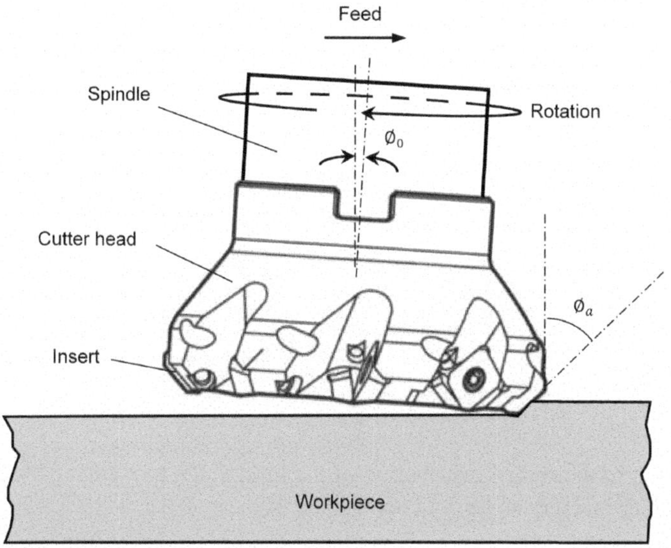

Fig. 3.48 Face milling with a large cutter head and a tilted spindle

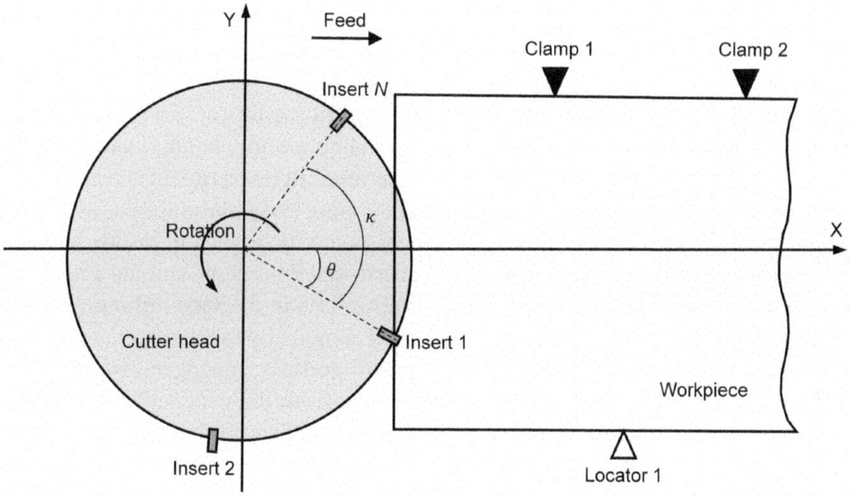

Fig. 3.49 Top view of a general face milling case

$$\begin{cases} x_1 = ft_c + R_I \cos(\theta + \Omega t_c) \\ y_1 = R_I \sin(\theta + \Omega t_c) \end{cases} \tag{3.18}$$

where R_I is the corrected radius of the circle formed by the contact point between the tool insert and the workpiece, and θ is the angle between the feed direction and the line extending from the first engaged insert to the center of the cutter head. This formulation captures the motion and path of the tool edge during the milling process, taking into account the rotational speed of the cutter and the linear feed rate.

Note that R_{I} is different from the radius of the cutter head R due to the spindle tilt, which can be approximately formulated as

$$R_{\mathrm{I}} = \frac{1+\cos\phi_0}{2}R \tag{3.19}$$

Assuming that all the inserts are evenly distributed on the circumference of the cutter head, the angle difference κ between insert i and insert $(i+1)$ (for $i = 1, 2, ..., N-1$) can be formulated as

$$\kappa = \frac{2\pi}{N} \tag{3.20}$$

According to the spatial distribution of the inserts, the tool-edge path generated by an arbitrary insert i can be obtained by

$$\begin{cases} x_i = ft_{\mathrm{c}} + R_{\mathrm{I}}\cos(\theta - (i-1)\kappa + \Omega t_{\mathrm{c}}) \\ y_i = R_{\mathrm{I}}\sin(\theta - (i-1)\kappa + \Omega t_{\mathrm{c}}) \end{cases} \tag{3.21}$$

Effect of Tool Conditions

Cutter runout is a critical tool condition that frequently arises in various machining operations that utilize rotary cutting tools, including milling, honing, and grinding [67, 68]. In the context of face milling, the occurrence of insert runout is particularly prevalent when employing a disc milling cutter equipped with multiple inserts. In the absence of insert runout, the engagement points between the inserts and the workpiece are considered to be uniformly distributed around the spindle's nominal axis, indicating an equal contribution from all inserts in the face milling process. However, the advent of insert runout introduces complexity to the material removal process due to the non-uniform radii at the insert-workpiece engagement points.

In the context of face milling operations, inserts featuring wiper edges are extensively employed to achieve machined surfaces with enhanced flatness [69, 70]. As depicted in Fig. 3.50, a standard square insert with four symmetric wiper edges is showcased. The insert possesses a thickness of l_{T}, four principal cutting edges each measuring l_{G} in length, and four wiper edges each measuring l_{W} in length. The angle where the adjacent principal cutting edge and the wiper edge meet is indicated by α. With the feed per insert being f_z at a steady feed rate f, the thickness of the undeformed chip is dependent on the lead angle ϕ and the feed per insert f_z. Furthermore, when the spindle is slightly tilted with an inclination of ϕ_0, the actual lead angle of the insert on the cutter head is calculated as $\phi_a = \phi + \phi_0$. Consequently, the thickness of the undeformed chip can be expressed as:

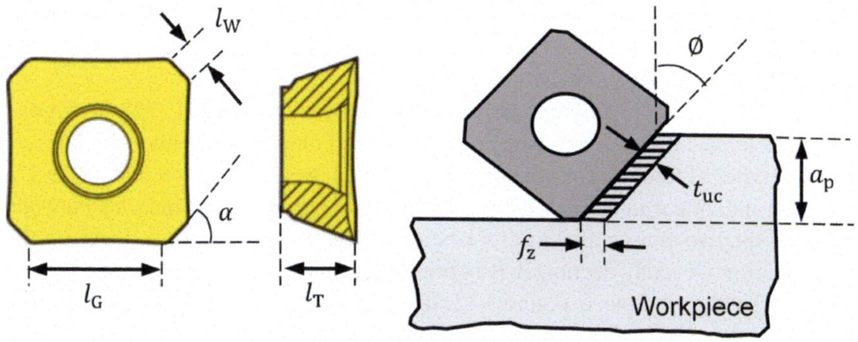

Fig. 3.50 The square insert with wiper edges and corresponding undeformed chip

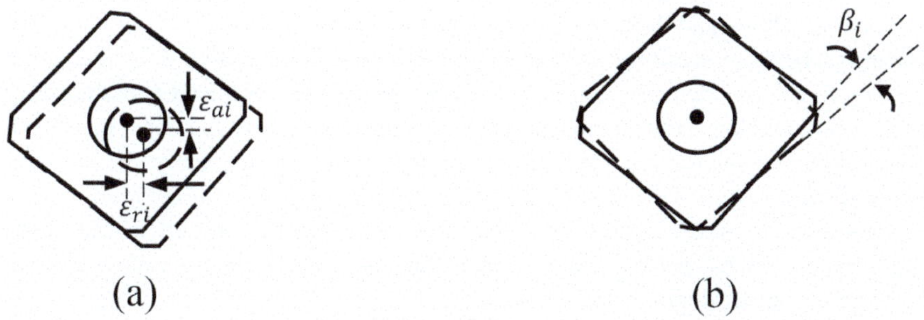

Fig. 3.51 Insert runout errors: **a** axial and radial runout errors and **b** rotational runout error

$$t_{uc} = f_z \cos \phi_a \tag{3.22}$$

As shown in Fig. 3.51, the radial, axial, and rotational runouts for insert i are represented by ε_{ri}, ε_{ai} and β_i respectively, for $i = 1, 2, \ldots, N$. These runouts collectively contribute to variations in chip removal, which in turn result in surface texture imperfections. To delve deeper into the impact of insert runout on surface quality, the radial and axial runouts are conceptualized as the state of insert-workpiece engagement points orbiting around an axis that is offset from the original spindle axis. Concurrently, the rotational runout error of an insert can be superimposed onto the spindle tilt, directly influencing the actual lead angle and the thickness of the undeformed chip. Thus, accounting for spindle tilt, the influence of insert runouts on chip removal can be articulated as:

$$\begin{cases} a'_{p\_i} = a_p + \varepsilon_{ai} \\ t'_{uc\_i} = (f_z - \varepsilon_{ri}) \cos(\phi + \phi_0 + \beta_i) \end{cases} \tag{3.23}$$

where the subscript i refers to the serial number of the inserts, a'_p is the actual axial cutting depth, and t'_{uc} is the thickness of the undeformed chip after considering the effects of insert runout and spindle tilt.

The surface texture is significantly influenced by tool wear [71, 72], hence this section examines the characteristics of undeformed chips, taking into account two common types of tool wear: local chipping and flank wear. As depicted in Fig. 3.52a and b, an individual undeformed chip i is demarcated by the boundaries between itself and the two consecutive inserts labeled as insert i and insert $i-1$. Along the feed direction (X-axis), section B-B is presumed to intersect the geometric central line (GCL) of the workpiece. Figure 3.52c displays section B-B of chip i and the angle γ_i between the wiper edge and the ideal horizontal plane. Additionally, Fig. 3.52d illustrates the impact of tool wear on chip removal, which is attributed to the reduction in the effective cutting area for insert i.

The angle between the wiper edge and the ideal horizontal plane causes the actual axial cutting depth and the area of section A-A to be non-uniform and vary with the distance L between section A-A and section B-B. As shown in Fig. 3.53, the actual axial cutting depth increases non-uniformly as one approaches the GCL of the cutter head. This variation is due to the changing engagement angle of the wiper edge with the workpiece, which affects the cutting geometry and, consequently, the material removal rate.

Denoting the intersection of chip edge trace and the GCL of cutter head along feed direction by B_0, as for an arbitrary reference point A_0 on the trace, the variation of the axial cutting depth is derived as

$$\Delta a_p(L_0) = \left(R_I - \sqrt{R_I^2 - L_0^2}\right) \tan \gamma_i \tag{3.24}$$

where L_0 is the projection distance from A_0 to the GCL of cutter head, ranging from 0 to $\frac{a_e}{2}$.

Consequently, the actual axial cutting depth of A_0 can be given by

$$a_p(L_0) = a'_{p\_i} - \left(R_I - \sqrt{R_I^2 - L_0^2}\right) \tan \gamma_i \tag{3.25}$$

where γ_i is jointly determined by spindle tilt, rotational insert runout, and insert geometry, formulated as $\gamma_i = \phi_0 + \beta_i + \alpha + \phi - \frac{\pi}{2}$.

To quantify the impact of tool wear on chip removal, the geometrical features of the effective cutting area for insert i are analyzed in the presence of typical tool wear, as illustrated in Fig. 3.54. A local coordinate system is set up in section B-B of chip i, and the maximum height wear of the wiper edge is defined as:

$$H_i = \max\{z_{i,b_i}\} \tag{3.26}$$

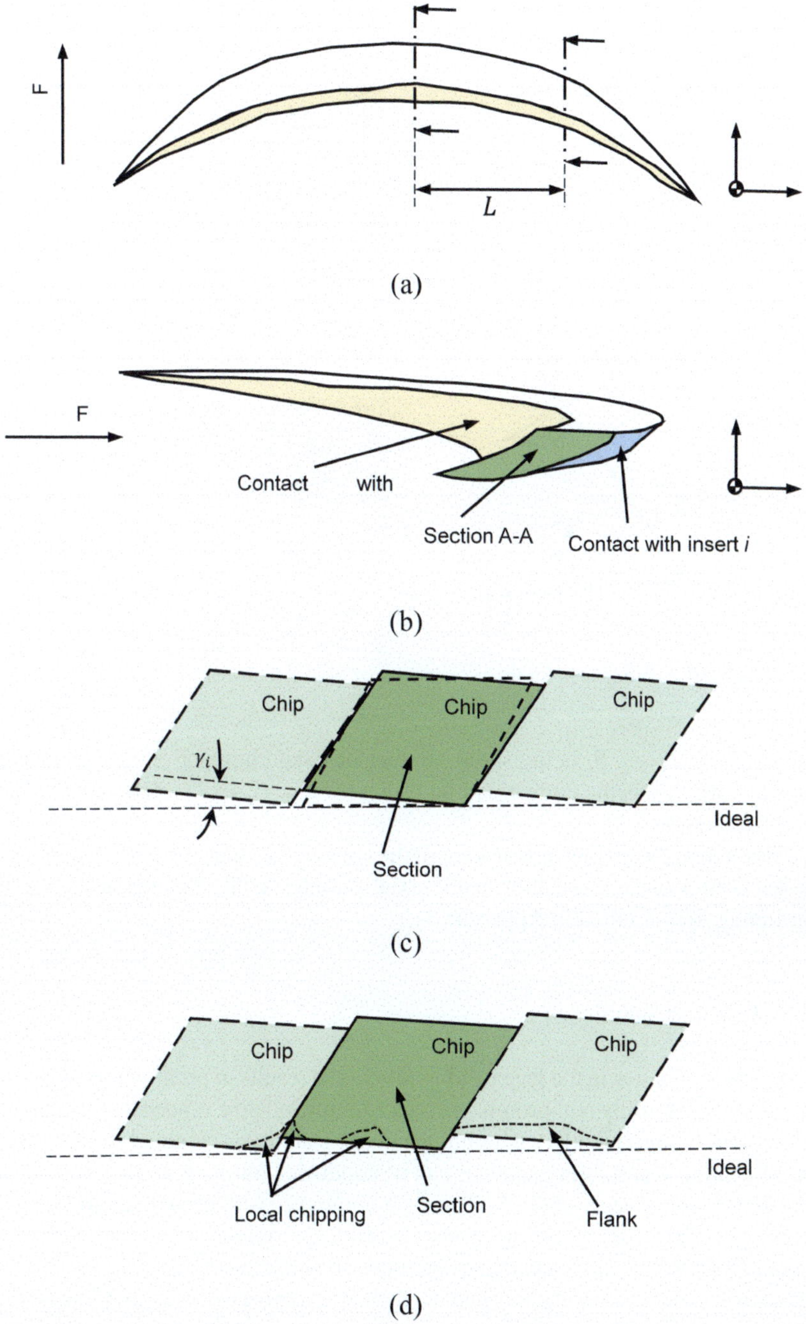

Fig. 3.52 The undeformed chip considering spindle tilt: **a** top view of chip i, **b** section A-A of chip i, **c** section B-B of chip i without tool wear, and **d** section B-B of chip i with different types of tool wear

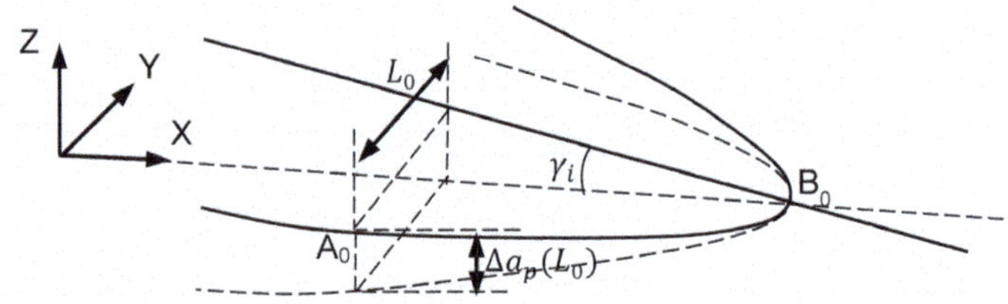

Fig. 3.53 Variation of axial cutting depth induced by the angle between wiper edge and ideal horizontal

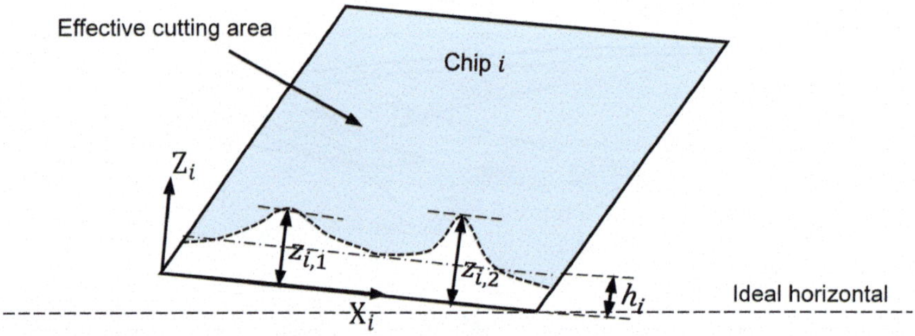

Fig. 3.54 Effective cutting area for insert i considering typical tool wear

where $b_i = 1, 2, \ldots, B_i$ is the serial number of local chipping on the wiper edge of insert i, and z_{i,b_i} is the corresponding ordinate value (Z-axis) relative to the local coordinate system.

Additionally, the overall height wear of the wiper edge, denoted as h_i, is significant as it leads to a direct reduction in the axial cutting depth. The methodology for determining h_i is detailed in Appendix 1.

Effect of Chatter Stability

When chatter emerges in the face milling process, it results in prominently abnormal tool marks and severely compromised surface texture quality, rendering the outcome unacceptable. Consequently, maintaining chatter stability is essential to prevent the onset of chatter. The stability analysis can be conducted utilizing the dynamic model of the cutter-workpiece system. Following a common assumption in chatter-related research [73–75], the face milling system is often approximated as a two-degree-of-freedom (2-DOF) vibration system to facilitate dynamic modeling.

Chatter stability is contingent upon the dynamic response of the vibration system, typically represented by the frequency response function (FRF). For a 2-DOF vibration system, the FRF at the chatter frequency ω_c can be formulated in matrix notation as:

$$[G(\mathrm{i}\omega_\mathrm{c})] = \begin{bmatrix} G_{xx}(\mathrm{i}\omega_\mathrm{c}) & G_{xy}(\mathrm{i}\omega_\mathrm{c}) \\ G_{yx}(\mathrm{i}\omega_\mathrm{c}) & G_{yy}(\mathrm{i}\omega_\mathrm{c}) \end{bmatrix} \tag{3.27}$$

where i is the imaginary unit. It should be noted that chatter frequencies can be of multiple orders and are generally distributed around the dominant modes of the vibration system, which can be determined through simulations or vibration tests.

Incorporating the regenerative effect, the regeneration displacement vector at any given time t during the machining process is represented by $\{\Delta(t)\}$. With the directional coefficient matrix $\boldsymbol{\alpha}$, which is a time-independent function of the radial immersion angle of the milling cutter, the dynamic cutting force model for the face milling process can be expressed as:

$$\{F(t)\} = \frac{N_\mathrm{e} a_\mathrm{p}}{4\pi} K_\mathrm{t} \boldsymbol{\alpha} \{\Delta(t)\} \tag{3.28}$$

where N_e denotes the number of inserts in contact with the workpiece, and K_t is the tangential cutting coefficient within the exponential instantaneous cutting force model [76].

The precision of the dynamic model is influenced by the radial cutting depth a_e and the axial cutting depth a_p. Particularly for workpieces featuring discontinuous surfaces, the actual radial cutting depth fluctuates with the cutting distance L_c, which is a function of the cutting duration t_c and the feed rate f. Additionally, the actual axial cutting depth undergoes changes throughout the face milling process, attributable to the impacts of spindle tilt, insert runout, and tool wear as previously discussed.

As depicted in Fig. 3.55, the top surface of a typical workpiece featuring multiple cylinder bores is being milled by a large disc milling cutter. The number of inserts in contact with the workpiece, N_e, changes as the cutter advances. The face milling process is characterized by its intermittent nature due to the discontinuities in the machined surface, with a step decrease in N_e occurring as the cutter encounters the cylinder bores, and an increase when it moves away. Given that $\{F(t)\}$ is a function of N_e, as per Eq. (3.29), the fluctuation in N_e is a critical factor for the accuracy of the model and cannot be disregarded. Building on the authors' prior research [77], the pattern of variation in N_e is intricate but calculable for general scenarios, and can be expressed as:

$$\left\lceil N_\mathrm{e} = \frac{N C_a(t)}{2\pi R} \right\rceil \tag{3.29}$$

where $C_a(t)$ is the total length of the cutting arcs at an arbitrary moment t within the range of $[0, t_\mathrm{c}]$, and the mathematical operator $\lceil\cdot\rceil$ means rounding up the filled value.

By accurately calculating the time-dependent total cutting arc length while taking into account the geometry of the workpiece, the dynamics model of the face milling process can be refined using Eq. (3.30). Consequently, the critical value of the axial cutting depth and its associated spindle speed can be determined by:

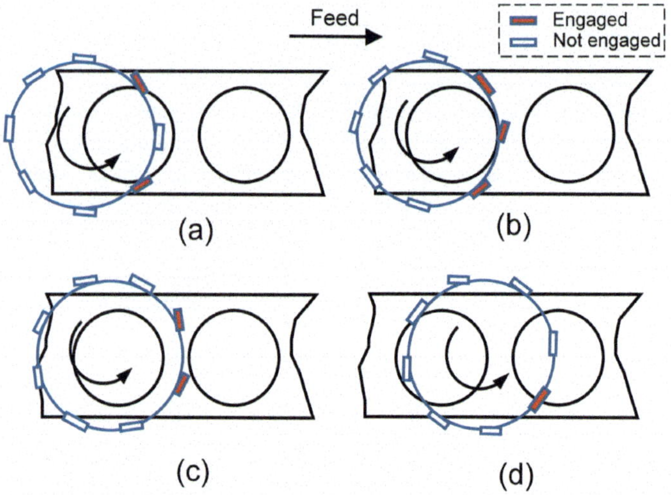

Fig. 3.55 Variation of cutting insert engagement: **a** $N_e = 2$, **b** N_e switches from 2 to 3, **c** N_e switches from 3 to 2, and **d** N_e switches from 2 to 1

$$\begin{cases} a_{p\_lim} = \frac{2\pi\left(\Lambda_R^2+\Lambda_I^2\right)}{N_e K_t \Lambda_R} \\ \Omega_{lim} = \frac{60}{N_e((2k_L+1)\pi - 2\arctan(\Lambda_I/\Lambda_R))} \end{cases} \tag{3.30}$$

where $k_L = 0, 1, 2\ldots$, indicating the successive stability lobes, while Λ_R and Λ_I denote the real part and imaginary part of the eigenvalue for the characteristic equation of regenerative chatter [77] respectively.

The comprehensive derivations for Λ_R and Λ_I are presented in Appendix 2. Utilizing Eq. (3.31), the stability lobe diagram (SLD) [78] for the face milling process that incorporates workpiece geometry can be constructed through iterative computations via programming, thereby ensuring cutting operations are free from chatter.

Evaluation Indexes Determination

Given that each insert on the cutter head engages in material removal at regular time intervals, the impact of tool conditions on surface texture and the associated evaluation indexes exhibit periodicity. By taking into account spindle tilt, insert geometry, insert runouts, and the general height wear of wiper inserts, the average peak-valley difference h_d can be ascertained from the physical model of face milling and expressed as:

$$h_d = \frac{\sum_{i=1}^{N}[t'_{uc\_i}\sin\gamma_i/\sin\alpha + h_i\cos(\phi+\phi_0+\beta_i)/\cos\alpha]}{N} \tag{3.31}$$

It should be noted that the actual thickness of the undeformed chip t'_{uc_i} is a function of ε_{ri} as per Eq. (3.24), indicating that the influence of insert runouts on h_d is attributed not only to the rotational insert runout but also to the radial one. In light of Eq. (3.32), since the intersection angle α between the general cutting edge and the wiper edge is constant once the insert geometry is set, for each insert, the peak-valley difference h_d is directly proportional to γ_i, h_i and t'_{uc_i}. Concurrently, h_d is inversely proportional to the sum of ϕ, ϕ_0 and β_i.

The wavelength spacing s_λ is mainly affected by the spindle tilt and rotational insert runout, leading to a slight deviation from the feed per insert. And it can be obtained by

$$s_\lambda = \frac{60f\sum_{i=1}^{N}\cos(\phi_0 + \beta_i)}{\Omega N^2} \tag{3.32}$$

Equation (3.32) clearly demonstrates the reliance of s_λ on various processing parameters. With the spindle tilt and the rotational runout of each insert held as constants throughout the face milling process, s_λ can be readily anticipated and regulated by modifying the feed rate f, spindle speed Ω, and the total number of inserts N prior to the machining operation.

As per Eq. (3.33), the height fluctuation h_f is largely a result of changes in the actual axial cutting depth. Incorporating the impact of insert runout, the expression becomes:

$$h_f = \frac{R[1 + \cos(\phi_0 + \beta_i)]}{2N}\sum_{i=1}^{N}\left\{1 - \sqrt{1 - \frac{a_e^2}{R^2[1 + \cos(\phi_0 + \beta_i)]^2}}\right\}\tan\gamma_i \tag{3.33}$$

where the radial cutting depth a_e is a variable that is contingent upon the cutting path, the cutter geometry, and the discontinuity of the machined surface.

The relationship of h_f with various processing parameters exhibits significant nonlinearity, making it challenging to regulate by simply adjusting a single processing parameter in a monotonic fashion, with the exception of γ_i, which is strongly positively correlated with the spindle tilt ϕ_0. To achieve an optimal machined surface, striking a balance among different evaluation indices is essential.

3.2.2.3 Physics-Informed Optimization Modeling

Within the scope of this section, the optimization of processing parameters, which encompasses N, Φ_0, a_p, f and Ω, is considered variable and adjustable within a feasible range, constrained by practical engineering limitations such as the load-bearing capacity of the machine tools. Concurrently, other influencing factors, including spindle stiffness and the material properties of the workpiece, are treated as constants or as variables that are contingent upon the processing parameters.

Objective Function

The principal objective of this research is to enhance surface texture quality, as quantified by both the proposed evaluation indexes and the ISO evaluation parameters. Given the absence of a direct expression linking ISO evaluation parameters to processing parameters, the objective function comprising h_d, s_λ and h_f is more aligned with the physics-informed optimization model. This approach leverages the explicit relationships with processing parameters derived from physical modeling (i.e., Eqs. (3.32)–(3.34)). Provided that the weight assignment is rationally determined, primarily based on expert knowledge or specific technical requirements, the multi-objective optimization problem can be converted into a single-objective optimization. The objective function can then be formulated as:

$$\min W = w_1 h_d + w_2 \frac{s_\lambda}{\lambda_f - s_\lambda} + w_3 h_f \tag{3.34}$$

where w_1,w_2 and w_3 are the corresponding weights and $w_1 + w_2 + w_3 = 1$.

In many scenarios, assigning appropriate weights to each objective is challenging, and sometimes the physical meanings of the objectives remain fundamentally different even after non-dimensional processing [79, 80]. Furthermore, both h_d and h_f significantly impact surface texture quality as it relates to ISO parameters, making them crucial factors. In contrast, s_λ primarily influences the selection of the cutoff wavelength λ_f for waviness filtering rather than the surface texture quality itself, rendering it of lesser importance. There is no necessity to deliberately pursue the minimum value of $\frac{s_\lambda}{\lambda_f - s_\lambda}$. That is, even if the weights for each objective could be ascertained, w_2 would be considerably smaller than w_1 and w_3. Consequently, in this research, it is suggested that the objective function be expressed in a bi-objective form and is given by:

$$\min W = [h_d, h_f] \tag{3.35}$$

Note that $\frac{s_\lambda}{\lambda_f - s_\lambda}$ is regarded as a supplementary objective, and it is deemed as acceptable for general cases when its value is less than 0.5.

Constraints

For the ease of automated control or manual adjustment, the variables in this physics-informed optimization model primarily encompass standard processing parameters, which include the spindle speed Ω, feed rate f, axial cutting depth a_p, spindle inclination ϕ_0, and the total number of inserts N. Consequently, the optimization model is governed by the following constraints.

Constraint 1: The selection of spindle speed must take into account the material properties of the workpiece and the load-bearing capacity of the machine tool. By defining the lower and upper limits of the spindle speed as Ω_{st} and Ω_{fi}, respectively, it follows that:

$$\Omega_{\mathrm{st}} < \Omega < \Omega_{\mathrm{fi}} \tag{3.36}$$

Constraint 2: To ensure the cutting stability of the face milling process, the axial cutting depth a_{p} must not exceed the critical axial cutting depth $a_{\mathrm{p\_lim}}(\Omega)$, which can be expressed as:

$$0 < a_{\mathrm{p}} < a_{\mathrm{p\_lim}}(\Omega) \tag{3.37}$$

Constraint 3: Establishing an excessively low feed rate can result in an unacceptably low material removal rate, affecting the efficiency of long-term and mass manufacturing. Additionally, the machine's feeding accuracy imposes a lower limit, f_{st}, on the selection of feed rate. In parallel with Constraint 1, the maximum feed rate, f_{fi}, is also constrained by the machine's load capacity. Therefore, the constraint for the feed rate f is delineated as:

$$f_{\mathrm{st}} < f < f_{\mathrm{fi}} \tag{3.38}$$

Constraint 4: To enhance the symmetry of cutting forces, it is a common practice in engineering to set the total number of inserts N to an even number, which significantly narrows the solution space for this problem. By defining N^* as the maximum number of inserts that can be mounted on the disc milling cutter, the constraint is expressed as:

$$N \in \{2, 4, \ldots, N^*\} \tag{3.39}$$

Optimization Solution

As a widely recognized multi-objective search-based algorithm, NSGA-II is characterized by its robust solution set convergence and swift computational efficiency, positioning it as a yardstick against which other multi-objective optimization algorithms are measured [63]. Consequently, NSGA-II is deployed to address the proposed bi-objective optimization model. The flowchart depicting the application of NSGA-II for optimization is presented in Fig. 3.56, where "*Gen*" and "*MaxGen*" represent the current generation number and the maximum generation limit, respectively.

The principal procedures of NSGA-II, which encompass rapid non-dominated sorting, computation of crowding distances, and preservation of elitism, are well-established concepts as documented in existing literature [61]. The essential aspects for the execution of NSGA-II are outlined below.

Variable Discretization: Given that all processing parameters under consideration, with the exception of N, are continuous, it is essential to establish an appropriate step size for the discretization of each continuous variable. Balancing engineering

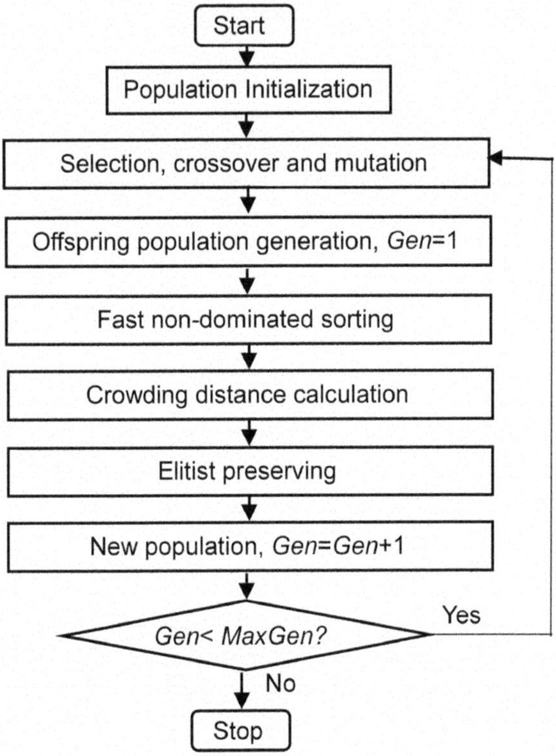

Fig. 3.56 Flow chart of NSGA-II utilized for optimization solution

practicality and computational efficiency, the step sizes for Φ_0, a_p, f and Ω are designated as 0.02°, 0.01 mm, 10 mm/min, and 50 rpm, respectively.

Stopping criteria: The solution process in this section is halted when the maximum number of generations, set at 1000, is reached. This signifies that after 1000 iterations, it is assumed that NSGA-II has converged upon the Pareto optimal solution set.

Pareto Compromise: An excessive number of Pareto optimal solutions, as represented by the Pareto front, can be impractical for guiding engineering applications. Thus, the population size per generation is capped at 100, which in turn limits the total number of Pareto optimal solutions. The selected solutions should ideally represent a balanced compromise to prevent any objective from being too extreme; that is, solutions in the central region of the Pareto front are generally preferred, except in specific cases.

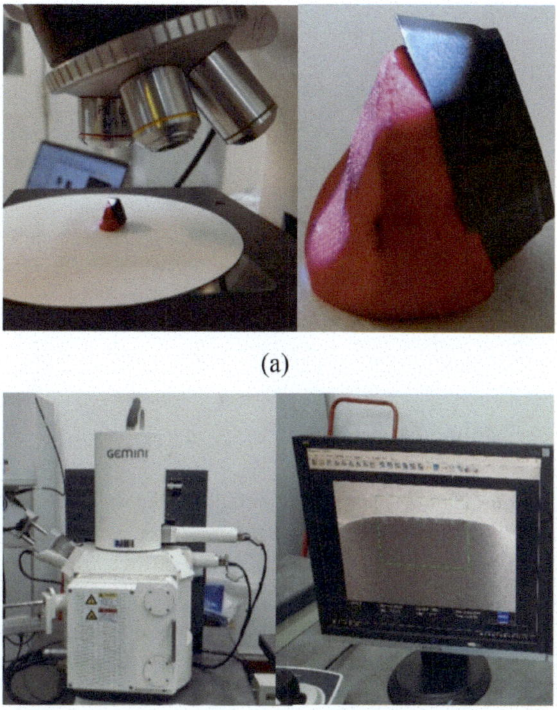

(a)

(b)

Fig. 3.57 Experimental setup for tool wear measurement, **a** the optical microscope, **b** the FE-SEM system

3.2.3 Case Study

3.2.3.1 Experimental Setup

Tool Conditions Measurement

SEEX1204ZZTN-M14 MK1500 inserts are employed in this case study. To ascertain the tool wear condition, which impacts surface finish quality, the wiper edges of a selection of inserts are examined using an optical microscope prior to the face milling experiments (refer to Fig. 3.57a). Due to the limitations in measurement resolution, precise height wear of the wiper edge cannot be solely determined from the optical microscope's data. Consequently, the morphology of the inserts in use is further scrutinized using field emission scanning electron microscopy (FE-SEM, Carl Zeiss Ultra 55, Germany) operated at 5 kV (refer to Fig. 3.57b). Additionally, once the inserts are fixed onto the disc cutter head, the insert runouts are manually gauged using a micron outside micrometer.

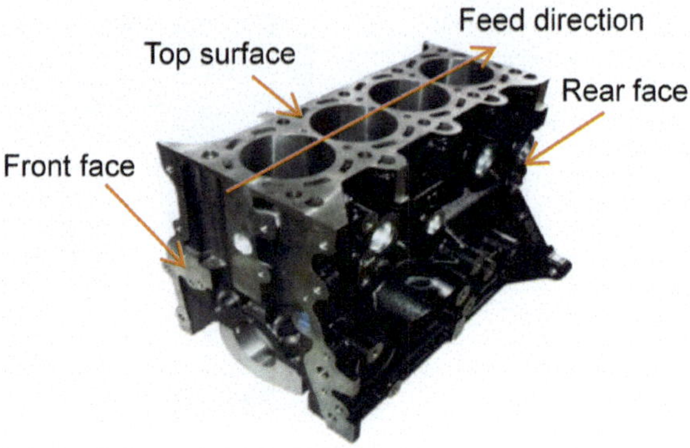

Fig. 3.58 Four-cylinder engine blocks utilized in face milling experiments

Face Milling Experiments

For this scenario, four-cylinder engine blocks of the B-12 type are utilized to ascertain the efficacy of the proposed methodology (as shown in Fig. 3.58). The top surface of the engine block, a critical joint surface for assembly, is machined in a single pass using a disc milling cutter in the face milling process. The texture quality of this surface is essential for the engine's sealing performance and reliability. Given that the engine block is constructed from gray cast iron HT250, surface hardening due to cutting temperatures is effectively circumvented during the material removal process [81, 82]. Otherwise, the occurrence of surface hardening would add further complexity to the explanation of surface texture imperfections, which is the primary focus of this section.

Taking into account the attributes of the workpiece geometry and the dynamics of the machining system, the SLD for the face milling process is derived and presented in Fig. 3.59. The axial cutting depth is explored within a range of 0.02–0.5 mm, the feed rate from 30 to 800 mm/min, the spindle speed from 200 to 2500 rpm, and the spindle inclination from 0.02 to 0.3°. Utilizing NSGA-II to address the proposed optimization model, the resulting Pareto optimal solution set is depicted in Fig. 3.60.

The Pareto optimal solution set provided by NSGA-II and the stability lobe diagram (SLD) of the face milling system inform four distinct sets of face milling experiments conducted on engine blocks. The specific processing parameters for each experiment are outlined in Table 3.9. Case (1) employs the original processing parameters typically used in practical manufacturing, primarily based on engineering experience. Cases (2) and (3) are derived from the Pareto optimal solution set. In contrast, case (4) follows the recommended processing parameter design from literature [39], serving as a benchmark to validate the advantages of the proposed approach, with consideration for the relevance and reproducibility of the methodology. The face milling experiments are carried out on engine blocks using the DMG-HSC-75 CNC machining center, as depicted in Fig. 3.61a, with a disc milling cutter measuring 160 mm in diameter and capable of accommodating up to 10 cubic boron nitride (CBN) inserts, as shown in Fig. 3.61b.

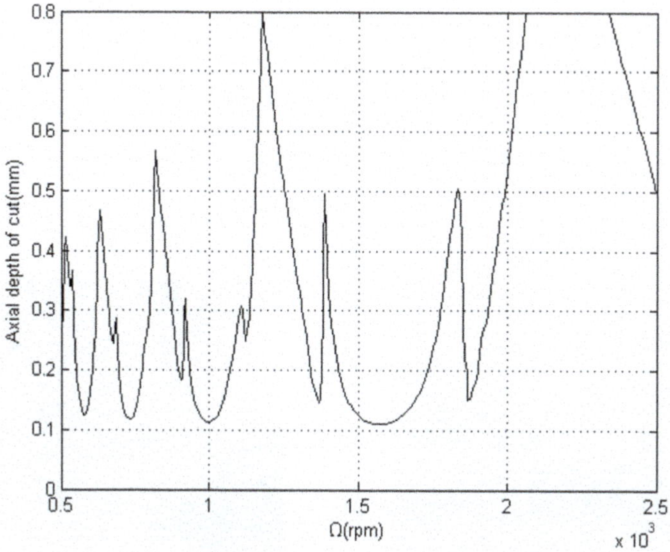

Fig. 3.59 SLD of the face milling system

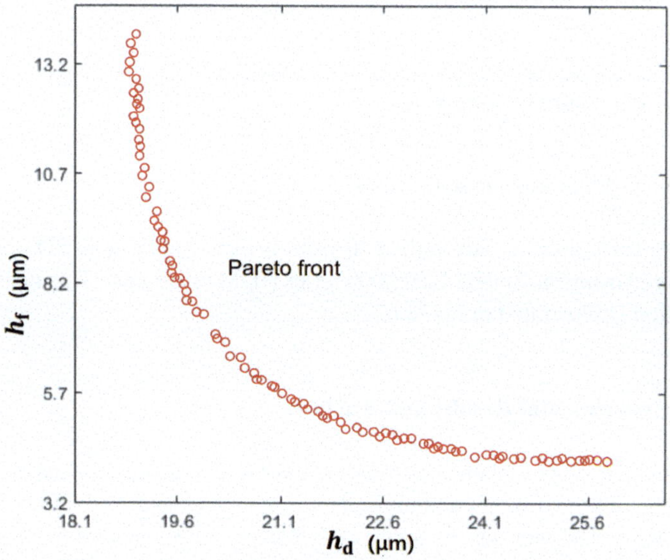

Fig. 3.60 Pareto optimal solution set given by NSGA-II

Table 3.9 Processing parameter design for face milling experiments

Case No	f (mm/min)	a_p(mm)	Ω (rpm)	N	Φ_0(°)
(1)	360	0.05	950	6	0.14
(2)	400	0.1	1350	8	0.1
(3)	630	0.12	1200	8	0.2
(4)	520	0.08	1100	8	0.25

(a)

(b)

Fig. 3.61 Experimental setup for face milling: **a** fixturing in the CNC machining center and **b** the disc milling cutter utilized in experiments

Surface Finish Measurement with HDM

The surface texture of the face milled surfaces is captured through 3D point cloud data, acquired using the ShaPix3D® 3000 series HDM instrument. The measurement configuration is illustrated in Fig. 3.62.

3.2.3.2 Experimental Results and Analysis

The optical microscope-obtained measurements of tool wear are depicted in Fig. 3.63. Based on the observed wear conditions, the inserts are categorized into two groups: new and worn. The new inserts serve as a reference for calculating height wear, while representative worn inserts are chosen for subsequent face milling experiments. Given that the face milling experiments are carried out under laboratory conditions over a relatively short period, the wear accumulated on the inserts is minimal and can be considered negligible.

The micrographs reveal that the coating on new inserts remains intact (refer to Fig. 3.63b), in contrast to the severely damaged coating on worn inserts (refer to Fig. 3.63c), which clearly illustrates the extent of flank wear on the wiper edges.

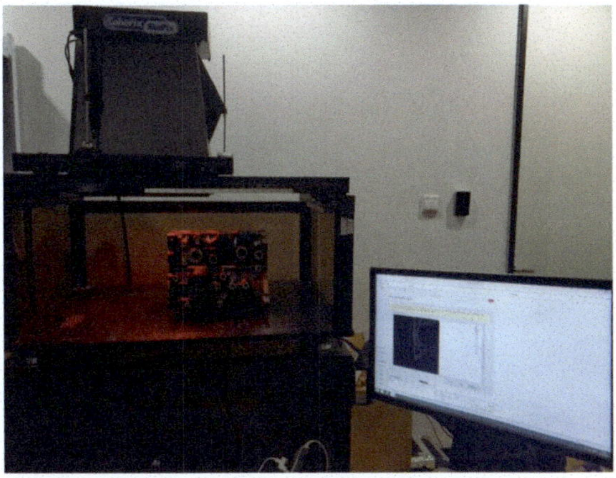

Fig. 3.62 Experimental setup for surface finish measurement with the HDM system

Additionally, to conduct a quantitative analysis of the tool wear condition, particularly the general height wear h_i, the FE-SEM system is employed, with the outcomes presented in Fig. 3.64. Observable local chipping on the wiper edges is evident, and h_i is determined by comparing the measured height against the nominal height of the wiper edge.

The values of h_i are presented in Table 3.10. The irregular variation in h_i across different inserts can be attributed to two potential factors. Initially, these eight inserts are randomly selected from a group of worn inserts that were initially identified using an optical microscope. Consequently, the sequence of h_i listed in Table 3.10 is neither in ascending nor descending order. Additionally, the specific usage duration for each insert may vary, leading to diverse wear conditions, as reflected by h_i.

Following the face milling experiments, a partial view of the top surface of an engine block is depicted in Fig. 3.65. The presence of discernible tool marks on the machined surfaces confirms the applicability of the proposed evaluation indexes and the optimization approach to this experimental case.

Figure 3.66 displays the HDM-acquired point cloud data, which has been converted into a two-dimensional image for enhanced visualization. Given that the disc milling cutter's diameter (160 mm) is marginally smaller than the maximum width of the machined top surface (171 mm), the assessment of surface texture is significantly impacted when examining local areas near the extremities of the cutting path. Consequently, the section focuses on the relatively central region of the surface texture, which has an average width of 109 mm and is defined by a cutting distance range of $75\,\text{mm} < L_c < 275\,\text{mm}$. This selected area is fully machined by the disc milling cutter and encompasses the geometric features of two complete cylinder bores located centrally within the cutting pass, providing an adequate representation of the workpiece's geometric characteristics.

Figure 3.66 provides a visual impression that the quality of the machined surfaces has been markedly enhanced by both the proposed model and the benchmark model. It is observed that the amplitude of height variation on the machined surface in case (4)

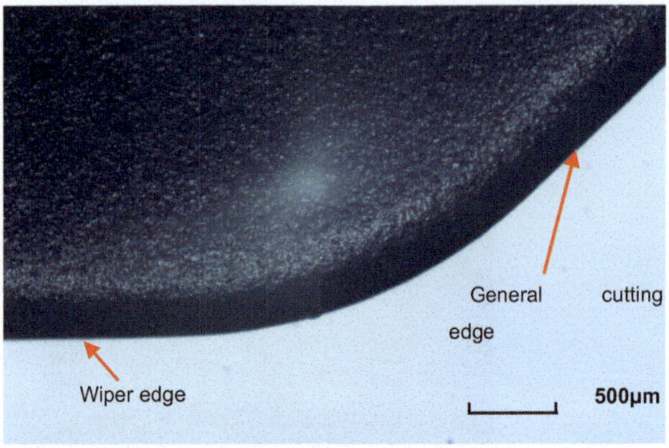

(a)

(b)

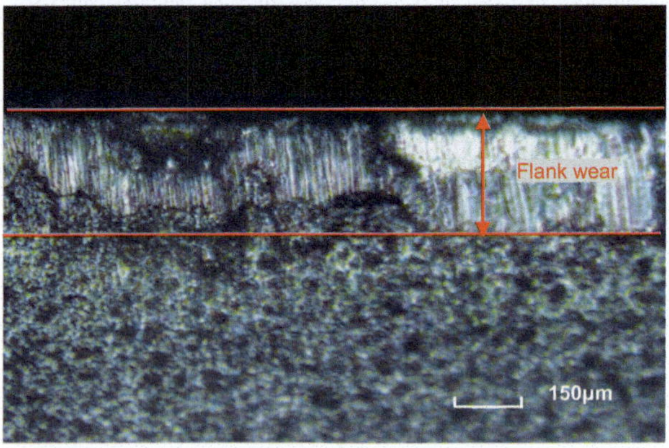

(c)

Fig. 3.63 Tool wear measured by the optical microscope: **a** overview of the insert corner, **b** micrograph of a brand-new wiper edge and **c** micrograph of a worn wiper edge

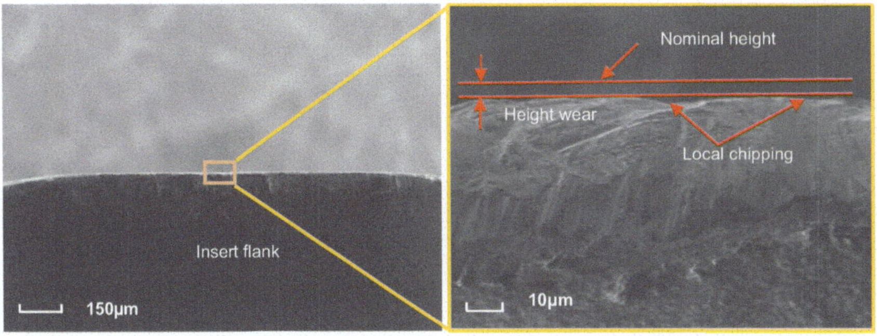

Fig. 3.64 Tool wear measured by the FE-SEM system

Table 3.10 Measurement results of general height wear of the wiper edges

Insert No.	1	2	3	4	5	6	7	8
h_i(μm)	2.3	4.2	0.2	0.8	3.3	7.5	5.3	3.6

Fig. 3.65 Partial view of the top surface of engine block after face milling

(refer to the color bar of Fig. 3.66d) is considerably lower than that in case (1) (refer to the color bar of Fig. 3.66a). Nonetheless, a typical central concave phenomenon is noted in case (4), signifying an improved h_d but a degraded h_f following optimization by the benchmark model. The potential cause for the degradation of h_f is analyzed as follows: Eq. (3.25) indicates that the variation amplitude of the actual axial cutting depth correlates positively with γ_i within a certain range. Should other parameters, including α, β_i, ϕ and L_0, remain constant, a larger ϕ_0 would result in a larger

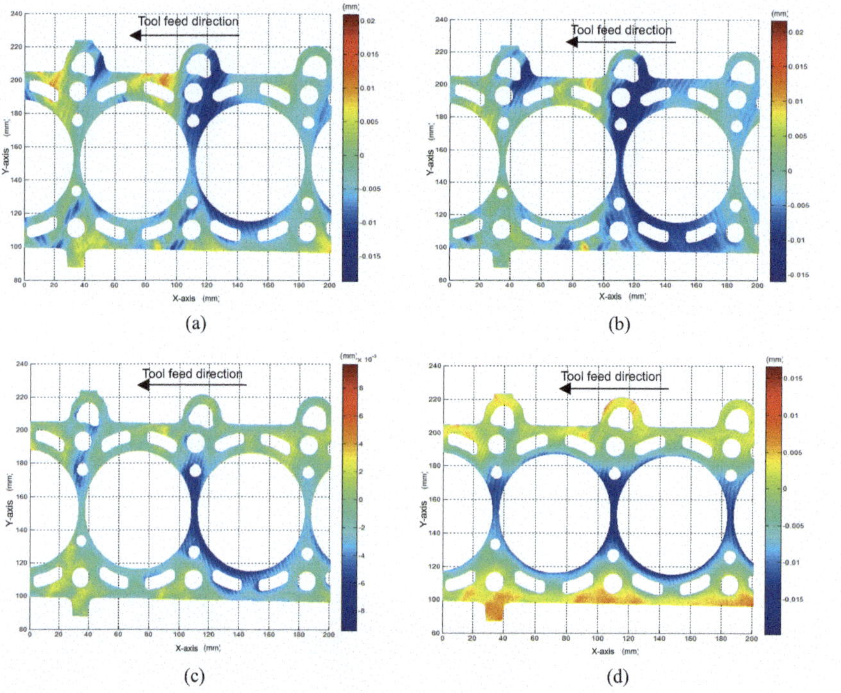

Fig. 3.66 HDM data transformed two-dimensional image: **a** data for case (1), **b** data for case (2) and **c** data for case (3) **d** data for case (4)

$\Delta a_p(L_0)$, and thus the height fluctuation of a single tool mark produced by insert i would increase, leading to a poorer h_f.

To conduct a thorough comparison of the efficacy between the proposed model and the benchmark model, it is essential to extract characteristics of tool marks, including waviness information, from the HDM data. Utilizing the HDM data preprocessing methods described in Sect. 2.1, the measured evaluation indexes for each scenario can be derived. Concurrently, with all processing parameters identified, the evaluation indexes can be theoretically determined based on the physical model outlined in Sect. 2.2. The comparative outcomes are presented in Table 3.11. Furthermore, ISO areal surface texture parameters are computed to assess the surface texture quality from an alternative viewpoint, as detailed in Table 3.12.

The analysis of the two tables yields several significant insights. Initially, the prediction errors for all evaluation indexes fall nearly within 15%, which is considered tolerable for practical engineering applications. It should be noted that the prediction errors for case (4) pertain to the proposed model, not the benchmark model, since the predictors were introduced in this section and the prediction values were derived from the physical model presented here. The sources of prediction errors can be attributed to two main aspects. Firstly, HDM measurement conditions and sensor noise may contribute to the prediction errors to a certain extent, which can be effectively managed within an acceptable range through proper maintenance

Table 3.11 Proposed evaluation indexes for different cases

Index	h_d			s_λ			h_f		
	Measured (μm)	Predicted (μm)	Error (%)	Measured (μm)	Predicted (μm)	Error (%)	Measured (μm)	Predicted (μm)	Error (%)
Case (1)	29.4	27.6	6.12	130.0	128.3	1.31	9.3	7.9	15.05
Case (2)	21.4	20.3	5.08	100.5	102.1	1.59	8.1	7.2	8.1
Case (3)	20.1	18.8	5.88	81.3	78.6	3.32	10.4	9.5	8.65
Case (4)	26.2	24.8	5.32	84.2	82.5	2.21	11.4	10.3	9.6

Table 3.12 ISO areal surface texture parameters for different cases

Case	S_q (μm)	S_p (μm)	S_v (μm)	S_z (μm)	S_a (μm)	S_{sk}	S_{ku}	S_{al}	S_{tr}
(1)	2.34	8.63	13.65	23.69	1.25	0.52	3.62	2.33	0.94
(2)	1.22	4.22	8.27	4.36	0.65	− 0.13	1.33	1.59	0.41
(3)	0.98	5.36	7.37	10.81	0.49	0.36	1.03	1.64	0.33
(4)	1.58	6.32	10.04	12.56	0.78	0.44	1.68	1.87	0.63

and adherence to standard operating procedures for HDM equipment. Secondly, despite the proposed model's attempt to closely mirror the actual physical scenario by accounting for multiple intrinsic factors, discrepancies between the theoretical model and the actual physical process persist, offering an area for potential refinement in future research endeavors.

Secondly, a comprehensive comparison of both the proposed evaluation indexes and the ISO parameters reveals that the proposed model outperforms the benchmark model in enhancing surface texture quality. This superiority is primarily due to the model's holistic consideration of the effects of spindle tilt, runout, tool wear, and chatter stability. In contrast, the benchmark model only accounts for the impact of spindle tilt on surface texture formation. Since the wavelength spacing s_λ for each case satisfies the condition $\frac{s_\lambda}{\lambda_f - s_\lambda}$, all cases are deemed acceptable in terms of s_λ. Thus, the efficacy of the proposed approach is further substantiated.

Thirdly, the processing parameter designs for case (2) and case (3), selected from the Pareto front provided by NSGA-II, each possess distinct advantages. For example, case (3) effectively reduces the average peak-valley difference h_d by employing a larger spindle inclination. However, the height fluctuation h_f in case (2) is significantly lower than in case (3). The experimental findings suggest that h_f tends to deteriorate with the use of a larger spindle inclination ϕ_0, which could markedly amplify the variation in the actual axial cutting depth during each revolution of a single insert. Nonetheless, by appropriately enlarging ϕ_0, the average peak-valley difference h_d may achieve a more favorable outcome according to Eq. (3.32). This observation aligns with the proposed guideline of Pareto compromise, emphasizing the necessity for a balanced approach in practical engineering when applying the proposed methodology.

3.2.4 Conclusion

An innovative HDM-based methodology for enhancing surface texture quality has been introduced in this section. A comprehensive physical model for face milling operations on workpieces with discontinuous surfaces has been developed, accounting for spindle tilt, tool conditions, and chatter stability. Three evaluation indexes, specific to HDM data and pertinent to surface texture marked by tool marks, have been defined and are quantifiably predictable through the proposed physical model. Employing these indexes as objectives, a physics-informed optimization model has been formulated to refine processing parameters aimed at improving the quality of face milled surfaces. The efficacy of this approach has been tested through a series of face milling experiments on a standard model of four-cylinder engine blocks. Tool conditions, including insert runouts and wear, were measured, and the machined surfaces were analyzed using HDM instruments to secure high-resolution surface texture data. The theoretical calculations of the evaluation indexes, based on the proposed physical model, closely match the measured outcomes, demonstrating the approach's capability for accurate surface texture quality prediction. Furthermore, the surface texture quality, as assessed by both the evaluation indexes and ISO standard parameters derived from the HDM data, has been markedly enhanced through the application of optimized processing parameters.

For future work, consideration of additional intrinsic factors influencing surface texture quality, such as residual stress and heat-induced surface hardening, should be integrated to enhance the precision and universality of the physical model.

Appendix 1

The general height wear of the wiper edge h_i can be generated by the following steps.

Step 1:	Check if the wiper edge of insert i is worn. If so, proceed to Step 2; if not, $h_i = 0$ and the program is ended
Step 2:	Initialization by setting $B_i = 1$ and $h_i = 0$
Step 3:	By continuously enlarging h_i till $h_i = H_i$, update h_i with the current value and set $B_i = B_i + 1$ once a new intersect is generated by the wear edge curve and the function line $z = h_i$
Step 4:	The program is ended when $h_i = H_i$

Appendix 2

Denoting the identity matrix by $[I]$, the characteristic equation of regenerative chatter is given by

$$\det\{[I] - \frac{2\pi\Lambda}{N_e}[G(i\omega_c)]\} = 0 \quad (3.40)$$

Therein, Λ is the eigenvalue and formulated by

$$\Lambda = \frac{N}{4\pi} a_p K_t \left(1 - e^{-i\omega_c\tau}\right) \quad (3.41)$$

where τ is the insert passing period given by $\tau = \frac{60}{\Omega N}$, while e is the base of natural logarithm and approximately equals to 2.718 in this research.

Evidently Λ is a complex number and thus can be rewritten as the sum of real part and imaginary part, namely, $\Lambda = \Lambda_R + i\Lambda_I$. By utilizing the Euler's formula [83], Λ_R and Λ_I can be calculated by

$$\begin{cases} \Lambda_R = \frac{N}{4\pi} a_p K_t (1 - \cos(\omega_c\tau)) \\ \Lambda_I = \frac{N}{4\pi} a_p K_t \sin(\omega_c\tau) \end{cases} \quad (3.42)$$

And Eq. (3.40) can be further derived as

$$\frac{\Lambda_I}{\Lambda_R} = \frac{\sin(\omega_c\tau)}{1 - \cos(\omega_c\tau)} \quad (3.43)$$

References

1. Marie C, Lasseux D (2006) Experimental leak-rate measurement through a static metal seal. J Fluids Eng 129(6):799–805
2. Malburg MC (2003) Surface profile analysis for conformable interfaces. J Manuf Sci Eng 125(3):624–627
3. Shao YP, Yin YX, Du SC, Xia TB, Xi LF (2018) Leakage monitoring in static sealing interface based on three dimensional surface topography indicator. ASME Trans Manuf Sci Eng 140:101003
4. Liao Y, Stephenson DA, Ni J (2012) Multiple-scale wavelet decomposition, 3D surface feature exaction and applications. ASME Trans Manuf Sci Eng 134(1):011005
5. Liu S, Jin S, Zhang XP, Chen K, Tian A, Xi LF (2019) A coupled model for the prediction of surface variation in face milling large-scale workpiece with complex geometry. ASME Trans Manuf Sci Eng 141(3):031009
6. Jin S, Liu S, Zhang XP, Chen K (2019) A unified prediction model of 3D surface topography in face milling considering multi-error sources. Int J Adv Manuf Technol 102(1):705–717
7. Li PY, Zhai YF, Huang SK, Wang QD, Fu WP, Yang HP (2017) Investigation of the contact performance of machined surface morphology. Tribol Int 107:125–134

8. Huang ZH, Shih AJ, Ni J (2006) Laser interferometry hologram registration for three-dimensional precision measurements. ASME Trans Manuf Sci Eng 128(4):887–896
9. Wang M, Xi LF, Du SC (2014) 3D surface form error evaluation using high definition metrology. Precis Eng 38(1):230–236
10. Du SC, Liu CP, Huang DL (2015) A shearlet-based separation method of 3D engineering surface using high definition metrology. Precis Eng 40:55–73
11. Wang M, Shao YP, Du SC, Xi LF (2015) A diffusion filter for discontinuous surface measured by high definition metrology. Int J Precis Eng Manuf 16(10):2057–2062
12. Shao YP, Wang K, Du SC, Xi LF (2018) High definition metrology enabled three dimensional discontinuous surface filtering by extended tetrolet transform. J Manuf Syst 49:75–92
13. Du SC, Liu T, Huang DL, Li GL (2018) A fast and adaptive bi-dimensional empirical mode decomposition approach for filtering of workpiece surfaces using high definition metrology. J Manuf Syst 46:247–263
14. Du SC, Huang DL, Wang H (2015) An adaptive support vector machine-based workpiece surface classification system using high-definition metrology. IEEE Trans Instrum Meas 64(10):2590–2604
15. Du SC, Liu CP, Xi LF (2015) A selective multiclass support vector machine ensemble classifier for engineering surface classification using high definition metrology. ASME Trans J Manuf Sci Eng 137(1):011003
16. Wang M, Du SC, Xi LF (2015) Predicting machined surface topography based on high definition metrology. In: 15th IFAC symposium on information control problems in manufacturing INCOM, pp 1013–1017
17. Shao YP, Du SC, Xi LF (2017) 3D machined surface topography forecasting with space-time multioutput support vector regression using high definition metrology. In: 37th Computers and information in enginnering conference, Cleveland, Ohio, USA, p V001T002A069
18. Du Shichang LX (2019) High definition metrology based surface quality control and applications. Springer, Singapore
19. Nguyen HT, Wang H, Hu SJ (2013) Characterization of cutting force induced surface shape variation in face milling using high-definition metrology. ASME Trans Manuf Sci Eng 135(4):041014
20. Nguyen HT, Wang H, Hu SJ (2014) Modeling cutter tilt and cutter-spindle stiffness for machine condition monitoring in face milling using high-definition surface metrology. Int J Adv Manuf Technol 70(5–8):1323–1335
21. Wang M, Ken T, Du SC, Xi LF (2015) Tool wear monitoring of wiper inserts in multi-insert face milling using three-dimensional surface form indicators. ASME Trans Manuf Sci Eng 137(3):031006
22. ISO (2013) Geometrical product specifications(GPS)-filtration Part 85: morphological areal filter: segmentation. International Organization for Standardization, Geneva, Switzerland. Standard No. ISO 16610-85
23. Bleau A, Leon LJ (2000) Watershed-based segmentation and region merging. Comput Vis Image Underst 77(3):317–370
24. Barré F, Lopez J (2000) Watershed lines and catchment basins: a new 3D-motif method. Int J Mach Tools Manuf 40(8):1171–1184
25. ISO (2012) Geometrical product specifications (GPS)-surface texture: areal-part2: terms, definitions and texture parameters. In: International Organization for Standardization, Geneva, Switzerland, Standard No. ISO 25178-2
26. Otsu N (1979) A threshold selection method from gray-level histograms. IEEE Trans Syst Man Cybern 9(1):62–66
27. Barber CB, Dobkin DP, Huhdanpaa H (1996) The quickhull algorithm for convex hulls. ACM Trans Math Softw 22(4):469–483
28. Dong WP, Sullivan PJ, Stout KJ (1992) Comprehensive study of parameters for characterizing 3D surface topography I: some inherent properties of parameter variation. Wear 159:161–171
29. Ryu SH, Choi DK, Chu CN (2006) Roughness and texture generation on end milled surfaces. Int J Mach Tools Manuf 46:404–412

30. Thompson RA (1992) On the doubly regenerative stability of a grinder: the effect of contact stiffness and wave filtering. J Manuf Sci Eng 114(1):53–60
31. Mohandesi JA, Rafiee MA, Maffi O, Saffarzadeh P (2007) Dependence of the yield and fatigue strength of the thread rolled mild steel on dislocation density. J Manuf Sci Eng 129(1):216–222
32. Mohammadtabar N, Bakhshi-Jooybari M, Gorji H, Jamaati R, Szpunar JA (2021) Effect of electric current pulse type on springback, microstructure, texture, and mechanical properties during V-bending of AA2024 aluminum alloy. J Manuf Sci Eng 143(1):011004
33. Yoo S, Walczyk DF (2007) A preliminary study of sealing and heat transfer performance of conformal channels and cooling fins in laminated tooling. J Manuf Sci Eng 129(2):388–399
34. Wang K, Li G, Du S, Xi L, Xia T (2020) State space modelling of variation propagation in multistage machining processes for variable stiffness structure workpieces. Int J Prod Res 15:1–20
35. Blunt L, Stout KJ (2000) Three-dimensional surface topography. Penton, London
36. Li G, Du S, Huang D, Zhao C, Deng Y (2019) Elastic mechanics-based fixturing scheme optimization of variable stiffness structure workpieces for surface quality improvement. Precis Eng 56:343–363
37. Villa A, Rossetto S, Levi R (1983) Surface texture and machining conditions. part 1: model building logic in view of process control. J Eng Ind 105(4):259–263
38. Villa A, Rossetto S, Levi R (1983) Surface texture and machining conditions. part 2: development of mathematical model based upon pattern recognition. J Eng Ind 105(4):264–269
39. Hadad M, Ramezani M (2016) Modeling and analysis of a novel approach in machining and structuring of flat surfaces using face milling process. Int J Mach Tools Manuf 105:32–44
40. Salisbury EJ, Domala KV, Moon KS, Miller MH, Sutherland JW (2001) A three-dimensional model for the surface texture in surface grinding, part 1: surface generation model. J Manuf Sci Eng 123(4):576–581
41. Salisbury EJ, Domala KV, Moon KS, Miller MH, Sutherland JW (2001) A three-dimensional model for the surface texture in surface grinding, part 2: grinding wheel surface texture model. J Manuf Sci Eng 123(4):582–590
42. Baek DK, Ko TJ, Kim HS (1997) A dynamic surface roughness model for face milling. Precis Eng 20:171–178
43. Kiss AK, Bachrathy D, Stepan G (2020) Effects of varying dynamics of flexible workpieces in milling operations. J Manuf Sci Eng 142(1):011005
44. Shen J, Xu P, Yu Y (2020) Dynamic characteristics analysis and finite element simulation of steel-BFPC machine tool joint surface. J Manuf Sci Eng 142(1):011006
45. Batsch M (2020) A novel method of obtaining honing tool profile for machining gears with profile modifications. J Manuf Sci Eng 142(9):091004
46. Du S, Xi L (2019) High definition metrology based surface quality control and applications. Springer, Singapore
47. Huang D, Du S, Li G, Wu Z (2017) A systematic approach for online minimizing volume difference of multiple chambers in machining processes based on high-definition metrology. J Manuf Sci Eng 139(8):081003
48. Huang D, Du S, Li G, Zhao C, Deng Y (2018) Detection and monitoring of defects on three-dimensional curved surfaces based on high-density point cloud data. Precis Eng 53:79–95
49. Shi J, Cao H, Maroju NK, Jin X (2020) Dynamic modeling of aerostatic spindle with shaft tilt deformation. J Manuf Sci Eng 142(2):021006
50. Li J, Kilic ZM, Altintas Y (2020) General cutting dynamics model for five-axis ball-end milling operations. J Manuf Sci Eng 142(12):121003
51. Yin Y, Shao Y, Wang K, Du S, Xi L (2020) Segmentation of workpiece surfaces with tool marks based on high definition metrology. J Manuf Process 57:268–287
52. Shao Y, Yin Y, Du S, Xia T, Xi L (2018) Leakage monitoring in static sealing interface based on three dimensional surface topography indicator. J Manuf Sci Eng 140(10):101003
53. Ren J, Wang H (2018) Surface variation modeling by fusing multi-resolution spatially nonstationary data under a transfer learning framework. J Manuf Sci Eng 141:011002

54. Wang B, Dong X, Wang Z, Wang Y, Hou Z (2020) MEMS-based ionization gas sensors for VOCs with array of nanostructured silicon needles. ACS Sensors 5(4):994–1101
55. Lu HS, Chang CK, Hwang NC, Chung CT (2009) Grey relational analysis coupled with principal component analysis for optimization design of the cutting parameters in high-speed end milling. J Mater Process Technol 209(8):3808–3817
56. Mukherjee I, Ray PK (2006) A review of optimization techniques in metal cutting processes. Comput Ind Eng 50:15–34
57. Wang H, Suriano S, Zhou L, Hu SJ (2009) High-definition metrology based spatial variation pattern analysis for machining process monitoring and diagnosis. In: Proceedings of the ASME 2009 international manufacturing science and engineering conference, vol 2, pp 471–480
58. Suriano S, Wang H, Hu SJ (2012) Sequential monitoring of surface spatial variation in automotive machining processes based on high definition metrology. J Manuf Syst 31(1):8–14
59. Tai BL, Stephenson DA, Shih AJ (2011) Improvement of surface flatness in face milling based on 3-D holographic laser metrology. Int J Mach Tools Manuf 51(6):483–490
60. Zhou L, Wang H, Berry C, Weng X, Hu SJ (2011) Functional morphing in multistage manufacturing and its applications in high-definition metrology-based process control. IEEE Trans Autom Sci Eng 9(1):124–136
61. Deb K, Agrawal S, Pratap A, Meyarivan T (2002) A fast and elitist multiobjective genetic algorithm: NSGA-II. IEEE Trans Evol Comput 6:182–197
62. Li H, Zhang Q (2009) Multiobjective optimization problems with complicated pareto sets, MOEA/D and NSGA-II. IEEE Trans Evol Comput 13:284–302
63. Liu Q, Tian Y, Wang C, Chekem FO, Sutherland J (2018) Flexible job-shop scheduling for reduced manufacturing carbon footprint. J Manuf Sci Eng 140(6):061006
64. ISO (2015) Geometrical product specifications (GPS)-filtration part 61: linear areal filters: gaussian filters. International Organization for Standardization, Geneva, Switzerland, Standard No. ISO 16610-61:2015
65. Wang M, Ken T, Du S, Xi L (2015) Tool wear monitoring of wiper inserts in multi-insert face milling using three-dimensional surface form indicators. J Manuf Sci Eng 137(3):031006
66. ISO (2015) Geometrical product specifications (GPS)-filtration part 40: morphological profile filters: basic concepts. International Organization for Standardization, Geneva, Switzerland, Standard No. ISO 16610-40:2015
67. Schmitz TL, Couey J, Marshb E, Mauntler N, Hughes D (2006) Runout effects in milling: surface finish, surface location error and stability. Int J Mach Tools Manuf 47:841–851
68. Kline WA, Devor RE (1983) The effect of runout on cutting geometry and forces in end milling. Int J Mach Tool Des Res 23(2–3):123–140
69. Smith GT (2008) Cutting tool technology industrial handbook. Springer, London
70. Liu S, Jin S, Zhang X, Chen K, Tian A, Xi L (2019) A coupled model for the prediction of surface variation in face milling large-scale workpiece with complex geometry. J Manuf Sci Eng 141(3):031009
71. Stephenson DA, Agapiou JS (1997) Metal cutting theory and practice manufacturing. CRC Press, New York
72. Liao Y, Stephenson DA, Ni J (2010) A multifeature approach to tool wear estimation using 3D workpiece surface texture parameters. J Manuf Sci Eng 132(6):1033–1041
73. Smith S, Tlusty J (1993) Efficient simulation programs for chatter in milling. CIRP Ann 42(1):463–466
74. Eksioglu C, Kilic ZM, Altintas Y (2012) Discrete-time prediction of chatter stability, cutting forces, and surface location errors in flexible milling systems. J Manuf Sci Eng 134(6):061006
75. Ding Y, Zhu L, Zhang X, Ding H (2010) A full-discretization method for prediction of milling stability. Int J Mach Tool Manuf 50:502–509
76. Wan M, Pan W, Zhang W, Ma Y, Yang Y (2014) A unified instantaneous cutting force model for flat end mills with variable geometries. J Mater Process Technol 214:641–650
77. Li G, Du S, Huang D, Zhao C, Deng Y (2019) Dynamics modeling-based optimization of process parameters in face milling of workpieces with discontinuous surfaces. J Manuf Sci Eng 141:101009

78. Cao H, Li B, He Z (2012) Chatter stability of milling with speed-varying dynamics of spindles. Int J Mach Tool Manuf 52:50–58
79. Zain AM, Haron H, Sharif S (2010) Prediction of surface roughness in the end milling machining using artificial neural network. Expert Syst Appl 37:1755–1768
80. Vivancos J, Luis C, Costa L, Ortiz J (2004) Optimal machining parameters selection in high speed milling of hardened steels for injection moulds. J Mater Process Technol 155(1):1505–1512
81. Franco P, Estrems M, Faura F (2008) A study of back cutting surface finish from tool errors and machine tool deviations during face milling. Int J Mach Tools Manuf 48:112–123
82. Franco P, Estrems M, Faura F (2004) Influence of radial and axial runouts on surface roughness in face milling with round insert cutting tools. Int J Mach Tools Manuf 44:1555–1565
83. Svetlik M, Radojicic M, Radovic S, Simic-Muller K (2018) Justifying Euler's formula through motion in a plane. Math Enthus 15(3):397–406
84. Shao YP, Yin YX, Du SC, Xi LF (2019) A surface connectivity based approach for leakage channel prediction in static sealing interface. J Tribol 141(6):062201
85. Wang B, Dong X, Wang Z, Wang Y, Hou Z (2020) The humidity-induced sensitivity amplification effect in an ionization gas sensor with silicon nanostructures. IEEE Electron Device Lett 41(6):908–911

Chapter 4
Profile Monitoring and Surface Classification

4.1 Circular and Cylindrical Profile Monitoring Considering Spatial Correlations

4.1.1 Introduction

Throughout the product's life cycle, from design to manufacturing and eventually recycling, the management of geometric specifications plays a pivotal role [1]. These specifications, which indicate the discrepancies between the actual and the intended geometric forms of machined components, have gained significant attention in recent times due to their substantial impact on the quality of mechanical parts. Key geometric characteristics, including circularity and cylindricity, are critical for quality control and significantly affect the sealing, dependability, and lifespan of products [2, 3].

Statistical process control (SPC) is widely utilized for managing geometric tolerances. Deng and Chin [4] investigated the factors influencing hole's circularity employing Taguchi techniques. Keshteli et al. [5] introduced a technique to determine the process capability index for circular profiles. Nevertheless, these approaches concentrate solely on the general aspects of circularity and cylindricity, leading to potential inaccuracies in the control process.

Advancements in measurement technology have facilitated an increase in measurement points, thereby offering more detailed insights into the shapes of geometric specifications. Consequently, the concept of profile monitoring has emerged. Profiling captures the interplay between explanatory and response variables, heavily reliant on extensive measurements. The scope of research in profile monitoring has broadened from linear [6, 7] to nonlinear [8–10], from single to multivariate [11, 12], from two-dimensional (2D) [13] to three-dimensional (3D) surfaces [14], from parametric [15, 16] to nonparametric approaches, and from single-channel to multichannel systems [17].

S. Du and L. Xi, *Quality Control for Variable Stiffness Structured Workpiece*,
https://doi.org/10.1007/978-981-96-5376-8_4

In the realm of circular and cylindrical profile monitoring, various methods that account for the distinctive attributes of these profiles have been introduced. Cho and Tu [18] outlined a harmonic decomposition technique for analyzing circular profile variations. Building upon this technique, Jiang et al. [19] assessed roundness from circular coordinate data obtained via a coordinate measuring machine (CMM), while Zhang et al. [20] adapted this approach to cylindrical parts. Ramaswami et al. [21] developed a novel circular feature sampling strategy to ascertain the most effective sample size and computed circular profile errors using the method of the minimum circumscribed circle. Principal component analysis [22–24] and independent component analysis [25] have been employed to decompose geometric specification components and investigate error origins. Additionally, machine learning techniques, including neural network algorithms [26] and Gaussian processes [27, 28], have been proposed for the detection of profile changes.

In a data-intensive manufacturing context, the proliferation of data leads to pronounced spatial correlations among measurement points. Colosimo et al. [29] investigated a spatial autoregressive regression (SARX) model to analyze the spatial correlations present in error terms. Soleimani, Noorossana, and Amiri [30] incorporated profile autocorrelation into linear profile monitoring, positing that error terms exhibit correlation which extends to the observations. Yu and Liu [31] introduced an innovative control chart aimed at detecting mean shifts in autocorrelated manufacturing processes. Addressing profiles' data that do not conform to the normal distribution assumption, Liu et al. [28] suggested a mixed-effect profile model for monitoring wafer thickness. Nevertheless, scant research has been directed towards scenarios characterized by spatially correlated observations.

The structure of a geometric specification can be represented as:

$$y = f(x) + \varepsilon \tag{4.1}$$

where y denotes the deviation from the nominal value, $f(x)$ signifies the system error, and ε represents the error term. Existing profile monitoring techniques have ascertained the value of $f(x)$ [18, 20], and several studies have taken into account the correlations within ε [30, 31]. However, the issue of correlations in y has received limited attention. Consequently, the objective of this research is to examine the spatial correlations of the observations and mitigate the impact of these correlations on the monitoring of circular and cylindrical profiles. Drawing from the spatial lag model (SLM) in spatial econometrics [32], and acknowledging the spatial correlations among observations, this section introduces a novel approach for the monitoring of circular and cylindrical profiles, integrated with control charts within the framework of SPC.

4.1.2 *The Proposed Method*

4.1.2.1 Overview of the Proposed Method

This section presents the novel approach for monitoring circular and cylindrical profiles. It addresses the spatial correlations inherent in the measurement points of industrial manufacturing processes by introducing a parametric model that accounts for these correlations. The model is designed to handle geometric specifications with spatial correlations, which are divided into three distinct components: spatial correlation error, systematic error, and random error. The spatial correlation error is modeled using the SLM, which captures the influence of neighboring points. Systematic error is attributed to consistent sources of error during the manufacturing process. Detailed descriptions of how to determine systematic error for circles and cylinders will be provided in Sects. 2.2 and 2.3, respectively. In this approach, the deviation of a point from its nominal value is composed of three elements: the influence of adjacent points, the systematic error, and the random error term, which can be formulated as:

$$y = \sum_{s=1}^{p} \rho_s W^{(s)} y + X'\beta + \varepsilon \tag{4.2}$$

where $X' = [x_1 x_2 \ldots x_r]$, represents the r independent regression variables, and $\beta = [b_1 b_2 \ldots b_r]$ represents the coefficients of these variables. $W^{(s)}$ is the s-order spatial weight matrix, where the weights for the adjacent s points are set to 1, and all other weights are 0. The error term ε follows a normal distribution with a mean of 0 and variance σ^2. ρ is an unknown lag coefficient that indicates the intensity of spatial correlations, with its range being between 0 and 1; a value close to 1 suggests strong correlations. To prepare the experimental data and obtain the y values, it is necessary to preprocess the data since the circular profile should be independent of the center position and the average radius measured on each profile [22]. The y value is calculated by subtracting the radius of the least square circle (LSC) from the measurements, with the LSC centered on the profile's geometric center.

The procedural flow of this methodology is depicted in Fig. 4.1, with the execution steps detailed below:

Step 1: Develop an appropriate parametric model for the profiles. For assessing spatial correlations, the Moran test can quantify the extent of such correlations [34]. Where no spatial correlations are detected, a standard ordinary least squares (OLS) regression model is utilized. Conversely, if spatial correlations are present, the LM test is applied to conduct more exhaustive diagnostic tests for autocorrelation. Should the spatial correlations among observations be pronounced, a model informed by the SLM should be formulated; if the spatial correlations within error terms are more pronounced, a model informed by the Spatial Error Model (SEM) should be formulated [35, 36].

Step 2: Determine the systematic error and estimate the model parameters based on the parametric model established in Step 1.

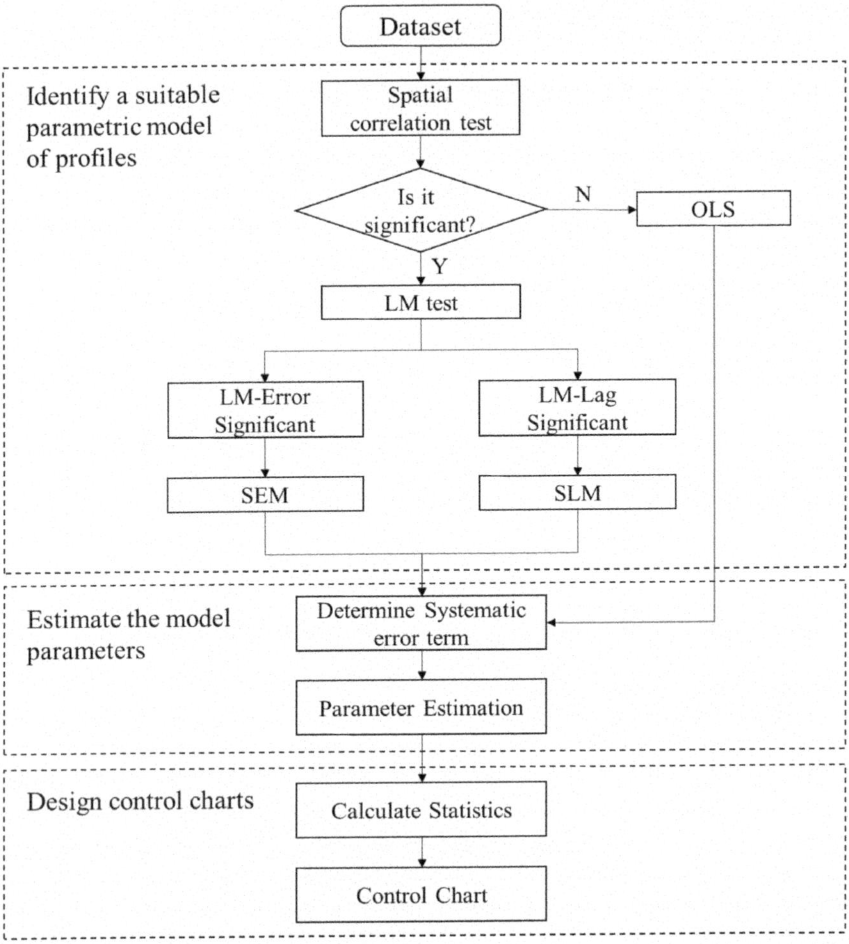

Fig. 4.1 Process flowchart for the proposed method

Step 3: Construct control charts for the parameters. Utilizing the control chart techniques from SPC, calculate the statistical parameters and design appropriate control charts for profile monitoring.

The specific monitoring procedures for circular and cylindrical profiles are elaborated in Sect. 4.1.2.2 and 4.1.2.3, respectively.

4.1.2.2 Circular Profile Monitoring

Suppose there are M circular profiles, with each profile $m(m = 1, 2 \ldots, M)$ consisting of N equally spaced observation points around the circumference. The deviation for each point $n(n = 1, 2, \ldots, N)$ within the mth profile is denoted by

y_{nm}, as illustrated in Fig. 4.2, where the abscissa represents the index n. Employing the SLM, the model for the circular profile can be formulated as follows:

$$y_{nm} = \sum_{s=1}^{p} \rho_{sm} W^{(s)} y_{nm} + X_n' b_m + \varepsilon_{nm} \tag{4.3}$$

Owing to factors such as machine tool spindle eccentricity, vibration, and ball wear, rotary components are prone to exhibit particular forms when being machined. Figure 4.3 displays various characteristic features of circular profiles.

As delineated in Section 2.1, the procedural steps for circular profile monitoring are outlined below:

Step 1: Develop an appropriate parametric model. Initially, the Moran test is employed to assess the level of spatial correlations. The outcome of this test is presented as the Moran index (Moran's I), indicating the extent of similarity between an observed value and its neighboring spatial elements. This test relies on the least squares estimation to derive the following statistic:

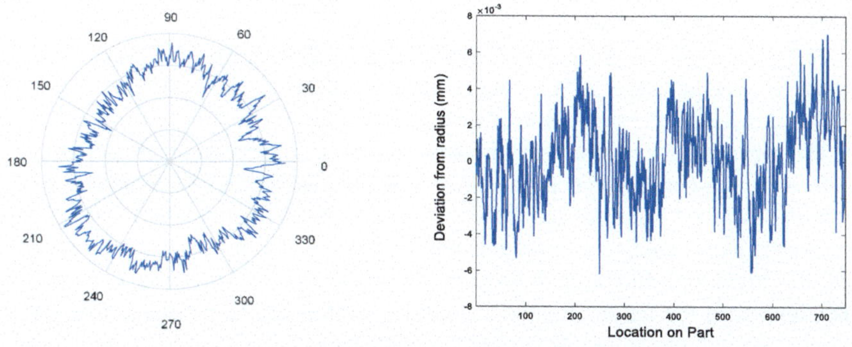

(a) (b)

Fig. 4.2 **a** y_j in polar coordinate system and **b** y_j in cartesian coordinate system

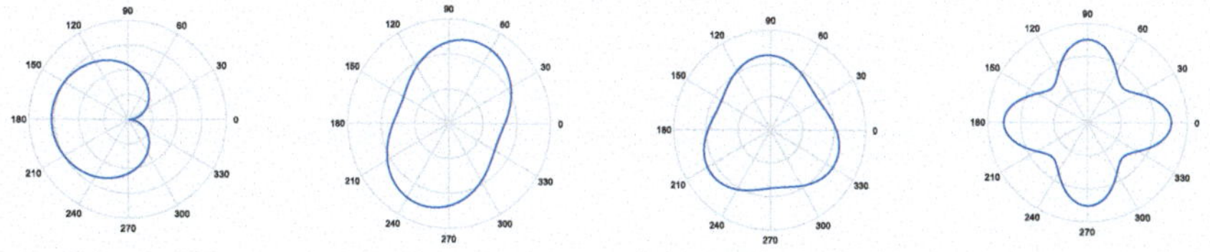

(a) Half-frequency spindle-motion error

(b) Bi-lobe spindle-motion error

(c) Tri-lobe spindle-motion error

(d) Four-lobe spindle-motion error

Fig. 4.3 Circular profiles of several typical features

$$
\mathrm{I} = \frac{n}{\sum_i \sum_j W_{ij}} \times \frac{\sum_i \sum_j W_{ij}\left(Y_i - \overline{Y}\right)\left(Y_j - \overline{Y}\right)}{\sum_i \left(Y_i - \overline{Y}\right)^2} \tag{4.4}
$$

where Y_i and Y_j represent the observations from spatial units; $\overline{Y}$ is the mean of Y; n denotes the total number of spatial units, and W signifies the spatial weight matrix.

A positive correlation in the data is indicated when Moran's I exceeds 0, with larger values signifying stronger spatial correlations. Conversely, a negative correlation is suggested when Moran's I is below 0, and the more negative the value, the greater the spatial disparity. Spatial data is considered to be approximately randomly distributed when Moran's I is near zero.

Following the Moran test as described in Sect. 4.1.2.1, step 1, the LM test is applied to conduct more extensive diagnostic tests for spatial correlation. The methodology for the LM test can be referenced from Breusch and Pagan [36]. The LM test provides two statistics, LM-Lag and LM-Error, which aid in deciding whether to opt for an SLM or SEM model. Should LM-Lag be more significant, a model informed by SLM is warranted; should LM-Error be more significant, a model informed by SEM is warranted. The SLM reflects the mutual influence among the response variables, whereas the SEM reflects the impact of neighboring unit errors on the observed values.

Given the pronounced spatial correlations among observations, the circular profile model is founded on the SLM, as depicted in Eq. (4.3).

Step 2. Identify the systematic error term.

The harmonic error commonly encountered in rotary parts during processing can be treated as a periodic function arising from the machine tool's rotational error. Damir's research [37] suggested that circular profiles can be simulated using Fourier harmonics, determining the minimum number of Fourier harmonics required to represent the circular profile through frequency component decomposition. Cho and Tu [18] confirmed the harmonic model's validity with a substantial dataset of actual circular and cylindrical profiles generated by turning and cylindrical grinding operations. Hii et al. [38] investigated the relationship between circular profiles and spindle motion error, assessing the reflection of spindle motion error on the circular profile. Colosimo et al. [29] identified potential spindle error forms during machining and their corresponding frequency components, characterizing each spindle error by specific amplitude and frequency components.

Consequently, the systematic component $X_n'b_m$ in Eq. (4.3) can be ascertained through the Fourier transform as specified in Eq. (4.5). Non-essential frequency components can be eliminated or isolated from the systematic component using Fourier harmonic separation.

$$
X_n'b_m = \sqrt{\frac{1}{N}}b_{0m} + \sqrt{\frac{2}{N}}\sum_{k=1}^{\frac{k}{2}}\left[b_{2k-1m}\cos(f_k(t-1)) + b_{2km}\sin(f_k(t-1))\right] \tag{4.5}
$$

Here, $f_k = k(2\pi/N)$; k represents the k th harmonic extracted through Fourier transformation; and t denotes the t th measurement point on the part. To ascertain the value of k, suitable frequency components are initially identified based on industrial expertise, complemented by experimental frequency decompositions. In practical applications, Damir [37] has empirically demonstrated that the second and third harmonics predominantly constitute the circular frequency components. By evaluating the relative magnitude of all amplitudes to the principal components, they proposed the following ranges for the order of Fourier harmonics: (1) for drilled surfaces: from the 2nd to the 6th harmonic; (2) for turned surfaces: from the 2nd to the 10th harmonic; (3) for cylindrical and centerless ground surfaces: from the 2nd to the 24th harmonic.

In experimental practice, one may consult the frequency components provided by Damir and integrate the outcomes of frequency decomposition to ascertain the harmonic components.

Step 3. Parameter Estimation.

Parameter estimation is executed subsequent to establishing the systematic error term and selecting an appropriate model. The parameters to be determined are ρ, β and ε. Given that ε is a random error distributed as $\varepsilon \sim \mathrm{N}(0, \sigma^2)$, the parameters for estimation are ρ, β and σ^2.. Initially, it is presumed in parameter estimation that spatial correlations can be depicted by a first-order spatial weight matrix (where s = 1).

The procedure for parameter estimation is as follows:

Step 3.1 Utilize the least-squares method to estimate parameters β and σ^2.

$$\hat{\beta} = \left(X'\mathrm{X}\right)^{-1} X'(y - \rho Wy) \tag{4.6}$$

$$\widehat{\sigma^2} = \frac{1}{n}(y - \rho Wy - X\beta)'(y - \rho Wy - X\beta) \tag{4.7}$$

Step 3.2 Estimate parameter ρ using the maximum likelihood method.

The likelihood function of the model is given as

$$\mathrm{L}\left(y|\rho, \beta, \sigma^2\right) = \frac{1}{2\pi\sigma^n}|I_n - \rho W|\exp\left\{-\frac{(y - \rho Wy - X\beta)'(y - \rho Wy - X\beta)}{2\sigma^2}\right\} \tag{4.8}$$

The logarithm of Eq. (4.8) is computed as

$$\begin{aligned} \ln\mathrm{L}\left(y|\rho, \beta, \sigma^2\right) &= -\ln 2\pi - \frac{n}{2}\ln\sigma^2 + \ln|I_n - \rho W| \\ &\quad - \frac{1}{2\sigma^2}(y - \rho Wy - X\beta)'(y - \rho Wy - X\beta) \end{aligned} \tag{4.9}$$

Input $\hat{\beta}$ and $\widehat{\sigma^2}$ into Eq. (4.9) to obtain the following relationship

$$\text{lnL}(y|\rho) = -\frac{n}{2}(1 + ln2\pi) - \frac{n}{2}\ln(e_0 - \rho e_1)'(e_0 - \rho e_1) + ln|I_n - \rho W| \quad (4.10)$$

A nonlinear optimization technique is deployed to optimize the likelihood function, yielding the estimated value $\hat{\rho}$ of ρ.

Step 3.3 Refine the estimates $\hat{\beta}$ and $\widehat{\sigma^2}$ utilizing the newly determined $\hat{\rho}$ from step 3.2.

Step 3.4 Iterate through steps 3.1–3.3 until the parameters reach convergence.

Once the parameter estimates for the profile are derived, if the residuals exhibit spatial correlations, the spatial weight matrix of order s + 1 should be incorporated, continuing this process until no spatial correlations are detected in the observations.

The derived parameter estimates can be expressed as

$$c'_m = \left[\rho'_m \beta'_m\right] = [\rho_{1m}\rho_{2m} \ldots \rho_{sm} b_{1m} b_{2m} \ldots b_{rm}] \quad (4.11)$$

which are normally distributed with a mean vector μ and a covariance matrix Σ.

Step 4. Design control chart.

Two control charts are designed in this step: One is the T^2 control chart for the parameters, the other is the individual control chart for the residuals. Assuming the false rate is α', the first type of error probability is $\alpha = 1 - \sqrt{1 - \alpha'}$.

The T^2 statistic is computed as

$$T_m^2 = (c_m - \mu)^T \Sigma^{-1} (c_m - \mu) \quad (4.12)$$

The upper control limit (UCL) is computed as

$$\text{UCL} = \frac{(m-1)^2}{m} \beta_{\alpha, n/2, (m-n-1)/2} \quad (4.13)$$

When the number of parts is large ($m > 100$), UCL $= \chi^2_{\alpha,n}$ can be used as an approximate control limit.

The residual statistic is given by

$$e_m = (I - \rho W) y_m - X b_m \quad (4.14)$$

The estimated variance of the residual is $\sigma_m^2 = \frac{e'_m e_m}{n-1}$.

A traditional Shewhart-type control chart can be used to monitor σ^2. The upper and lower control limits are given as

$$\begin{aligned} \text{UCL} &= \frac{\sigma^2}{n-1} \chi^2_{\frac{\alpha}{2}, n-1} \\ \text{CL} &= \sigma^2 \\ \text{LCL} &= \frac{\sigma^2}{n-1} \chi^2_{(1-\frac{\alpha}{2}), n-1} \end{aligned} \quad (4.15)$$

Upon establishing the control chart limits, the quality attributes of circular profiles can be surveilled. Should the statistical measures of the observation points exceed these control limits, it indicates that the production process is not within control parameters, necessitating further investigation.

4.1.2.3 Cylindrical Profile Monitoring

A cylindrical profile encompasses multiple circular profiles within a three-dimensional context. Within the cylindrical coordinate framework, a third dimension for height measurement is appended to the two-dimensional polar coordinate system, denoted by z. Along the axial direction, measurement points are evenly distributed, with points sharing the same z-value forming a layer; each layer is treated as an individual circular profile. In the radial direction, the position of a point is indicated by θ. Consequently, the radius of a measurement point within the cylindrical coordinate system can be determined by z and θ, and is formulated as $r(z, \theta)$.

In contrast to the circular profile, the cylindrical profile is subject to both radial and axial errors. The radial error, akin to the circular profile, is characterized by the Fourier series. Along the axial dimension, factors such as spindle and workpiece center misalignment, workpiece deflection, and thermal expansion contribute to axial errors [39]. Zhang et al. [20] examined the impact of diverse error sources on the geometric error of cylindrical components and demonstrated that the axial errors induced in cylindrical workpieces by manufacturing error sources can be expressed using the Legendre polynomial. The Legendre polynomial, being complete and orthogonal, ensures the coefficients' independence. Hence, the Legendre polynomial closely aligns with the axial direction error of cylindrical characteristics, allowing the axial error to be depicted by the Legendre polynomial as shown in Eq. (4.16).

$$
\begin{aligned}
P_0(z) &= 1 \\
P_1(z) &= z \\
P_2(z) &= \frac{1}{2}(3z^2 - 1) \\
P_3(z) &= \frac{1}{2}(5z^3 - 3z) \\
P_4(z) &= \frac{1}{8}(35z^4 - 30z^2 + 3) \\
&\cdots
\end{aligned}
\tag{4.16}
$$

Examining the synthesis of Legendre polynomials and Fourier functions enables the derivation of cylindrical profiles with diverse geometries. Glenn [40] employed the product of axial and radial functions to characterize radius errors. Zhang et al. [20] dissected the potential configurations of axial errors, validating that form errors could be articulated by a polynomial in z. Furthermore, they harnessed the Legendre-Fourier (L-F) polynomial to encapsulate the systematic error term of the cylindrical profile. The L-F polynomial, which is the result of multiplying Fourier and Legendre functions, serves to model cylindrical errors and is detailed in Eq. (4.17).

$$\begin{aligned} r(z,\theta) &= r_0 + \Delta r(z,\theta) \\ &= \sum A_{0j}P_j(z) + \sum\sum\left[A_{ij}P_j(z)\cos(i\theta) + B_{ij}P_j(z)\sin(i\theta)\right] \end{aligned} \tag{4.17}$$

where $z \in [-1, 1]$; $\theta \in [-\pi, \pi]$;

The jth Legendre polynomial function, $P_j(z)$, signifies the deviation along the z-axis; $\cos(i\theta)$ denotes the deviation in the x-axis direction, while $\sin(i\theta)$ denotes the deviation in the y-axis direction. Together, $\cos(i\theta)$ and $\sin(i\theta)$ constitute the radial variation of the cylinder, which is represented by F_i. The average radius of the workpiece, F_i, can be determined by the summation $\sum A_{0j}P_j(z)$. The shape error at the point (z, θ), denoted as Δr, is calculated through the product of the Fourier and Legendre functions.

The coefficients within the L-F polynomial are determined by the formula,

$$A_{ij} = \frac{\sum_{k=1}^{t}\sum_{l=1}^{u} r_{kl}P_j(z_k)\cos(i\theta_l)}{\sum_{k=1}^{t}\sum_{l=1}^{u} P_j^2(z_k)\cos^2(i\theta_l)} \tag{4.18}$$

$$B_{ij} = \frac{\sum_{k=1}^{t}\sum_{l=1}^{u} r_{kl}P_j(z_k)\sin(i\theta_l)}{\sum_{k=1}^{t}\sum_{l=1}^{u} P_j^2(z_k)\sin^2(i\theta_l)} \tag{4.19}$$

where k signifies the order of a point along the z-axis; t represents the total count of points along the z-axis; l is the order of the radial azimuth, and u is the total count of radial azimuths. The value of A_{0j} can similarly be derived from Eq. (4.18).

Table 4.1 and Fig. 4.4 shows several typical features of cylindrical profiles, where P_jF_i represents the order of the L-F polynomial.

Consistent with the circular profile monitoring process outlined in Sect. 4.1.2.2, the cylindrical profile monitoring procedure can be delineated as follows:

Step 1. As with the circular spatial correlations test in Sect. 4.1.2.2, step 1, both Moran and LM tests are conducted to assess cylindrical spatial correlations and to select an appropriate model.

Step 2. Identify the systematic error term. Given that the average radius has been subtracted from the observed values, the initial component on the right side of Eq. (4.17) should be disregarded in determining the systematic error term. Consequently, the systematic error term is expressed as

$$X_n'b_m = \sum\sum\left[A_{ij}P_j(z)\cos(i\theta) + B_{ij}P_j(z)\sin(i\theta)\right] \tag{4.20}$$

The orders of i and j, which correspond to the Fourier function and Legendre polynomial respectively, can be initially specified based on industrial expertise and further refined through the frequency analysis of measurements. Damir [37] suggested significant harmonics for cylindrical surfaces, yet did not provide the

Table 4.1 The geometric shape and error source of several L-F polynomial

Fourier i	Legendre j	Fourier function F_i	Legendre polynomial P_j	Cylindrical radius feature r_{ij}	Geometric shape	Error source
0	1	0	z	z	Axial taper	Misalignment of spindle/work centers
0	2	0	$(3z^2-1)/2$	$(3z^2-1)/2$	Axial bump	Workpiece deflection
1	0	$\cos(\theta),\ \sin(\theta)$	1	$\cos(\theta) + \sin(\theta)$	Axial eccentricity	Spindle rotation
1	1	$\cos(\theta),\ \sin(\theta)$	z	$z(\cos(\theta) + \sin(\theta))$	Axial tilt	Chuck clamping
1	2	$\cos(\theta),\ \sin(\theta)$	$(3z^2-1)/2$	$[(3z^2-1)/2](\cos(\theta) + \sin(\theta))$	Banana shape	Workpiece deflection and spindle rotation
2	0	$\cos(2\theta),\ \sin(2\theta)$	1	$\cos(2\theta) + \sin(2\theta)$	Bi-lobe circumflex	Spindle rotation

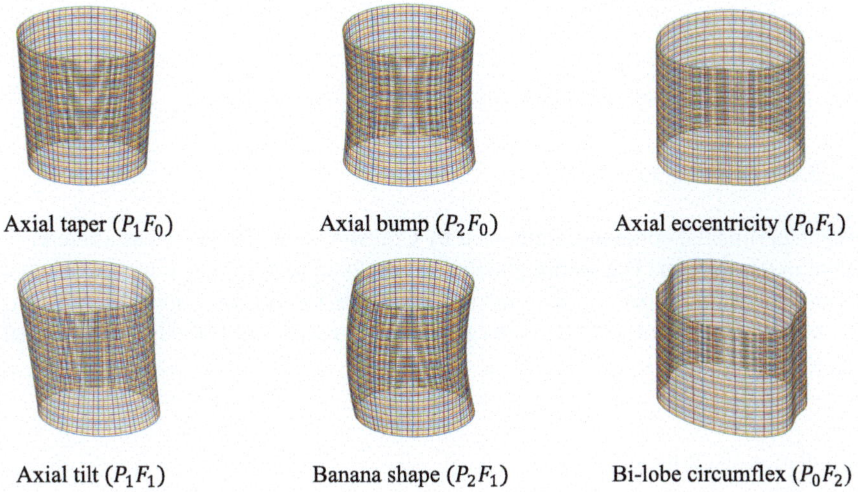

Fig. 4.4 Cylindrical profiles of several typical features

corresponding polynomial values. Colosimo et al. [27] identified six fundamental components crucial for characterizing cylindrical surfaces. These six polynomial orders are adopted as the fundamental elements to depict cylindrical profile errors in experimental settings. The relevant polynomials are presented in Table 4.2.

Table 4.2 The basic components of L-F polynomial

k	Order of Legendre polynomial	Order of Fourier function	L-F
1	0	2	$\cos 2\theta + \sin 2\theta$
2	0	3	$\cos 3\theta + \sin 3\theta$
3	1	0	z
4	1	2	$z(\cos 2\theta + \sin 2\theta)$
5	1	3	$z(\cos 3\theta + \sin 3\theta)$
6	2	0	$(3z^2\text{-}1)/2$

Step 3. Parameter estimation for the chosen model is conducted, taking into account the spatial nature of the cylinder, which affects the weight matrix values differently than in circular profiles. Once the cylindrical measurements are aligned, the weight matrix can be defined by rook-based contiguity (assigning a weight of 1 to points adjacent in four directions to a given point) or queen-based contiguity (assigning a weight of 1 to points adjacent in eight directions to a given point).

Step 4. The process for devising control charts for cylindrical profiles mirrors that of circular profile monitoring, as detailed in Sect. 4.1.2.2 The T^2 and residual statistics are determined using Eqs. (4.12) and (4.14), respectively, with control limits derived from Eqs. (4.13) and (4.15).

4.1.3 Numerical Simulation

4.1.3.1 Circular Profile

The experimental components detailed by Colosimo et al. [29] serve as a reference point in this section. The section involved 100 C20 cylinders, which were machined from an original diameter of 30 mm down to a target diameter of 26 mm. A CMM was utilized to sample a total of 748 points across each circular profile for the experimental parts.

Simulation Experiment

The simulation data is handled following the circular profile monitoring procedure outlined in Sect. 4.1.2.2. The steps are as follows:

Step 1. Develop an appropriate parametric model.

Initially, the spatial correlations within each profile are assessed using the Moran and LM tests. For practical implementation, the Matlab spatial statistics toolbox 'jplv7' can be deployed for these tests. The outcomes of these tests are summarized in Table 4.3. As indicated by the table, the Moran test reveals strong spatial correlations among the observations. Furthermore, the LM test results, where LM-Lag is more

Table 4.3 Results of Moran and LM test

Statistics	Statistical value	*P* value
Moran-istat	19.0004	0.000
LM (lag)	139.3217	0.000
LM (error)	13.5151	0.000

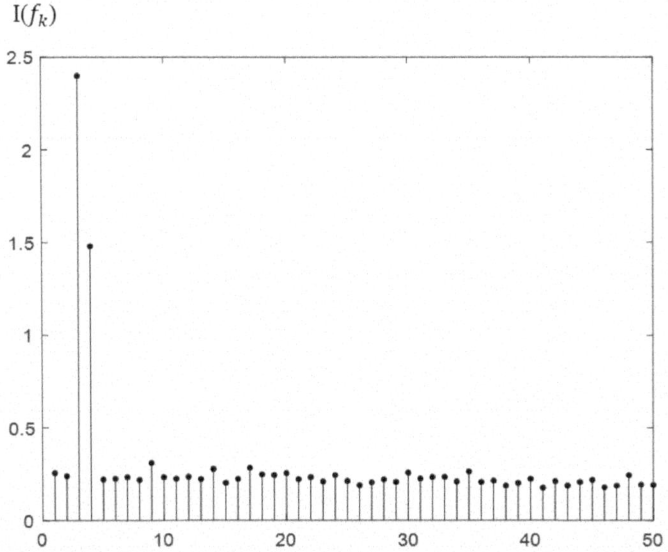

Fig. 4.5 Fourier decomposition map for the former 50 harmonics of simulated circular profiles

pronounced, confirm the presence of significant correlations in the observations. Consequently, it is justified to develop a model based on the SLM.

Step 2. Identify the systematic error term

During this phase, harmonic separation is utilized for decomposition purposes. The first Fourier harmonic b_{0m} (where $k = 0$) indicates the circle's center position, and the second Fourier harmonic b_{1m} (where $k = 1$) corresponds to the nominal radius of the sampled profiles. Given that the experimental circle is centered at the origin and the nominal radius has been factored out from the observations, both b_{0m} and b_{1m} are null. The value of k is derived from Fourier decomposition, and by considering the first 50 harmonics as depicted in Fig. 4.5, it is evident that the 2nd and 3rd harmonics exhibit significant amplitudes while the others are negligible and can be disregarded. Thus, the systematic error term in this scenario is represented by the 2nd and 3rd harmonics, and is expressed as

$$\begin{aligned} X_n' b_m = \sqrt{2/N}[&b_{3\text{m}} \cos(f_2(t-1)) + b_{4\text{m}} \sin(f_2(t-1)) \\ &+ b_{5\text{m}} \cos(f_3(t-1)) + b_{6\text{m}} \sin(3(t-1))] \end{aligned} \tag{4.21}$$

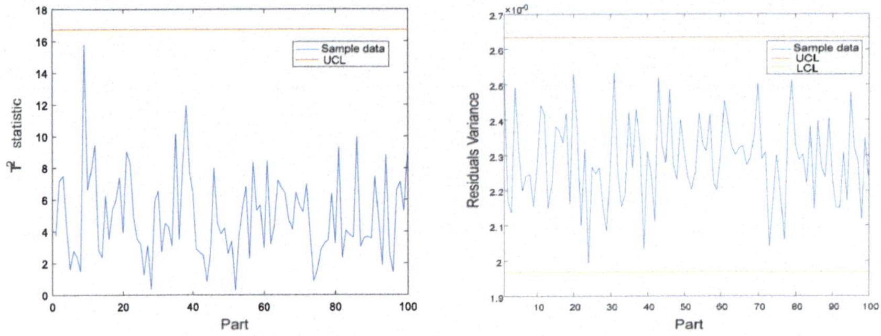

(a) T^2 control chart (b) σ^2 control chart

Fig. 4.6 Control charts of 100 circular profiles in phase I

Step 3. Parameter Estimation

Initially, a first-order correlation matrix is employed for parameter estimation. The parameters ρ and β are determined as described in Sect. 2.2 and arranged as $c'_m = \left[\rho'_m \beta'_m\right] = [\rho_m b_{1m} b_{2m} b_{3m} b_{4m}]$. Subsequently, the Moran test is applied to evaluate the spatial correlations of the residuals, continuing until the test statistics no longer indicate significance.

Step 4. Control Chart Design

The T^2 statistics for these 100 profiles are computed using Eq. (4.12). With a false alarm rate of $\alpha' = 1\%$ for each control chart, the probability of a type I error is $\alpha = 1 - \sqrt{1 - \alpha'} = 0.5013\%$. The upper control limit is set at UCL $= \chi^2_{0.005013,5} =$ 16.7496. The control limits for σ^2 are derived using Eqs. (4.14) and (4.15). Figure 4.6 displays the T^2 and σ^2 control charts for the 100 simulated circular profiles in phase I.

Method Comparison

This subsection compares the proposed method with existing approaches using the dataset from Colosimo et al. [29], focusing on common techniques for circular profile monitoring.

Out-of-roundness (OOR) measures the proximity of a cross-section to an ideal circle, assessing circularity accuracy. The LSC method is a prevalent and straightforward approach for determining OOR values; it calculates the OOR by finding the difference between the maximum and minimum distances from the actual profile to the LSC. Shewhart control charts can be used to monitor each circular profile's corresponding OOR value. If the OOR value does not exceed the upper control limit, the circular feature is deemed to meet the specifications.

The Location Control Chart (LOC), introduced by the Boeing Commercial Airplane Group [33], involves compiling measurement data from each specific location and tracking these measurements by plotting them within a shared coordinate system. Following the registration of profiles—a process detailed by Colosimo and Pacella [22]—the Shewhart control chart is implemented for each observation point under stable conditions. If the observed shape is within control, the data points should fall within the control limits with a specified level of probability. The control limits for the LOC are determined based on the mean and standard deviation of the measurement points at corresponding positions.

SARX, as introduced by Colosimo et al. [29], is utilized for the monitoring of circular profiles. This approach is rooted in the SEM, which accounts for the impact of spatial correlations on the error terms. It posits that each profile's observations are composed of a systematic component and an error component. Deviating from Eq. (4.2), the SEM-based representation of observations is articulated as follows:

$$\begin{aligned} y &= X'\beta + v \\ v &= \lambda W v + \varepsilon \end{aligned} \tag{4.22}$$

where $X'\beta$ denotes the systematic component; v signifies the error term; W is the weight matrix, and λ is the spatial correlation coefficient. Drawing from regression techniques, the authors proposed a residual correlation model that integrates with control charts. This technique takes into account the influence of spatial correlations on the error terms and scrutinizes the monitoring process for circular profiles. The distinction between the SLM and SARX models lies in how spatial correlations are depicted—in SLM, they appear as a distinct term, whereas in SARX, they are incorporated within the error term. These methodologies, while founded on different models, yield distinct outcomes. For comparative analysis, both T2 and residual control charts are implemented within these approaches; the variations lie in the statistical values and the control limits applied.

Principal Component Analysis (PCA) is a viable technique for monitoring circular profiles. Colosimo and Pacella [22] employed PCA to detect systematic patterns within roundness profiles, while Wang et al. [24] utilized it for the surveillance of multimode near-circular shape profiles. PCA serves as a comparative method, focusing on the monitoring of the T^2 statistics of principal components and the squared prediction error (SPE). Only principal components accounting for more than 80% of the total variance are retained.

This subsection details the performance comparison of these methods in phase II. The objective of phase II is the early detection of changes during the monitoring process. The monitoring methods are assessed based on the average run length (ARL), which denotes the number of parts inspected before the first out-of-control signal is detected. Given that the false-alarm probabilities for these methods are set at 1% in phase I, 1000 circular profiles with identical parameter distributions are simulated to establish the control limits for each method. To assess performance in phase II, simulations are conducted under three distinct out-of-control scenarios: spindle motion error, bi-lobe error, and tri-lobe error.

(1) Spindle motion error: $y_n(p) + \sqrt{\frac{2}{p}}\delta_1 \sin\left(\frac{1}{2}\theta_p\right)$;

Table 4.4 Mean of ARL results for 1000 simulations in phase II

	δ	OOR	LOC	SARX	PCA	SLM-based
Spindle	$\delta_1 = 0.005$	49.41	63.33	23.48	66.67	**23.17**
	$\delta_1 = 0.01$	54.23	60.44	13.88	18.18	**12.2**
	$\delta_1 = 0.05$	53.96	4.63	**1.37**	5.71	1.53
	$\delta_1 = 0.1$	78.48	1.02	1.23	2.06	**1**
Bi-lobe	$\delta_2 = 0.005$	66.83	10.54	4.86	76.92	**4.3**
	$\delta_2 = 0.01$	18.42	**1.06**	1.28	19.23	1.2
	$\delta_2 = 0.05$	1	1	1	1.78	1
	$\delta_2 = 0.1$	1	1	1	1	1
Tri-lobe	$\delta_3 = 0.005$	61.53	10.58	5.14	52.63	**4.95**
	$\delta_3 = 0.01$	10.77	**1.05**	1.23	7.69	1.32
	$\delta_3 = 0.05$	1	1	1	1.38	1
	$\delta_3 = 0.1$	1	1	1	1	1

(2) Bi-lobe error: $y_n(p) + \sqrt{\frac{2}{p}}\delta_2\left[c_{1n}\cos\left(2\theta_p\right) + c_{2n}\sin\left(2\theta_p\right)\right]$;

(3) Tri-lobe error: $y_n(p) + \sqrt{\frac{2}{p}}\delta_3\left[c_{3n}\cos\left(3\theta_p\right) + c_{4n}\sin\left(3\theta_p\right)\right]$.

In phase II, simulations for each scenario are conducted 1000 times to determine the average ARL. The ARL outcomes of these methods under the three specified conditions are detailed in Table 4.4. The performance comparison is derived from simulating profiles based on the models outlined for phase II and calculating the ARL using the control limits established during phase I.

Table 4.4 illustrates that the proposed method records the lowest ARL value in the majority of instances. The LOC and SARX methods also exhibit strong performance in the bi-lobe and tri-lobe scenarios. When the change magnitude, represented by δ, is minor, indicating a subtle alteration, the SLM-based and SARX approaches outperform the other methodologies. The LOC method also shows satisfactory performance in scenarios where the change is more pronounced. The OOR technique is the least effective, as the OOR value fails to represent the profile's original shape accurately. The PCA's performance is underwhelming; it lacks sensitivity to minor deviations. The results in Table 4.4 clearly indicate that methods accounting for spatial correlations significantly outperform conventional methods.

4.1.3.2 Cylindrical Profile

Simulation Experiment

At this juncture, a simulation of 100 cylinders is carried out utilizing the parameters specified in Table 4.5. Within each cylindrical profile, 10 strata are sampled, comprising 98 measurement points per stratum.

Table 4.5 Polynomial orders in the simulation

Order of Legendre polynomial	Order of Fourier component	x_{kn}
0	0	$P_0(z)$
1	2	$P_1(z)\cos(2\theta)$
1	2	$P_1(z)\sin(2\theta)$
2	0	$P_2(z)$

Table 4.6 Mean of ARL results in phase II

	δ	OOR	LOC	SARX	PCA	SLM-based
Axial error	$\delta_1 = 0.01$	50.47	39.29	20.51	77.78	**15.1**
	$\delta_1 = 0.05$	43.15	24.67	1.83	46.67	**1.61**
	$\delta_1 = 0.1$	27.51	5.82	**1.32**	27.77	1.35
Radial error	$\delta_2 = 0.01$	54.73	31.94	5.62	83.34	**5.38**
	$\delta_2 = 0.05$	48.69	**1.63**	2.83	28.57	2.64
	$\delta_2 = 0.1$	10.93	**1**	2.64	1.60	2.76
Both	$\delta_3 = 0.01$	63.14	36.23	19.07	66.67	**18.28**
	$\delta_3 = 0.05$	68.59	15.87	1.63	39.16	**1.59**
	$\delta_3 = 0.1$	46.16	2.91	**1.13**	18.18	1.21

Following the circular profile monitoring protocol, the Moran and LM tests are initially conducted on the observations to determine an appropriate spatial correlation model. Subsequently, the parameters and T^2 statistics are computed. Ultimately, control charts for the T^2 statistics and residuals are devised to ascertain the monitoring outcomes. The simulation findings are detailed in the subsequent subsection.

Method Comparison

Drawing parallels to the circular profile scenario, the quintet of methods—OOR, LOC, SARX, PCA, and SLM-based—can be deployed for the monitoring of cylindrical profiles. During phase I, a simulation of 1000 cylindrical profiles is executed to ascertain the control limits for these methods at a false-alarm probability of 1%. Consequently, these control limits serve as benchmarks for phase II. For assessing performance in phase II, simulations incorporate axial and radial errors, as delineated by the subsequent models:

(1) Axial error: $\delta_1 P_2(\cos(2\theta) + \sin(2\theta))$;
(2) Radial error: $\delta_2 P_0(\cos(4\theta) + \sin(4\theta))$;
(3) Axial and radial errors: $\delta_3 P_2(\cos(4\theta) + \sin(4\theta))$.

In this simulation, each scenario is replicated 1000 times to determine the average ARL for phase II. The average ARL for these methods is presented in Table 4.6.

Table 4.6 indicates that the SLM-based and SARX methods outperform the others substantially. When changes are pronounced, the LOC method yields results comparable to the SLM-based and SARX methods, all of which yield satisfactory outcomes relative to the OOR and PCA methods. The PCA shows insensitivity in the face of minor changes, and the OOR method ranks as the least effective among the quintet. Broadly speaking, the SLM-based method records the lowest ARL value in most instances. It is also observable from Table 4.6 that axial errors are more readily detected than radial errors.

4.1.4 *Case Study*

This section details the performance validation of the proposed method using cylinder bores from B12 series engines, as depicted in Fig. 4.7a. The cylindrical features within engine cylinder bores are commonly produced through honing processes. The experimental arrangement is illustrated in Fig. 4.7b, where the cylindrical hole is created by the honing head's vertical and rotational motion. The bore's standard diameter is 71 mm, with actual diameters being measured via a coordinate measuring machine. The experiment includes twelve engines—comprising six that meet specifications and six that do not, the latter being intentionally honed to distort the cylinder shape—resulting in a total of 48 bores. For each bore, ten layers with 98 measurement points per layer are sampled utilizing the CMM.

4.1.4.1 Circular Profile

In phase I, eighty-four qualified circles were selected for establishing phase I objects and deriving control limits. The OOR, LOC, SARX, PCA, and SLM-based techniques were applied to these 84 circular profiles. As depicted in Fig. 4.8, the historical samples were examined to discern the process behavior while in control,

(a) B12 engine with four-cylinder bores

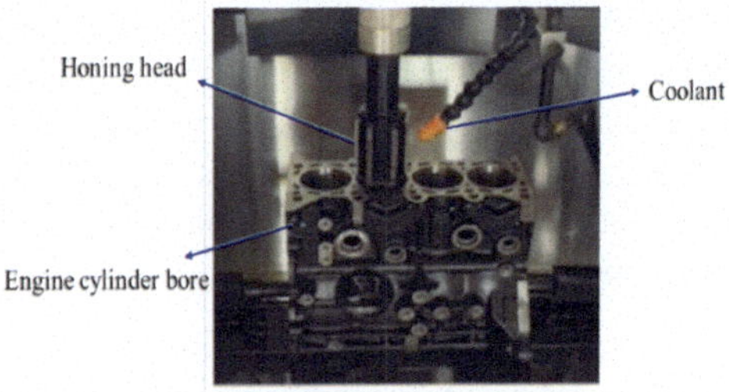

(b) Bore honing experimental system

Fig. 4.7 The measurement of cylinder bores

with control limits being calculated accordingly. The LOC method flagged eleven shapes as out of control, exceeding the acceptable Type I error rates of the other methods. Consequently, the LOC method's stringent criteria render it unsuitable for this scenario.

In phase II, a comparison of the detection rates of these methods was conducted using 120 circular profiles from defective engine bores. The OOR, LOC, SARX, and SLM-based control charts for these profiles are presented in Fig. 4.9, with Table 4.7 detailing the number of detected unqualified parts by each method. Figure 4.9 and Table 4.7 indicate that the SARX and SLM-based methods detected a significantly higher number of unqualified parts compared to the OOR, PCA, and LOC methods. The LOC method's false-alarm probability was excessively high in phase I, confirming its inapplicability. The SARX and SLM-based methods yielded comparable results, likely due to their consideration of spatial correlations, while the OOR and PCA methods were minimally effective in detecting out-of-control conditions in this context.

In this case, the SLM-based method exhibits a markedly higher detection rate in comparison to the OOR and PCA methods. Additionally, the SLM-based method outperforms the SARX method, which supports the spatial correlation test findings that the impact of spatial correlation in the observations is substantial. However, the overall detection rates of these methods remain quite low, making the identification of unqualified parts challenging. The potential reason for this phenomenon could be that the selected samples, which are unqualified cylinders, have circularities that are still within control limits.

4.1.4.2 Cylindrical Profile

Parallel to the monitoring of circular profiles, the same quintet of methods—OOR, LOC, SARX, PCA, and SLM-based—are deployed for the surveillance of cylindrical profiles. An assessment of 48 cylinders is conducted to gauge the efficacy of these techniques. Figure 4.10 illustrates the radial deviations for 24 compliant and 24 non-compliant cylinder bores.

Referring to the controlled machined surfaces, the objective of control charting is to identify any potential out-of-control shapes within this collection of items. The control limits for the control charts in phase I are derived from the in-control cylindrical profiles and serve as a benchmark for phase II. As depicted in Fig. 4.11, the control charts indicate that the sample profiles generally fall within the control limits, with the exception of the 16th part in the SARX method and the 17th part in the SPE chart of PCA. Given the limited sample size in this section, the error rate for these two methods stands at 4.17%, which is significant and should not be disregarded. Nonetheless, these methods continue to be employed for comparative purposes in phase II.

In phase II, the efficacy of these methods is assessed based on their detection rates, with a comparison of the number of out-of-control parts identified. Figure 4.12

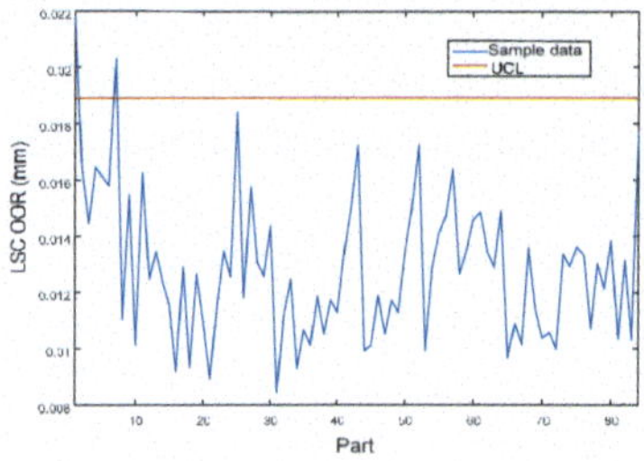

(a) Control chart of OOR method.

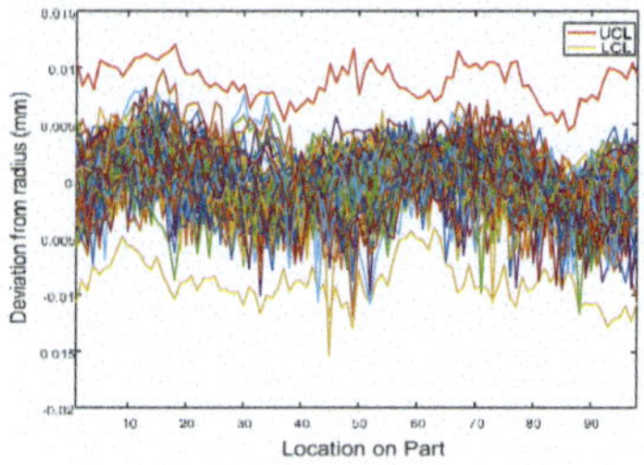

(b) Control chart of LOC method.

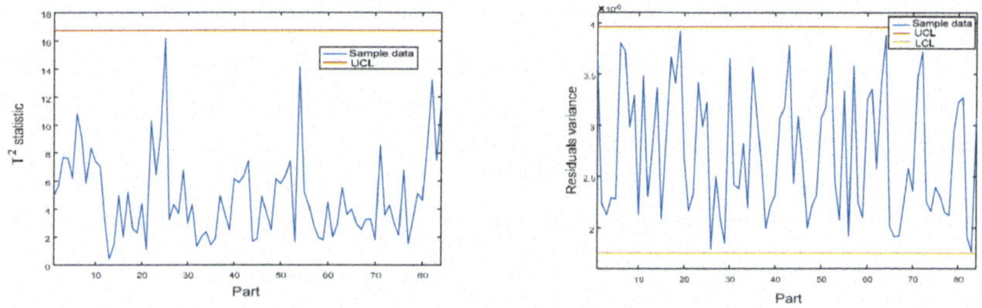

(c) T^2 and residual control charts for the SARX method.

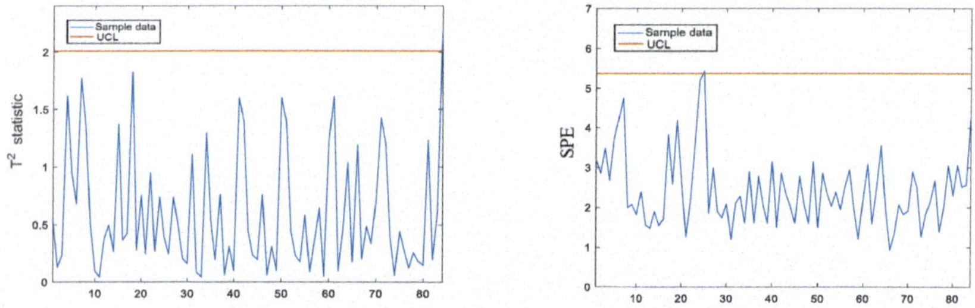

(d) T^2 and residual control charts for the PCA method.

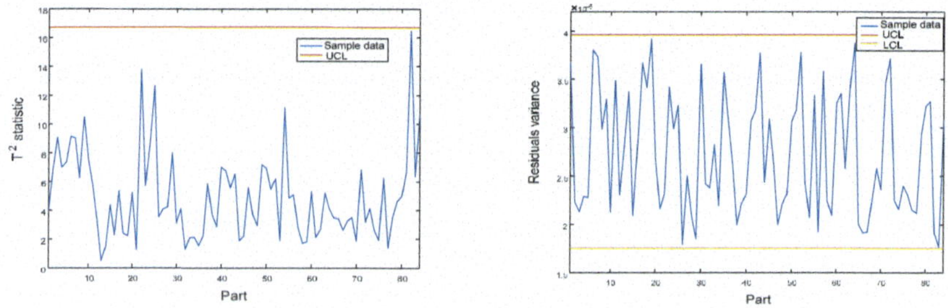

(e) T^2 and residuals control charts for the SEM-based method.

Fig. 4.8 Circular profile control charts for the five methods in phase I

Table 4.7 Number (rate) of unqualified parts detected using the five methods on Circular profile

Number of unqualified parts	OOR	LOC	SARX	PCA	SLM-based
	2	30	45	4	47
Detected (%)	1.67	25	37.5	3.33	39.2

displays the control charts for 24 out-of-control cylindrical profiles during phase II, while Table 4.8 lists the count of unqualified parts detected by each method.

Analysis of Fig. 4.12 and Table 4.8 reveals that the SLM-based method achieves the highest detection rate, closely followed by the SARX method. The OOR method ranks the lowest in effectiveness. The PCA outcomes mirror those of LOC, with the exception of a single point exceeding the control limit in phase I.

For cylindrical profiles, the SLM-based method outperforms SARX, with an 8.3% enhancement in detection rate, suggesting that the spatial correlations considered are more reflective of the actual conditions.

The distinction between the SARX and SLM-based approaches primarily hinges on the T^2 statistics and residual values, as both methods mitigate the impact of spatial correlations. The T^2 control chart is designed to track the coefficients of spatial correlations and systematic errors. Given that the two methods portray spatial correlations differently, their respective T2 statistics vary. In Fig. 4.12, during phase II, the T^2 control chart of the SLM method is able to identify two additional out-of-control parts: the third and twenty-third points. An additional benefit of the SLM-based method is evident in the T^2 control chart of phase I, Fig. 4.11, where the 16th T^2 statistic of the SARX method significantly exceeds the control limits, leading to a mistaken conclusion that a controlled part is out of control.

Furthermore, the detection rate for cylindrical profiles significantly surpasses that observed for circular profiles. A potential explanation for this discrepancy is the relative ease of identifying axial changes in cylinders compared to radial changes; additionally, the samples in question are cylinders that are deemed unqualified, with their circularities maintained within control limits.

4.1.4.3 Error Form Analysis

Error form analysis assists engineers in diagnosing the sources of errors in machined components and enhancing the manufacturing techniques. By eliminating the impact of random errors and scrutinizing the systematic errors of out-of-control components, the root cause of the error can be identified and rectified during production. In this scenario, the diagram of the pertinent error form is depicted in Fig. 4.13, revealing an error pattern characterized by radial half-frequency and axial tilt. Drawing from the machining error analysis of cylindrical parts conducted by Zhang et al. [20], radial half-frequency errors may stem from whirling or a defective ball bearing, while axial tilt errors could be attributed to misaligned spindle or work centers. Utilizing the

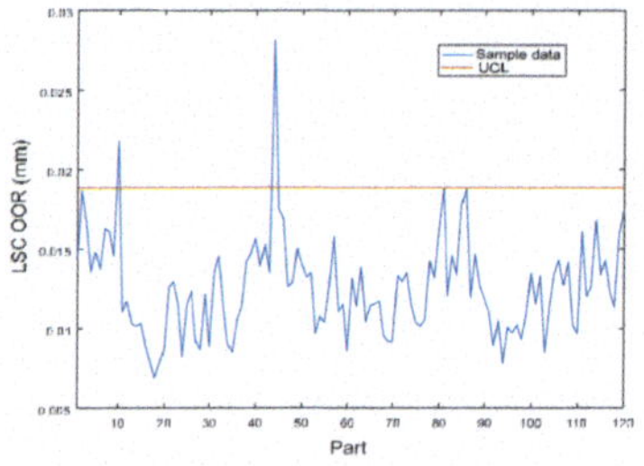

(a) Control chart of OOR method.

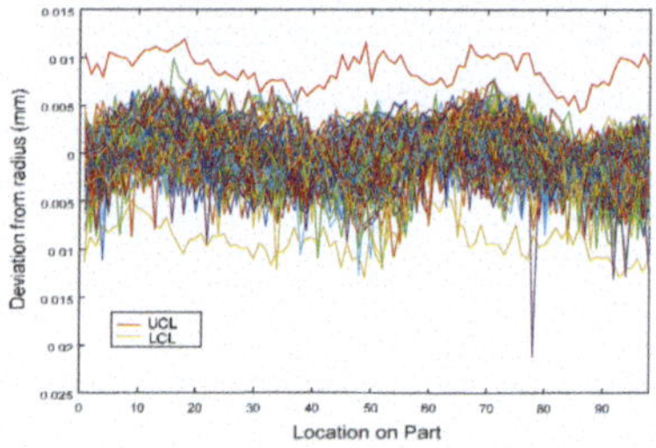

(b) Control chart of LOC method.

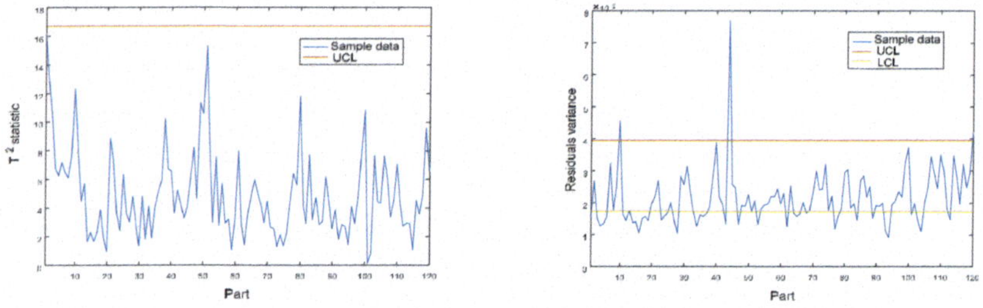

(c) T^2 and residual control charts for the SARX method.

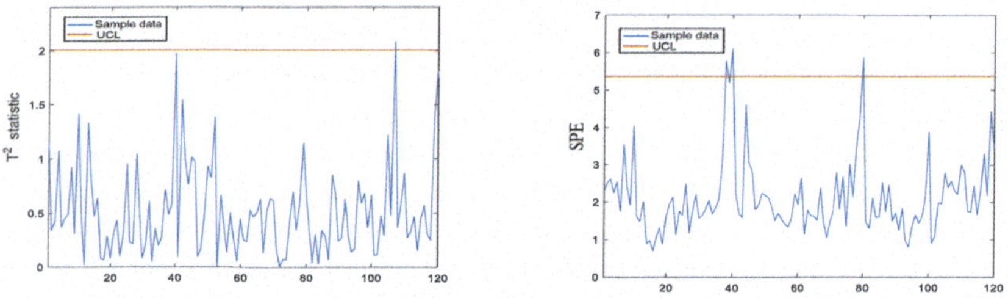

(d) T^2 and residual control charts for the PCA method.

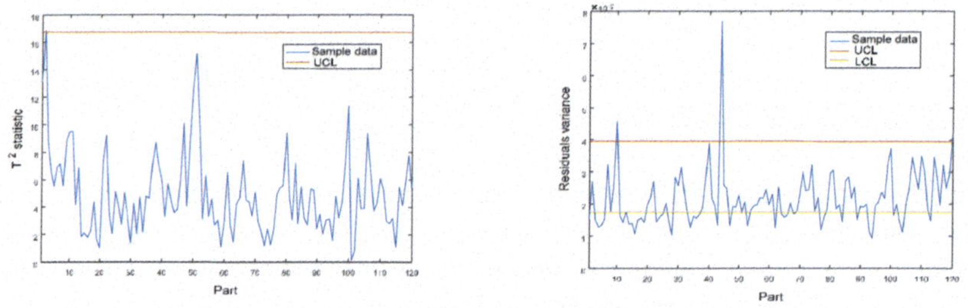

(e) T^2 and residuals control charts for the SEM-based method.

Fig. 4.9 Circular profile control charts for the five methods in phase II

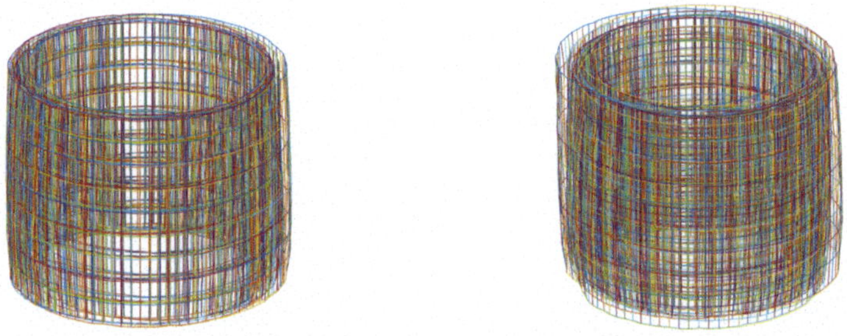

(a) Radial errors for 24 qualified cylinder profiles.

(b) Radial errors of for unqualified cylinder profiles.

Fig. 4.10 Radial errors of cylindrical profiles

analysis outcomes, the probable causes of the machining errors can be ascertained, allowing for targeted improvements in the processing stage.

4.1.5 Conclusions

Traditionally, geometric profile monitoring has encountered challenges due to the spatial correlations inherent in measurement points. Existing methodologies include OOR, LOC, PCA, and SARX, with the first three neglecting spatial correlations and SARX accounting for them within the error term. This approach examines the spatial correlations of observations and introduces a SLM-based method for monitoring circular and cylindrical profiles. Utilizing engine cylinder bores as a case study, the detection rates of the SLM-based method show a 1.7% enhancement in circular profiles and an 8% improvement in cylindrical profiles over SARX. Furthermore, these outcomes surpass those of other methodologies. The PCA method proves ineffective in this context, and its principal components, potentially a blend of various error shapes, can be challenging or impossible to decipher. Although both SLM-based and SARX models incorporate spatial correlations, they represent distinct spatial interaction effects; SLM reflects the mutual influence among response variables, whereas SARX reflects the impact of adjacent units' errors on the observed value. The phase II case study results indicate that methods considering spatial correlations outperform conventional methods, and they also suggest that the spatial correlations in observations align more closely with actual conditions.

Additionally, through the modeling and analysis of systematic error terms, the SLM-based approach can offer insights into the likely sources of machining errors in practical manufacturing and suggest potential avenues for enhancement.

Several avenues for future investigation present themselves:

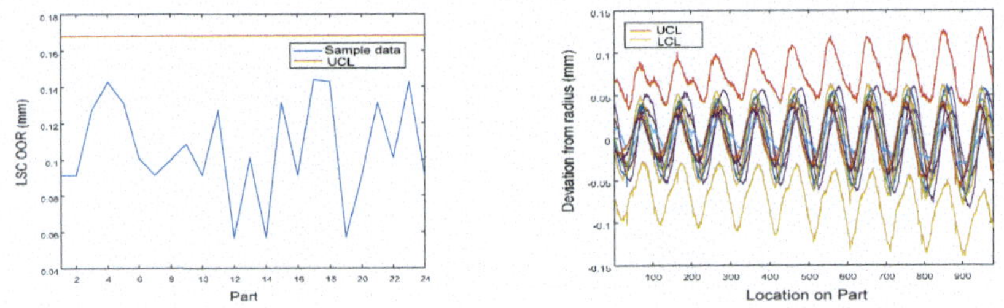

(a) Control charts of OOR method (b) Control charts of LOC method

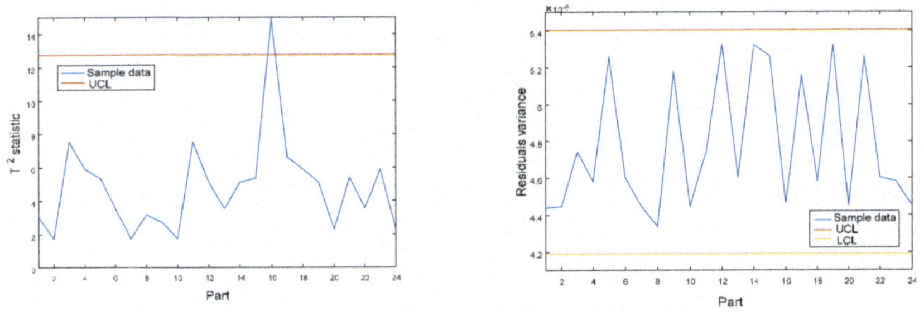

(c) T^2 and residual control charts for the SARX method

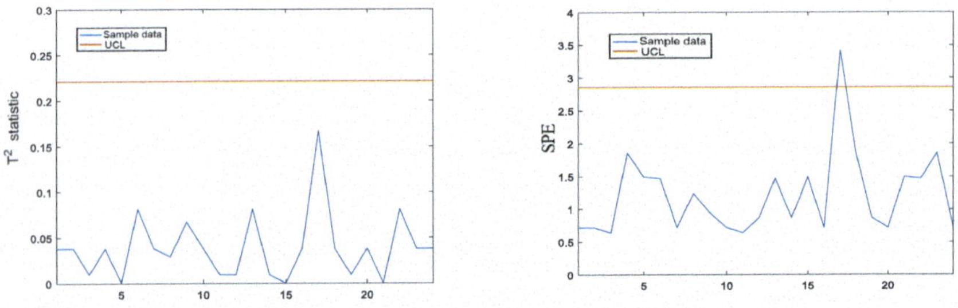

(d) T^2 and residual control charts for the PCA method

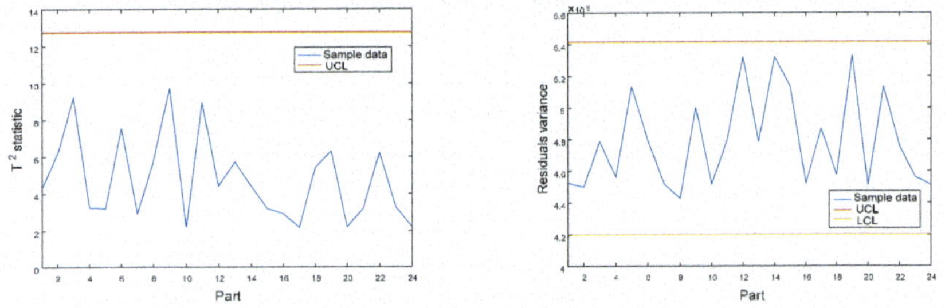

(e) T^2 and residual control charts for the SLM-based method

Fig. 4.11 Cylindrical profile control charts for the five methods in phase I

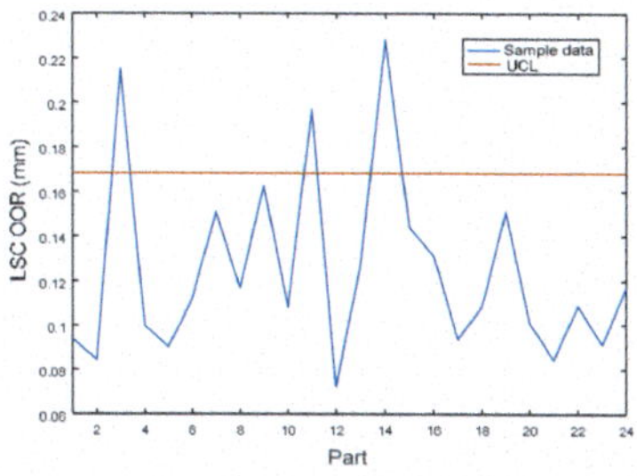

(a) Control charts of OOR method

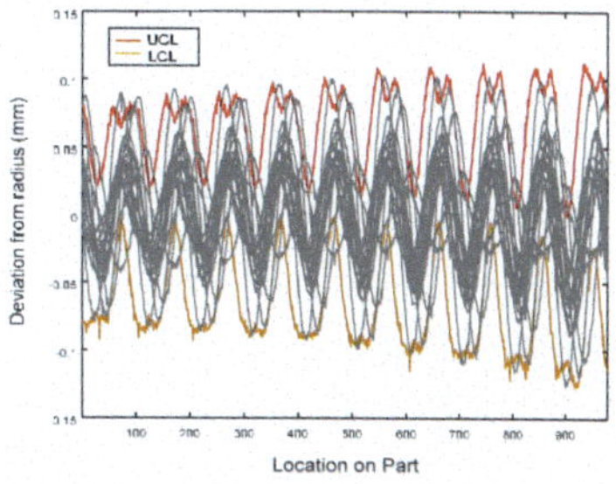

(b) Control charts of LOC method

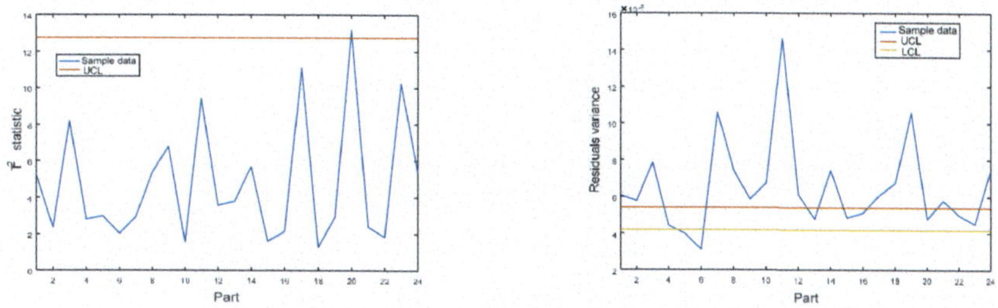

(c) T^2 and residual control charts for the SARX method

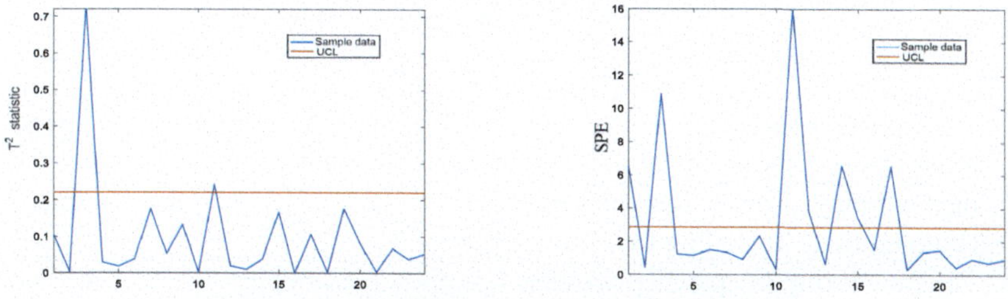

(d) T^2 and residual control charts for the PCA method

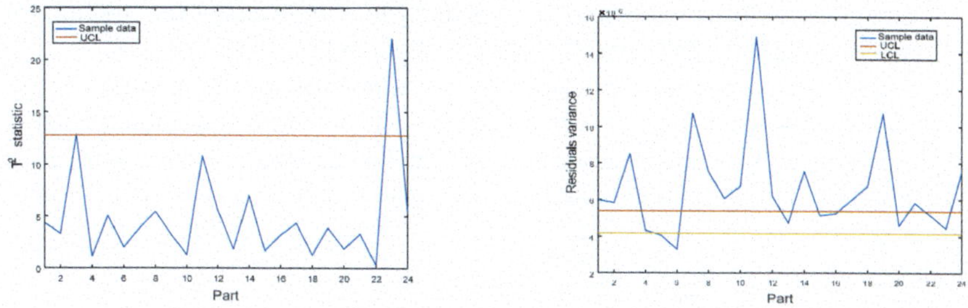

(e) T^2 and residual control charts for the SLM-based method

Fig. 4.12 Cylindrical profile control charts for the five methods in phase II

Table 4.8 Number (rate) of unqualified parts detected using the five methods on Cylindrical profile

Number of unqualified parts	OOR	LOC	SARX	PCA	SLM-based
	3	7	18	7	20
Detected (%)	12.5	29.2	75	29.2	83.3

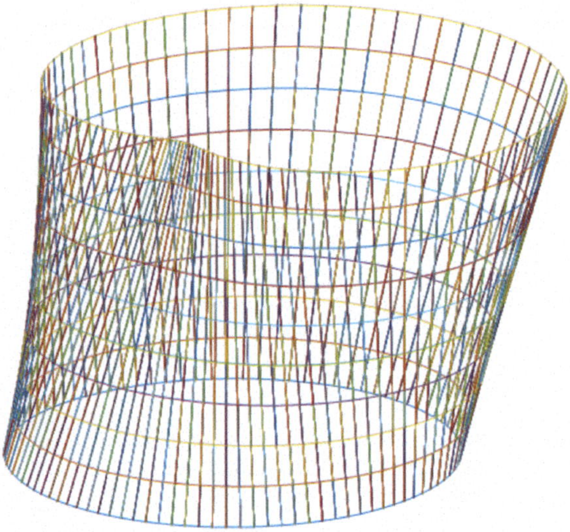

Fig. 4.13 Schematic of unqualified cylinders

(1) This section focuses on circular and cylindrical profiles as examples for examining the monitoring of geometric specifications with spatial correlations. The methodology proposed is versatile and can be applied to other geometric configurations, including free-form surfaces.
(2) The monitoring of spatially correlated profiles is a complex area that warrants further exploration. This approach offers a foundation for constructing models of spatially correlated profiles and could be integrated with other control chart techniques, such as exponentially weighted moving-average and cumulative sum control charts, to detect minor shifts in geometric specifications.
(3) The methodology can be expanded to nonparametric profile monitoring with spatial correlations. In cases where a profile is too intricate for parameterization, the parametric model form could be substituted with a nonparametric alternative.

4.2 A Novel Parallel Classification Network for Classifying Three-Dimensional Surface with Point Cloud Data

4.2.1 Introduction

With the progression of manufacturing technologies, 3D surfaces are becoming more prevalent in contemporary industrial applications. The integrity of these surfaces significantly influences the overall product quality. For instance, the interior surface of an engine cylinder head's combustion chamber can substantially impact engine performance, including the compression ratio. Surface classification stands as a highly effective method for assessing surface quality and is an essential step in the product quality control process.

Conventional quality control focuses on the critical product characteristics (KPCs) of manufactured components, including flatness [41, 42], roundness [43, 44], and surface profile [45]. These KPCs are gathered and processed by measurement devices such as calipers, CMMs, or digital cameras [46]. Yet, with the recent evolution of measurement technology, high-definition metrology (HDM) techniques, including laser scanners, stereo cameras, and structured light systems, have increasingly been implemented for data collection. HDM is capable of acquiring millions of data points within seconds, with measurement precision potentially as accurate as $\pm$ 1 μm [47]. Figure 4.14 illustrates the measurement principle of HDM and a case study of the measurement process for an engine block surface.

In contrast, conventional measurement approaches capture only specific points of interest, failing to capture the full contour of a surface. Figure 4.15 presents a comparative analysis of CMM and HDM measurement techniques applied to an engine

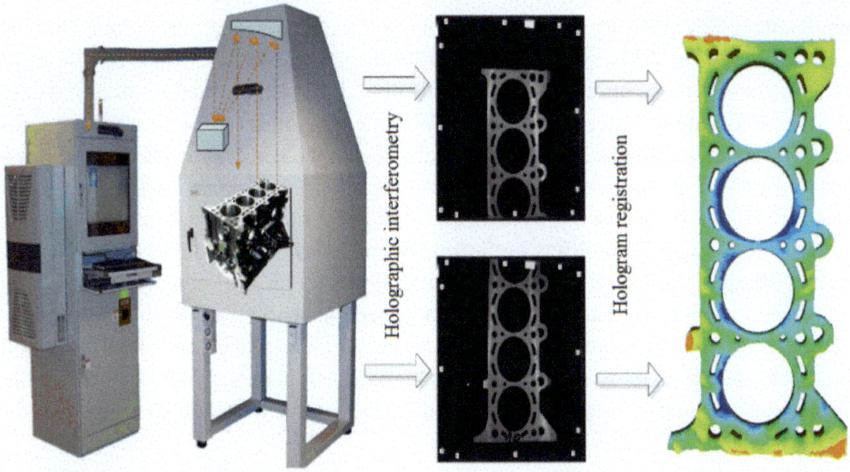

Fig. 4.14 Measurement by HDM

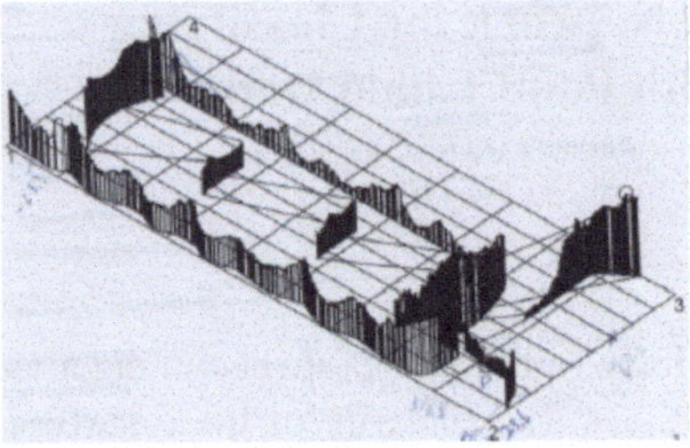

(a) CMM measurement result

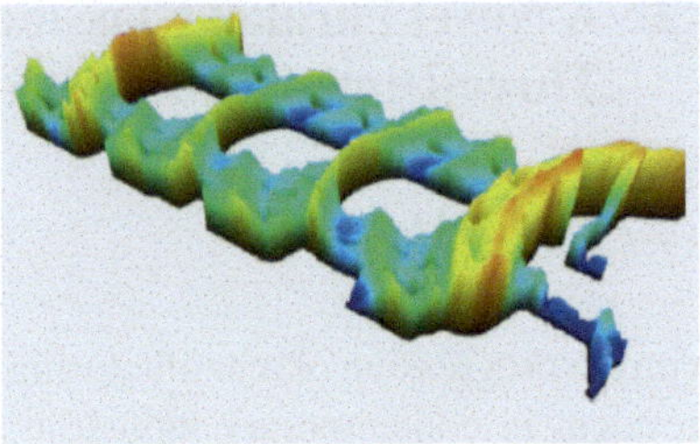

(b) HDM measurement result

Fig. 4.15 Comparison of different surface measurement techniques

block surface. HDM's point cloud data collection surpasses CMM in accurately representing the comprehensive 3D topographical details of the entire surface.

Regarding point cloud data, while KPCs are employed to identify discrepancies, they fail to leverage the complete information contained within the point cloud. This discrepancy between the data and the processing techniques has led to the issue of "abundant data but scarce information" [48]. Numerous studies have been undertaken to address profile monitoring and defect detection utilizing point cloud data. Du et al. [49, 50] introduced a suite of point cloud filtering techniques leveraging support vector machines and shearlet methods to isolate various surface components. Wang, Xi, and Du [47] transformed point cloud data into grayscale images for assessing surface form errors. However, these investigations have been limited to flat surfaces.

Alternative methodologies have been put forth for monitoring discrepancies between point cloud data and the nominal surface using control charts. Zang and Qiu [51, 52] determined the translation and rotation matrices for both measured and nominal point clouds to assess the surface quality of 3D prints. This technique is applicable to any arbitrary 3D surface with regular or sparse measurements. Stankus and Castillo-Villar [53] introduced a multivariate generalized likelihood ratio control chart for monitoring curved surfaces. Wells et al. [46] transformed point cloud data into a NURBS model and employed the EWMA control chart to monitor the NURBS model parameters. However, these approaches necessitate access to a known nominal surface or CAD model, which may not be accessible in certain scenarios. Wang and Tsung [54] and Wells et al. [55] utilized a Q-Q plot method for point cloud monitoring. The Q-Q plot simplifies high-dimensional data into linear profiles to assess the distribution similarity between data sets. Osada et al. [56] and Laga et al. [57] employed distribution histograms to characterize the shape features of 3D surfaces. Huang et al. [14] segmented a 3D surface into smaller planes and then evaluated the surface based on these segments' characteristics. While these methods are viable for 3D surface quality assessment, feature extraction for intricate 3D shapes might overlook details of local variations.

Addressing the limitations of current methods in their applicability to arbitrary 3D surfaces, a pioneering deep learning (DL) strategy is introduced to handle point cloud data and assess the quality of 3D surfaces, which are segmented into distinct sub-regions with unique attributes.

DL has recently become a potent framework for managing high-density data, with its layered architecture adept at extracting salient features from raw inputs [58]. In the realm of point cloud data, recent DL advancements have predominantly targeted object classification, employing techniques such as Volumetric-based methods [59, 60], multi-view approaches [61, 62], and PointNet [63, 64]. To our knowledge, no DL techniques have yet been developed for the quality classification of 3D surfaces. Unlike object classification, this task demands a heightened focus on the nuanced features of objects, crucial for identifying localized quality deficiencies in workpieces that might otherwise evade detection in a holistic assessment.

This novel methodology incorporates a systematic framework featuring a parallel classification network (PCN) for the quality classification of 3D surfaces. Within this framework, a 3D surface undergoes registration and is segmented into multiple sub-regions. The DL-driven PCN then extracts and processes the features of these sub-regions to determine the overall quality classification of the 3D surface.

4.2.2 The Proposed Method

4.2.2.1 Overview and Framework

A 3D surface can be perceived as an assembly of smooth, integrated sub-regions, each exhibiting distinct traits attributable to design intent or manufacturing processes, thereby influencing the overall surface quality to varying degrees. Within the proposed technique, the 3D surface is partitioned into distinct sub-regions based on point characteristics. A cutting-edge neural network, the PCN, which encompasses both a parallel neural network (PNN) and a deep neural network (DNN), is crafted to distill the features of these sub-regions and to classify the 3D surface quality.

The architecture of the proposed method is depicted in Fig. 4.16, with the process encompassing the subsequent stages:

Step 1: Segmentation of the 3D surface region. During this phase, the 3D surface is divided into multiple sub-regions via point cloud registration and segmentation techniques. Point characteristics such as normal vectors and curvature are utilized to delineate the surface into regions.

Step 2: Classification of the surface. The PCN, comprising PNN and DNN modules, is employed to extract the sub-region features. The PNN is tasked with extracting features from various sub-regions, which are then linearly integrated and processed through the DNN's fully connected layers. The output layer subsequently yields the classification outcome.

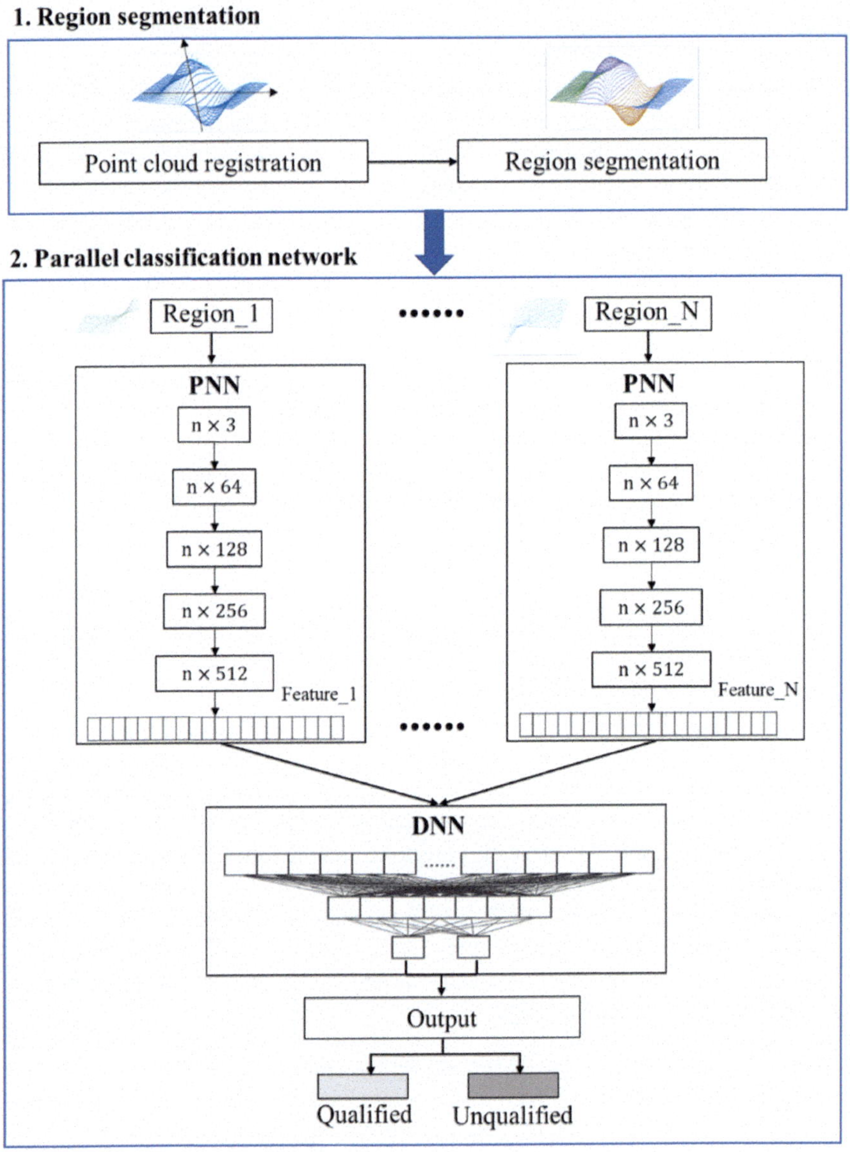

Fig. 4.16 Framework of the proposed method

4.2.2.2 Region Segmentation of 3D Surface

Surface partitioning is executed to break down a complex surface into distinct sub-regions, where points sharing similar attributes are grouped into a single region. Typically, at the boundary where two regions meet, there is a significant variation in point characteristics [65], allowing for the differentiation of points on either side of the boundary into separate regions. The segmentation process includes steps of point cloud registration and region segmentation, detailed in the subsequent two subsections.

Point Cloud Registration

Point cloud registration aligns the point clouds of various samples within a unified coordinate system through rotational and translational adjustments. During this phase, the principal component analysis (PCA) technique is deployed for the registration of point clouds across different samples.

The process begins with translating the centroid of a point cloud sample to the coordinate system's origin, resulting in the derivation of a translation matrix. Subsequently, the principal component directions for the sample are determined, with the first three principal components being aligned with the X-axis, Y-axis, and Z-axis to establish the rotation matrix. Ultimately, the point cloud samples are realigned using these translation and rotation matrices. These steps are repeated for all point cloud samples, standardizing the coordinate systems and achieving registration of point clouds from various samples.

Region Segmentation

In contemporary product manufacturing, surface design is commonly executed through computer-aided design and computer-aided manufacturing (CAD/CAM) systems. Within CAD/CAM, product development encompasses a multitude of surface combinations, each with distinct geometries. The practical processing or application of these surfaces varies, leading to a variety of quality assessment criteria, each with a unique impact on the overall quality. Segmenting the surface into coherent natural point clusters is vital for shape analysis and the classification of point clouds. Consequently, segmentation stands as a pivotal phase in the processing of point clouds.

In the context of region segmentation, the acquired point cloud data is partitioned into distinct, non-overlapping regions that share uniform characteristics, leveraging the attribute features of the points. This segmentation facilitates deep learning-based classification by minimizing the disorder among points and enabling the neural network to accurately extract features specific to each sub-region.

Common approaches to point cloud segmentation fall into three categories: edge-based, region-based, and cluster-based methods [66]. The initial two categories are characterized by hard partitioning, being unambiguous, and having a heightened sensitivity to noise. In this section, Fuzzy C-Means (FCM) clustering is employed for the segmentation of point cloud data. FCM clustering exhibits robustness against noise, as it employs an objective function to ascertain the degree of membership of each point to a specific cluster.

In the FCM algorithm, the point cloud dataset $P = \{p_1, p_2, \ldots, p_n\}$ is partitioned into c subsets, with each subset represented by a $c \times n$ matrix U. The objective function is articulated as:

$$J_m(\mathrm{U}, \mathrm{v}) = \sum_{i=1}^{n} \sum_{k=1}^{c} (\mu_{ik})^m (d_{ik})^2 \tag{4.23}$$

where μ_{ik} is the membership degree of the ith data point p_i in the kth cluster; m is the fuzzifier parameter; d_{ik} is the feature-weighted distance between the ith data point p_i and the kth cluster center v_k, encompassing spatial distance, the disparity in normal vectors, and the variance in curvature.

The optimal fuzzy partitioning is achieved by minimizing the objective function, following these procedural steps:

Step 1: Establish the cluster count c, set the parameter value m, and initialize the cluster centers V;

Step 2: Compute or refresh the membership matrix U, where $\mu_{ik} = \left[\sum_{j=1}^{c} \left(\frac{d_{ik}}{d_{ij}}\right)^{\frac{2}{m-1}}\right]^{-1}$;

Step 3: Compute or refresh the cluster centers V, with $v_i = \frac{\sum_{i=1}^{n} (\mu_{ik})^m (d_{ik})^2}{\sum_{i=1}^{n} (\mu_{ik})^m}$;

Step 4: Iterate between steps 2 and 3 until the discrepancy between the new and original centroids falls below a predetermined threshold.

In the FCM algorithm, the cluster quantity is predefined. For objects with a straightforward composition or those comprised of a limited number of curved surfaces, the number of sub-regions can be manually specified. When the composition of the curved surface is intricate and not easily discerned, the number of sub-regions is ascertained using the elbow method. This method involves graphing the explained variation against the number of clusters and identifying the curve's elbow point as the optimal cluster count [67].

4.2.2.3 Parallel Classification Network

Following region segmentation, the surface is partitioned into multiple sub-regions, with the PCN employed for feature extraction and the subsequent classification of the 3D surface.

The architecture of the classification network is depicted in Fig. 4.17, comprising two modules: PNN and DNN. The PNN facilitates feature extraction from sub-regions by converting multidimensional arrays into lower-dimensional vectors. Within the DNN, a concatenated layer aggregates the features from all sub-regions, which are then processed through fully connected layers to yield the classification outcome.

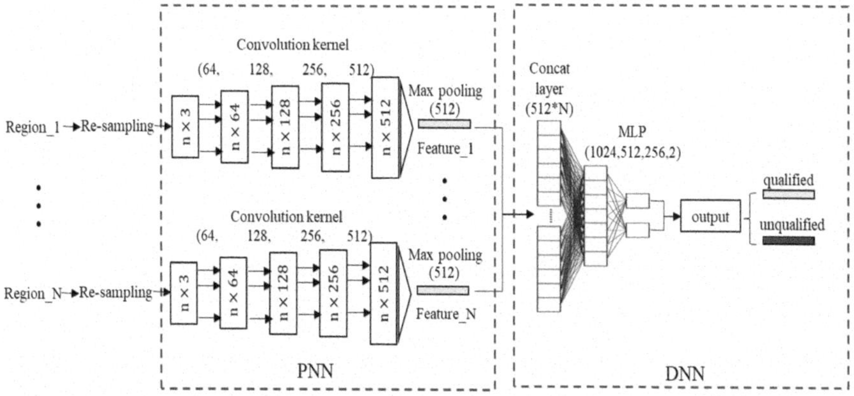

Fig. 4.17 The structure of the PCN

Table 4.9 The structure of the PNN

Parameter	L_1	L_2	L_3	L_4	Max pooling
Input	$n \times 3$	$n \times 64$	$n \times 128$	$n \times 256$	$n \times 512$
Conv	(3,64,1)	(64,128,1)	(128,256,1)	(256,512,1)	—
BatchNorm	64	128	256	512	—
Activation function	✓	✓	✓	—	—
Output	$n \times 64$	$n \times 128$	$n \times 256$	$n \times 512$	512

Feature Extraction Based on PNN

Prior to the PNN stage, resampling of the points within each sub-region is conducted to ensure uniformity in the number of input points across the same region for all samples. This standardization is crucial as neural networks process input data in batches, necessitating that data vectors be of identical dimensions. The farthest point sampling (FSP) technique is employed in this context to uniformly sample points within each region, ensuring that the count of points in equivalent regions across different samples is consistent and that the points are evenly dispersed.

The architecture of the PNN is detailed in Table 4.9, which includes four convolutional layers and a single pooling layer. Within the convolutional layers, the point cloud undergoes one-dimensional (1D) convolution with kernels. The dimensions of these kernels for each layer are specified in Table 4.9. As an illustration, in the initial layer, the input data has dimensions of n × 3, with n representing the number of points in a sub-region, each delineated by a 3D coordinate. The kernel size is 3 × 1, and with 64 kernels, the output data dimension becomes n × 64. Subsequently, the data undergoes normalization and non-linear transformation through BatchNorm and an Activation function.

Table 4.10 The structure of the DNN

Parameter	L_1	L_2	L_3	L_4
Input	$N \times 512$	1024	512	256
Linear	($N \times 512$,1024)	(1024,512)	(512,256)	(256,2)
BatchNorm	1024	512	256	—
Activation function	✓	✓	✓	—
Dropout	✓	✓	✓	—
Output	1024	512	256	2

Following the sequence of four convolutional layers, a feature vector of 512 dimensions is produced. Each point's 3D features are thus expanded to a 512-dimensional representation. A max pooling layer then isolates 1D features by identifying the peak values among the point features. Consequently, the PNN yields an output of 512 features that encapsulate the characteristics of a sub-region.

Feature Connection and Classification Based on DNN

The DNN, with its layered architecture, is capable of dimension reduction through non-linear transformations. Within the DNN module, the features derived from N sub-regions are amalgamated linearly by the concatenation layer, followed by processing through a multi-layer perceptron (MLP). The MLP consists of multiple fully connected layers that synthesize the features and convey them to the ultimate classification layer.

As depicted in Table 4.10, the DNN comprises four linear layers, each fully connected, with neurons in one layer linking to all neurons in the preceding layer. Within these linear layers, the initial number indicates the count of input neurons, while the subsequent number denotes the count of output neurons. The output data are subjected to normalization and non-linear activation, further complemented by BatchNorm and Activation functions, alongside a dropout mechanism to mitigate overfitting. The DNN's final layer is responsible for delivering the classification outcome.

4.2.3 Simulation Study

4.2.3.1 Problem Description and Analysis

In engineering practice, product surfaces are typically a combination of planar and natural quadric surfaces. For the purposes of this simulation, a 3D surface integrating cylindrical, spherical, and planar elements is constructed. As shown in Fig. 4.18,

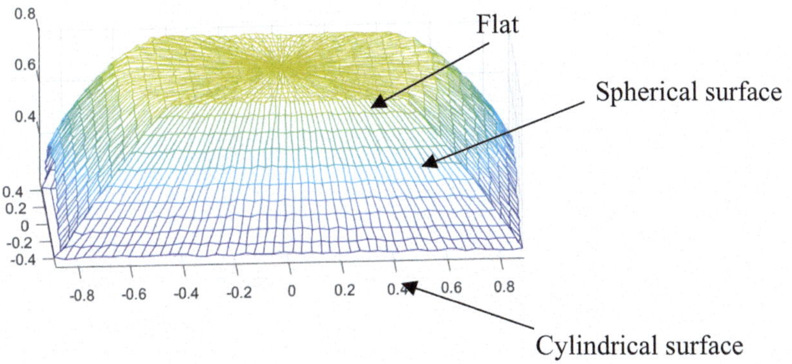

Fig. 4.18 The qualified surface $g_0(x)$

the random errors of the designed qualified surface follow a normal distribution: $\varepsilon \sim N(0, \sigma^2), \sigma^2 = 0.01$.

The four unqualified samples are designed as:

(a) Global deviation, $\sigma^2 = 0.02, 0.03, 0.05$.
(b) Deviation in the cylindrical surface, $\sigma^2 = 0.02, 0.03, 0.05$.
(c) Deviation in the spherical surface, $\sigma^2 = 0.02, 0.03, 0.05$.
(d) Deviation in the flat, $\sigma^2 = 0.02, 0.03, 0.05$.

These four deviation states are shown in Fig. 4.19.

Here are the four distinct scenarios outlined in the simulation:

1. Case 1: Training and test datasets, denoted as $g_0(x)$ and $g_1(x)$. Both datasets feature globally deviated defective samples, with 20 instances per unique σ^2 value. Consequently, each dataset comprises 80 samples.
2. Case 2: Training dataset: $g_0(x)$ and $g_1(x)$. Test set: $g_0(x)$, $g_2(x)$, $g_3(x)$ and $g_4(x)$. The training dataset's defective samples exhibit global deviation, while the test dataset's defective samples show local deviation. Each deviation case, characterized by a distinct σ^2 value, includes 20 samples. The training and test datasets consist of 80 and 200 samples, respectively.
3. Case 3: Training dataset: $g_0(x)$, $g_2(x)$, $g_3(x)$ and $g_4(x)$. Test set: $g_0(x)$ and $g_1(x)$. The training dataset's defective samples are locally deviated, while the test dataset's defective samples are globally deviated. Each deviation case, with a unique σ^2 value, contains 20 samples. The training and test datasets contain 200 and 80 samples, respectively.
4. Case 4: Training and test datasets: $g_0(x)$, $g_1(x)$, $g_2(x)$, $g_3(x)$ and $g_4(x)$. Both datasets encompass all conditions, with 260 samples each.

The simulated surface consists of five continuous sub-surfaces, with empirical evidence suggesting that segmentation into five sub-regions yields effective results without over-segmentation. To validate this claim, the simulation compares segmentation outcomes for 1–6 sub-regions. Moreover, an excessive number of sub-regions

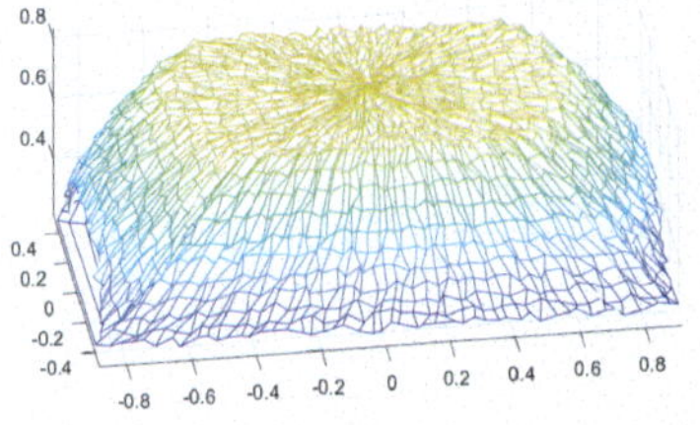

(a) Deviation in global space, $g_1(x)$

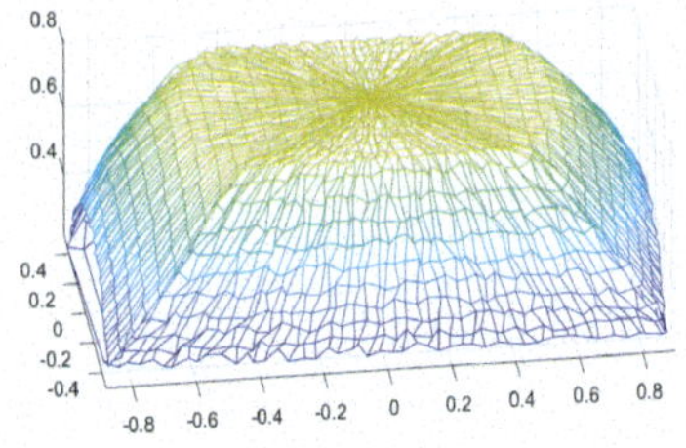

(b) Deviation in the cylindrical surface, $g_2(x)$

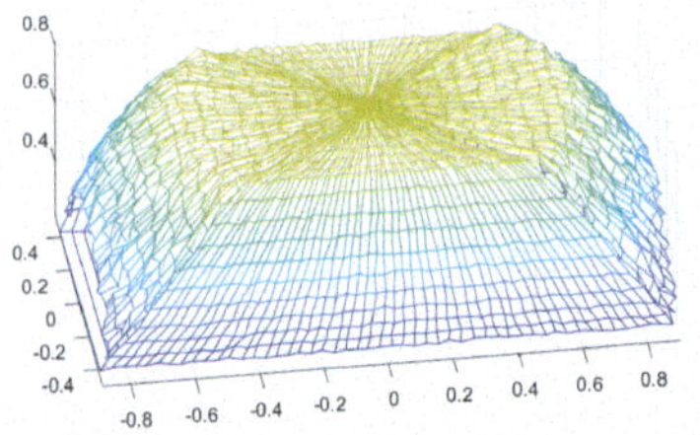

(c) Deviation in the spherical surface, $g_3(x)$

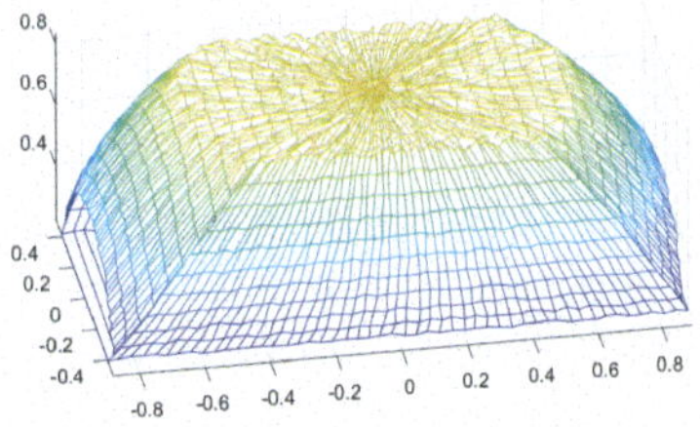

(d) Deviation in flat, $g_4(x)$

Fig. 4.19 The four deviation states

during point cloud segmentation can lead to varying segmentation outcomes for samples under high noise conditions. To mitigate such discrepancies, the cluster center matrix for the samples is archived within the training set to maintain consistency in the spatial alignment of regions across different samples. Figure 4.20 illustrates the segmentation outcomes for varying sub-region counts.

The classification network's hyperparameters play a pivotal role in shaping the learning process and the ultimate classification outcomes. An examination of their impact on classification precision and the rate at which the loss value diminishes is essential. Hyperparameter values, as established through experimental determination, are capable of delivering robust performance across diverse scenarios. Table 4.11 presents the specific hyperparameter settings employed in this section.

The classification accuracy of the PCN with 1–6 sub-regions and in various cases was determined respectively as shown in Table 4.12.

The classification outcomes are markedly influenced by the quantity of sub-regions; an increased number does not inherently outperform a reduced count. The scenario featuring five sub-regions achieves the highest accuracy in classification. Given that the simulation surface is segmented into five sub-surfaces, it is logical that the segmentation aligning with the true composition of the surface yields the best results. This finding aligns with theoretical expectations. Effective segmentation of regions enhances the precision of the classification network.

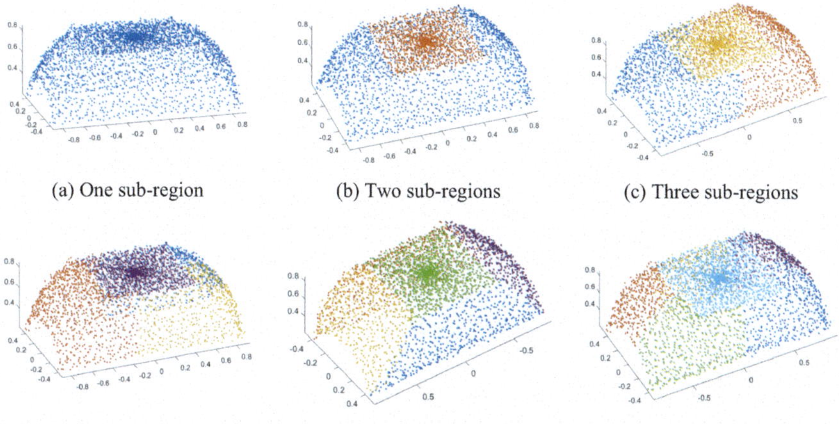

(a) One sub-region (b) Two sub-regions (c) Three sub-regions

(d) Four sub-regions (e) Five sub-regions (f) Six sub-regions

Fig. 4.20 Segmentation results under different numbers of sub-regions

Table 4.11 The values of hyperparameters in the PCN

Hyperparameter	Value
Dropout rate	0.3
Epoch	120
Learning rate	0.001
Activation function	Tanh

Table 4.12 The classification accuracy of PCN in simulation

Number of sub-regions	Case 1	Case 2	Case 3	Case 4	Average
1	0.767	0.589	0.731	0.664	0.688
2	0.875	0.701	0.845	0.762	0.796
3	0.913	0.706	0.874	0.787	0.820
4	0.924	0.712	0.866	0.801	0.826
5	0.957	0.758	0.925	0.843	0.871
6	0.936	0.720	0.872	0.819	0.837

The findings reveal that Case 2 exhibits the least classification precision. In this scenario, global deviation is used to train for partial deviation, leading to a comparatively weaker accuracy. However, when the segmentation is divided into five subregions, there is a notable enhancement in classification precision. The outcomes from Case 2 and Case 3 demonstrate the model's capacity for generalization.

Subsequent discussions will contrast the proposed approach with alternative methodologies to underscore its efficacy.

4.2.3.2 Comparison and Discussion

Research on 3D surface quality classification utilizing point cloud data is scarce, with a particular dearth of studies focusing on deep learning-based methodologies. Since PNN is predominantly employed for feature extraction, the comparison is drawn with techniques that extract features from point clouds. Current methods for feature extraction classify, assess, or monitor 3D surfaces based on the point cloud's normal vectors, curvature, and other descriptors. These descriptors capture the geometric and topological characteristics of the 3D surface's shape and are categorized into global and local features [68]. The following section details global and local feature methodologies, providing two exemplary methods for comparative analysis.

(1) Global Feature

Derived from the entire point cloud dataset, the global feature method encapsulates the geometric properties within a singular feature vector. Osada et al. [56] pioneered the quantification of 3D model geometry using a shape function, statistically computing its probabilities and employing a distribution histogram to encapsulate shape characteristics. Building on this foundation, Laga et al. [57] quantified model dissimilarities by comparing their distribution histograms through a dissimilarity measure, with the χ^2 statistics serving as descriptors for this metric.

The procedural application of this technique involves the following steps: Initially, construct the histogram of global features for each training set sample. Subsequently, conduct a least squares fitting on the histograms of all qualified training samples to derive a statistical distribution curve. For the test set, calculate the χ^2 statistics, which represent the divergence between each sample's histogram and the distribution curve. A sample is deemed unqualified if its statistic exceeds a predefined threshold.

(2) Local Feature

Local features encompass statistical data of vertices, distributions of normal, projection, and other elements pertinent to vertex feature description. Descriptors are often formed using histograms or statistics of local geometric characteristics such as normal and curvature. Huang et al. [14] partitioned a 3D surface into several smaller regions, approximating each to a plane and utilizing wavelet packet entropy and normal vectors to characterize the features of these regions. They also introduced three distinct control charts to oversee three quality parameters; a sample is flagged as unqualified if any parameter exceeds the control limit.

These methodologies are established and potent for classifying 3D surfaces. The simulation outcomes are presented in Table 4.13.

As indicated by the data in Table 4.13, the proposed approach achieves the highest classification accuracy in the majority of scenarios, surpassing the local feature method, with the global feature method ranking the least effective among the trio. The global feature method struggles with detail discrimination and often overlooks local features, leading to a failure in identifying non-conforming samples. This is particularly evident in Case 2 and Case 4, where the global feature method is nearly incapable of accurately classifying the substantial number of non-conforming samples.

Table 4.13 Comparison of classification accuracy

Method	Case 1	Case 2	Case 3	Case 4	Average
PCN	**0.957**	0.758	**0.925**	**0.843**	**0.871**
Global feature [57]	0.663	0.485	0.763	0.542	0.613
Local feature [14]	0.738	**0.775**	0.813	0.804	0.783

The local feature method segments the surface into multiple smaller areas, extracting more comprehensive information than the global feature method. However, its use of three concurrent control charts can mistakenly classify conforming samples as non-conforming. Consequently, the local feature method exhibits greater classification accuracy when the proportion of conforming products is relatively low.

In contrast, the PCN method segments the point cloud into meaningful sub-regions and manually extracts deviation-related feature information from the point cloud. This approach significantly enhances classification accuracy compared to other methods, further validating the strong generalization capabilities of the proposed method.

4.2.4 Case Study

This section delves into a case study examining the classification of the inner surface quality of an engine cylinder head's combustion chamber to evaluate the efficacy of the proposed method. The inner surface of the engine cylinder head's combustion chamber constitutes a 3D surface, the contours of which can influence the chamber's volume and, consequently, the engine's performance. Manufactured through casting, the inner surface of the combustion chamber is subject to wear due to the mold's exposure to high temperatures and prolonged contact with the casting material.

In contemporary manufacturing, the inner surface quality is assessed indirectly through mold inspection utilizing a CMM. During this process, the measurement area and point count are selected manually, and mold quality is appraised based on engineering judgment. This approach, which measures the mold at discrete intervals, is prone to overlooking issues in batches of components, potentially leading to significant economic damage.

Employing HDM technology enables the acquisition of point cloud data for the inner surface in a matter of minutes, facilitating direct measurement of this surface. Consequently, any discrepancies in the cylinder head combustion chamber's processing can be swiftly detected through analysis of the inner surface's point cloud data. This enables prompt mold adjustments to be made, thereby ensuring the maintenance of product quality.

In this experimental setup, HDM machinery is utilized to measure sixteen B12 series engines from an automotive manufacturing facility, capturing point cloud data of the cylinder head combustion chambers. Figure 4.21 illustrates the measurement

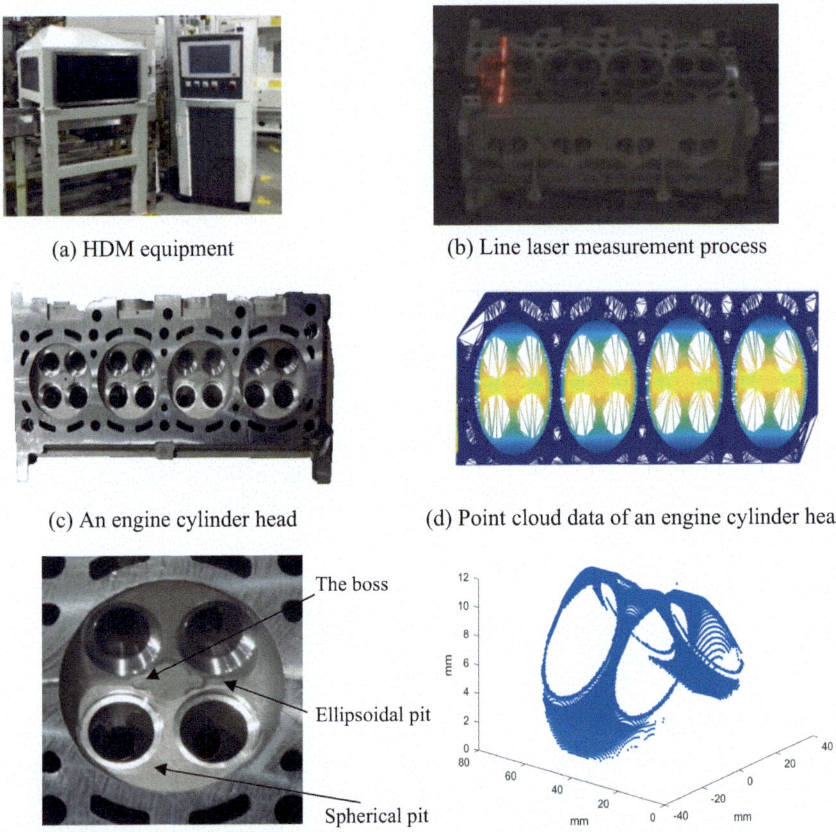

(a) HDM equipment (b) Line laser measurement process

(c) An engine cylinder head (d) Point cloud data of an engine cylinder head

(e) A cavity of engine cylinder head (f) Point cloud data of a cavity

Fig. 4.21 The measurement process and results

procedure for the cylinder heads. Figure 4.21a depicts the measurement apparatus, while Fig. 4.21b details the measurement process. The cylinder heads undergo scanning from left to right by a line laser, with each laser line comprising 1280 data points and 2000–3000 lines being recorded per cylinder head. Figure 4.21c and d present the point cloud data for an engine cylinder head, whereas Fig. 4.21e and f display the point cloud data for a single cavity.

4.2.4.1 Result Analysis

64 combustion chambers were examined using HDM machinery, with 24 of these samples identified as non-conforming due to mold wear, as determined by manual inspection. These samples were evenly allocated between the training and test datasets.

Following the point cloud measurement, each cavity contained approximately 90,000 points, indicating a rich set of sampling data. To augment the training dataset, a random selection of 70,000 points was extracted from each sample. Each point cloud was sampled twice, effectively doubling the size of the training dataset. Consequently, the training dataset comprised 64 samples, while the test dataset included 32 samples.

Taking a sample cavity as a case study, the preprocessing and segmentation procedures for combustion chambers are described. Measurement interference from ambient light and surface reflection of the workpiece can introduce outliers. Given the high-density nature of the HDM-derived point cloud, these outliers are conspicuous and can be detected with relative ease. The spatial distance from neighboring points within the point cloud is employed to remove these outliers (Huang et al. 2018). The process begins by calculating the average distance of the K-nearest neighbor points for each data point. Subsequently, the 3σ rule is applied to filter out outliers. Points with an average distance exceeding the threshold are classified as outliers and are removed. As depicted in Fig. 4.22, the points enclosed in the red circle are identified as outliers and are subject to elimination.

Analogous to the simulated surface, the cavity consists of three distinct surface types. The elbow method is employed in this context to ascertain the optimal number of sub-regions for the combustion chamber surface. As illustrated in Fig. 4.23, the curve indicates an inflection point when the sub-region count is set to five. Consequently, the number of sub-regions is established as five, leading to the segmentation outcomes depicted in Fig. 4.24.

Following the methodology outlined, each sample undergoes sampling, registration, and segmentation, after which they are subjected to training and evaluation using the PCN. Figure 4.25 delineates the training and testing workflow. The resultant accuracy for the test dataset is recorded at 0.848.

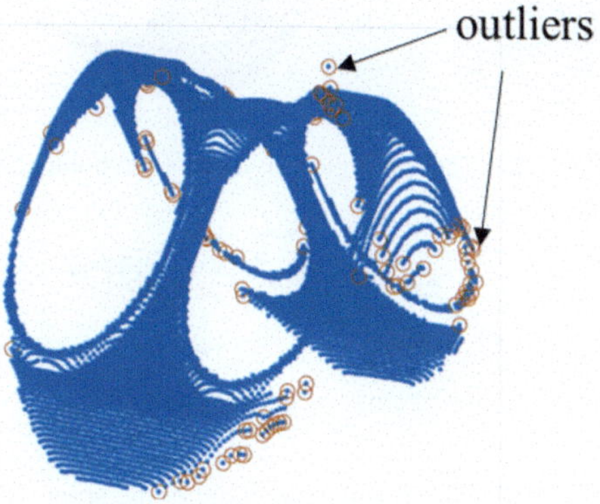

Fig. 4.22 The outliers of the point cloud

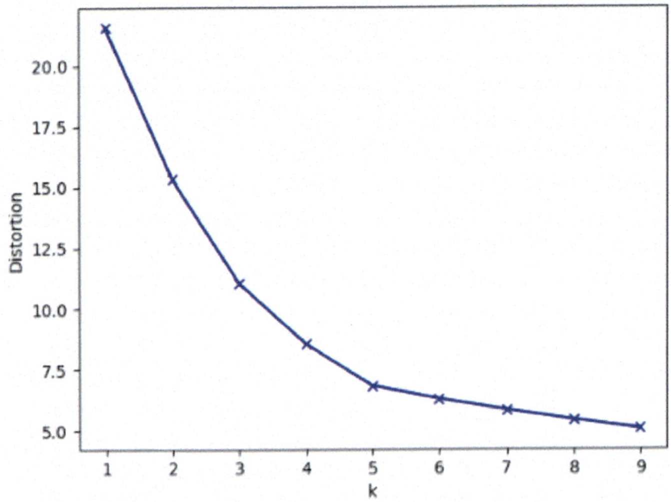

Fig. 4.23 The elbow curve of the number of sub-regions

Fig. 4.24 The sub-region division of the point cloud

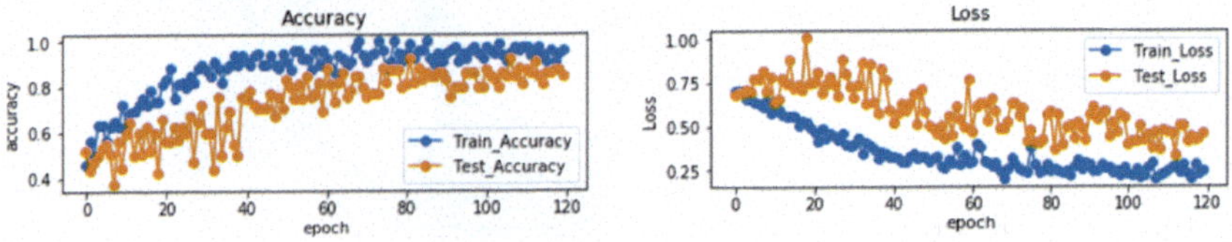

Fig. 4.25 The classification result of the proposed method

4.2.4.2 Comparison and Discussion

The identical training and test datasets are employed to evaluate the performance of the PCN in comparison with both global and local feature methodologies. The comparative outcomes are presented in Table 4.14.

Table 4.14 illustrates that the proposed approach achieves the highest classification accuracy, with the local feature method ranking second. The global feature method records the lowest accuracy. These findings align with the simulation outcomes. The proposed method demonstrates a superiority of 6.7% over the global feature method and 22.3% over the local feature method in terms of accuracy. The proposed method maintains robust performance for the classification of intricate cylinder head surfaces.

4.2.5 Conclusion

The PCN method, introduced in this section, leverages deep learning for the categorization of 3D surface quality from point cloud data. Pioneering in its application of deep learning to point cloud-based surface quality assessment, this approach encompasses both regional segmentation and quality evaluation. Regional segmentation clusters points with analogous attributes within the same region, while the quality classification employs an innovative network, PCN, designed to directly mine the features from point cloud datasets and classify 3D surfaces.

The method outperforms two prevalent point cloud feature extraction techniques, demonstrating superior outcomes in both simulated and real-world scenarios. In the simulations, various training and testing datasets were assembled for evaluation. The findings indicate that effective point cloud segmentation significantly enhances the neural network's feature extraction capabilities. Comparative analysis reveals that PCN achieves the highest classification accuracy in the majority of instances, highlighting its robust generalization capabilities. In the real-world application, the engine cylinder head combustion chamber's internal surface quality was classified, and the proposed method demonstrated a notably higher classification accuracy compared to alternative approaches. Collectively, these findings underscore the enhanced classification efficacy achieved through the segmentation and feature extraction of subregions over global information processing. Furthermore, the automatic extraction of sub-region features by neural networks surpasses manually defined local features in terms of 3D surface quality classification.

Table 4.14 The classification results of the methods

Method	Accuracy
PCN	**0.848**
Global feature (Laga et al. 2019)	0.625
Local feature (Huang et al. 2018)	0.781

The findings suggest that the proposed method is adept at classifying 3D surfaces utilizing point cloud data. The potential for acquiring substantial datasets in manufacturing environments could further refine classification outcomes. Several avenues for future investigation are identified:

(1) This section utilizes solely offline data for surface quality classification. Enhancing the algorithm's precision and reliability for online detection is essential. Integrating with the swift measurements of HDM devices, this approach could facilitate rapid detection of production condition changes that affect surface quality, enabling real-time 3D surface quality classification.
(2) Harnessing 3D point cloud data presents a considerable research challenge. This method offers a framework for the extraction of detailed information from high-density point cloud data and can be integrated with other point cloud technologies to achieve precise and expeditious processing of point cloud data.

4.3 Geometrical Deviation Modeling and Monitoring of 3D Surface Based on Multi-output Gaussian Process

4.3.1 Introduction

Deviations in geometry, defined as the discrepancies between a measured element and its theoretical counterpart, significantly influence the precision and quality of components, including attributes such as straightness, flatness, circularity, and cylindricity. As advancements in processing and measurement technologies continue, the quality of 3D surface processing has emerged as a critical area of study. These geometrical deviations are key targets for the control of 3D surfaces and have a substantial impact on the sealing integrity, reliability, and lifespan of products [69].

Assessing the quality of 3D surfaces requires initial acquisition of measurement data. The prevalent technique involves the section curve approach derived from three-coordinate surface measurements [70]. Nonetheless, this technique suffers from low efficiency, and the restricted section curves fail to comprehensively represent the entire surface's processing quality, potentially overlooking local defects. As non-contact measurement technology progresses, HDM technology is capable of gathering vast datasets that capture the complete geometrical information of manufactured surfaces. Consequently, the inference, reconstruction, and monitoring of 3D surfaces from extensive point cloud data have emerged as innovative areas of research.

In numerous applications, the assessment of process or product quality is defined by a functional link between the response and explanatory variables. The constancy of these connections over time is scrutinized and observed through the lens of SPC. Woodall [71] encapsulated these techniques under the umbrella of profile monitoring. The genesis of profile monitoring studies was anchored in a straightforward linear regression model. To capture more intricate patterns, Kazemzadeh et al. [10] and Zou et al. [72] introduced polynomial and multiple regression models, respectively.

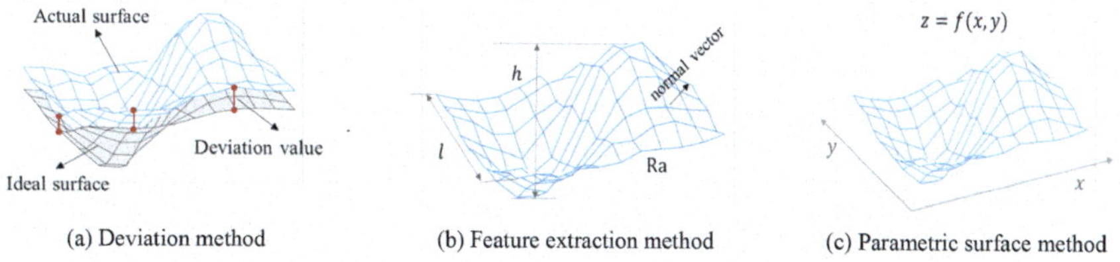

Fig. 4.26 Schematic diagram of geometric deviations evaluation of three methods

Colosimo et al. [29] ventured into a SARX model to analyze circular profiles. Zhao et al. [44] expanded this model's application to cylindrical surfaces. These monitoring endeavors can be broadened to encompass the quality assessment of various surface attributes, including roundness, flatness, and cylindricity.

The majority of profile monitoring studies have concentrated on 2D profiles that are captured by regression models. Wells et al. [46] proposed that these monitoring endeavors could be expanded into 3D realms with the support of point cloud data. Yet, the inherent complexity of 3D surfaces means that profiles are not readily definable by a single equation. Consequently, in contrast to the assessment methods for 3D surface textures, research into the geometrical deviations of 3D surfaces remains relatively constrained.

The methodologies for modeling and monitoring 3D surfaces, as depicted in Fig. 4.26, can be categorized into three distinct approaches: the deviation approach, the feature-based extraction technique, and the parametric surface methodology.

(i) Deviation method. This technique assesses 3D surface accuracy by determining the discrepancy between actual measurements and the intended surface geometry. For a precise deviation assessment, it is imperative to align the measured surface accurately with the design surface to ensure congruence. Commonly, the least squares or minimum area methods are employed to derive the rigid body transformation matrix necessary for this alignment [73]. Zang and Qiu [51, 52] determined the translational and rotational matrices of the measured data relative to the nominal point cloud for 3D printing surface quality monitoring. This approach is viable for 3D surfaces with both dense and sparse measurement points. Stankus and Castillo-Villar [53] segmented the point cloud into regions of interest (ROIs) and computed the average deviation for each ROI. Mehrad et al. [74] and Wells et al. [46] utilized B-spline for 3D surface reconstruction, with the surface's geometric deviations indicated by the positions of the spline's control points. The deviation approach aims to achieve optimal alignment between measured and theoretical surfaces, with the deviation's magnitude hinging on the precision of the registration process. This computation is applicable to surfaces with a known CAD model or a set of theoretical data points.

(ii) The Feature Extraction Technique. Quality of a surface can, in certain instances, be articulated through pertinent macro or micro-geometrical indicators such as length, height, roughness, and normal vectors [75]. Yet, these attributes are acutely susceptible to local imperfections and fail to capture the qualitative and quantitative variations in the contour of the measured surface in relation to the theoretical surface. Addressing this limitation, Kase et al. [76] introduced a dual approach for assessing the geometric deviation of 3D surfaces, encompassing both local and global evaluations. The local assessment quantifies alterations in principal curvatures between the CAD model and the actual surface, while the global assessment leverages cumulative normal vectors to describe surface segments. Wang and Tsung [54] along with Wells et al. [55] employed the quantile–quantile (Q-Q) plot technique for 3D surface monitoring. This method translates high-dimensional data into linear profiles to ascertain the congruence of data distributions. Osada et al. [77] and Laga et al. [78] utilized distribution histograms to depict the morphological traits of 3D surfaces. However, the transition from point clouds to distribution maps may result in the loss of details regarding local variations on intricately shaped 3D surfaces. Huang et al. [14] segmented a 3D surface into planar sub-regions and assessed surface quality by extracting the normal vectors and wavelet packet entropy from these segments. In summary, the feature extraction method does not precisely mirror surface morphology, impeding a comprehensive characterization of geometric deviations.

(iii) The Parametric Surface Approach. This technique constructs a regression model to capture the contours of a 3D surface. Colosimo et al. [27] employed a Gaussian process (GP) to represent a cylindrical surface, assessing surface quality through deviation values at control points. Zhao et al. [79] utilized height values as the dependent variable for the x and y coordinates of the 3D surface, employing Gaussian process regression (GPR) for modeling. The parametric surface approach develops a regression model for each measured surface, with 3D surface geometrical deviations monitored via the model's parameters. The benefit of surface parameterization is that it circumvents the need for precise surface alignment, captures the full spectrum of surface shape information, and offers a valuable reconstruction technique for reverse engineering endeavors. Many contemporary parametric techniques treat surface data as a unique profile with a single response variable and a pair of explanatory variables. However, this presumption is restrictive and does not account for the three-dimensional freedom present in actual measurements.

To address the constraints inherent in parametric modeling, del Castillo et al. [80] employed manifold learning to project 3D surfaces onto a 2D plane and introduced a Geodesic Gaussian process (GGP) to capture the relationship between the projected surfaces and their three-dimensional counterparts. This technique reconstructs the underlying surface through 2D manifold learning, providing a more accurate reflection of the spatial correlation between points on the surface. A comparable methodology is the Gaussian process latent variable model (GPLVM), proposed by Lawrence [81]. The GPLVM incorporates an additional set of latent variables, characterizing the data through a nonlinear dimensionality reduction framework. The underlying premise of these strategies is that high-dimensional datasets may inherently reside on a low-dimensional, nonlinear manifold that is the target of the modeling effort [82].

Conventional techniques for dimensionality reduction aim to condense high-dimensional datasets into lower-dimensional representations. In the context of 3D surfaces, frequently utilized reduction methods like GPLVM, Locally Linear Embedding (LLE), and isometric mapping are all planar parameterization techniques, and their primary limitation lies in the inability to be directly applied to closed surface models.

For manufacturing surfaces, geometric representations of surfaces are typically characterized by closed, genus-0 forms, namely, spheres that have been distorted. For these models, the sphere serves as the most intuitive domain for parametrization. Consequently, with respect to closed surfaces, the sphere can be seen as the foundational surface, and spherical mapping is utilized to dimensionally reduce the three-dimensional surface. This approach ensures that the points within the lower-dimensional space maintain a rational and consistent structure, thereby facilitating an efficient mapping outcome for closed surfaces.

Additionally, according to Sun et al. [83], the measurements derived from machined surfaces through non-contact methods adhere to a Gaussian distribution and exhibit significant spatial correlations. Accounting for this spatial correlation, the connection between the latent spatial data and the actual surface can be effectively modeled utilizing a Gaussian process.

The objective of this research is to investigate a regression model suitable for closed surfaces. Leveraging spherical mapping and multi-output Gaussian process (MOGP), a novel approach termed spherical multi-output Gaussian process (S-MOGP) has been developed for 3D surface modeling. Concurrently, to address the extensive structure inherent in point cloud data, a compactly supported covariance matrix function is engineered to alleviate computational demands. Furthermore, the monitoring of geometrical deviations is achieved by assessing the discrepancies between the actual measured surface and the surface's predicted values.

The merits of the proposed approach include: (i) the 3D surface model is constructed via spherical mapping, utilizing angular distance in place of Euclidean and geodesic distances, making it apt for closed and non-extendable surfaces. (ii) A regression model grounded in MOGP is formulated to capture the 3D surface's form, with a holistic representation of the x, y, and z axes, while accounting for the spatial correlations among measurement points. (iii) This technique is applicable for both modeling and monitoring point clouds within large-scale datasets.

4.3.2 Related Prior Work

4.3.2.1 Spherical Mapping

Surface parameterization facilitates a one-to-one correspondence between a surface and a designated parameter domain, effectively transferring one surface to another.

This parameterization is articulated through a vector-valued function $\phi: \Omega \rightarrow S$, which projects the 3D surface $\Omega \subset \mathbb{R}^3$ onto the 2D parameter domain $S \subset \mathbb{R}^2$.

The process of transferring a surface to the parameter domain entails reducing specific forms of distortion. Length distortion, angle distortion, and area distortion are predominant, aligning with Isometric, Conformal, and Equiareal mapping techniques, respectively. In practical applications, surface parameterization takes into account the topology of the data produced, seeking to minimize one form of distortion or a blend of various distortions [84].

A 3D geometric model is often depicted as a sphere that has been altered; hence, the sphere emerges as the most intuitive parameterized domain for 3D surfaces [86]. For a given surface M, spherical mapping is executed by assigning each vertex a parameterization, establishing a continuous, reversible mapping from the surface to the sphere, denoted as $\phi: M \rightarrow S$. The parametric approach to spherical surfaces remains consistent across translations, rotations, and scaling, making it well-suited for the portrayal of point clouds [87].

Point cloud measurements are stored within the Cartesian coordinate system of three-dimensional space, with each point defined by a triplet of coordinates (x, y, z). When discrepancies exist between the measurement coordinate systems, the alignment of the measured object is achieved through rotational and translational adjustments (as referenced in the PCA-based coarse registration by Yao et al. [88]). Within the spherical coordinate framework, each point is characterized by a pair of angular values (u, v), with the elevation $u \in [-\pi/2, \pi/2]$ and the azimuth $v \in [-\pi, \pi]$. The spherical coordinates corresponding to (x, y, z) are denoted as

$$r(u, v) = (\mathrm{R}\cos u \cos v, \mathrm{R}\cos u \sin v, \mathrm{R}\sin u) \tag{4.24}$$

Consequently, the coordinates (x, y, z) of a 3D surface can be depicted using the duo of parameters (u, v).

In the case of spherical surfaces, the process begins with spherical fitting to ascertain the sphere's approximate center, followed by translating the sphere's center to the coordinate system's origin. Figure 4.27 illustrates the concept of spherical mapping.

In the realm of spherical parameterization for intricate surfaces, various scholars have put forth distinct approaches to map 3D surfaces onto a spherical domain. These methods prioritize establishing a one-to-one correspondence between the points of the original surface and the spherical parameter domain, eschewing the equal area or equal arc length mapping approaches. As a result, the conformal method is often employed for the parameterization of the spherical surface.

Employing the conformal parameterization technique, Shen et al. [84] and Choi et al. [89] executed spherical mapping by resolving a sequence of Laplace equations.

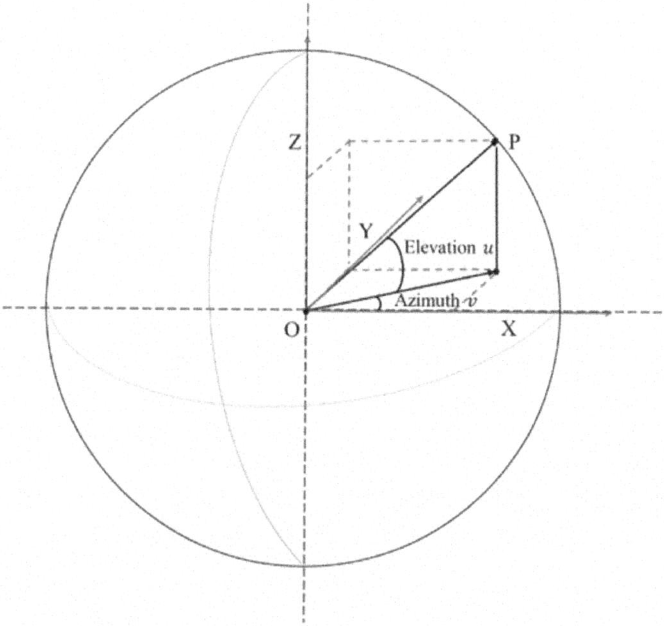

Fig. 4.27 Spherical domain

4.3.2.2 Gaussian Process Regression

GP is a stochastic process consisting of an infinite number of random variables, each following a Gaussian distribution over a continuous domain. GPR is a non-parametric approach that leverages the GP as a prior for conducting regression analyses on datasets. From a functional perspective, GPR aims to identify a subset of functions from an extensive collection that aligns with the observed data. Utilizing prior knowledge, the Bayesian framework and the properties of the Gaussian distribution are employed to determine the distribution of this function set. This distribution is then harnessed for the prediction of forthcoming data.

Consider a dataset comprising N paired observations: $D = \{(x_n, y_n)|n = 1, \ldots, N\}$, where x and y represent the input and output vectors, respectively. Within the framework of GPR, the relationship is expressed as $\boldsymbol{y} = f(\boldsymbol{x}) + \boldsymbol{\varepsilon}$, with $\boldsymbol{\varepsilon}$ following a normal distribution $N\left(0, \sigma_n^2\right)$, σ_n^2 denoting the variance of the noise. The function $f(\boldsymbol{x})$ represents the underlying latent function, indicating that any finite set of these latent variables is multivariate Gaussian-distributed, denoted as

$$f(\boldsymbol{x}) \sim GP\left(m(\boldsymbol{x}), k\left(\boldsymbol{x}, \boldsymbol{x}'\right)\right) \tag{4.25}$$

where $m(\boldsymbol{x})$ serves as the mean function, often assumed to be zero for simplicity. $k(\boldsymbol{x}, \boldsymbol{x}')$ is the covariance function that characterizes the relationship between the variables.

In training observations $\{x_i, y_i\}_{i=1}^n$, the input data is $\boldsymbol{x} = \{x_1, \ldots, x_n\}^T \in \mathbb{R}^{1\times n}$, and the output data is $\boldsymbol{y}$. The joint prior distribution of the test points $\boldsymbol{x}_*$ and the predictive targets $\boldsymbol{f}_*$ is

$$\begin{bmatrix} \boldsymbol{y} \\ \boldsymbol{f}_* \end{bmatrix} \sim \mathrm{N}\left(0, \begin{bmatrix} K(\mathrm{X}, \mathrm{X}) + \sigma_n^2 I & K(\mathrm{X}, \mathrm{X}_*) \\ K(\mathrm{X}_*, \mathrm{X}) & K(\mathrm{X}_*, \mathrm{X}_*) \end{bmatrix}\right) \tag{4.26}$$

where K(X, X) is the covariance matrix with the element $\mathrm{K}_{ij} = k(x_i, x_j)$. X_* is the matrix of test inputs, and X is the matrix of the training inputs.

The covariance matrix comprises elements that represent the correlation between each pair of x values. This matrix is symmetric and positive semi-definite (PSD), and its computation is facilitated by the kernel function.

After observing the training sample, the posterior distribution of the test output function is

$$\boldsymbol{f}_* | \mathrm{X}, \boldsymbol{y}, \mathrm{X}_* \sim \mathrm{N}(\overline{\boldsymbol{f}}_*, \mathrm{cov}(\boldsymbol{f}_*)) \tag{4.27}$$

where

$$\overline{\boldsymbol{f}}_* = \mathrm{K}(\mathrm{X}_*, \mathrm{X})\left[\mathrm{K}(\mathrm{X}, \mathrm{X}) + \sigma_n^2 I\right]^{-1} \boldsymbol{y} \tag{4.28}$$

$$\mathrm{cov}(\boldsymbol{f}_*) = \mathrm{K}(\mathrm{X}_*, \mathrm{X}_*) - \mathrm{K}(\mathrm{X}_*, \mathrm{X})\left[\mathrm{K}(\mathrm{X}, \mathrm{X}) + \sigma_n^2 I\right]^{-1} \mathrm{K}(\mathrm{X}, \mathrm{X}_*) \tag{4.29}$$

In the context of multivariate GP, the aforementioned techniques remain applicable. The distinction lies in the fact that the kernel function in multivariate GP captures the correlation between two vectors.

The essence of the GP model is its covariance function, also known as the kernel function, which dictates the covariance of the function values. Different covariance functions yield distinct regression outcomes in GP models, yet all must adhere to the semi-definite form. Typical kernel functions employed include the Gaussian kernel, Matern kernel, squared exponential kernel, and periodic kernel, among others.

4.3.2.3 Multi-output Gaussian Process Regression

In the realm of multi-output regression, traditional GPR is limited to modeling individual outputs and fails to account for inter-output correlations. Several researchers have explored leveraging these correlations to develop a MOGP model, thereby enhancing predictive accuracy [85]. Below is a succinct overview of MOGP.

For training points $X = \{x_1, \ldots, x_n\}^T$, there are D-dimensional output $\mathrm{Y} = (y_1, \ldots, y_D)^T$. Function $f(\boldsymbol{x})$ are assumed to follow GP as $f(\boldsymbol{x}) \sim GP(0, \mathrm{K}(\boldsymbol{x}, \boldsymbol{x}'))$, and the covariance function is

$$\mathrm{K}(\boldsymbol{x},\boldsymbol{x}') = \begin{bmatrix} k_{11}(\boldsymbol{x},\boldsymbol{x}') \cdots k_{1\mathrm{D}}(\boldsymbol{x},\boldsymbol{x}') \\ \vdots \ddots \vdots \\ k_{\mathrm{D}1}(\boldsymbol{x},\boldsymbol{x}') \cdots k_{\mathrm{DD}}(\boldsymbol{x},\boldsymbol{x}') \end{bmatrix} \in \mathbb{R}^{D\times D} \tag{4.30}$$

The joint distribution of multiple outputs can be written as

$$\begin{bmatrix} Y \\ \boldsymbol{f}_* \end{bmatrix} \sim \mathrm{MN}\left(0, \begin{bmatrix} \mathrm{K(X, X)} + \sigma_n^2 I & \mathrm{K(X, X_*)} \\ \mathrm{K(X_*, X)} & \mathrm{K(X_*, X_*)} \end{bmatrix}\right) \tag{4.31}$$

where $K(\mathrm{X}, \mathrm{X}) \in \mathbb{R}^{nD\times nD}$ is the covariance matrix.

Given the training set X and the output observations Y, the posterior distribution of the tested output function is: $\boldsymbol{f}_*|\mathrm{X}, \mathrm{Y}, \mathbf{x}_* \sim \mathcal{N}(\overline{\boldsymbol{f}}_*, \mathrm{cov}(\boldsymbol{f}_*))$. The prediction mean and prediction variance are given as

$$\overline{\boldsymbol{f}}_* = \mathrm{K(X_*, X)}\left[\mathrm{K(X, X)} + \sigma_n^2 I\right]^{-1}\mathrm{Y} \tag{4.32}$$

$$\mathrm{cov}(\boldsymbol{f}_*) = \mathrm{K(X_*, X_*)} - \mathrm{K(X_*, X)}\left[\mathrm{K(X, X)} + \sigma_n^2 I\right]^{-1}\mathrm{K(X, X_*)} \tag{4.33}$$

4.3.3 The Proposed Method

4.3.3.1 Model Description

The S-MOGP approach, which integrates spherical mapping with MOGP, has been developed for 3D surface modeling. This methodology commences with the parameterization of the 3D surface utilizing spherical mapping. Subsequently, the surface's geometry is represented through MOGP. Ultimately, the T^2 and Q statistical measures are employed to track the geometrical deviations of the 3D surface. A schematic representation of the key procedural steps is depicted in Fig. 4.28.

The proposed methodology encompasses three distinct stages:

Step 1: Spherical Parameterization. Initially, the coordinates (m_x, m_y, m_z) of the surface measurement points are aligned. Subsequently, the surface is parameterized through spherical mapping, with the coordinate values expressed as two parameters: elevation (u) and azimuth (v). The center of the sphere used for fitting is set at the origin.

Step 2: MOGP Model Estimation. It is posited that the parametric surface $S(u, v)$ on $\mathbb{R}^2$ follows a Gaussian distribution, with the output function being a multivariate Gaussian process. The joint distribution of the training observations and

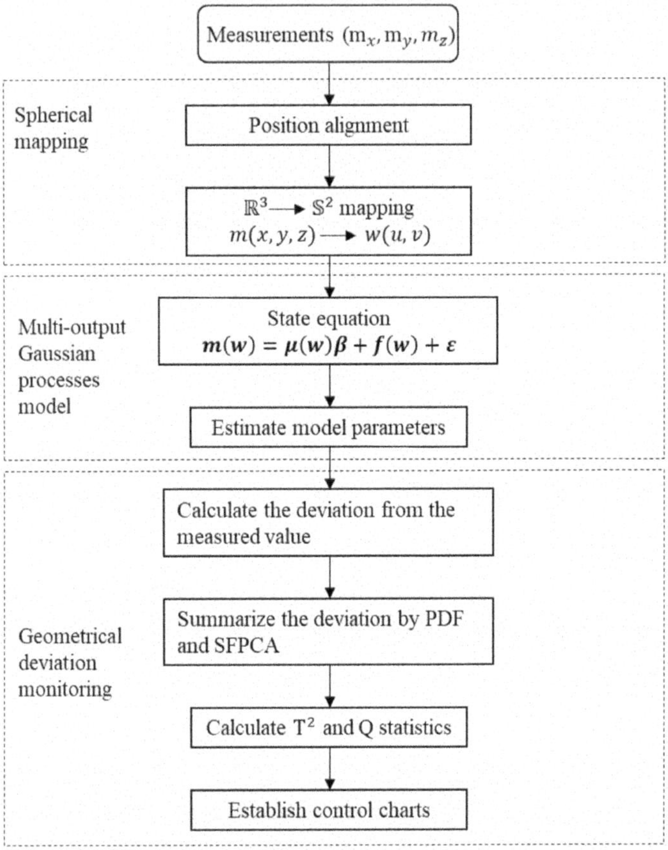

Fig. 4.28 Flowchart of the proposed method

the prediction targets is derived using Bayesian inference. Utilizing MOGP, the 3D surface model is formulated, and the model parameters are determined.

Step 3: Geometric Deviation Surveillance. A set of deviation values is obtained by calculating the distances from the measured values. Thereafter, the probability density function (PDF) and simplicial functional principal component analysis (SFPCA) are applied to encapsulate the deviation characteristics. T^2 and Q control charts are constructed to monitor the geometric deviations.

4.3.3.2 Multi-output Gaussian Processes Model

For each measured coordinate within the vector $m = \left(m_x, m_y, m_z\right) \subset \mathbb{R}^3$, spherical mapping establishes a transformation $f : \mathbb{R}^3 \rightarrow \mathbb{S}^2$. Consequently, a surface situated in three-dimensional space can be parameterized by a pair of parameters (u, v). The separation between any two points is not calculated using Euclidean distance but is instead represented by angular measurements.

Parameterize the measuring point (m_x, m_y, m_z) on the spherical surface. Each point is represented by two parameters:

$$\mathrm{m}(w) = \begin{cases} m_x = x(u, v) \\ m_y = y(u, v) \\ m_z = z(u, v) \end{cases}, \quad w = (u, v) \subset \mathbb{S}^2 \tag{4.34}$$

Construct the parameter equation for the surface: the input set is $w = \{u_i, v_i\}_{i=1}^{n}$, which means that each measurement point is associated with a parameter value via spherical mapping; the output set consists of the three-dimensional coordinates of the measured values $m = \left\{m_{x_i}, m_{y_i}, m_{z_i}\right\}_{i=1}^{n}$, with n representing the total number of measurement points. The model proposed is comprised of three components: the mean (fixed effect), deviation (correlation effect), and noise (random effect), expressed as

$$\boldsymbol{m}(\boldsymbol{w}) = \boldsymbol{\mu}(\boldsymbol{w})^{\mathrm{T}}\boldsymbol{\beta} + \boldsymbol{f}(\boldsymbol{w}) + \boldsymbol{\varepsilon} \tag{4.35}$$

where $\boldsymbol{\mu}(\boldsymbol{w}) = \mathrm{diag}\left(\mu_x(\boldsymbol{w}), \mu_y(\boldsymbol{w}), \mu_z(\boldsymbol{w})\right)$ is a block diagonal matrix with the blocks being $\left\{\mu_x = \cos u \cos v, \mu_y = \cos u \sin v, \mu_z = \sin u\right\}$, $\boldsymbol{\beta} = [\beta_1, \beta_2, \beta_3]^{\mathrm{T}}$ is the coefficient vector, $\boldsymbol{f}(\boldsymbol{w})$ is a multivariate random field presumed to adhere to MOGP, and $\boldsymbol{\varepsilon}$ conforms to a normal distribution $N\left(0, \sigma^2 \boldsymbol{I}\right)$.

To address the model, the covariance structure must be examined. Firstly, not every kernel function is appropriate for spherical applications. A process on the sphere is considered stationary if its covariance function is a function solely of the spherical angle. Huang et al. [90] assessed the suitability of covariance functions on the sphere, based on spherical angle distances, across various models. Secondly, the processing of extensive spatial datasets presents significant computational hurdles. A compactly supported function, which is zero beyond a certain range, can generate a sparse covariance matrix, thus simplifying the computation for large datasets.

Consequently, a non-separable covariance structure is engineered by integrating the spherical covariance function with a compactly supported function.

Considering two points $w_1(u_1, v_1)$ and $w_2(u_2, v_2)$ on $\mathbb{S}^d$, the spatial separation is defined by the spherical distance along the great circle of $\mathbb{S}^d$ that connects w_1 and w_2; specifically, within our model where $w = (u, v)$, implying $d = 2$,, the distance is denoted as

$$\theta(w_1, w_2) = \arccos\{\sin u_1 \sin u_2 + \cos u_1 \cos u_2 \cos(v_1 - v_2)\} \tag{4.36}$$

Subsequently, a covariance model with compact support is considered, and its efficacy on the sphere is evaluated. A non-separable covariance model with compact support can be formulated through the convolution method as detailed in Du et al. [91] Theorem 3,

$$C_{ij}(x) = \begin{cases} \int_0^1 (u - \|x\|)_+^{\nu} g_{ij}(u)du, & \|x\| \le 1, \\ 0, & \|x\| > 1, x \in \mathbb{R}^d, \end{cases} \quad i,j = 1, \ldots, m \tag{4.37}$$

where $g(x)$ serves as the indicator function of a compact subset in $\mathbb{R}^d$, and the function $g_{ij}(x)$ is continuous across the interval [0,1].

For dimensions where $d \ge 2$, Du et al. [91] present an m-variate elliptically contoured random field with both direct and cross covariances in Theorem 4, and subsequently, Ma et al. [92] validated the covariance's applicability on the spherical $\mathbb{S}^d$. In the context of our model, with $d = 2$, the direct and cross covariances are derived as

$$C_{ii}(\theta) = \begin{cases} 2\alpha_i^2 \arccos(\tau_i(\theta)) - 2\alpha_i^2 \tau_i(\theta)\left(1 - \tau_i^2(\theta)\right)^{\frac{1}{2}}, & \theta \le 2\alpha_i \\ 0, & \theta > 2\alpha_i \end{cases} \tag{4.38}$$

$$C_{ij}(\theta) = \begin{cases} \pi \min\left\{\alpha_i^2, \alpha_j^2\right\}, & \theta \le |\alpha_i - \alpha_j| \\ \alpha_i^2 \arccos(\tau_i(\theta)) - \alpha_i^2 \tau_i(\theta)\left(1 - \tau_i^2(\theta)\right)^{\frac{1}{2}} & \\ +\alpha_j^2 \arccos\left(\tau_j(\theta)\right) - \alpha_j^2 \tau_j(\theta)\left(1 - \tau_j^2(\theta)\right)^{\frac{1}{2}}, & |\alpha_i - \alpha_j| < \theta \le \alpha_i + \alpha_j \\ 0, & \theta > \alpha_i + \alpha_j \end{cases} \tag{4.39}$$

where $\tau_i(\theta) = \left(\alpha_i^2 - \alpha_j^2 + \theta^2\right)/(2\alpha_i\theta)$ and $\tau_j(\theta) = \left(\alpha_j^2 - \alpha_i^2 + \theta^2\right)/\left(2\alpha_j\theta\right)$, $i \ne j$, $i, j = 1, 2, 3$.

Considering $\lambda = \{\beta, \alpha, \sigma\}$ as the set of hyper-parameters in the proposed model, the log-likelihood function is

$$L = -\frac{3n}{2}\log(2\pi) - \frac{1}{2}\log\left|C + \sigma^2\mathbf{I}\right| - \frac{1}{2}(\boldsymbol{m} - \boldsymbol{\mu\beta})^{\mathrm{T}}\left(C + \sigma^2\mathbf{I}\right)^{-1}(\boldsymbol{m} - \boldsymbol{\mu\beta}) \tag{4.40}$$

By solving the hyperparameters, the 3D surface can be modeled from the measurement points.

4.3.3.3 Surface Monitoring

In the realm of non-contact measurement, dense point cloud data often presents high-dimensionality, complicating the direct application of conventional control charts. This approach involves monitoring the 3D surface by assessing the discrepancy between the actual measured surface and the predicted values obtained from the S-MOGP model. To effectively compare these geometric discrepancies, a suitable summary of the deviation values is essential. While sample statistics such as mean and variance could be utilized, this would result in significant information loss. Menafoglio et al. [93] suggested summarizing the PDF of deviation values as a method to balance information retention with simplification. In line with this strategy, the deviation values of 3D surfaces are encapsulated using the PDF and principal component analysis (PCA), and monitored through T^2 and Q control charts.

The S-MOGP model is constructed based on a nominal sample dataset. For the jth measurement sample within the range ($j = 1, \ldots, M$), the deviation value of that sample is determined by the formula.

$$d_{ij} = \left\| m_{ij} - \hat{z}_{ij} \right\| \tag{4.41}$$

where $\hat{z}_{ij}$ represents the predicted value, m_i denotes the actual coordinates with $i = 1, \ldots, N_j$, and N_j signifies the count of measurement points for the jth sample.

Drawing on the approach outlined by Scimone et al. [94], the PDF can be approximated using the Bernstein polynomial estimator. The PDF estimator $\hat{f}_j$, which is derived from the empirical cumulative distribution function (CDF) F_j, is expressed as

$$\hat{f}_j(x) = N_j \sum_{k=0}^{N_j-1} \left[\left(F_j\left(\frac{k+1}{N_j}\right) - F_j\left(\frac{k}{N_j}\right) \right) b_{k,N_j}(x) \right] \tag{4.42}$$

with N_j being the total number of points in the point cloud dataset. The k-th Bernstein polynomial b_{k,N_j} is defined as $b_{k,N_j}(x) = N_j k x^k (1-x)^{N_j-k}$, where x is the variable mapped onto the interval [0, 1].

Subsequently, SFPCA is applied to extract the eigenvalues-eigenfunctions (λ_i, ζ_i) from the sample covariance operator, along with the corresponding scores on the ith principal component, denoted as

$$S_{ij} = \hat{f}_j - \hat{f}, \zeta_i \tag{4.43}$$

where $\hat{f} = \frac{1}{M}\sum_{j=1}^{M} \hat{f}_j$ is the sample mean.

Choose a number $K \in \{1, \ldots, N\}$ of principal components that account for 98% of the dataset's total variability. Two distinct statistical measures are utilized: the T^2 statistic for detecting anomalies within the K selected components and the Q statistic for assessing prediction error. These statistics are determined by

$$T_j^2 = \sum_{i=1}^{K} \frac{\left(S_{ij}\right)^2}{\lambda_i} \tag{4.44}$$

$$Q = \hat{f}_j^* - \hat{f}_j, \hat{f}_j^* - \hat{f}_j \tag{4.45}$$

where $\hat{f}_j^* = \sum_{i=1}^{K} z_{ij}\zeta_i$ is the reconstruction of the jth density curve which retaining the first K principal components.

Two multivariate control charts are performed to monitor the T^2 and Q statistics. The upper control limits of T^2 and Q statistics are

$$\mathrm{UCL}_\alpha\left(T^2\right)=\frac{(N-1)^2}{N}q_{\mathrm{Beta}\left(1-\alpha,\frac{K}{2},(N-k-1)/2\right)} \tag{4.46}$$

where q_{Beta} denotes the quantile of the Beta distribution and α is the Type I error probability.

$$\mathrm{UCL}_\alpha(Q)=\theta_1\left\{1-\theta_2 h_0\left(\frac{1-h_0}{\theta_1^2}\right)+\frac{\sqrt{z_\alpha\left(2\theta_2 h_0^2\right)}}{\theta_1}\right\}^{\frac{1}{h_0}} \tag{4.47}$$

where z_α is the $(1-\alpha)$ quantile of the standard normal distribution, $h_0=1-\frac{(2\theta_1\theta_3)}{3\theta_2^2}$, $\theta_r=\sum_{j=K+1}^{\infty}\lambda_j^r$, $r=1,2,3$.

Following the aforementioned procedures, the discrepancy between the actual measured sample and the nominal surface model can be tracked using Hotelling's T^2 and Q control charts.

4.3.4 Simulation

4.3.4.1 Simulation Surface Prediction

Within this section, simulations of three surfaces with varying geometries are created to assess the efficacy of the proposed method in surface modeling. Points on these surfaces are segregated into sets for training and testing. The 3D surface's parameter model is constructed utilizing the training set, and predictions are subsequently made at the test points. The model's performance is gauged by computing the mean square error (MSE) between the predicted and actual values:

$$\mathrm{MSE}=\sqrt{\frac{\sum_{i=1}^{n}\left|\hat{z}_i-m_i\right|^2}{n}} \tag{4.48}$$

where $\hat{z}_i$ represents the predicted value, m_i the true coordinates, n the count of points for prediction, and $|\cdot|$ signifies the Euclidean distance.

In the conventional GP model, the x and y coordinates of the measurement point are treated as explanatory variables, while the z coordinate is considered the response variable. del Castillo et al. [80] modified the GP kernel function by substituting the Euclidean distance with the geodesic distance to model 3D surfaces, thereby more accurately capturing the correlation between points. For the simulation of surface prediction, comparisons are made among the standard GP approach, the GGP

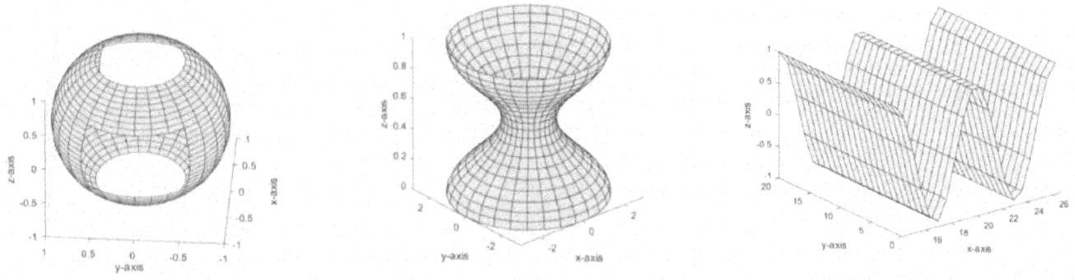

(a) Sphere with holes (b) Deformed cylinder (c) Sinusoidal surface

Fig. 4.29 Three simulated surfaces

method [80], and the GPLVM method [81]. The simulation includes three distinct surface types—a sphere with four holes, a distorted cylinder, and a sinusoidal surface, depicted in Fig. 4.29—for comparative analysis.

1. Spherical Surface with Apertures. A sphere, radius $R = 1$, featuring four apertures is generated to assess the method's capability in modeling surfaces that contain holes.
2. Modified Cylinder. A cylinder is defined through the profile function: $2 + \alpha \times \cos(u)$ to evaluate the method's performance on semi-enclosed surfaces; α represents the deformation parameter, with $\alpha = 1$ corresponding to the standard state.
3. Sinusoidal Surface. A sinusoidal surface is defined by the profile function: $\beta \times \sin(u)$ to test the method's effectiveness on non-enclosed, flexible surfaces; β is the amplitude factor, and $\beta = 1$ in the standard condition.

For each synthetic surface, an error term with a mean of 0 and variances of $\sigma^2(\sigma^2 = 0.01, 0.05, 0.1)$ is introduced in the x, y, and z directions. Each scenario is run 50 times to determine the average MSE. Within each scenario, 400 points are created on the surface, with 300 being randomly chosen for fitting and the remaining 100 designated for prediction. Table 4.15 presents the outcomes of a series of simulations across the four methods.

As depicted in Table 4.15, the proposed method achieves superior predictive accuracy for surfaces with holes and deformed cylinders. These surfaces are essentially variations of a standard sphere. Consequently, the proposed method, which more precisely captures the correlation between points, outperforms other methods in predictive accuracy. In the case of the sinusoidal surface, the S-MOGP method's performance lags behind the GGP and GPLVM methods. This is attributed to the fact that GGP and GPLVM unfold the sinusoidal surface onto a plane, reducing distortion for flexible surfaces. Spherical mapping, on the other hand, introduces significant distortion for such surfaces.

Nonetheless, the proposed method still surpasses the conventional GP model. The simulation results across the three distinct surface types indicate that the proposed method is particularly advantageous for closed and semi-closed surfaces and is less

Table 4.15 Prediction results for three simulation surfaces

Shape	σ^2	MSE (S-MOGP)	MSE (GP)	MSE (GGP)	MSE (GPLVM)
Sphere with holes	0.01	0.0130	0.6732	0.4417	0.5466
	0.05	0.0438	0.7035	0.6733	0.6670
	0.1	0.0803	0.7321	0.7485	0.7044
Deformed cylinder	0.01	0.0241	0.5997	0.5443	0.4284
	0.05	0.0677	0.7479	0.7229	0.6400
	0.1	0.1032	0.7825	0.7383	0.7296
Sinusoidal surface	0.01	0.0332	0.1224	0.0089	0.0274
	0.05	0.0741	0.1932	0.0312	0.0707
	0.1	0.1439	0.2130	0.0674	0.1365

sensitive to noise, maintaining proximity to the true surface even under high noise conditions.

4.3.4.2 Simulation Surface Monitoring

Utilizing the identical surface model as outlined in Sect. 4.3.4.1, the geometrical deviations of surfaces are monitored, with control results assessed through the ARL value. A lower ARL value indicates greater sensitivity of the control chart to deviations and a superior method. Additionally, to evaluate the performance of the proposed method, the Q-Q plot technique is employed for comparative analysis.

A Q-Q plot is employed to determine if two datasets originate from the same distribution. This method graphs the expected values of a specified distribution against the sample data's quantiles. Linearity in the resulting plot suggests that the sample data conforms to the specified distribution. Wells et al. [55] evaluated the Q-Q plot's monitoring parameters, encompassing the intercept, slope, and two categories of residual values. Their findings indicated that monitoring the intercept and slope generally yields the most effective results. Consequently, in this simulation, the Q-Q plot's intercept and slope are tracked for comparative analysis of the methods.

For the purpose of tracking the alterations in form and noise levels of the simulated surfaces, 500 samples are generated for each simulation surface under normal operating conditions (Phase I) to establish control parameters and the UCL. A type I error rate of $\alpha = 0.005$ is assumed. During Phase II, the process is perturbed 100 times, with 500 samples per simulation, to compute the mean and standard deviation of the ARL for each method under comparison.

For the trio of simulated surfaces, with the form factors $R = 1, \alpha = 1, \beta = 1$ and the noise factor $\sigma^2 = 0.01$ set as the baseline conditions, the proposed method and the Q-Q plot approach are implemented to evaluate and compare the ARLs during Phase II. The outcomes of these simulations are depicted in Fig. 4.30.

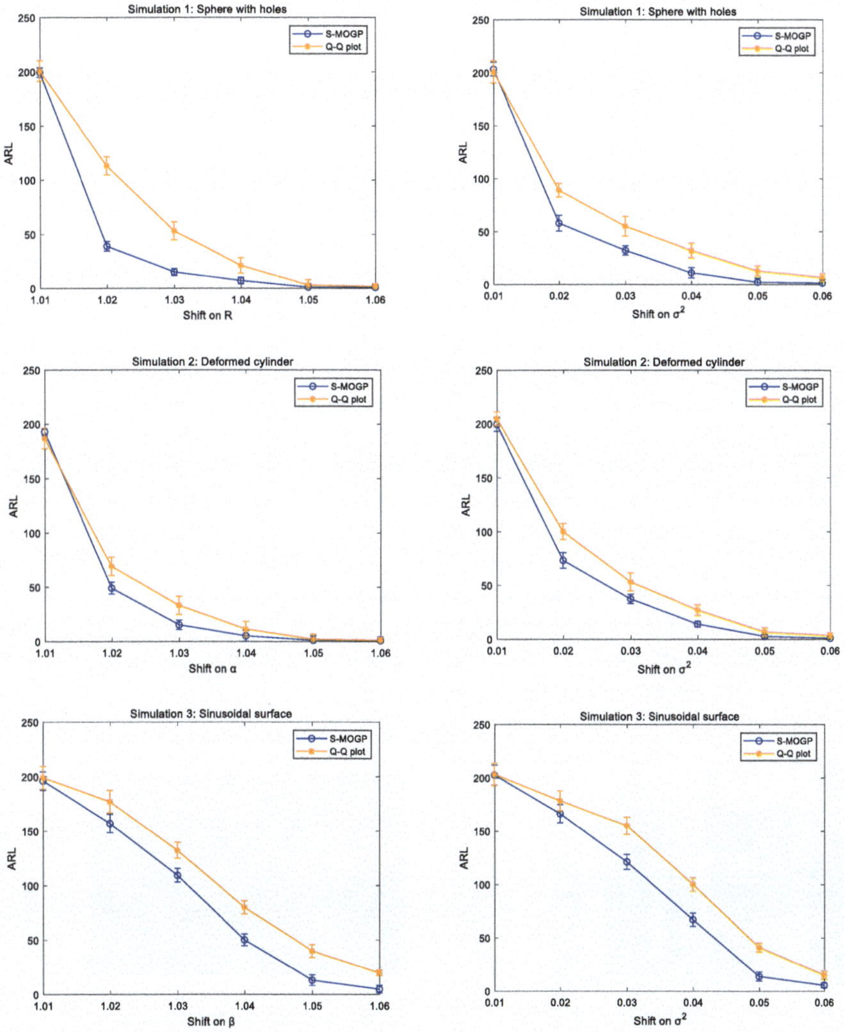

Fig. 4.30 ARLs and 95% confidence intervals in simulated surfaces 1, 2, 3 for S-MOGP and Q-Q plot methods

Figure 4.30 illustrates the ARLs for both the proposed method and the Q-Q plot method across various scenarios. The proposed method outperforms the Q-Q plot in monitoring efficacy. This approach is capable of analyzing the deviation profile shapes, yielding superior results compared to the monitoring of the Q-Q plot's slope and intercept.

This technique effectively and comprehensively captures the information of the 3D surface. It demonstrates a more robust monitoring capability than the Q-Q plot method and can detect process anomalies more swiftly.

4.3.5 Case Study

The combustion chamber within the engine's cylinder head presents a complex 3D surface. Defects on this surface can impact the chamber's volume, consequently affecting engine performance. Hence, investigating the surface quality of the combustion chamber is of paramount importance.

Conventional methods rely on three-coordinate measurements, with the measurement area and point count selected manually, which limits the representation of the combustion chamber's overall contour. These methods also suffer from long measurement times, leading to interval-based measurements that can overlook batch deviations.

Advanced HDM technology, with its rapid measurement capabilities, provides high-density point cloud data that accurately captures the entire surface contour. Figure 4.31 depicts the HDM measurement process for the cylinder head's combustion chamber, where a line laser scans the chamber. In this experiment, the cylinder heads from eight B12 series engines at an automotive manufacturing facility were measured, with four combustion chambers per cylinder head, totaling 32 chambers. One cylinder head was found to be deformed due to mold wear. The proposed method was employed to model and monitor these combustion chambers.

The measurement outcomes indicate that the combustion chamber is roughly hemispherical with apertures. Spherical mapping is applied to the point cloud of each chamber, parameterized by (u, v).

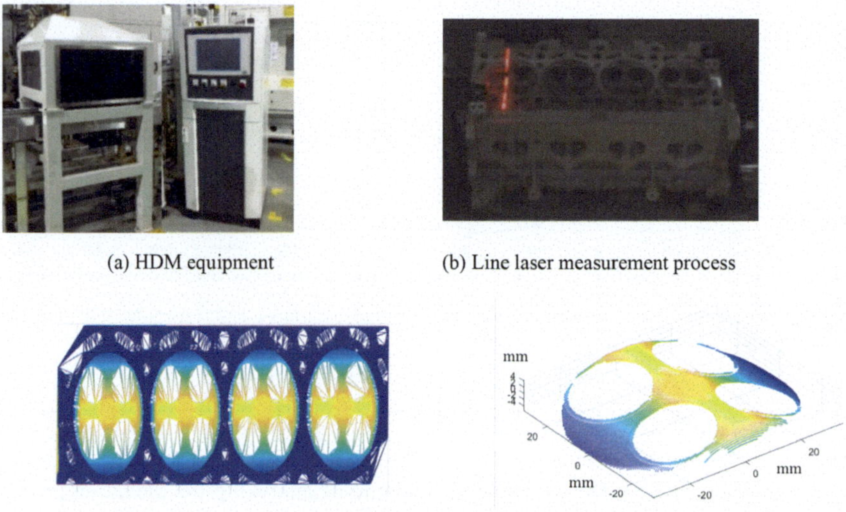

(a) HDM equipment

(b) Line laser measurement process

(c) Point cloud data of an engine cylinder head

(d) Point cloud data of a chamber

Fig. 4.31 The measurement process and measured point cloud data

Initially, the impact of the number of fitting points on model quality is examined. The dataset from a combustion chamber comprises 9094 points. Subsets of 500, 1500, 3500, and 8000 points are randomly extracted as fitting points (n_{fit}), with 1000 points reserved for prediction.

In the proposed method, the covariance structure with compact support facilitates the processing of large-scale datasets. For the comparative GGP method, del Castillo et al. [80] employed sparse matrix techniques to manage inverse and determinant operations. This approach is also utilized for the GPLVM model to assess the model's fitting performance under large-scale measurement points.

Table 4.16 shows the MSE values of the three methods with different numbers of fitting points.

When evaluating the proposed S-MOGP method against the GP, GGP, and GPLVM methods, it is found that the S-MOGP method yields the most accurate predictive outcomes. Additionally, for surfaces featuring holes, the geodesic distance calculated by the GGP deviates significantly from the actual values, resulting in less accurate predictions compared to the conventional Euclidean GP model. The GPLVM model's performance lies intermediate between the GP and GGP. In this scenario, spatial correlations are more effectively captured in spherical space rather than in geodesic or Euclidean spaces.

Subsequently, the inner surfaces of the 32 combustion chambers are monitored. Among these, the combustion chambers from the 29th to 32nd cylinder heads are identified as defective due to wear-related issues.

The initial 16 combustion chambers are designated as Phase I, and subsequently, T^2 and Q control charts are utilized to oversee the subsequent 16 chambers, as illustrated in Fig. 4.32. The final quartet of cavity surfaces exhibit geometric deviations attributable to wear, which are detectable by the proposed method.

Ultimately, the Q-Q plot method is employed for comparative analysis. Within this approach, each combustion chamber measurement is juxtaposed with the initial qualified sample to construct a Q-Q plot. The Q-Q plot's slope and intercept are subjected to monitoring via a Shewhart control chart. Figure 4.33 displays the control chart for the Q-Q plot method. Figure 4.33a presents the slope values of the Q-Q plot; a slope value near 1 signifies minimal divergence from the qualified sample, hence the slope control chart is depicted with only a lower control limit. Figure 4.33b depicts the intercept values of the Q-Q plot, with red lines representing the control limits. In this scenario, the last four samples are deemed unqualified, yet the method only detects one unqualified sample.

Table 4.16 The MSE of three methods

n_{fit}	MSE (S-MOGP)	MSE (GP)	MSE (GGP)	MSE (GPLVM)
500	0.1007	0.3820	0.5168	0.3977
1500	0.0632	0.1284	0.3570	0.1446
3500	0.0619	0.1445	0.3696	0.1218
8000	0.0501	0.1109	0.3143	0.1169

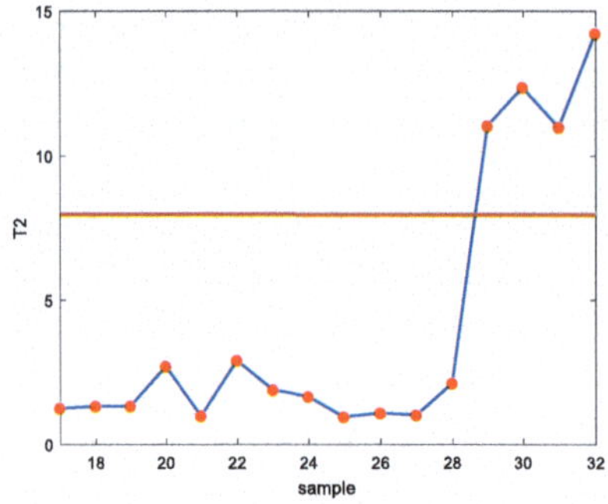

(a) T^2 control chart

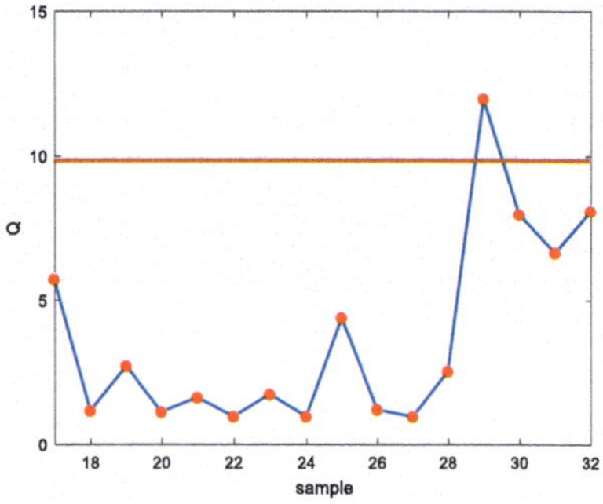

(b) Q control chart

Fig. 4.32 The control chart of S-MOGP

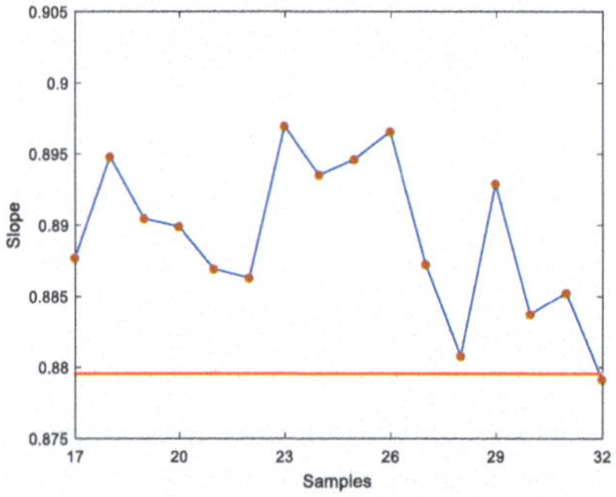

(a) Slope value of Q-Q plot

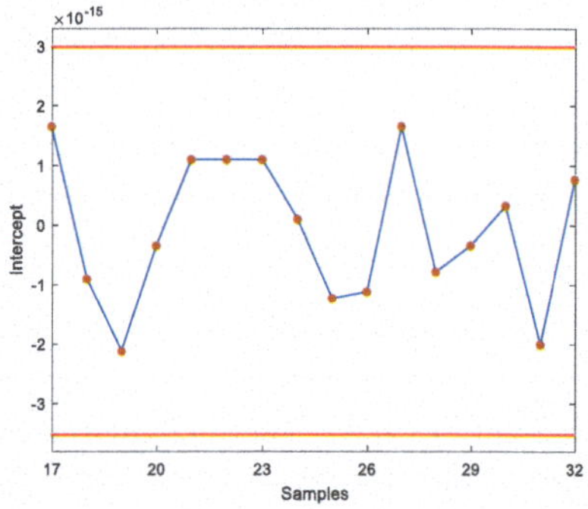

(b) Intercept value of Q-Q plot

Fig. 4.33 The control chart of Q-Q plot

The experimental outcomes demonstrate that, in comparison to the GP, GGP, and GPLVM methods, the proposed approach achieves superior fit and is capable of effectively modeling the inner surfaces of the engine cylinder head's combustion chambers. Although the chamber's inner surface is not closed, its approximate hemispherical shape results in minimal distortion during spherical mapping, making it well-suited for the proposed method. During monitoring, the proposed method can efficiently detect changes in surface geometry, enabling swift adjustments to the manufacturing process to maintain product quality. The practical efficacy of the method is confirmed through both the fitting error and the monitoring outcomes.

4.3.6 Conclusions

The presented S-MOGP technique, grounded in spherical mapping and MOGP, is designed for the modeling and monitoring of 3-dimensional surfaces. This approach translates 3D coordinates into spherical parameters via spherical mapping, constructs a state equation based on MOGP, and offers control charts for the surveillance of geometric deviations, utilizing spherical distance in place of Euclidean and geodesic distances. The method's efficacy for 3D surface modeling and monitoring is substantiated through simulations and practical applications.

In contrast to GP, GGP, and GPLVM methods, the proposed approach exhibits the lowest prediction error for closed and semi-closed surfaces. In practical scenarios, the method significantly enhances prediction accuracy. It takes into account the spatial correlations across the x, y, and z axes comprehensively. For spherical and surfaces with multiple apertures, the method relies not on the spatial distance between points but on angular values, which facilitates a more accurate modeling of 3D surfaces.

In terms of surface monitoring, the proposed method is adept at swiftly detecting process anomalies. Relative to the Q-Q plot method, it can effectively identify surface changes, offering crucial feedback for part processing to uphold product quality.

To summarize, the S-MOGP method for 3D surface modeling and monitoring is proposed. It holds significant advantages over existing methods, particularly for non-developable surfaces, including closed and multi-hole surfaces. In conjunction with HDM technology, this method finds application in the processing of 3D surface point cloud data. Future research directions include:

(1) Currently, analysis is limited to offline data. The method could be integrated with online HDM rapid measurement for real-time 3D surface monitoring, reflecting online processing information.
(2) While the method can monitor surface shape changes, it does not pinpoint specific defect locations. By integrating image data and additional processing techniques, further exploration could enhance the method's capability to identify surface defect locations.

References

1. Yan X, Ballu A (2018) Tolerance analysis using skin model shapes and linear complementarity conditions. J Manuf Syst 48:140–156
2. Zhang X, Wang X, Wang D, Yao Z, Xi L, Wang X (2016) Methodology to improve the cylindricity of engine cylinder bore by honing. J Manuf Sci Eng 139(3):031008–031008–10
3. Shao Y, Yin Y, Du S, Xia T, Xi L (2018) Leakage monitoring in static sealing interface based on three dimensional surface topography indicator. J Manuf Sci E T ASME 140(10):101003
4. Deng C-S, Chin J-H (2004) Hole roundness in deep-hole drilling as analysed by Taguchi methods. Int J Adv Manuf Technol 25(5–6):420–426
5. Nemati Keshteli R, Kazemzadeh RB, Amiri A, Noorossana R (2014) Functional process capability indices for circular profile. Qual Reliab Eng Int 30(5):633–644

6. Kang L, Albin SL (2018) On-line monitoring when the process yields a linear profile. J Qual Technol 32(4):418–426
7. Zou C, Zhang Y, Wang Z (2006) A control chart based on a change-point model for monitoring linear profiles. IIE Trans 38(12):1093–1103
8. Ding Y, Zeng L, Zhou S (2018) Phase I analysis for monitoring nonlinear profiles in manufacturing processes. J Qual Technol 38(3):199–216
9. Williams JD, Woodall WH, Birch JB (2007) Statistical monitoring of nonlinear product and process quality profiles. Qual Reliab Eng Int 23(8):925–941
10. Kazemzadeh RB, Noorossana R, Amiri A (2008) Phase I monitoring of polynomial profiles. Commun Stat-Theory Methods 37(10):1671–1686
11. Noorossana R, Eyvazian M, Vaghefi A (2010) Phase II monitoring of multivariate simple linear profiles. Comput Ind Eng 58(4):563–570
12. Guo W, Shao C, Kim TH, Hu SJ, Jin J, Spicer JP, Wang H (2016) Online process monitoring with near-zero misdetection for ultrasonic welding of lithium-ion batteries: an integration of univariate and multivariate methods. J Manuf Syst 38:141–150
13. Wang K, Jiang W, Li B (2016) A spatial variable selection method for monitoring product surface. Int J Prod Res 54(14):4161–4181
14. Huang DL, Du SC, Li GL, Zhao C, Deng YF (2018) Detection and monitoring of defects on three-dimensional curved surfaces based on high-density point cloud data. Precis Eng 53:79–95
15. Zeng L, Neogi S, Zhou Q (2014) Robust phase I monitoring of profile data with application in low-E glass manufacturing processes. J Manuf Syst 33(4):508–521
16. Yang W, Zou C, Wang Z (2017) Nonparametric profile monitoring using dynamic probability control limits. Qual Reliab Eng Int 33(5):1131–1142
17. Wang H, Kababji H, Huang Q (2009) Monitoring global and local variations in multichannel functional data for manufacturing processes. J Manuf Syst 28(1):11–16
18. Cho N, Tu J (2001) Roundness modeling of machined parts for tolerance analysis. Precis Eng 25(1):35–47
19. Jiang Q, Feng H-Y, OuYang D, Desta MT (2006) A roundness evaluation algorithm with reduced fitting uncertainty of CMM measurement data. J Manuf Syst 25(3):184–195
20. Zhang XD, Zhang C, Wang B, Feng SC (2005) Unified functional tolerancing approach for precision cylindrical components. Int J Prod Res 43(1):25–47
21. Ramaswami H, Acharya SB, Kovvur Y, Anand S (2006) A multivariate statistical analysis of sampling uncertainties in geometric and dimensional errors for circular features. J Manuf Syst 25(2):77–94
22. Colosimo BM, Pacella M (2007) On the use of principal component analysis to identify systematic patterns in roundness profiles. Qual Reliab Eng Int 23(6):707–725
23. Pacella M, Colosimo BM (2018) Multilinear principal component analysis for statistical modeling of cylindrical surfaces: a case study. Qual Technol Quant Manag 15(4):507–525
24. Wang K, Li J, Tsung F (2018) Registration-free monitoring of multimode near-circular shape profiles. Qual Reliab Eng Int 34(4):529–542
25. Fathizadan S, Niaki STA, Noorossana R (2017) Using independent component analysis to monitor 2-D geometric specifications. Qual Reliab Eng Int 33(8):2075–2087
26. Pacella M, Semeraro Q (2011) Monitoring roundness profiles based on an unsupervised neural network algorithm. Comput Ind Eng 60(4):677–689
27. Colosimo BM, Cicorella P, Pacella M, Blaco M (2014) From profile to surface monitoring: SPC for cylindrical surfaces via Gaussian processes. J Qual Technol 46(2):95–113
28. Liu JP, Jin R, Kong ZJ (2018) Wafer quality monitoring using spatial Dirichlet process based mixed-effect profile modeling scheme. J Manuf Syst 48:21–32
29. Colosimo BM, Semeraro Q, Pacella M (2008) Statistical process control for geometric specifications: on the monitoring of roundness profiles. J Qual Technol 40(1):1–18
30. Soleimani P, Noorossana R, Amiri A (2009) Simple linear profiles monitoring in the presence of within profile autocorrelation. Comput Ind Eng 57(3):1015–1021
31. Yu J, Liu J (2010) LRProb control chart based on logistic regression for monitoring mean shifts of auto-correlated manufacturing processes. Int J Prod Res 49(8):2301–2326

32. Anselin L (1988) Spatial econometrics: methods and models, 1st edn. Springer, Netherlands
33. Boeing Commercial Airplane Group, Quality Assurance Department. Advanced quality system tools, AQS D1–9000–1. http://www.boeingsuppliers.com/supplier/d1-9000-1.pdf. Accessed July 2018
34. Cliff A, Ord K (1972) Testing for spatial autocorrelation among regression residuals. Geogr Anal 4(3):267–284
35. Anselin L, Bera AK, Florax R, Yoon MJ (1996) Simple diagnostic tests for spatial dependence. Reg Sci Urban Econ 26(1):77–104
36. Breusch TS, Pagan AR (1980) The Lagrange multiplier test and its applications to model specification in econometrics. Rev Econ Stud 47(1):239–253
37. Damir MNH (1979) Approximate harmonic models for roundness profiles. Wear 57(2):217–225
38. Hii K-F, Vallance RR, Grejda RD, Marsh ER (2004) Error motion of a kinematic spindle. Precis Eng 28(2):204–217
39. Weihua N, Zhenqiang Y (2012) Cylindricity modeling and tolerance analysis for cylindrical components. Int J Adv Manuf Technol 64(5–8):867–874
40. Glenn P (1984) Set of orthogonal surface error descriptors for near-cylinder optics. Opt Eng 23:384–390
41. Wang AD, Wang KB, Tsung FG (2014) Statistical surface monitoring by spatial-structure modeling. J Qual Technol 46(4):359–376
42. Zhang LM, Wang KB, Chen N (2016) Monitoring wafers' geometric quality using an additive Gaussian process model. IIE Trans 48(1):1–15
43. Colosimo BM, Mammarella F, Petro S (2010) Quality control of manufactured surfaces. Front Stat Qual Control 9:55–70
44. Zhao C, Du SC, Deng YF, Li GL, Huang D (2020) Circular and cylindrical profile monitoring considering spatial correlations. J Manuf Syst 54:35–49
45. Jin YA, Li H, He Y, Fu JZ (2015) Quantitative analysis of surface profile in fused deposition modelling. Addit Manuf 8:142–148
46. Wells LJ, Dastoorian R, Camelio JA (2021) A novel NURBS surface approach to statistically monitor manufacturing processes with point cloud data. J Intell Manuf 32(2):329–345
47. Wang M, Xi LF, Du SC (2014) 3D surface form error evaluation using high definition metrology. Precis Eng-J Int Soc Precis Eng Nanotechnol 38(1):230–236
48. Choudhary AK, Harding JA, Tiwari MK (2009) Data mining in manufacturing: a review based on the kind of knowledge. J Intell Manuf 20(5):501–521
49. Du SC, Liu CP, Huang DL (2015) A shearlet-based separation method of 3D engineering surface using high definition metrology. Precis Eng-J Int Soc Precis Eng Nanotechnol 40:55–73
50. Du SC, Liu CP, Xi LF (2015) A selective multiclass support vector machine ensemble classifier for engineering surface classification using high definition metrology. J Manuf Sci Eng-Trans Asme 137(1)
51. Zang YY, Qiu PH (2018) Phase I monitoring of spatial surface data from 3D printing. Technometrics 60(2):169–180
52. Zang YY, Qiu PH (2018) Phase II monitoring of free-form surfaces: an application to 3D printing. J Qual Technol 50(4):379–390
53. Stankus SE, Castillo-Villar KK (2019) An Improved multivariate generalised likelihood ratio control chart for the monitoring of point clouds from 3D laser scanners. Int J Prod Res 57(8):2344–2355
54. Wang KB, Tsung F (2005) Using profile monitoring techniques for a data-rich environment with huge sample size. Qual Reliab Eng Int 21(7):677–688
55. Wells LJ, Megahed FM, Niziolek CB, Camelio JA, Woodall WH (2013) Statistical process monitoring approach for high-density point clouds. J Intell Manuf 24(6):1267–1279
56. Osada R, Funkhouser T, Chazelle B, Dobkin D (2002) Shape distributions. Acm T Graphic 21(4):807–832
57. Laga H, Guo Y, Tabia H, Fisher RB, Bennamoun M (2019) Global shape descriptors in 3D shape analysis. Wiley, pp 65–91

58. Li X, Zhang W, Ding Q, Sun JQ (2020) Intelligent rotating machinery fault diagnosis based on deep learning using data augmentation. J Intell Manuf 31(2):433–452
59. Maturana D, Scherer S (2015) VoxNet: A 3D convolutional neural network for real-time object recognition. IEEE Int C Int Robot 922–928
60. Ning FW, Shi Y, Cai ML, Xu WQ (2020) Various realization methods of machine-part classification based on deep learning. J Intell Manuf 31(8):2019–2032
61. Wu ZR, Song SR, Khosla A, Yu F, Zhang LG, Tang XO, Xiao JX (2015) 3D ShapeNets: a deep representation for volumetric shapes. Proc Cvpr IEEE 1912–1920
62. Su H, Maji S, Kalogerakis E, Learned-Miller E (2015) Multi-view convolutional neural networks for 3D shape recognition. IEEE I Conf Comp Vis 945–953
63. Qi CR, Su H, Mo KC, Guibas LJ (2017) PointNet: deep learning on point sets for 3D classification and segmentation. Proc Cvpr Ieee 77–85
64. Qi CR, Yi L, Su H, Guibas LJ (2017) PointNet plus plus: deep hierarchical feature learning on point sets in a metric space. Adv Neural Inf Process Syst 30(Nips 2017):30
65. Di Angelo L, Di Stefano P (2015) Geometric segmentation of 3D scanned surfaces. Comput Aided Des 62:44–56
66. Nurunnabi A, Belton D, West G (2012) Robust segmentation in laser scanning 3D point cloud data. In: 2012 International conference on digital image computing techniques and applications (Dicta)
67. Syakur MA, Khotimah BK, Rochman EMS, Satoto BD (2018) Integration K-means clustering method and elbow method for identification of the best customer profile cluster. Iop Conf Ser-Mat Sci 336
68. Lopez GL, Negron APP, Jimenez AD, Rodriguez JR, Paredes RI (2017) Comparative analysis of shape descriptors for 3D objects. Multimed Tools Appl 76(5):6993–7040
69. Shao YP, Du SC, Tang HT (2021) An extended bi-dimensional empirical wavelet transform based filtering approach for engineering surface separation using high definition metrology. Measurement 178
70. Rajamohan G, Shunmugam MS, Samuel GL (2011) Effect of probe size and measurement strategies on assessment of freeform profile deviations using coordinate measuring machine. Measurement 44(5):832–841
71. Woodall WH (2007) Current research on profile monitoring. Production 17(3):420–425
72. Zou CL, Tsung FG, Wang ZJ (2007) Monitoring general linear profiles using multivariate exponentially weighted moving average schemes. Technometrics 49(4):395–408
73. Lin H, Keller F, Stein M (2020) Influence and compensation of CMM geometric errors on 3D gear measurements. Measurement 151
74. Mehrad V, Xue D, Gu P (2013) Prediction of surface reconstruction uncertainties for freeform surface inspection. Measurement 46(8):2682–2694
75. Hagen H, Hahmann S, Schreiber T (1995) Visualization and computation of curvature behaviour of freeform curves and surfaces. Comput Aided Des 27(7):545–552
76. Kase K, Makinouchi A, Nakagawa T, Suzuki H, Kimura F (1999) Shape error evaluation method of free-form surfaces. Comput Aided Des 31(8):495–505
77. Osada R, Funkhouser T, Chazelle B, Dobkin D (2002) Shape distributions. ACM Trans Graph 21(4):807–832
78. Laga H, Guo Y, Tabia H, Fisher R, Bennamoun M (2019) 3D Shape analysis: fundamentals, theory and application. Wiley
79. Zhao C, Lv J, Du SC, Deng YF (2019) Geometric error modeling and monitoring of the 3D Surface by Gaussian correlation model. In: IEEE international conference on industrial engineering and engineering management (IEEM), pp 1022–1025
80. del Castillo E, Colosimo BM, Tajbakhsh SD (2015) Geodesic Gaussian processes for the parametric reconstruction of a free-form surface. Technometrics 57(1):87–99
81. Lawrence N (2005) Probabilistic non-linear principal component analysis with Gaussian process latent variable models. J Mach Learn Res 6:1783–1816
82. Rasmussen CE, Nickisch H (2010) Gaussian processes for machine learning (GPML) toolbox. J Mach Learn Res 11:3011–3015

83. Sun XF, Rosin PL, Martin RR, Langbein FC (2009) Noise analysis and synthesis for 3D laser depth scanners. Graph Models 71(1–6):34–48
84. Shen L, Makedon F (2006) Spherical mapping for processing of 3D closed surfaces. Image Vis Comput 24(7):743–761
85. Liu HT, Cai JF, Ong YS (2018) Remarks on multi-output Gaussian process regression. Knowl-Based Syst 144:102–121
86. Praun E, Hoppe H (2003) Spherical parametrization and remeshing. Acm Trans Graph 22(3):340–349
87. Brechbühler C, Gerig G, Kübler O (1995) Parametrization of closed surfaces for 3-d shape description. Comput Vis Image Underst 61(2):154–170
88. Yao ZW, Zhao QX, Li XF, Bi QS (2021) Point cloud registration algorithm based on curvature feature similarity. Measurement 177
89. Choi GPT, Ho KT, Lui LM (2016) Spherical conformal parameterization of genus-0 point clouds for meshing. SIAM J Imag Sci 9(4):1582–1618
90. Huang CF, Zhang HM, Robeson SM (2011) On the validity of commonly used covariance and Variogram functions on the sphere. Math Geosci 43(6):721–733
91. Du J, Ma CS (2013) Vector random fields with compactly supported covariance matrix functions. J Stat Plan Inference 143(3):457–467
92. Ma CS (2016) Stochastic representations of isotropic vector random fields on spheres. Stoch Anal Appl 34(3):389–403
93. Menafoglio A, Grasso M, Secchi P, Colosimo BM (2018) Profile monitoring of probability density functions via simplicial functional PCA with application to image data. Technometrics 60(4):497–510. https://doi.org/10.1080/.00401706.2018.1437473
94. Scimone R, Taormina T, Colosimo BM, Grasso M, Secchi P (2021) Statistical modeling and monitoring of geometrical deviations in complex shapes with application to Additive Manufacturing. Technometrics 1–34. https://doi.org/10.1080/00401706.2021.1961870

Chapter 5
Surface Characterization and Evaluation

5.1 High Definition Metrology Enabled Three Dimensional Discontinuous Surface Filtering by Extended Tetrolet Transform

5.1.1 Introduction

During the last three decades, the correlation between surface topography and mechanical manufacturing has garnered growing attention. Surface topography serves as a crucial intermediary between the manufacturing process of a component and its intended functionality [1]. To meet stricter tolerances and enhanced performance criteria, it is imperative for engineers to extend their focus beyond dimensions, shapes, and peak values to include the specification of various surface functional attributes. A surface can be perceived as a linear combination of flaws, form error, and surface texture, which encompasses both waviness and roughness [2]. Furthermore, a surface can be decomposed into distinct frequency components ranging from finer to coarser scales, namely roughness, waviness, and form. This categorization has been acknowledged since the inception of surface texture measurement. The rationale behind such a classification is that these surface characteristics at varying scales have distinct origins and exert different impacts on the functionality of parts. They are capable of reflecting the diverse functions of workpieces, such as vibration, tool wear, chatter, and sealing [3–5]. Malburg et al. introduced an innovative approach for analyzing surface profiles in relation to the comfortable properties of components like gaskets, seals, and bushings, which are widely used in engine and hydraulic applications [6]. Bottiglione et al. [7, 8] suggested a method where surface topography parameters, the applied load, and the geometry of the seal were the primary factors influencing the leakage mechanism, which includes the formation of leak paths and the calculation of leakage.

S. Du and L. Xi, *Quality Control for Variable Stiffness Structured Workpiece*,
https://doi.org/10.1007/978-981-96-5376-8_5

Surface filtering, a vital component of surface topography analysis, involves the segregation of a surface profile into its constituent elements: form, waviness, and roughness. This technique has been extensively utilized in surface metrology, where a well-established system has been developed. In the realm of 3D surface analysis, studies have introduced an areal spline filtering method [9, 10]. The integration of wavelet transform into engineering surface texture analysis has been explored, with Fu et al. [11] comparing various wavelet bases. The utility of wavelets was underscored through their multi-resolution analysis capabilities on surface profiles. Jiang et al. [12] put forth a lifting wavelet representation for the characterization of surface topography. Raja et al. provided an overview of prevalent filter techniques, encompassing both traditional and advanced approaches such as 2RC, Gaussian, spline, morphological, wavelet, regression filters, and robust regression filters [13].

The majority of surface filtering studies have focused on continuous surfaces, yet there is a significant presence of discontinuous surfaces in engineering applications, such as engine block faces featuring cylinder holes, bolt holes, and cooling holes. Traditionally, surfaces are measured using coordinate measuring machines (CMM), which, due to their low-density sampling, fail to capture the complete surface topography. Consequently, there is a requirement for a full-inspection, high-density sampling instrument capable of measuring large and discontinuous surfaces. Recently, a new measuring instrument known as non-contact high definition metrology (HDM) has emerged [14]. The Coherix ShaPix (ShaPix® Surface DetectiveTM, Ann Arbor, MI) is utilized in the HDM measurement process, employing a laser holographic interferometry technique that can capture the entire surface height of discontinuous objects with holes and empty areas, rather than just a localized region. This technology generates millions of data points over an area of 300 mm × 300 mm in mere seconds, providing a comprehensive representation of the measured surface with a 150 μm lateral resolution in the x–y plane and a 1 μm accuracy in the z-axis. Figure 5.1 illustrates the HDM device alongside a measured engine block face. HDM has facilitated the development of various studies, including 3D surface topography evaluation [15], filtering [16, 17], classification [18, 19], forecasting [20], and monitoring methods [21–24], all aimed at the pre-control of manufacturing processes.

Limited research has addressed filtering on discontinuous surfaces using HDM. Zhang et al. [25] employed a Gaussian (DoGs) filter bank to classify workpiece surfaces by extracting features from HDM data. Liao et al. [26] introduced a method using biorthogonal wavelets to decompose 3D surfaces into multi-scale subsurface and also explored the correlation between wavelet scales and surface characteristics. On discontinuous surfaces, edge distortion typically arises from holes and boundaries. Specifically, edge distortion represents the manifestation of singularities. In two-dimensional discontinuous surfaces, edge distortion is a point singularity, while in three-dimensional discontinuous surfaces, it corresponds to line and surface singularities. The discrete wavelet transform (DWT) has become a prevalent tool in image processing for its effectiveness in analyzing point singularities, which can address edge distortion. However, DWT is inadequate for capturing geometric features with line and surface singularities, as its basis functions are fixed in horizontal, vertical, and

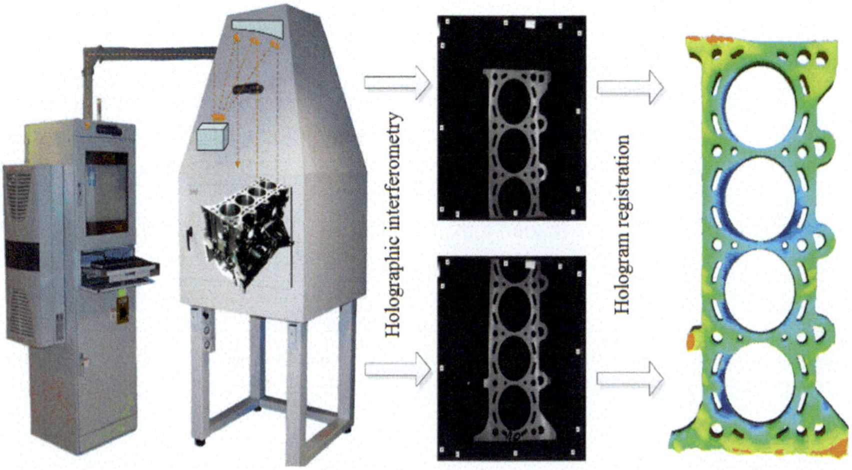

Fig. 5.1 Measurement by HDM on an engine block face

diagonal directions and do not adapt to image geometric structures. Consequently, there is a need to develop an optimal filtering technique tailored for 3D discontinuous surface.

The emerging academic field of multi-scale geometric analysis (MGA) in harmonic analysis is dedicated to finding the most effective way to represent high-dimensional data, with potential applications in surface filtering. The tetrolet transform, a recent addition to MGA, is a non-redundant, adaptive geometric wavelet transform that is noted for its small support, symmetry, and orthogonality [27, 28]. These characteristics enable the tetrolet transform to avoid pseudo-Gibbs artifacts, while maintaining anisotropic edges and features, offering a straightforward yet highly efficient method for image processing and the compression of real data arrays. However, the standard tetrolet transform, while adept at representing the original surface, falls short in depicting the decomposed and reconstructed subsurface. To address the challenges of discontinuous surface filtering, an extension of the tetrolet transform is warranted. The current methods' shortcomings are encapsulated by the following points:

- Surface topography is intrinsically linked to mechanical properties and functional characteristics, with various surface components indicating different functions of workpieces, such as vibration, tool wear, chatter, and sealing [3–5].
- Filtering is crucial for extracting surface components. Conventional filtering techniques like Gaussian and spline filters are suitable for continuous surfaces but are not applicable to discontinuous surfaces due to issues with boundary distortion or end effects.
- Engineering practice frequently encounters discontinuous surfaces, such as those found on engine block faces, engine desk faces, and automatic transmission

valve joint surfaces. There is a pressing need for a filtering method tailored to discontinuous surfaces.

To the authors' best knowledge, no filtering method utilizing the tetrolet transform has been developed for discontinuous surfaces employing HDM. Capitalizing on advancements in measurement technology, the primary contribution of the proposed approach is the introduction of an innovative filtering technique for discontinuous surfaces that leverages an extended tetrolet transform. This novel filtering technique effectively addresses the issue of edge effects typically associated with discontinuous surfaces when traditional filtering methods are applied. It precisely separates the low-frequency and high-frequency components of discontinuous engineering surfaces, eliminating edge distortions. With the accurate extraction of surface components, the method can accurately reflect the machining process and the functionality of the part.

5.1.2 *Brief Introduction of Tetrolet Transform*

Tetrolet transform is founded on the principle of tetrominoes, a concept introduced by Golomb [29]. Tetrominoes are configurations created by combining four individual unit squares, with each square sharing an edge with at least one other square. There exist five distinct varieties of tetrominoes, known as free tetrominoes, which do not account for rotations and reflections (refer to Fig. 5.2). It is evident that an image of dimensions $N \times N$, where N is an even number, can be constructed from these fundamental tetrominoes. Notably, there are 117 distinct ways to cover a board using tetrominoes.

The tetrolet concept parallels that of the wedgelet, utilizing Haar functions on the edges. For an image $\mathbf{a} = (a[m_1, m_2])_{(m_1,m_2)\in I}$ with $I = \{0, \ldots, N-1\} \subset \mathbb{Z}^2$, $N = 2^J$, $J \in \mathbb{N}$, it can be divided into a series of 4×4 blocks. In a 4×4 block, the four tetrominoes are labeled with $\{I_0, I_1, I_2, I_3\}$, and the four indices within each subsets I_v are associated with the scalar set $\{0, 1, 2, 3\}$ through a one-to-one mapping L. For any tetromino I_v, the discrete basis functions are defined as:

$$\phi_{I_v}[m_1, m_2] := \begin{cases} 1/2 & (m_1, m_2) \in I_v \\ 0 & \text{otherwise} \end{cases} \tag{5.1}$$

and

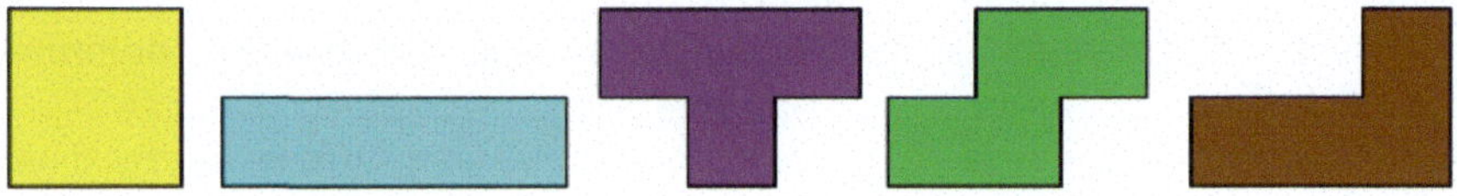

Fig. 5.2 The five free tetrominoes

$$\psi^l_{I_v}[m_1, m_2] := \begin{cases} \epsilon[l, L(m_1, m_2)] & (m_1, m_2) \in I_v \\ 0 & \text{otherwise} \end{cases} \tag{5.2}$$

for $l = 1, 2, 3$. Due to the underlying tetromino support, $\psi^l_{I_v}$ are called tetrolets and ϕ_{I_v} is the corresponding scaling function. In the tetrolet definition, the function values $\varepsilon[l, L(m_1, m_2)]$ entries from Haar wavelet transform matrix

$$W := (\epsilon[l, m])^3_{l,m=0} = \frac{1}{2}\begin{pmatrix} 1 & 1 & 1 & 1 \\ 1 & 1 & -1 & -1 \\ 1 & -1 & 1 & -1 \\ 1 & -1 & -1 & 1 \end{pmatrix} \tag{5.3}$$

Similar to the wavelet transform, the tetrolet transform breaks down an image into low-frequency components and high-frequency tetrolet components at a defined level. The key distinction lies in the tetrolet transform's approach of segmenting the image into 4×4 blocks and selecting the most suitable tetromino partition that aligns with the block's image geometry from the 117 available options [27]. The tetrolet transform's k level decomposition process is detailed below:

Step 1: The image is segmented into 4×4 blocks, each of which is covered by any set of four free tetrominoes.
Step 2: Adhering to the principle of the sparsest tetrolet representation, the most optimal coverage is identified within each block from the 117 possible configurations.
Step 3: The low-frequency components and high-frequency tetrolet components of each block are reorganized into a 4×4 block based on the count of deviations.
Step 4: The low-frequency components and high-frequency tetrolet components are recorded.
Step 5: Repeat steps 1–4 on the low-frequency components for subsequent decompositions until reaching level k.

Through an appropriate decomposition, the low-frequency components and high-frequency tetrolet components of the original image at each level can be extracted.

5.1.3 The Proposed Method

5.1.3.1 Overview of the Proposed Method

This section introduces an innovative filtering technique for discontinuous surfaces, leveraging the extended tetrolet transform. The approach surpasses conventional methods in three key areas: (1) It addresses line and surface singularities in 3D discontinuous surfaces that lead to edge distortions, capitalizing on the small support, symmetry, orthogonality, and adaptive geometrical structure. (2) The method incorporates an edge detector to preemptively identify surface boundaries and holes, facilitating the application of varied decomposition techniques. (3) It effectively isolates and reconstructs subsurfaces from the decomposition, each representing distinct surface components. The methodology encompasses 3D discontinuous surface measurement, point cloud conversion, edge detector generation, extended tetrolet transform, and the assessment of 3D characterization parameters. The framework is depicted in Fig. 5.3. The process unfolds as follows:

Step 1: High-definition measurement (HDM) is utilized to capture the 3D discontinuous surface, yielding millions of points. The point cloud is then transformed into a grayscale image that encapsulates the entire surface [15].
Step 2: Edge detector generation. A detector is developed to pinpoint the boundaries and holes of the discontinuous surface in advance. Decomposition strategies are tailored to areas with and without edges to mitigate end effects.
Step 3: Extended tetrolet transform. This transform is compartmentalized into three stages: decomposition, shrinkage, and reconstruction. The subsurfaces derived from this process establish a correlation between the decomposition level and the actual surface components, enabling the precise separation of the discontinuous surface into its form and waviness components.
Step 4: 3D characterization parameters evaluation. Advanced 3D characterization parameters, including height, shape, and spatial parameters, are computed to assess the filtered surface components for a comprehensive surface topography analysis.

5.1.3.2 Surface Measurement and Point Cloud Conversion

To accurately capture surface details, high definition metrology is employed for surface measurement, offering an effective basis for surface topography analysis. Shapix's precise measurement technique generates a substantial dataset, comprising millions of points that define the surface. Wang et al. [15] propose a preprocessing technique that translates the extensive point cloud into a gray image that preserves height and position information. This approach offers numerous benefits for subsequent analysis. The Z-direction's lower and upper specification limits, SL and SU, are set between − 30 and 30 μm. HDM's lateral resolution restricts its ability to measure roughness, with the data focusing solely on 3D surface form and waviness. Figure 5.4 illustrates two examples of measured surfaces: the engine cylinder block and head surfaces.

A key distinction in filtering techniques lies between discontinuous and continuous surfaces, with the former's surface boundaries being predetermined. To ensure distortion-free filtering of discontinuous surfaces, an edge detector $\mathbf{w}$ is crucial for delineating these boundaries. The extended tetrolet transform employs distinct decomposition strategies for areas with and without edges to mitigate end effects.

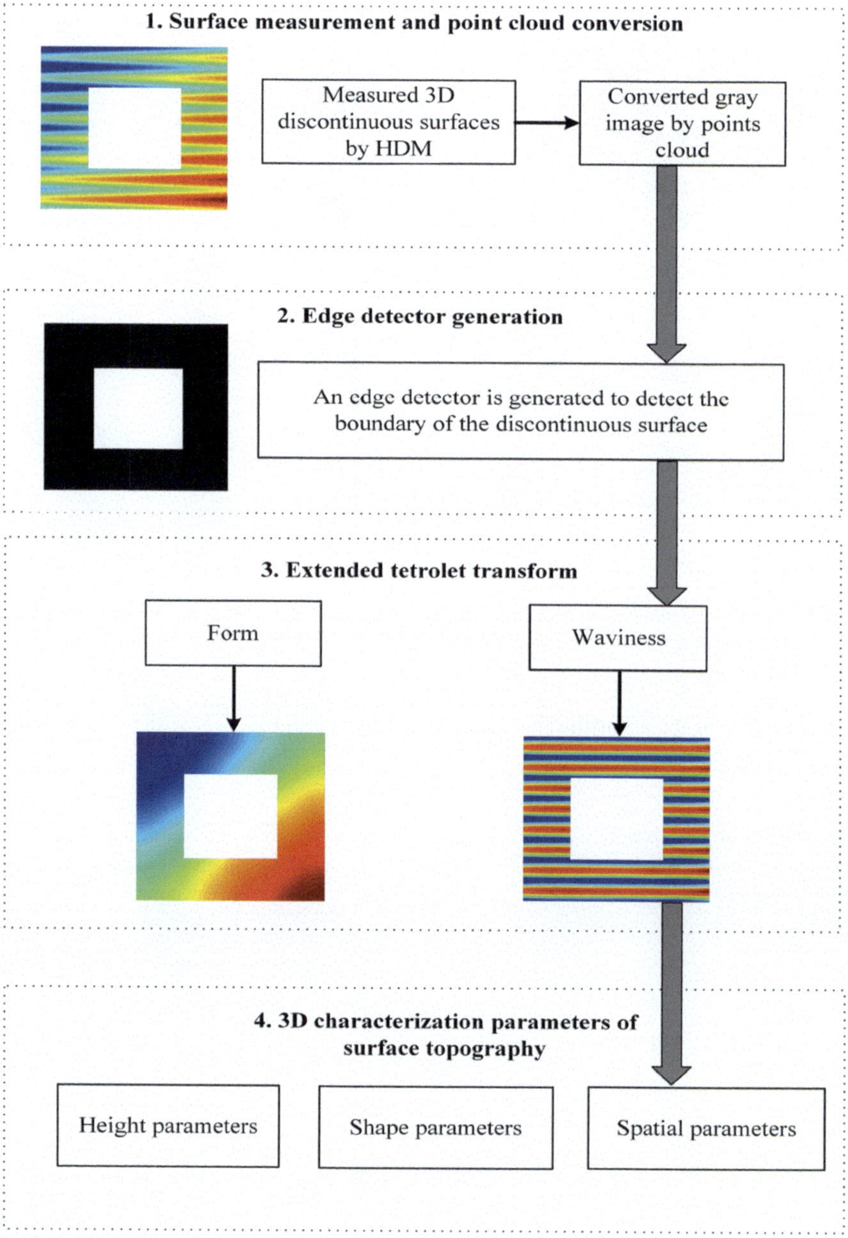

Fig. 5.3 The framework of the proposed method with four steps

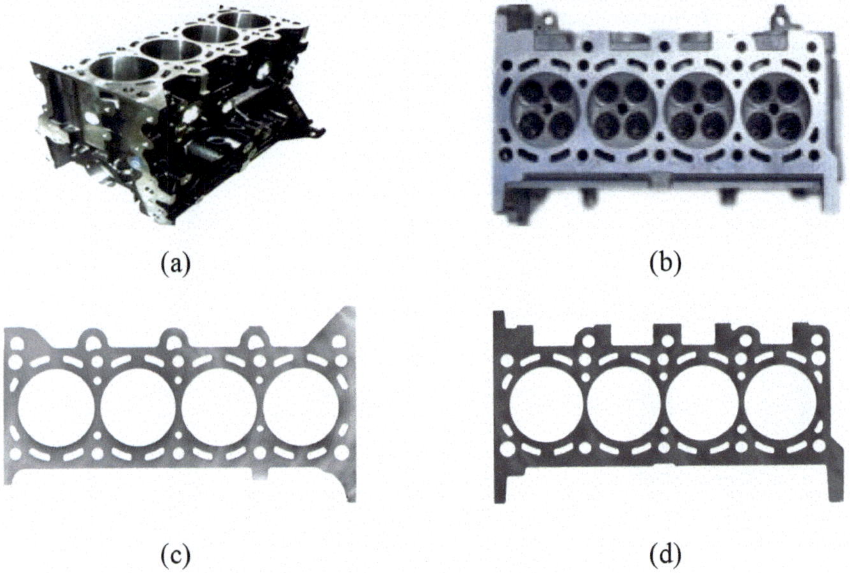

Fig. 5.4 **a** Engine cylinder block surface, **b** engine cylinder head surface, **c** the converted gray image of the block surface, **d** the converted gray image of the head surface

The structure of the algorithm in question is depicted in Fig. 5.5, with the following detailed steps outlined.

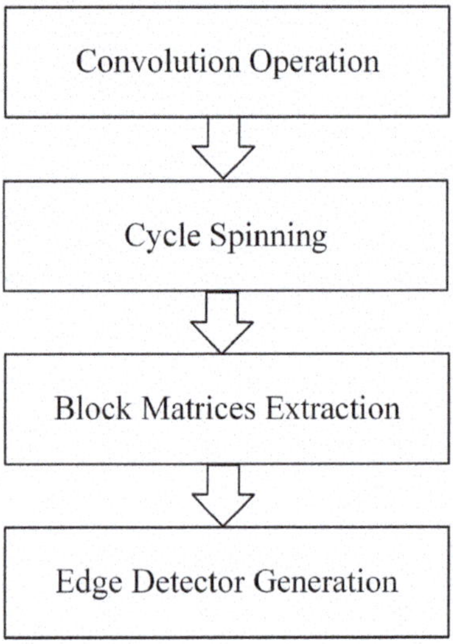

Fig. 5.5 The architecture of the developed algorithm

Step 1: Convolution operation

For a discontinuous surface $\mathbf{a}=[a(m_1, m_2)]_{m_1,m_2=0}^{N-1}$ can be represented by a matrix of size $N \times N$, the convolution sum $\mathbf{b}$ is calculated as:

$$\mathbf{b} = \mathbf{a} * \mathbf{h_1} + \mathbf{a} * \mathbf{h_2} \tag{5.4}$$

where the symbol $*$ means convolution operation, $\mathbf{b}$ has the same sizes with $\mathbf{a}$. The convolution kernels $\mathbf{h_1}$ and $\mathbf{h_2}$ are defined as:

$$\mathbf{h_1} = \begin{pmatrix} 1 & -1 \end{pmatrix}, \mathbf{h_2} = \begin{pmatrix} 1 & -1 \end{pmatrix}^T \tag{5.5}$$

The regions where the pixel dimensions undergo swift alterations are of paramount importance, marking the locations known as singular points on the surface. The convolution process is designed to identify these singular points by scanning the entire surface. The presence of boundaries and holes becomes evident through the pattern of these singular points' distribution.

Step 2: Cycle Spinning

The convolution sum $\mathbf{b}$ is also a matrix of size $N \times N$, let the elements of the Nth column and the Nth row of the matrix $\mathbf{b}$ be equal to 0. The new matrix $\mathbf{g_0}$ is defined as:

$$\mathbf{g_0} = \begin{pmatrix} (\mathbf{B_1})_{(N-1)\times(N-1)} & \mathbf{0}_{(N-1)\times 1} \\ \mathbf{0}_{1\times(N-1)} & \mathbf{0}_{1\times 1} \end{pmatrix} \tag{5.6}$$

Upon the identification of boundaries and holes, cycle spinning is implemented to address the absence of translational invariance and to mitigate the pseudo-Gibbs effects observed in the surface. This technique serves to smoothen the surface and minimize distortions. Specifically, it rearranges the sequence of surface pixels, which in turn alters the distribution of singular points across the surface, leading to a reduction in oscillation amplitude. The operator of cycle spinning $CS_{p,q}$ is defined as:

$$CS_{p,q}(g_0(m_1, m_2)) = g_0(\mathrm{mod}(m_1 + p, N), \mathrm{mod}(m_2 + q, N)) \tag{5.7}$$

where p, q are the numbers of shifts in column and row respectively.

Then the updated matrix $\mathbf{g}$ is defined as:

$$\mathbf{g} = CS_{0,-1}(CS_{-1,0}(\mathbf{g_0})) \tag{5.8}$$

Step 3: Block Matrices Extraction

Conforming to the aforementioned procedures, the dimensions of the processed surface remain unchanged from the original, which is not aligned with the block structure required for the tetrolet transform. To accommodate the 4 × 4 block structure inherent to the tetrolet transform, the surface must be downsized by a factor of four in both the x and y axes, achieved by selecting one out of every four points. Consequently, the resultant matrix A is then partitioned into block matrices. A new matrix $\mathbf{u_0}$ is calculated as:

$$\mathbf{u_0} = \mathbf{g} * \mathbf{K_4} \tag{5.9}$$

where $\mathbf{K_4}$ is a matrix of size 4 × 4 and the size $N_{u_0} \times N_{u_0}$ of $\mathbf{u_0}$ depends on $\mathbf{g} \cap \mathbf{K_4}$, $N_{u_0} = N - 4 + 1 = N - 3$. The matrix $\mathbf{K_4}$ is given as:

$$\mathbf{K_4} = \begin{pmatrix} 1 & 1 & 1 & 1 \\ 1 & 1 & 1 & 1 \\ 1 & 1 & 1 & 1 \\ 1 & 1 & 1 & 1 \end{pmatrix} \tag{5.10}$$

Let $\mathbf{u_0}=\{u_0(i,j), i = 0, \ldots, N_{u_0} - 1, j = 0, \ldots, N_{u_0} - 1\}$, the extracted block matrix $\mathbf{u}$ is defined as:

$$\mathbf{u}=\{u(i',j'), i' = 0, \ldots, N_u - 1, j' = 0, \ldots, N_u - 1\} \tag{5.11}$$

where $u(i',j') = u_0(4i', 4j')$ and $N_u = \lfloor (N_{u_0} - 1)/4 \rfloor$

Step 4: Edge Detector Generation

In this step, the edge detector's value is ascertained based on a predefined threshold, which is typically derived from the advanced knowledge of the original surface characteristics. Let $\mathbf{w}$ be the edge detector of the surface $\mathbf{a}$, it is defined as:

$$\mathbf{w}=\{w(i',j'), i' = 0, \ldots, N_u - 1, j' = 0, \ldots, N_u - 1\} \tag{5.12}$$

$$w(i',j') = \begin{cases} 0, & |u[i',j']| > \theta \\ 1, & |u[i',j']| \le \theta \end{cases} \tag{5.13}$$

where θ is the threshold to determine whether the pix point belongs to the surface or not. If the pix point belongs to the surface, $w(i',j') = 0$, otherwise $w(i',j') = 1$.

Example 1 Referencing the discontinuous surface segment depicted in Fig. 5.6, the edge detector w is produced by the aforementioned algorithm with $\theta = 0$. The boundary of the surface is clearly delineated.

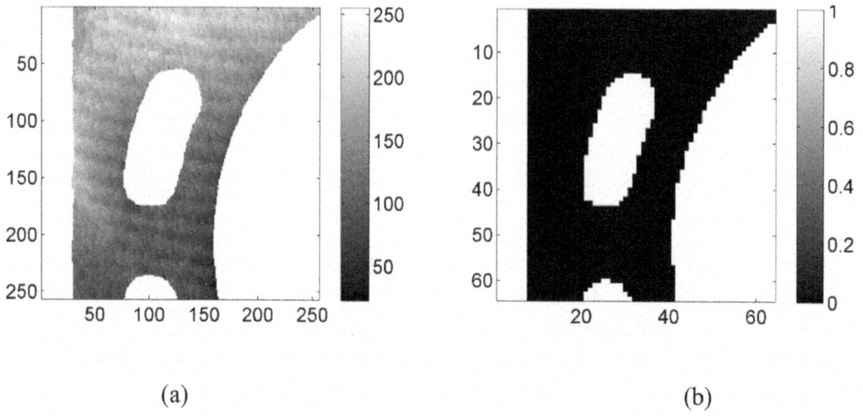

Fig. 5.6 **a** A discontinuous surface, **b** the edge detector

5.1.3.3 Extended Tetrolet Transform

The Idea of the Extended Tetrolet Transform

The ISO 16610-29 [30] has recognized a wavelet-based filtering technique for engineering surfaces. Sharing similar attributes with the tetrolet transform, which is a non-redundant, adaptive geometric wavelet transform, it allows for the decomposition of an original surface **a** into low-pass approximation components (low-frequency coefficients) a_i and high-pass multiscale detail components (high-frequency tetrolet coefficients) d_i. Denoting a_0 as the original surface, the logical tetrolet decomposition at J levels is as follows:

$$\begin{aligned} a_0 &= a_1 + d_1 \\ &= a_2 + d_2 + d_1 \\ &= a_3 + d_3 + d_2 + d_1 \\ &= \cdots \\ &= a_J + d_J + d_{J-1} + \cdots + d_1 \end{aligned} \quad (5.14)$$

Consequently, the multiresolution aspect of the tetrolet transform involves the creation of a sequence of approximations for the surface. The detailed steps are depicted in Fig. 5.7.

Inverting the ladder structure, the ith subsurface M_i is reconstructed using the reconstructed low-frequency coefficients $\widetilde{a}_i$ and the reconstructed high-frequency tetrolet coefficients $\widetilde{d}_i$. Hence, the i th subsurface M_i is formulated as:

$$M_i = \widetilde{a}_i + \widetilde{d}_i \quad (5.15)$$

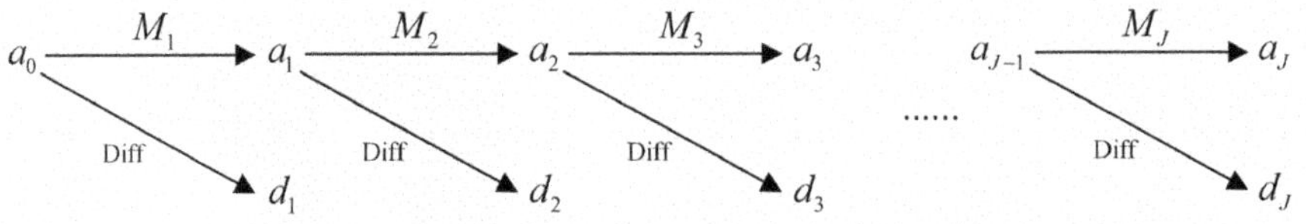

Fig. 5.7 A multiresolution separation using the extended tetrolet transform

As previously discussed, the conventional tetrolet transform does not accommodate the representation of decomposed and reconstructed subsurface. The down-sampling step in the decomposition phase reduces the dimensions of the low-frequency coefficients and high-frequency tetrolet coefficients to half of their previous size, making it impossible for the subsurface to match the original surface's dimensions. To address this for discontinuous surface filtering, the tetrolet transform is enhanced. The enhanced tetrolet transform is segmented into three components: decomposition, shrinkage, and reconstruction. A flowchart of the enhanced tetrolet transform algorithm is presented in Fig. 5.8, with the detailed procedures described below.

Module 1: Decomposition

For a discontinuous surface where $\mathbf{a}$=$[a(m_1, m_2)]_{m_1,m_2=0}^{N-1}$, r levels decomposition is applied. The decomposition rationale is equally applicable to the extended tetrolet transform, allowing the low-pass approximation component $a_{r-1}[m_1, m_2]$ to be further decomposed as follows:

$$a_{r-1}[m_1, m_2]=\sum_{l=1}^{3}\sum_{v=0}^{3} d_r^l[m_1, m_2]\psi_{I_v}^l[m_1, m_2] + \sum_{v=0}^{3} a_r[m_1, m_2]\phi_{I_v}[m_1, m_2] \tag{5.16}$$

where $I = \{I_v|v = 0, 1, 2, 3\} = \{(m_1, m_2) : m_1, m_2 = 0, \ldots N - 1\} \subset \mathbb{Z}^2$, $r = 1, 2, \ldots J - 1, N = 2^J, J \in \mathbb{N}$.

Step 1: In the rth level, the low-pass approximation part a_{r-1} is divided into blocks Q of size 4×4.

Step 2: Each block Q has 117 tetromino tillings $k = 1, \ldots, 117$. The tetromino subsets $I_v^{(k)}(v = 0, 1, 2, 3)$ is ordered by the pixel size, and the corresponding mapping $L(m', n') = \{0, 1, 2, 3\}$, $m', n' \in I_v^{(k)}$.

(a) When the edge detector $w(i', j') = 0$, it indicates that the pixel point is part of the surface and not an edge. In such cases, the Haar wavelet transform is applied to each tile k, yielding the low-frequency coefficients $a_r^{(k)}$ and the high-frequency tetrolet coefficients $d_r^{l,(k)}$ for $l = 1, 2, 3$ as detailed below:

$$a_r^{(k)} = a_r^{(k)}[v]|_{v=0}^{3} \text{ with } a_r^{(k)}[v] = \sum_{(m',n')\in I_v^{(k)}} \epsilon[0, L(m', n')]a_{r-1}[m', n'] \tag{5.17}$$

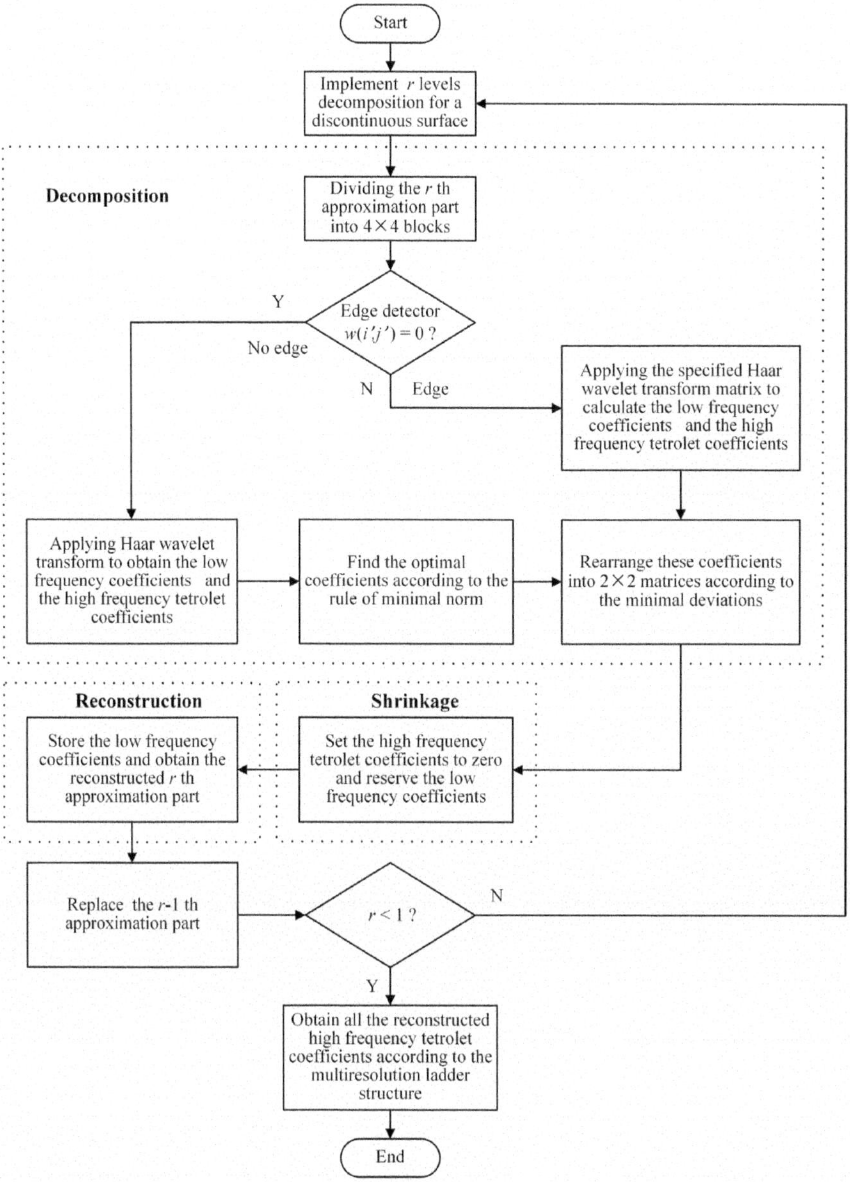

Fig. 5.8 Extended tetrolet transform algorithm

$$d_r^{l,(k)} = d_r^{l,(k)}[v]\big|_{v=0}^{3} \text{ with } d_r^{l,(k)}[v] = \sum_{(m',n')\in I_v^{(k)}} \epsilon[l, L(m',n')]a_{r-1}[m',n'] \tag{5.18}$$

According to the rule of minimal norm which satisfies:

$$k^* = \arg\min \sum_{l=1}^{3}\sum_{v=0}^{3} \left|d_r^{l,(k)}[v]\right| \tag{5.19}$$

The optimal low frequency coefficients and high frequency tetrolet coefficients $\left[a_r^{(k^*)}, d_r^{1,(k^*)}, d_r^{2,(k^*)}, d_r^{3,(k^*)}\right]$ are found.

(b) Given that the edge detector $w(i',j') = 1$, it signifies that the pixel point is at the edge and not part of the surface. For such blocks Q that meet the tetromino tiling $k = 1$, the low-frequency coefficients $a_r^{(k)}$ and the high-frequency tetrolet coefficients $d_r^{l,(k)}$ for $l = 1, 2, 3$ are derived using the designated Haar wavelet transform matrix **ha**.

$$\mathbf{ha} = \begin{pmatrix} \mathbf{ha}_1 & \mathbf{ha}_2 & \mathbf{0} & \mathbf{0} \\ \mathbf{ha}_3 & \mathbf{ha}_4 & \mathbf{0} & \mathbf{0} \\ \mathbf{0} & \mathbf{0} & \mathbf{ha}_1 & \mathbf{ha}_2 \\ \mathbf{0} & \mathbf{0} & \mathbf{ha}_3 & \mathbf{ha}_4 \end{pmatrix} = \left\{ha_{ij} | i, j = 1, 2\ldots, 16\right\} \tag{5.20}$$

$$a_r^{(k)} = \left\{ \sum_{j=1}^{16} ha_{ij} \times a_{r-1}[m',n'] | i = 1, 5, 9, 13 \right\} \tag{5.21}$$

$$d_r^{l,(k)} = \left\{ \sum_{j=1}^{16} ha_{ij} \times a_{r-1}[m',n'] | i = 2, 3, 4, 6, 7, 8, 10, 11, 12, 14, 15, 16 \right\} \tag{5.22}$$

where $\mathbf{ha}_1 = \frac{1}{2}\begin{pmatrix} 1 & 1 & 0 & 0 \\ 1 & 1 & 0 & 0 \\ 1 & -1 & 0 & 0 \\ 1 & -1 & 0 & 0 \end{pmatrix}$, $\mathbf{ha}_2 = \frac{1}{2}\begin{pmatrix} 1 & 1 & 0 & 0 \\ -1 & -1 & 0 & 0 \\ 1 & -1 & 0 & 0 \\ -1 & 1 & 0 & 0 \end{pmatrix}$, $\mathbf{ha}_3 = \frac{1}{2}\begin{pmatrix} 0 & 0 & 1 & 1 \\ 0 & 0 & 1 & 1 \\ 0 & 0 & 1 & -1 \\ 0 & 0 & 1 & -1 \end{pmatrix}$

and $\mathbf{ha}_4 = \frac{1}{2}\begin{pmatrix} 0 & 0 & 1 & 1 \\ 0 & 0 & -1 & -1 \\ 0 & 0 & 1 & -1 \\ 0 & 0 & -1 & 1 \end{pmatrix}$.

Step 4: Determine the deviations from the tiling pattern, and then organize the optimal low-frequency coefficients $a_r^{(k^*)}$ and high-frequency tetrolet coefficients $d_r^{l,(k^*)}$ into a 2×2 matrix based on the smallest deviations observed.

i.e. $a_r = \mathrm{R}(a_r^{(k^*)}) = \begin{bmatrix} a_r^{(k^*)}[0], \ a_r^{(k^*)}[2] \\ a_r^{(k^*)}[1], \ a_r^{(k^*)}[3] \end{bmatrix}$ and $d_r^l = \mathrm{R}(d_r^{l,(k^*)}) = \begin{bmatrix} d_r^{l,(k^*)}[0], \ d_r^{l,(k^*)}[2] \\ d_r^{l,(k^*)}[1], \ d_r^{l,(k^*)}[3] \end{bmatrix}$.

Module 2: Shrinkage

Once the decomposition is complete, a shrinkage step is implemented on the resulting coefficients. Utilizing the globally recognized hard threshold function $S_\alpha(x)$ within the extended tetrolet transform, the threshold parameter α is contingent upon the specific number of the most significant tetrolet coefficients that are maintained.

Step 5: Zero out the high-frequency tetrolet coefficients while retaining the low-frequency coefficients. The specific function is detailed as follows:

$$S_\alpha(d_r^{l,(k^*)}) = \begin{cases} d_r^{l,(k^*)}, & \left|d_r^{l,(k^*)}\right| \geq \alpha \\ 0, & \left|d_r^{l,(k^*)}\right| < \alpha \end{cases} \tag{5.23}$$

where $\alpha = M$ and M is a significantly large positive constant that exceeds the absolute values of all high-frequency tetrolet coefficients.

Module 3: Reconstruction

To maintain the approximation segment at a size equivalent to the original surface area **a**, the rth approximation segment a_r must be reconstructed. The process of reconfiguring the coefficients can be referred to as the inverse extended tetralet transform.

Step 6: The reconstructed equation is listed below:

$$\widetilde{a_r}[m_1, m_2] = \sum_{l=1}^{3}\sum_{v=0}^{3} S_\alpha(d_r^l[m_1, m_2])\psi_{I_v}^l[m_1, m_2] + \sum_{v=0}^{3} a_r[m_1, m_2]\phi_{I_v}[m_1, m_2] \tag{5.24}$$

where $I = \{I_v | v = 0, 1, 2, 3\} = \{(m_1, m_2) : m_1, m_2 = 0, \ldots N-1\} \subset \mathbb{Z}^2, N = 2^J$, $J \in \mathbb{N}$.

Step 7: Store the reconstructed $\widetilde{a_r}$, and replace the low-pass approximation part a_{r-1} by a_{r-2} until $r < 1$.

Step 8: Obtain all the reconstructed high frequency tetrolet coefficients $\widetilde{d_r^l}$ according to the multiresolution ladder structure, such as $\widetilde{d_r^l} = \widetilde{a}_{r-1} - \widetilde{a}_r$.

Transformation Between Decomposition Levels and Surface Components

Surface filtering primarily deals with the differentiation of topographical elements. These elements are typically identified by their wavelength or frequency, as reflected in the terminology of form, waviness, and roughness. Form encompasses all low-frequency components, whereas waviness and roughness represent high-frequency or short-wavelength components that are inherent to every engineering surface. Consequently, a surface can be bifurcated into low-pass and high-pass segments.

Example 2 A 3D engineering surface:

$$S(x, y) = L(x, y) + H(x, y) \tag{5.25}$$

where $L(x, y)$ is the low-pass part which represents the form components, and $H(x, y)$ is the high-pass part which represents the waviness and roughness components. It is clear that it has $L(x, y) = F(x, y)$ and $H(x, y) = W(x, y) + R(x, y)$, where $F(x, y)$ is the form, $W(x, y)$ is the waviness and $R(x, y)$ is the roughness.

Given that the HDM data encompasses solely the 3D surface form and waviness details, the original surface adheres to the subsequent convention:

$$\begin{aligned} S(x, y) &= L(x, y) + H(x, y) \\ &= F(x, y) + W(x, y) \end{aligned} \tag{5.26}$$

For elucidating the general perception of surface components across various decomposition levels, it is essential to establish a link between these levels and the actual surface components. In a manner akin to "transmission bandwidth", a surface variable $M_{i,j}$ is formulated by determining the disparity between the ith subsurface M_i and the jth subsurface M_j.

$$\begin{aligned} M_{i,j} &= M_i - M_j \\ &= \tilde{a}_i + \tilde{d}_i - \tilde{a}_j - \tilde{d}_j \\ &= \tilde{d}_{j-1} + \tilde{d}_{j-2} + \cdots + \tilde{d}_i \quad (i < j) \end{aligned} \tag{5.27}$$

where the level i is corresponding to the cutoff wavelength λ_i, and the level j is corresponding to the cutoff wavelength λ_j. They satisfy that $\lambda_i = \Delta s \times 2^{i+1}$ and $\lambda_j = \Delta s \times 2^{j+1}$ in the extended tetrolet transform, Δs is the sampling interval.

According to ISO 4287, the surface components such as roughness, waviness, and form are delineated by specific cutoff wavelengths λ_s, λ_c, and λ_f. Specifically, λ_s signifies the boundary where the very short wave components meet the roughness, λ_c marks the boundary between the roughness and waviness, and λ_f indicates the boundary where the waviness transitions into form components.

As previously stated, the HDM data encompasses solely the 3D surface form and waviness details. Consequently, the surface component with wavelengths between λ_c and λ_f is classified as waviness, while the remaining part is attributed to the surface form. In this context, let $\lambda_1 = \lambda_c$ and $\lambda_j = \lambda_f$ represent the wavelengths corresponding to level j, which is determined by the specified cutoff wavelength λ_f. The suggested values for λ_f are detailed in ISO 16610-29 [30], including 2.5 um, 8 um, 25 um, 80 um, 250 um, 0.8 mm, 2.5 mm, 8 mm, 25 mm, and so forth. The surface waviness $W(x, y)$ and form $F(x, y)$ can be derived accordingly.

$$\begin{aligned} W(x, y) &= M_{1,j} = M_1 - M_j \\ &= \widetilde{d}_{j-1} + \widetilde{d}_{j-2} + \ldots + \widetilde{d}_i \quad (j > 1) \end{aligned} \tag{5.28}$$

$$\begin{aligned} F(x, y) &= M_1 - W(x, y) = M_j \\ &= \widetilde{a}_j + \widetilde{d}_j \\ &= \widetilde{a}_{j-1} \quad (j > 1) \end{aligned} \tag{5.29}$$

where M_1 is the reconstructed original surface.

5.1.3.4 3D Characterization Parameters of Surface Topography

For a more precise assessment of the proposed filtering method's efficacy, areal surface texture parameters are utilized to appraise the outcomes of the filtering process. While numerous organizations have defined profile parameters for surface characterization, and many have been incorporated into standards, these profile parameters alone cannot furnish comprehensive and dependable data on 3D surface topography. In contrast, areal parameters present a compelling and practical alternative. Drawing from the current areal surface texture standard ISO 25178-2 [31], a selection of key parameters is made, encompassing height parameters (S_q, S_p, S_v, S_z, S_a), shape parameters (S_{sk}, S_{ku}), and spatial parameters (S_{al}, S_{tr}).

(1) Height parameters

The parameter S_q, denoting the root mean square height, is specified as:

$$S_q = \sqrt{\frac{1}{A} \iint_A z^2(x, y) dxdy} \tag{5.30}$$

where A is a definition area and $z(x, y)$ is the height value of the surface at position x, y.

The maximum height parameter S_z is defined as:

$$S_z = S_p + S_v \tag{5.31}$$

where S_p is maximum peak height parameter and S_v is the maximum pit height parameter within the definition area A.

The arithmetical mean height parameter S_a is defined as:

$$S_a = \frac{1}{A}\iint_A |z(x, y)|dxdy \tag{5.32}$$

(2) Shape parameters

The skewness of height distribution parameter S_{sk} is defined as:

$$S_{sk}=\frac{1}{S_q^3}\left[\frac{1}{A}\iint_A z^3(x, y)dxdy\right] \tag{5.33}$$

The kurtosis of height distribution parameter S_{ku} is defined as:

$$S_{ku}=\frac{1}{S_q^4}\left[\frac{1}{A}\iint_A z^4(x, y)dxdy\right] \tag{5.34}$$

(3) Spatial parameters

The spatial parameter S_{al} is the horizontal distance of $f_{ACF}(\tau_x, \tau_y)$ which has the fastest decay to η.

$$S_{al} = \min_{\tau_x,\tau_y \in R} \sqrt{\tau_x^2 + \tau_y^2}\ (R = \{(\tau_x, \tau_y) : f_{ACF}(\tau_x, \tau_y) \le \eta\}) \tag{5.35}$$

where $f_{ACF}(\tau_x, \tau_y)$ is the autocorrelation function and η is a specified constant with $\eta \in (0, 1)$. The default value is $\eta = 0.2$.

The surface texture aspect ratio parameter S_{tr} is defined as:

$$S_{tr} = \frac{\min\limits_{\tau_x,\tau_y \in R} \sqrt{\tau_x^2 + \tau_y^2}}{\max\limits_{\tau_x,\tau_y \in R} \sqrt{\tau_x^2 + \tau_y^2}}\ (R = \{(\tau_x, \tau_y) : f_{ACF}(\tau_x, \tau_y) \le \eta\}) \tag{5.36}$$

5.1.4 Numerical Simulation

Gaussian filter [32], which is the prevailing standard in filtering techniques, is endorsed in ISO 16610-21, superseding the 2RC filter. However, the Gaussian filter is limited to continuous surfaces due to significant edge effects. The Spline filter, capable of mitigating edge distortion, is also recommended in ISO 16610-22 [33].

These filters have been extended to areal Gaussian and areal Spline filters for 3D surfaces [9, 34]. In contrast, the proposed extended tetralet transform offers effective filtering solutions for both continuous and discontinuous surfaces. To demonstrate the superior efficiency of the proposed method, a comparison is made among the three methods concerning their filtering characteristics and outcomes.

5.1.4.1 Filtering Properties

To ensure effective surface filtering, the extended tetralet transform adheres to several key properties: (1) Finite Impulse Response (FIR), with a finite tetromino support, (2) linear phase, where the compact and symmetrical nature of the tetrolet prevents distortion in filtered profiles, and (3) shape amplitude transmission, which ensures that a high-pass filter passes almost all wavelengths below a specific cut-off without loss of amplitude and significantly reduces wavelengths above the cut-off, with the inverse being true for the low-pass filter.

It is widely recognized that an ideal filtering method should exhibit a sharp amplitude transmission curve near the cut-off wavelength [32–34]. This curve signifies that the filter allows 50% transmission at the cut-off wavelength, implying a preference for a steeper amplitude transmission profile in surface filtering. Figure 5.9 depicts the low-pass and high-pass amplitude transmission characteristics of the areal Gaussian filter, areal spline filter, and the extended tetralet transform, with cut-off wavelengths in both x and y directions set at 0.8mm. For a more detailed comparison, Fig. 5.10 displays the one-dimensional low-pass amplitude transmission characteristics of these three filters. It is evident that the extended tetralet transform has the steepest amplitude transmission profile, indicating its effectiveness in distinguishing surface topography features.

A synthetic continuous surface is created using the equation for $S(x, y) = 0.05x + 0.05y - 0.5\cos(2\pi x/6.4)$. Evidently, the cosine component represents the surface waviness, while the remaining component represents the surface form. The sampling interval is 0.2 mm in both the x and y directions, with the simulated surface dimensions being 51.2 mm × 51.2 mm. The surface has 256 × 256 sampling points, and the wavelength of the surface waviness is 6.4 mm. The continuous surface is depicted in Fig. 5.11a. Conversely, the discontinuous surface is derived by removing segments from the continuous surface, as illustrated in Fig. 5.11b.

For both synthetic surfaces, the areal Gaussian filter, areal spline filter, and extended tetralet transform are utilized to differentiate between form and waviness. The cutoff wavelength λ_f for the areal Gaussian and areal spline filters is set at 25 mm, denoted as $\lambda_f = 25$ mm. For the extended tetralet transform, the wavelength cutoff is $\lambda_f = 0.2 \times 2^{6+1} = 25.6$ mm (nearly equivalent to 25 mm) with a decomposition level of $i = 6$. Consequently, the waviness is denoted as $W(x, y) = M_{1,6} = M_1 - M_6 = \widetilde{d}_5 + \widetilde{d}_4 + \widetilde{d}_3 + \widetilde{d}_2 + \widetilde{d}_1$ and the form as $F(x, y) = M_1 - W(x, y) = M_6 = \widetilde{a}_6 + \widetilde{d}_6 = \widetilde{a}_5$. Figure 5.12 displays the filtering outcomes for both continuous and discontinuous surfaces using the areal Gaussian filter, while Fig. 5.13 presents the same for the areal spline filter. Additionally,

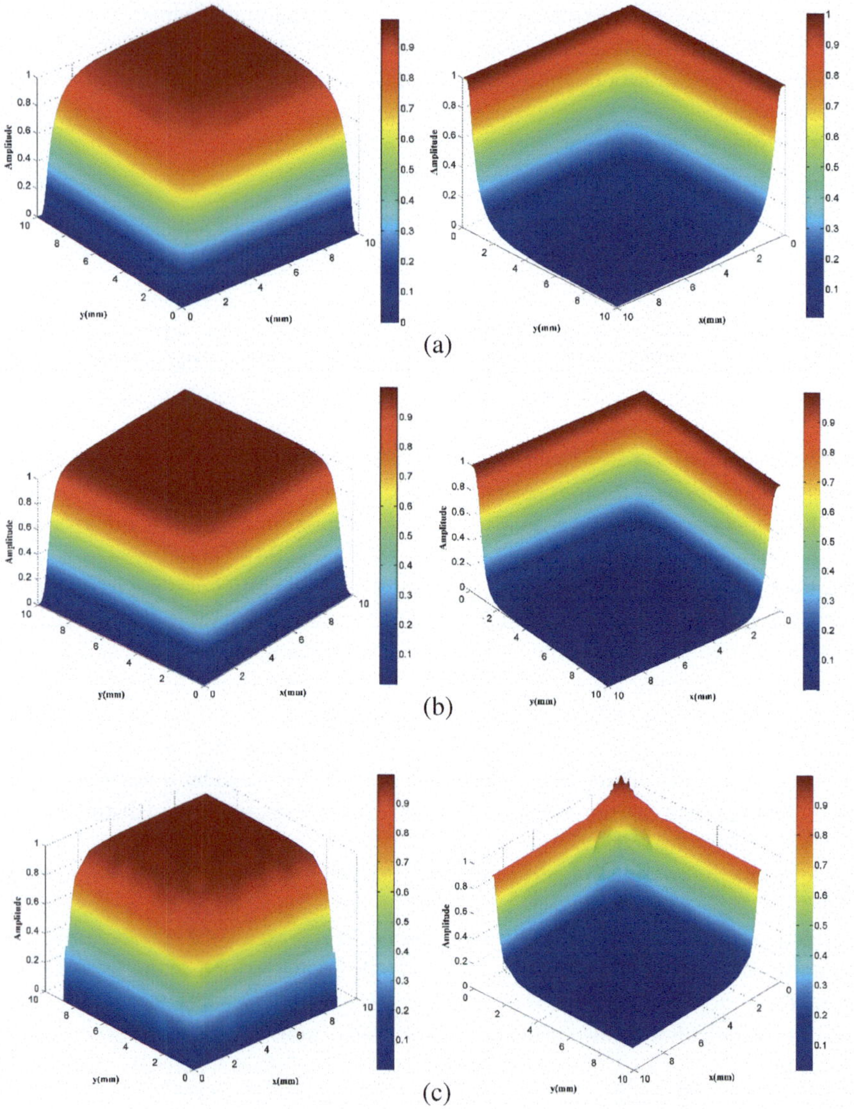

Fig. 5.9 **a** Amplitude transmission characteristics of areal Gaussian filter, **b** amplitude transmission characteristics of areal spline filter, **c** amplitude transmission characteristics of the extended tetrolet transform

Fig. 5.14 depicts the filtered results for both surfaces using the extended tetralet transform. In Fig. 5.12, edge distortions are noticeable in the areal Gaussian filter results for both continuous and discontinuous surfaces, particularly at boundaries and holes. To mitigate these edge effects and assess surface parameters, a half-cutoff wavelength is typically excluded from the start and end of the surface when employing

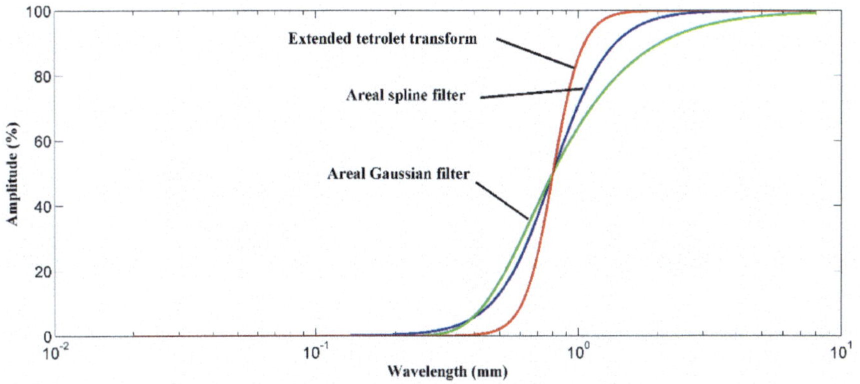

Fig. 5.10 Low-pass amplitude transmission characteristics of the three filters

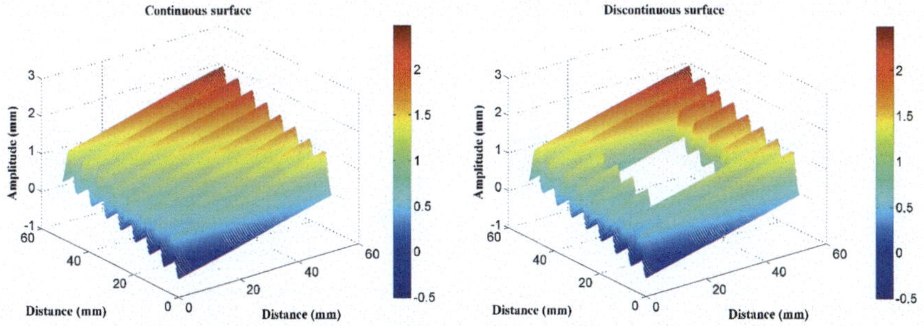

Fig. 5.11 **a** Continuous surface, **b** discontinuous surface

the areal Gaussian filter [34]. In accordance with ISO 16610-21 and 16610-61 [32, 34], the areal Gaussian filtered results for the continuous surface are regarded as the benchmark filtered outcomes, as shown in Fig. 5.15a.

Using the standard filtered results as a benchmark, for the continuous surface, it is evident that the form and waviness extracted by the areal spline filter and the extended tetralet transform closely resemble those obtained from the cutoff areal Gaussian filter (refer to Figs. 5.13a, 5.14a and 5.15a). However, for the discontinuous surface, significant discrepancies are observed in the filtering outcomes. In the case of areal Gaussian filtering, pronounced edge distortions are present at both the boundary and the hole. After accounting for the end effect, the remaining cutoff areal Gaussian filtered results for the discontinuous surface are insufficient for parameter evaluation, as depicted in Fig. 5.15b. In the areal spline filtering, while the end effect is mitigated at the boundary, it persists near the holes (see Fig. 5.13b), a phenomenon less prevalent with the extended tetralet transform (see Fig. 5.14b). Specifically, the form and waviness are substantially distorted near the holes by both the areal Gaussian and areal spline filters, yet the extended tetralet transform maintains accurate separation of the surface components.

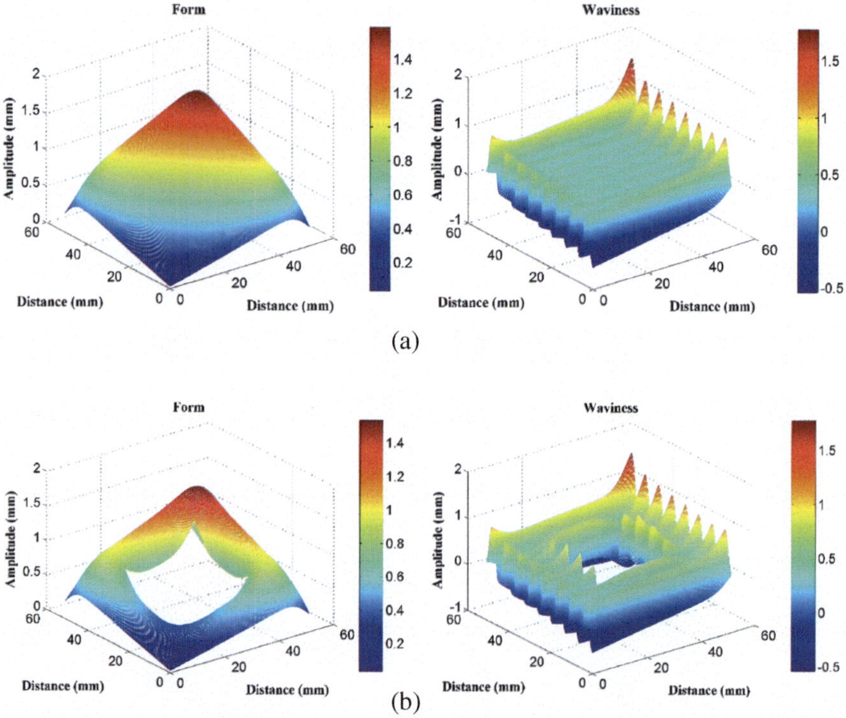

Fig. 5.12 Results of areal Gaussian filter, **a** continuous surface, **b** discontinuous surface

Building on this, the 3D characterization parameters outlined in Sect. 3.5 are utilized for a quantitative assessment of the filtered surface components. The cutoff areal Gaussian filtered results from the continuous surface are accepted as the standard, and the corresponding standard waviness parameters are detailed in Table 5.1. In accordance with ISO 25178-2 [31], the evaluation of filtered waviness parameters necessitates calculation within a defined area, the side length of which should be twice the cutoff wavelength of waviness. However, for the discontinuous surface, the defined area from the cutoff areal Gaussian filtered results is inadequate for parameter evaluation. Despite this, the discontinuous surface, created by removing sections from the continuous surface, does not affect the parameter values within a defined area. Consequently, the waviness parameters from the cutoff areal Gaussian filtered results of the continuous surface are used as the reference for both continuous and discontinuous surfaces. Tables 5.2 and 5.3 compare the waviness parameters derived using the areal spline filter and the extended tetralet transform for both surfaces. The differences are computed using the formula:

$$\mathrm{Diff}_1 = \frac{\mathrm{par(areal\ spline\ filter)} - \mathrm{par(cutoff\ areal\ Gaussian\ filter)}}{\mathrm{par(cutoff\ areal\ Gaussian\ filter)}} \tag{5.37}$$

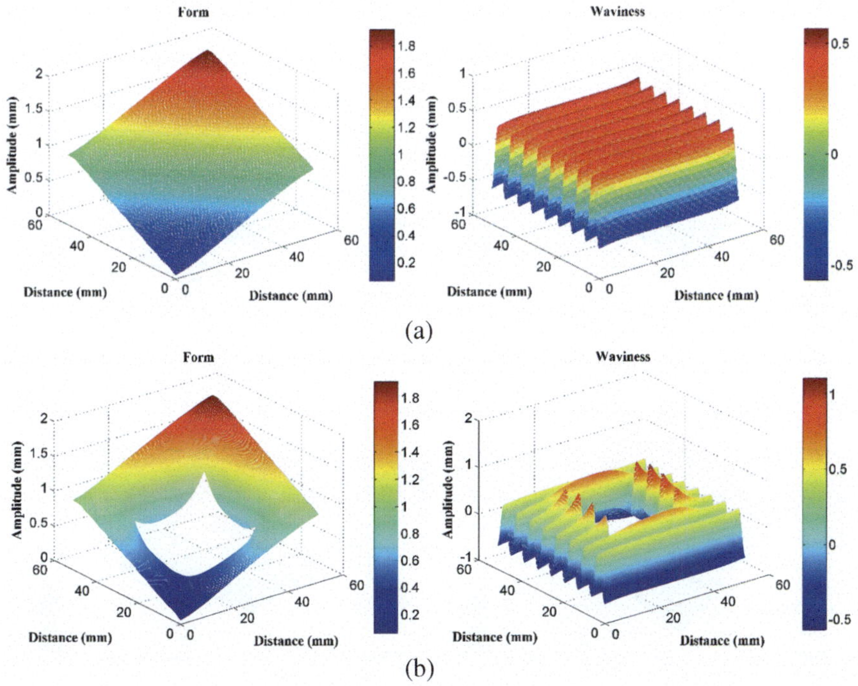

Fig. 5.13 Results of areal spline filter, **a** continuous surface, **b** discontinuous surface

$$\text{Diff}_2 = \frac{\text{par(extended tetrolet transform)} - \text{par(cutoff areal Gaussian filter)}}{\text{par(cutoff areal Gaussian filter)}} \tag{5.38}$$

where Diff_1 and Diff_2 represent the deviations from the filtered results by the areal spline filter and the extended tetralet transform, respectively, and $\text{par}(\cdot)$ signifies the respective surface parameters.

In this context, a deviation of 10% is deemed acceptable, signifying the absence of edge distortions. Table 5.2 shows that the waviness parameter discrepancies for the continuous surface across the cutoff areal Gaussian filter, areal spline filter, and the extended tetralet transform are within 2%, suggesting that all three filtering techniques are appropriate for continuous surfaces. For the discontinuous surface detailed in Table 5.3, the waviness parameter deviations for the areal spline filter exceed 20%, whereas for the extended tetralet transform, they remain below 7%. This indicates that the proposed extended tetralet transform is effective in separating the components of the discontinuous surface without introducing edge distortions.

To assess the degree of similarity among the surface components filtered by the three aforementioned methods, Pearson's correlation coefficient r is employed [35, 36]. It is determined by the formula:

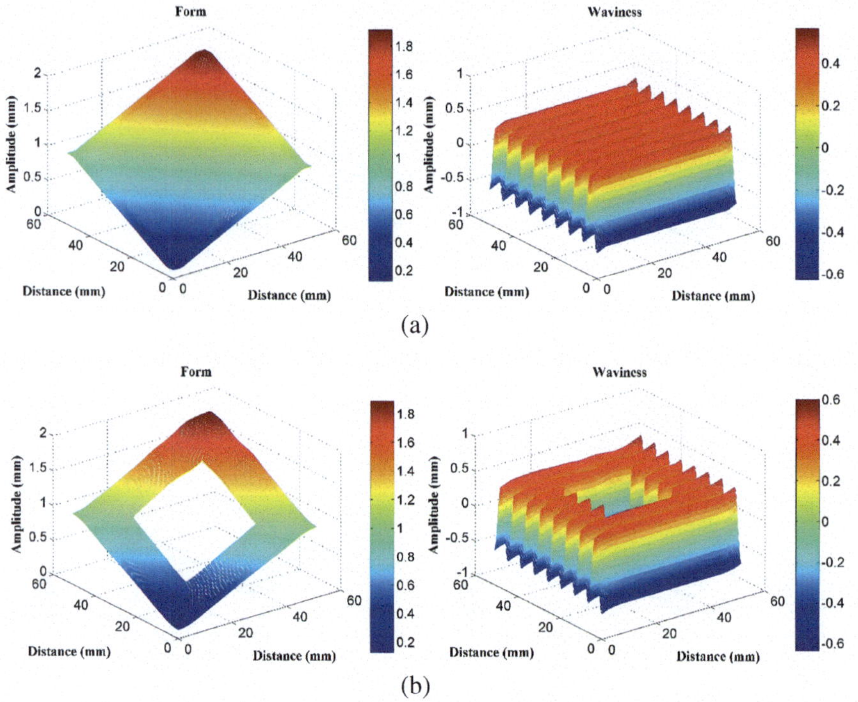

Fig. 5.14 Results of the extended tetrolet transform, **a** continuous surface, **b** discontinuous surface

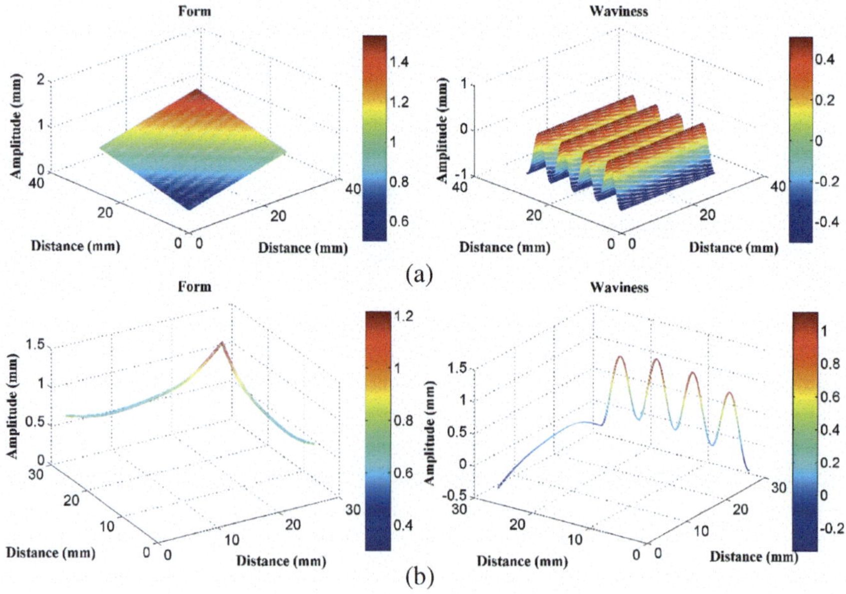

Fig. 5.15 Results of cutoff areal Gaussian filter, **a** continuous surface, **b** discontinuous surface

Table 5.1 The standard values of waviness parameters

The cutoff areal Gaussian filtered results of the continuous surface					
S_q/mm	0.357	S_p/mm	0.520	S_v/mm	0.484
S_z/mm	1.004	S_a/mm	0.322	S_{sk}	0.066
S_{ku}	1.493	S_{al}	7.066	S_{tr}	0.220

Table 5.2 The comparisons of the waviness parameters in the continuous surface

Simulation	Continuous surface		Diff_1	Diff_2
Method	Areal spline	Extended tetrolet	Areal spline (%)	Extended tetrolet (%)
S_q/mm	0.354	0.357	– 1.02	0.04
S_p/mm	0.528	0.524	1.52	0.71
S_v/mm	0.491	0.485	1.45	0.33
S_z/mm	1.019	1.009	1.44	0.48
S_a/mm	0.319	0.323	– 1.03	0.04
S_{sk}	0.066	0.066	– 0.36	– 0.20
S_{ku}	1.494	1.492	0.08	– 0.02
S_{al}	7.073	7.071	0.10	0.07
S_{tr}	0.222	0.221	0.81	0.26

Table 5.3 The comparisons of the waviness parameters in the discontinuous surface

Simulation	Discontinuous surface		Diff_1	Diff_2
Method	Areal spline	Extended tetrolet	Areal spline (%)	Extended tetrolet (%)
S_q/mm	0.464	0.358	29.98	0.15
S_p/mm	0.786	0.547	51.24	5.14
S_v/mm	0.585	0.513	20.96	6.08
S_z/mm	1.371	1.060	36.59	5.54
S_a/mm	0.422	0.325	30.79	0.78
S_{sk}	0.092	0.063	38.25	– 4.88
S_{ku}	1.991	1.497	33.40	0.29
S_{al}	8.711	7.073	23.28	0.10
S_{tr}	0.279	0.222	26.57	0.81

$$r(X,Y) = \frac{Cov(X,Y)}{\sqrt{Var(X)Var(Y)}} = \frac{\sum_m \sum_n (X_{mn} - \overline{X})(Y_{mn} - \overline{Y})}{\sqrt{(\sum_m \sum_n \sum (X_{mn} - \overline{X})^2)(\sum_m \sum_n (Y_{mn} - \overline{Y})^2)}} \tag{5.39}$$

Table 5.4 Correlation coefficient of the continuous surface

$r(F_{c\_Standard}, F_{c\_Spline})$	0.9999	$r(F_{c\_Standard}, F_{c\_Tetrolet})$	1.000
$r(W_{c\_Standard}, W_{c\_Spline})$	0.9999	$r(W_{c\_Standard}, W_{c\_Tetrolet})$	1.000

Table 5.5 Correlation coefficient of the discontinuous surface

$r(F_{disc\_Spline}, F_{c\_Spline})$	0.9622	$r(F_{disc\_Tetrolet}, F_{c\_Tetrolet})$	0.9994
$r(W_{disc\_Spline}, W_{c\_Spline})$	0.9418	$r(W_{disc\_Tetrolet}, W_{c\_Tetrolet})$	0.9990

where X and Y are two matrices of size $m \times n$, and $\overline{X}$ and $\overline{Y}$ are the mean values of the respective matrices. The correlation coefficient $r \in [0, 1]$, with values closer to 1 indicating a stronger similarity. Table 5.4 presents the correlations among the different filtering techniques, where $F_{c\_Standard}$, F_{c_Spline}, $F_{c\_Tetrolet}$ and $W_{c\_Standard}$, W_{c_Spline}, $W_{c\_Tetrolet}$ represent the filtered form and waviness of the continuous surface using the cutoff areal Gaussian filter, areal spline filter, and extended tetrolet transform, respectively. Table 5.5 compares the correlations between the filtered results of the continuous and discontinuous surfaces, with F_{disc_Spline}, $F_{disc\_Tetrolet}$ and W_{disc_Spline}, $W_{disc\_Tetrolet}$ denoting the filtered form and waviness of the discontinuous surface using the same filters. The high correlation coefficients evident in Tables 5.4 and 5.5 further substantiate the suitability of the extended tetrolet transform for filtering discontinuous surfaces.

5.1.5 *Case Study*

5.1.5.1 Case Study I

A steel grooved plate, characterized as a discontinuous surface with pronounced edges, is depicted in Fig. 5.16. The actual plate surface is illustrated in Fig. 5.16a, while the corresponding gray-scale image is presented in Fig. 5.16b. The processing time required for the extended tetralet transform is contingent upon the dimensions of the HDM data for the component. With computational efficiency and the square area requirement of the extended tetralet transform in mind, a square section of the plate is arbitrarily chosen, as indicated in Fig. 5.16b.

The dimensions of the chosen surface measure 51.2 mm × 51.2 mm, with a sampling interval of 0.2 mm. The sampling points total 256 in both the X and Y directions. According to the roughness and waviness specifications $Ra = 1.6$ um, $Rz = 6.3$ um, and $Wt = 2$ um, the cutoff wavelengths for roughness are 0.08 and 0.8 mm, while those for waviness are 0.8 and 8 mm [37]. Given that the HDM data encompasses only the 3D surface form and waviness, the cutoff wavelength is set at $\lambda_f = 8$ mm. For the extended tetrolet transform, the corresponding cutoff wavelength

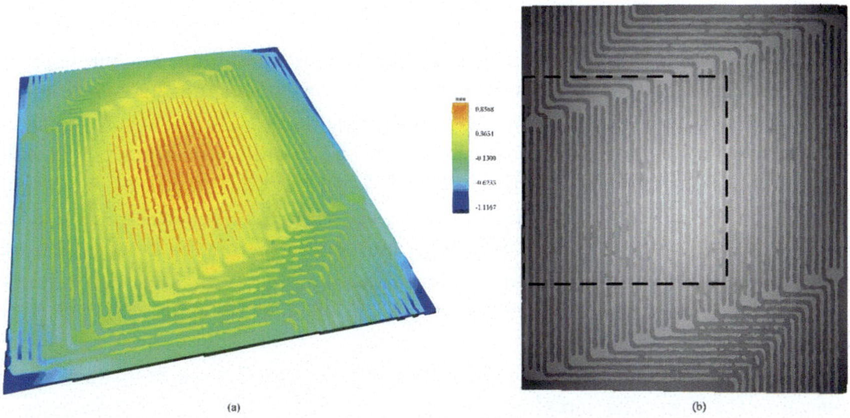

Fig. 5.16 Discontinuous surface, **a** steel plate surface, **b** the converted gray image

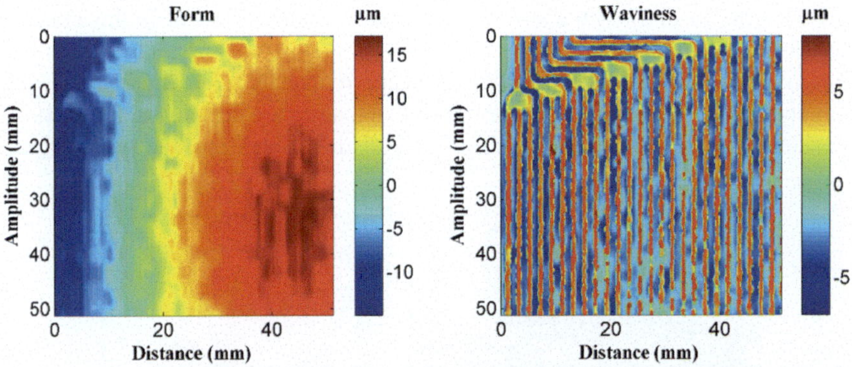

Fig. 5.17 Plate surface in 2D view

is $\lambda_f = 0.2 \times 2^{4+1} = 6.4$ mm (nearly equivalent to 8 mm) with a specific decomposition level $i = 4$. Thus, the waviness is denoted as $W(x, y) = M_{1,4} = M_1 - M_4 = \widetilde{d}_3 + \widetilde{d}_2 + \widetilde{d}_1$ and the form as $F(x, y) = M_1 - W(x, y) = M_4 = \widetilde{a}_4 + \widetilde{d}_4 = \widetilde{a}_3$. The selected area contains approximately 0.07 million points, with a computational time of 8.6 s. The filtering outcomes are displayed in Fig. 5.17. It is evident that the form and waviness are fully separated, with no distortions at the boundaries or deep edges. The precisely filtered waviness reveals an acceptable surface quality, suggesting a stable machining process and the absence of surface irregularities.

5.1.5.2 Case Study II

The second case study focuses on the top surface of an engine cylinder block, as depicted in Fig. 5.18a. The engine cylinder block is made of Cast iron FC250 and serves as a critical sealing surface within the automotive powertrain. Its production involves rough milling, semi-finish milling, and finish milling. The milling was performed on an EX-CELL-O machining center using a face milling cutter with a diameter of 200 mm, equipped with 15 cutting inserts and 3 wiper inserts. Quaker 370 KLG cutting fluid was utilized in the process. The milling parameters included a speed of 816.4 m/min, a milling depth of 0.5 mm, and a feed rate of 3360 mm/min. Due to the constraints of the measurement range, the top surface was evenly divided into two sections (upper and lower) for measurement, with the results from both sections subsequently merged by the measurement system. The comprehensive measurement outcome is akin to Fig. 5.1, and the corresponding gray image is presented in Fig. 5.18b.

Each segment of the top surface measures 102.4 mm × 102.4 mm, with a sampling interval of 0.2 mm. The sampling points are 512 in both the X and Y directions. According to the roughness and waviness specifications $Ra = 3.2\,\text{um}, Rz = 12.5\,\text{um}$, and $Wt = 10\,\text{um}$, the cutoff wavelengths for waviness are set at 2.5 and 25 mm [37]. The corresponding cutoff wavelength is $\lambda_f = 0.2 \times 2^{6+1} = 25.6\,\text{mm}$ (nearly equivalent to 25 mm) with a specific decomposition level $i = 6$. In this context, the waviness is denoted as $W(x, y) = \widetilde{d}_5 + \widetilde{d}_4 + \widetilde{d}_3 + \widetilde{d}_2 + \widetilde{d}_1$ and the form as $F(x, y) = \widetilde{a}_6 + \widetilde{d}_6 = \widetilde{a}_5$. Each part contains approximately 0.3 million points, with a computational time of 31.6 s. The filtering outcomes are depicted in Fig. 5.19. It is evident that the tool marks are completely distinguished from the surface form,

(a) (b)

Fig. 5.18 Discontinuous surface, **a** top surface of engine cylinder block, **b** the converted gray image with no distortions observed at the boundaries or holes. The precise filtering results suggest that the surface quality is satisfactory and the machining process is consistent.

Leakage remains a critical issue in engine production, potentially causing engine overheating, loss of compression, and diminished power output. Predominantly, leaks manifest at the junction between the engine cylinder block and the head. On this surface, form and roughness are accommodated by the flexibility of the gasket material. However, waviness with peak-to-peak undulations is not accommodated, leading to localized contact points and leakage [6]. Therefore, the separation of waviness from the top surface of the engine cylinder block using the proposed filtering method can pinpoint potential leakage areas.

An additional top surface of an engine cylinder block is displayed in Fig. 5.20. Despite the block passing the leakage test, leakage persisted. The surface's tooling marks were extracted using the extended tetrolet transform, and the waviness parameters from eight specific positions in Fig. 5.20 are detailed in Table 5.6.

Evidently, the waviness parameters at positions 4 and 5 exhibit anomalies. Additionally, the enlarged views of these two positions are depicted in Fig. 5.21. The alternating red and blue colors indicate tooling marks with significant peak-to-peak

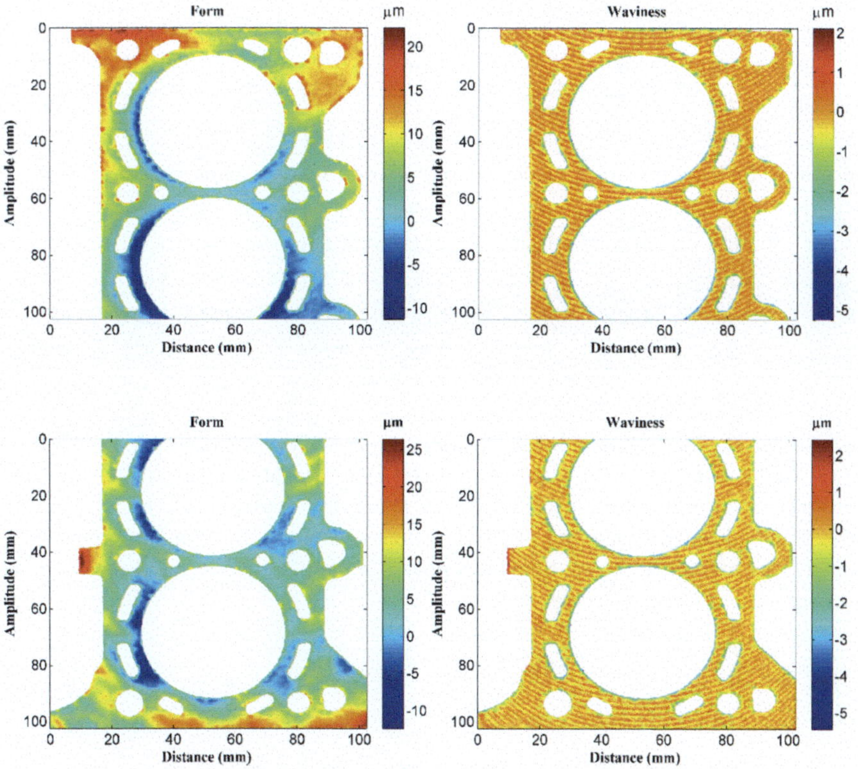

Fig. 5.19 Top surface of engine cylinder block

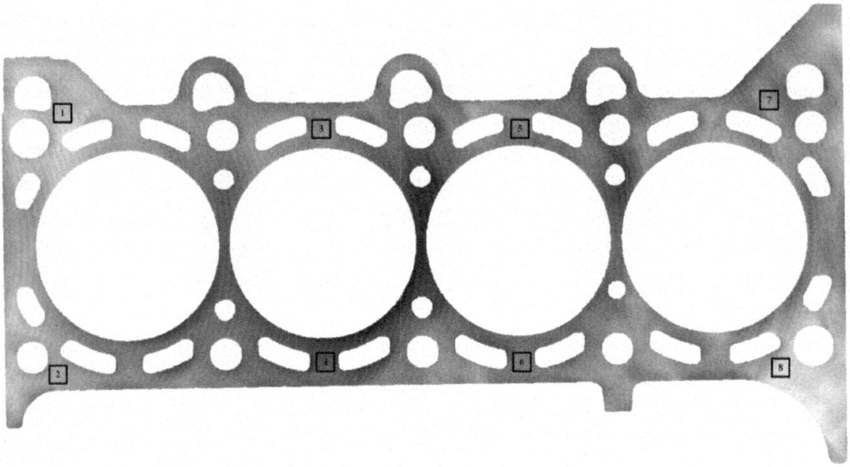

Fig. 5.20 Leakage surface

Table 5.6 Waviness parameters of each position

Position	1	2	3	4	5	6	7	8
$S_q/\mu m$	0.764	0.974	0.979	3.256	1.156	1.005	0.837	0.669
$S_p/\mu m$	2.926	3.222	2.894	7.619	4.672	3.348	3.058	2.439
$S_v/\mu m$	2.749	3.228	5.191	13.634	5.861	4.221	2.519	2.460
$S_z/\mu m$	5.675	6.450	8.085	21.254	10.533	7.569	5.578	4.898
$S_a/\mu m$	0.597	0.762	0.736	2.150	0.792	0.708	0.651	0.527
S_{sk}	0.085	− 0.040	− 0.544	− 1.325	− 1.156	− 0.158	0.397	− 0.096
S_{ku}	3.392	3.212	4.786	5.78	7.202	5.077	3.641	3.423
S_{al}	3.606	4.123	4.123	4.123	4.123	4.123	4.123	4.123
S_{tr}	0.806	0.825	0.972	0.206	0.294	1.144	0.922	0.972

variations. The likelihood of leakage is directly related to the peak-to-peak values of these marks. This suggests that positions 4 and 5 could be areas prone to leakage, with position 4 being the most susceptible. Consequently, to assess the risk of leakage, it is essential to scrutinize the precisely filtered waviness in critical regions. Ultimately, the extended tetrolet transform proves to be effective in the domain of 3D discontinuous surface filtering.

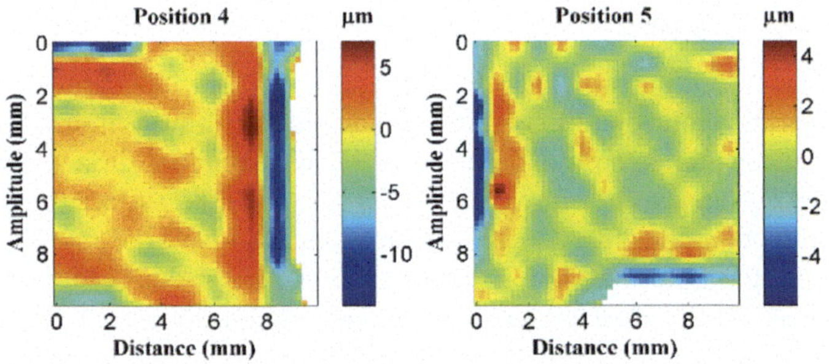

Fig. 5.21 **a** Waviness of position 4, **b** waviness of position 5

5.1.6 Discussion

To ensure stringent surface quality, the precision of surface attributes such as form, waviness, and roughness for high-precision workpieces is paramount for maintaining quality standards. The edges and apertures of discontinuous engineering surfaces are pivotal areas that encompass a wealth of surface data, significantly influencing factors like tooling, markings, vibrations, distortions, and heat treatment during machining. Surface filtering for the precise extraction of these components is crucial for analyzing manufacturing processes and functional performance. A filtering approach that addresses the edge effect, a common issue with traditional methods on discontinuous surfaces, is thus essential.

The enhancements of the proposed filtering technique are as follows:

(1) The method, characterized by a compact support, symmetry, orthogonality, and adaptive geometric structure, effectively addresses line and surface singularities in 3D discontinuous surfaces, thereby mitigating edge distortions.
(2) It incorporates an edge detector to preemptively identify surface boundaries and holes, enabling the application of various decomposition techniques.
(3) The method adeptly extracts and reconstructs the decomposed subsurface, which correspond to distinct surface components.

Despite the minimal block effect in the extended tetrolet transform due to the nature of tetrominoes, future optimization of the method through smoothing techniques is anticipated. Once the precise surface components are ascertained, they can facilitate the diagnosis of the underlying causes of surface defects, allowing for timely identification of product flaws. Furthermore, they can highlight surface irregularities, enabling engineers to swiftly adjust manufacturing processes to prevent the progression of non-conforming workpieces into subsequent, more costly stages of machining or assembly. However, there have been limited efforts to explore the correlation between surface components and the functional behavior of components, which presents a future research avenue.

5.1.7 Conclusions

A novel filtering approach utilizing the extended tetrolet transform has been introduced, specifically tailored for the analysis of 3D discontinuous surface topography. This method encompasses the processes of 3D discontinuous surface measurement, conversion to point cloud data, edge detection, application of the extended tetrolet transform, and the assessment of 3D characterization parameters. In comparison with the areal Gaussian and areal spline filters, the proposed method has been confirmed to excel in 3D surface filtering without introducing edge distortions, particularly for surfaces that are discontinuous in nature. Both simulated and experimental discontinuous surfaces have confirmed the method's efficacy as a robust tool for 3D discontinuous surface filtering. Furthermore, the filtered surface data can be utilized to compute 3D surface parameters, which are indicative of surface quality and functional performance. Given the demonstrated effectiveness of the extended tetrolet transform, it is anticipated to outperform the Gaussian and spline filters that are currently standardized by ISO. The proposed filtering method is poised to become a valuable asset in the analysis of 3D discontinuous surface topography, especially as the demand for analyzing such surfaces grows in the field of engineering.

5.2 A Surface Connectivity Based Approach for Leakage Channel Prediction in Static Sealing Interface

5.2.1 Introduction

Leaks are a significant issue, potentially resulting in severe security breaches and substantial economic damage. Within the industrial sector, preventing leakage is a critical aspect and a key functional behavior that affects product quality performance. Seals are extensively employed to prevent leakage, serving to fill the gap between rigid material components [31]. The primary types of seals are static and dynamic. A static seal involves pressing two rough surfaces together, a technique that is prevalent and significant in engineering applications, such as the flange assembly interfaces in pipeline construction and the mating surfaces between engine heads and blocks in automotive powertrain manufacturing. Despite good sealing performance, leakage failures still occur in many engineering applications. Consequently, it is imperative to model and diagnose leakage in static sealing interfaces to ensure superior product quality and minimize economic losses.

Investigations into leakage can be categorized into three distinct stages: ascertaining the likelihood of leakage, pinpointing the leakage pathways, and quantifying the leak rate. Throughout the past decades, a multitude of theoretical and experimental studies have concentrated on addressing this issue. Theoretically, Lebeck initially introduced the "waviness theory," attributing the waviness of the end face to the leakage in sealing interfaces [32]. More recently, Robbe-Valloire et al. developed a model based on waviness motifs to analyze surfaces and compute leak rates [33]. Persson et al. introduced the critical-junction theory for estimating the leak rate of static seals [34–36]. Building upon Persson's approach, Bottiglione et al. advanced a theoretical framework to account for multiple leakage paths in flat seals [37, 38]. Subsequently, Persson et al. [39] introduced an effective-medium theory to estimate leak rates in rubber seals, further elucidating the leakage mechanism. Perez-Rafols et al. [40] formulated a numerical model for calculating leakage in metallic seals. In terms of experimental studies, Marie et al. were the first to elucidate fluid leakage through rough metal contacts, confirming that intermediate surface waviness is a significant factor in leakage [41, 42]. Through a series of experiments, Persson et al. [43, 44] determined that the leak rate of seals is highly influenced by the skewness of the height probability distribution of rough sealing surfaces. Vladescu et al. designed a novel apparatus to emulate a compliant sealing interface, characterizing the effects of contact pressure, roughness, and surface energy on liquid leakage [45].

Drawing from these studies, it is evident that the leakage in static sealing interfaces is contingent upon contact pressure, the sealing element, and surface topography. The majority of prior research has focused on the impact of contact pressure and sealing elements, with surface topography being complex and variable, making it challenging to control during machining. Furthermore, the post-assembly interface topography is not readily measurable due to limitations in measurement technology. Consequently, understanding the role of surface topography in static seal leakage is still in its nascent stages. Malburg pioneered a profile filtering technique to elucidate the effects of two-dimensional surface topography on leakage [46]. Liao et al. established correlations between waviness and sealing performance, identifying potential leak paths from significant tooling marks [47]. Sun et al. constructed a leakage channel model based on the asperity contact of sealing interfaces [48]. Zhang et al. detailed an approach for calculating leak channels and rates between metallic sealing surfaces using three-dimensional finite element contact analysis [49]. Ren et al. introduced a lattice graph model for diagnosing potential leak areas in surface assembly [50]. Shao et al. proposed a leakage monitoring technique for static sealing interfaces utilizing three-dimensional surface topography indicator [51]. Collectively, these studies have demonstrated a robust correlation between surface topography and leakage. However, theoretical contact mechanics and finite element analyses rely on approximations that may deviate from actual conditions. Experimental studies have yet to uncover the multiscale impact of surface topography on leakage. Recently, Liu et al. introduced a connectivity characterization method to explore the link between surface topography and sealing performance, offering a novel perspective on leakage [52]. Nevertheless, the connectivity index definition was incomplete, and the mechanism by which surface connectivity affects leakage was not elucidated. Moreover, to the best of the authors' knowledge, there is a significant deficiency in research assessing surface connectivity. Thus, surface topography remains a central focus in leakage and sealing performance studies, with surface connectivity emerging as a novel research avenue to understand leakage channels. The limitations of current methodologies are encapsulated as follows:

- Most theoretical studies, including contact mechanics and finite element analyses, are predicated on simplified assumptions and fail to account for the complete actual surface topography due to the absence of suitable measurement tools.
- Experimental research has not been able to reveal the multiscale effects of surface topography on leakage, attributed to the presence of multiscale surfaces.
- Previous studies have seldom considered the spatial distribution and surface connectivity of the sealing interface, which are pertinent to the leakage channel, and the influence of surface connectivity on the leakage channel remains obscure.

Over the past few years, there has been a marked acceleration in the advancement of precision measurement technologies. A state-of-the-art noncontact measurement device, known as HDM, has been implemented [53]. This instrument is capable of capturing millions of data points in a matter of seconds, providing a comprehensive representation of the 3D surface topography. The device boasts a measurement resolution of 150 μm in the x–y plane and 1 μm in the z-axis. A depiction of the HDM apparatus and a sample HDM-measured engine block surface can be found in Fig. 5.1. Leveraging the high precision and comprehensive nature of these measurements, numerous studies have been conducted on 3D surface topography to gain in-depth insights into manufacturing processes [18, 54–59].

Utilizing the innovative measurement platform, the primary contribution lies in the introduction of a surface connectivity-based methodology to forecast potential leakage paths via surface topography analysis. Initially, the method established in prior research [51] is employed to generate the contact surface, which serves as the sealing interface. This contact surface is delineated by integrating the virtual gasket surface with the waviness surface. Subsequently, with spatial connectivity in mind, a set of leakage parameters are introduced, encompassing both connectivity and correlation parameters, to characterize the attributes of the contact surface. Ultimately, a cutting-edge prediction algorithm is crafted to pinpoint the potential leakage channels on the surface.

5.2.2 Brief Introduction of Connectivity

The concept of connectivity in mathematics is extensively applied within both topological and graph-theoretic framework [60]. Traditional connectivity is crucial for image processing and analysis and has seen widespread application across various domains. To bridge the gap between topological and graph-theoretic connectivity, Serra [61] introduced a formal definition of connectivity classes within a comprehensive lattice structure, grounded in mathematical morphology. Below, foundational theoretical concepts are reviewed and detailed.

(1) **Connectivity class**

Definition: Let Z be a collection of points and X a subset within Z. If for any two arbitrary points within $a, b \in X$, there exists a continuous function ϕ mapping from the interval [0, 1] to X that meets conditions $\phi(0) = a$ and $\phi(1) = b$, then X is classified as a connectivity class.

That is, the continuous function ϕ may represent a path extending from a to b, entirely contained within X.

In this definition, the continuous function ϕ is contingent upon the adjacency of neighboring points. Consequently, the principle of adjacency between proximate points directly influences the structure of the connectivity class.

(2) **Adjacency principle**

Consider a set X classified as a connectivity class, which may exhibit 4-connected or 8-connected in accordance with the principles of 4-adjacency (d_4) and 8-adjacency (d_8). These principles outline distinct definitions of distance, denoted as

$$d_4 = |x_1 - x_2| + |y_1 - y_2| \tag{5.40}$$

$$d_8 = \max\{|x_1 - x_2|, |y_1 - y_2|\} \tag{5.41}$$

where (x_1, y_1) and (x_2, y_2) are any two points within X. Figure 5.22 illustrates three distinct points, $a(2, 3)$, $b(3, 3)$ and $c(3, 2)$. According to Eq. (5.40), with $d_4(a, b) = 1$ and $d_4(a, c) = 2$, a and b are adjacent, while a and c are not. Conversely, Eq. (5.41) indicates $d_8(a, b) = d_8(a, c) = 1$, rendering both a, b and a, c as adjacent points.

In alignment with the 4-adjacency (d_4) principle, Fig. 5.23 provides an example of a connectivity class. It showcases two distinct sets, X_1 and X_2, each comprising a number of points, where X_{21}, X_{22} are subsets that belong to X_2. Then it has:

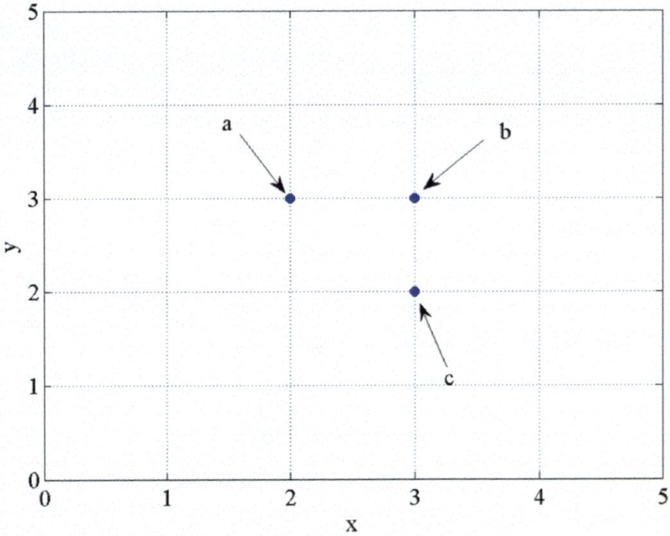

Fig. 5.22 Examples of different adjacency principles

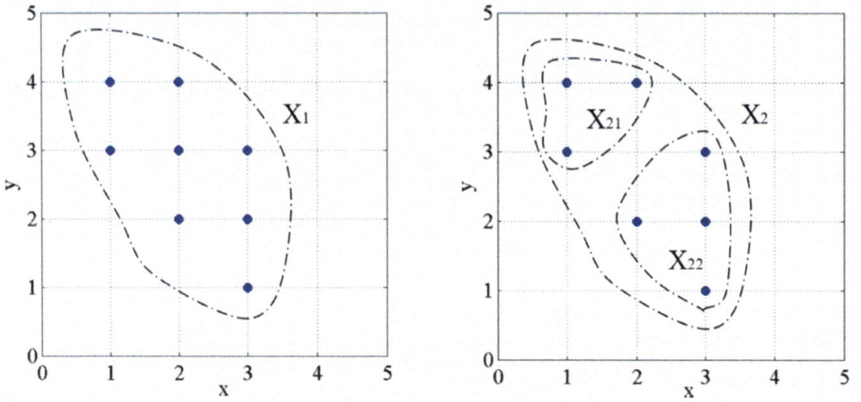

Fig. 5.23 Examples of connectivity class

$$X_1 = \{(1, 4), (1, 3), (2, 2), (2, 3), (2, 4), (3, 1), (3, 2), (3, 3)\} \tag{5.42}$$

$$X_2 = \{(1, 4), (1, 3), (2, 2), (2, 4), (3, 1), (3, 2), (3, 3)\} \tag{5.43}$$

$$X_{21} = \{(1, 4), (1, 3), (2, 4)\} \tag{5.44}$$

$$X_{22} = \{(2, 2), (3, 1), (3, 2), (3, 3)\} \tag{5.45}$$

Under the 4-adjacency principle, the continuous mapping ϕ signifies a pathway connecting any two arbitrary points. It is evident that such a pathway consistently exists within X_1, yet fails to satisfy the criteria in X_2. Consequently, X_1 qualifies as a connectivity class, whereas X_2 does not. In the case of the combined subsets X_{21}, X_{22}, both are recognized as connectivity classes. Here, the 4-adjacency principle is applied, with ϕ being interpreted as a continuous path between any two points within the subsets.

(3) Lattice operators

Lattice-based operations are crucial for assessing connectivity. Within the comprehensive lattice structure, four fundamental lattice operations are identified: dilation, erosion, opening, and closing. They are defined as:

Dilation:

$$\delta_B(X) = X \oplus B \tag{5.46}$$

Erosion:

$$\varepsilon_B(X) = X \ominus B \tag{5.47}$$

Opening:

$$\gamma_B(X) = X \circ B = \delta_B(\varepsilon_B(X)) = (X \ominus B) \oplus B \tag{5.48}$$

Closing:

$$\beta_B(X) = X \cdot B = \varepsilon_B(\delta_B(X)) = (X \oplus B) \ominus B \tag{5.49}$$

For the input connectivity class X, a compact convex structuring element B is utilized, with $\oplus$ and $\ominus$ representing Minkowski addition and subtraction, respectively. In this context, a spherical structuring element is employed.

(4) **Multiscale operators**

To measure the level of connectivity, lattice operators are expanded into their multiscale counterparts. The structuring element B is upgraded to a multiscale variant rB ($rB = \{rb : b \in B, r \geq 0\}$). Consequently, multiscale dilation, multiscale erosion, multiscale opening, and multiscale closing can be defined iteratively as follows:

Multiscale dilation:

$$\delta_B^r(X) = X \oplus rB = \delta_B(\delta_B^{r-1}(X)) = \cdots = \delta_B(\delta_B \ldots \delta_B(\delta_B(X))) \tag{5.50}$$

Multiscale erosion:

$$\varepsilon_B^r(X) = X \ominus rB = \varepsilon_B(\varepsilon_B^{r-1}(X)) = \cdots = \varepsilon_B(\varepsilon_B \ldots \varepsilon_B(\varepsilon_B(X))) \tag{5.51}$$

Multiscale opening:

$$\gamma_B^r(X) = X \circ rB = \delta_B^r(\varepsilon_B^r(X)) \tag{5.52}$$

Multiscale closing:

$$\beta_B^r(X) = X \cdot rB = \varepsilon_B^r(\delta_B^r(X)) \tag{5.53}$$

where $\delta_B^0(X) = X$, $\varepsilon_B^0(X) = X$ and $r = 1, 2, 3 \ldots, r \in \mathbb{N}_+$.

(5) **Conditional operators**

Given the expansive nature of the dilation operator, the original connectivity class X may surpass its boundaries when dilation is applied iteratively without constraints. To circumvent this issue, the dilation operator is confined within a designated mask. Consequently, a novel non-translational invariance dilation operator, termed conditional dilation, is introduced as follows:

Conditional dilation:

$$\delta_B(X|Y) = (X \oplus B) \cap Y \tag{5.54}$$

where Y is a mask of X.

5.2.3 *The Proposed Method*

This section outlines the proposed methodology, which is composed of three key components: contact surface generation, definition of leakage parameters, and prediction of leakage channels. The architecture of this approach is depicted in Fig. 5.24. Module 1 establishes the contact surface by integrating the virtual gasket surface with the waviness surface. Here, the methodology introduced by Shao et al. [51] is utilized to create the contact surface that indicates potential leakage. Module 2 introduces a set of leakage parameters, encompassing connectivity and correlation parameters, to characterize the attributes of the contact surface. Module 3 develops an innovative algorithm for predicting leakage channels, based on the defined leakage parameters. A leakage index is constructed to pinpoint the surface's potential leakage paths. Detailed descriptions of Module 1 are provided in Sect. 3.1, with Modules 2 and 3 being discussed in Sects. 3.2 and 3.3, respectively.

5.2.3.1 Contact Surface Generation

Drawing from the findings in references [42, 46, 51], the intermediate surface component waviness was identified as the predominant factor influencing leakage. Consequently, the methodology from a prior study [51] is implemented to produce the contact surface that mirrors potential leakage paths. This contact surface is derived

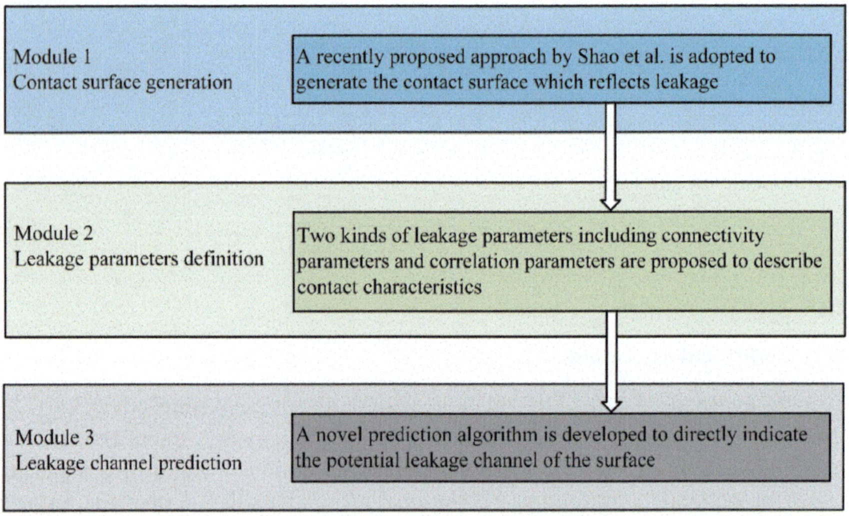

Fig. 5.24 The framework of the proposed method with three modules

by combining the virtual gasket surface with the waviness surface. A schematic representation of the approach is displayed in Fig. 5.25, with the subsequent detailed procedures outlined below.

Step 1: Engineering surface measurement and filtration

Based on the precision measurement platform, the entire and accuracy engineering surface information is measured by HDM. It is noted that the measured HDM data don't contain the three-dimensional roughness information with the limitation of the resolution in z-axis. For example, Fig. 5.26 shows the measured results of engine cylinder head and block surfaces.

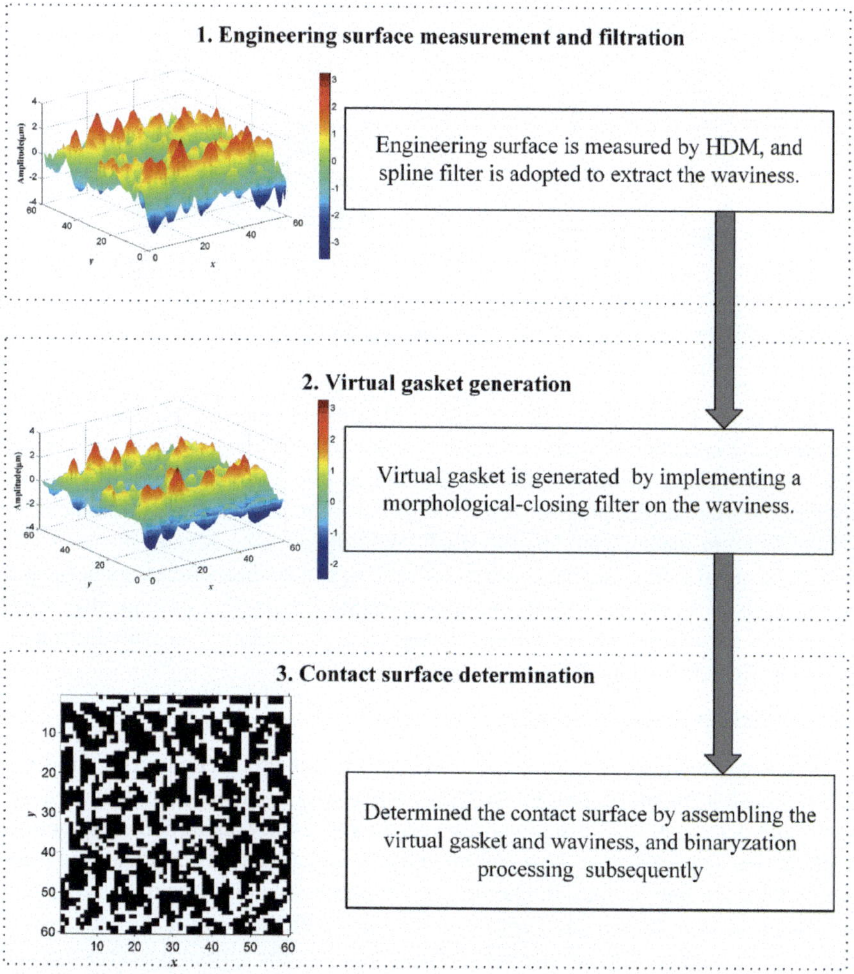

Fig. 5.25 The architecture of the contact surface generation approach

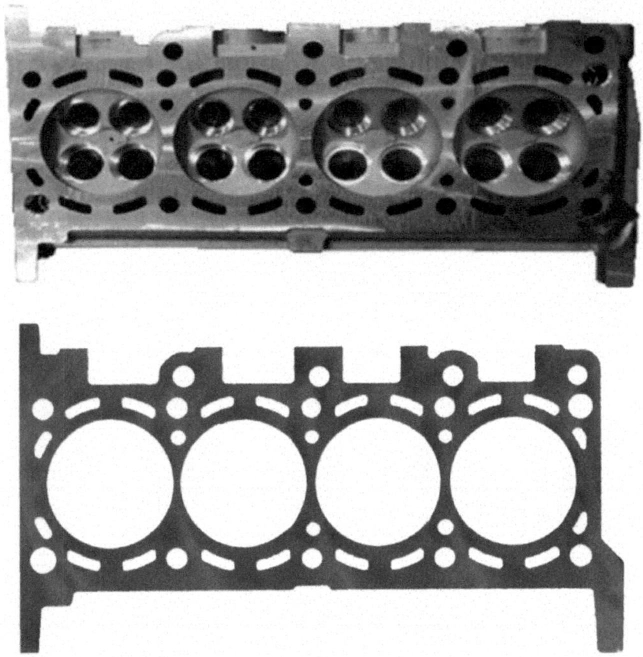

Fig. 5.26 **a** Engine cylinder head surface, **b** engine cylinder block surface

In the field of surface metrology, surface filtration is predominantly employed to elucidate the influence of various surface components [62]. The spline filter, a linear filter extensively utilized, is explicitly defined in ISO 16610-22 [63]. The profile spline filter enables the precise separation of two-dimensional surface components across different scales, from large to small. With the ongoing advancements in ISO 16610, the areal spline filter has been introduced to isolate three-dimensional surface components, with its publication anticipated in ISO 16610-62. Consequently, waviness can be accurately extracted from the original high-resolution measured surfaces, as clearly depicted in Fig. 5.27.

Step 2: Virtual gasket generation

Gasket seal constitutes a critical sealing method in numerous engineering products that demand stringent sealing integrity. As previously discussed, the actual topography and deformation of assembled gaskets are not readily quantifiable in practical applications. In response, a concept termed the virtual gasket was introduced in prior research to facilitate contact simulation [51]. Within the realm of geometrical product specification, morphological filters are commonly applied to capture the geometrical attributes of engineering surfaces [64]. Consequently, the virtual gasket is created by applying a morphological-closing filter to the most recently filtered waviness. Illustrations of a virtual gasket profile and surface are presented in Fig. 5.28.

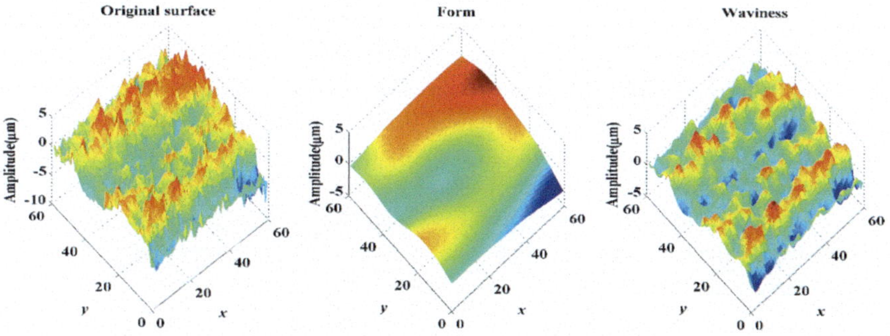

Fig. 5.27 Filtered results of areal spline filter

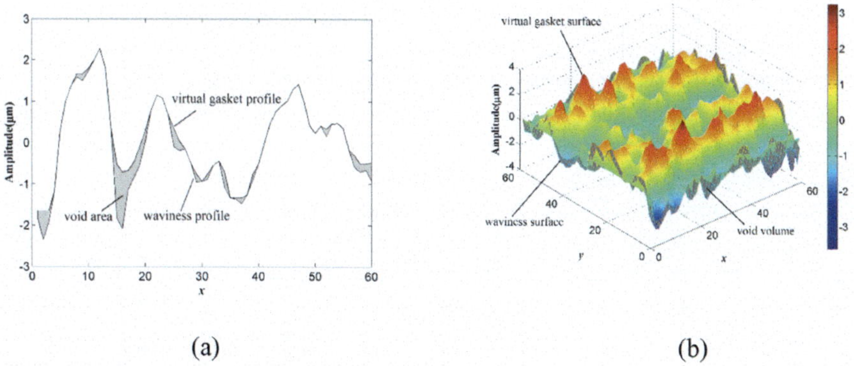

Fig. 5.28 **a** Virtual gasket profile, **b** virtual gasket surface

Step 3: Contact surface determination

With the waviness $SW(x, y)$ and the virtual gasket $SC(x, y)$ in hand, the subsequent phase involves the creation of the contact surface. In a study previously conducted [51], three areal leakage parameters—CAP (contact area percentage), VV(void volume), and $SWvoid$ (relative void volume)—were introduced as metrics to indicate the occurrence of leakage. A concise overview of these parameters is provided below. The definitions of these three areal leakage parameters are as follows:

$$CAP = \sum_{S} [if\,(SC_i(x, y) = SW_i(x, y))1\; else\; 0]/S \tag{5.55}$$

$$VV = \iint_{S} (SC(x, y) - SW(x, y))dxdy \tag{5.56}$$

$$SWvoid = VV/S \tag{5.57}$$

where S denotes the nominal area of the surface.

Given the autonomy of *SWvoid*, it serves as a valid indicator for monitoring leakage. Specifically, if the leakage parameter *SWvoid* on a surface exceeds the predetermined threshold *SWvc*, the surface is identified as a leakage surface. Consequently, upon identifying a leakage surface, it becomes imperative to further delineate the leakage channel. It is evident that the interaction between the virtual gasket and waviness is intricately linked to the leakage channel. Let Z represent the contact surface, formed by integrating the virtual gasket and waviness. Based on the contact status between the two surfaces, the contact surface Z can be treated as a binary image. Then, it can be defined as:

$$Z = \begin{cases} 0 \; SC(x, y) \leq SW(x, y) \text{ contact} \\ 1 \; SC(x, y) > SW(x, y) \text{ no contact} \end{cases} \tag{5.58}$$

where $Z = 0$ means total contact and $Z = 1$ means no contact. $SW(x, y)$ and $SC(x, y)$ correspond to the z-direction values of the waviness and virtual gasket surfaces, respectively, encapsulating the height data of the surfaces.

It should be acknowledged that the determination of the contact surface hinges on the comparison of the z-direction values between the virtual gasket surface and the waviness surface. The height data of these surfaces follow a random distribution pattern. Drawing from probability and statistical theory, the dimensions of the sampling areas do not influence the distribution of surface heights. Concurrently, the contact area percentage at a specified average normal pressure is independent of the system size for a given surface [65], yet the volume of surface data and the computational efficiency are contingent upon the size of the sampling areas. Thus, while the sampling area's size does not alter the distribution of the contact surface, it does impact the computation time, and can be determined based on engineering practices.

Following the outlined procedures, Fig. 5.29 presents a standard depiction of the contact surface. Observably, each grid square representing a contact area is either black or white (with black indicating full contact and white signifying no contact). The contact surface is essentially a binary image devoid of height information, and all subsequent processes rely on this contact surface.

5.2.3.2 Leakage Parameters Definition

With the contact surface secured, subsequent efforts concentrate on characterizing the contact attributes, which are pertinent to the leakage channel. Connectivity, a concept hailing from topological geometrical theory, significantly impacts lubrication, surface distribution, and sealing performance [50, 52, 66]. The connectivity can differ between surfaces, even when they share identical amplitude parameters. To the best of the authors' knowledge, there is a notable scarcity of research assessing surface connectivity. To date, no specific standard or systematic method exists for

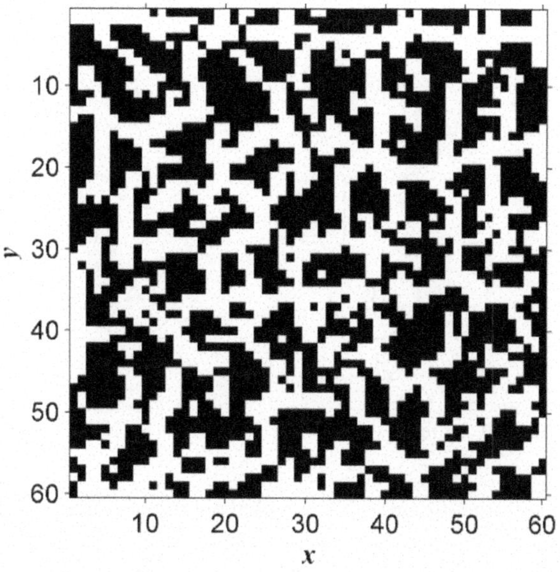

Fig. 5.29 Contact surface

evaluating surface connectivity. Hence, this section defines a suite of leakage parameters grounded in connectivity. These proposed parameters are categorized into connectivity parameters, which assess the connectivity measure of an individual connectivity class, and correlation parameters, which denote the interplay among various connectivity classes.

Connectivity Parameters

Broadly speaking, the contact surface Z is depicted by a matrix that comprises a collection of discrete points within discrete spaces. Consequently, the notion of connectivity essentially pertains to the geometrical interplay among these discrete points.

As illustrated in Fig. 5.29, the surface contains multiple connectivity classes, each with distinct shapes and sizes. For any given connectivity class, there is invariably a most constricted path, which can be considered the class's weakest segment. Additional exemplary cases are presented in Fig. 5.30. Classes A_1, A_2 and A_3 represent three distinct connectivity classes, with their sizes satisfying $\mu(A_1) > \mu(A_2) > \mu(A_3)$ under the conditions $R_1 > R_2$ and $h_1 > h_2$. To differentiate these connectivity classes, a measure of connectivity for an individual class is essential. In essence, the connectivity measure is designed to assess the most confined pathway.

Tzafestas [60] has detailed various methods for measuring connectivity, such as multiresolution connectivity measure, generalized connectivity measure, erosion-based connectivity measure, and adjunctional connectivity measure. Building on

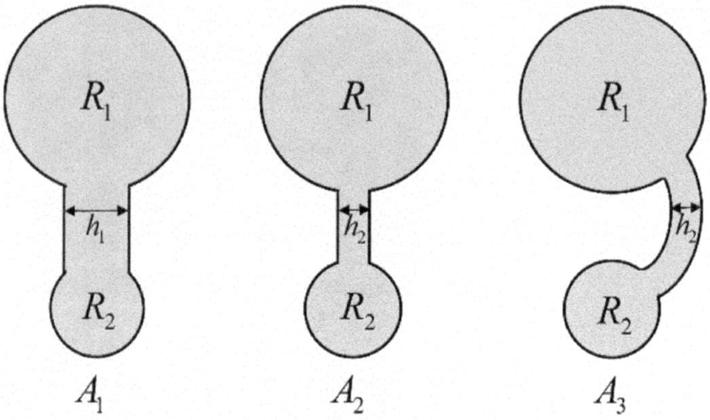

Fig. 5.30 Different connectivity classes

Tzafestas' findings, an extended operator known as multiscale conditional operators and a modified connectivity measure referred to as the average adjunctive multiscale connectivity function $\overline{\mu}_\alpha$ have been developed. Let X be a subset of Z, and suppose X constitutes a connectivity class. The definitions for these two parameters are outlined below.

(1) **Multiscale conditional operators**

Adhering to the rationale of multiscale operators, the multiscale conditional dilation is delineated as follows:

Multiscale Conditional Dilation:

$$\delta_B^r(X|Y) = \delta_B(\delta_B^{r-1}(X|Y)|Y) = \ldots = \delta_B(\delta_B(\ldots\delta_B(\delta_B(X|Y)|Y)\ldots|Y)|Y) \tag{5.59}$$

where X denotes the connectivity class and Y serves as the mask for X. B represents a compact convex structuring element, and r signifies the scale with $r = 1, 2, 3\ldots, r \in \mathbb{N}_+$. Through this approach, account is taken of both the length and width of the most constricted path within the connectivity class.

(2) **Average adjunctive multiscale connectivity function**

Let $\alpha = (\varepsilon_B, \delta_B)$ be an adjunction on $P(E)$. The adjunctive multiscale connectivity function μ_α is specified as:

$$\mu_\alpha(X, s) = \exp(-\lambda r_\alpha(X, s)) \tag{5.60}$$

where $r_\alpha(X, s) = inf\{r \in \mathbb{N}_+ : \delta_B^r(\varepsilon_B^s(X)|X) = 1\}$, and $\lambda > 0$ is a parameter indicating the rate of the exponential function. λ is not a constant value; it is used to modify the function's slope and varies across different applications. Per prior research and practice [60], $\lambda = 0.05$ is deemed appropriate. $s \in [0, s_{\max}]$ represents the scale, with $s_{\max}$ denoting the maximum scale applicable, defined by:

$$s_{\max} = \left\{ s_{\max} \in \mathbb{N}_+ : n(\varepsilon_B^{s_{\max}}(X)) = 0\,, X \in P(E) \right\} \tag{5.61}$$

where $n(X)$ is the count of connectivity classes within X. Consequently, the average adjunctive multiscale connectivity function $\overline{\mu}_\alpha$ is determined as:

$$\overline{\mu}_\alpha = \frac{1}{s_{\max}} \sum_{s=0}^{s_{\max}} \mu_\alpha(X\,, s)\, \overline{\mu}_\alpha \in [0,\, 1] \tag{5.62}$$

According to the definition, the proposed connectivity function $\overline{\mu}_\alpha$ encompasses both the complete geometrical characteristics (shape and size) and incorporates multi-scale information. Specifically, $\overline{\mu}_\alpha$ reflects the rate at which a connectivity class becomes disconnected or disappears. When $\overline{\mu}_\alpha \to 0$, X is "nearly disconnected." Conversely, when $\overline{\mu}_\alpha \to 1$, X is "fully connected." In essence, a larger $\overline{\mu}_\alpha$ signifies a larger connectivity class size. In practice, $\overline{\mu}_\alpha$ assesses the "width" and "length" of the "narrowest" path within X. For instance, the connectivity classes A_1, A_2 and A_3 exhibit varying "width" and "length" as shown in Fig. 5.30. Given the conditions $R_1 > R_2$ and $h_1 > h_2$, it is evident that $\mu(A_1) > \mu(A_2) > \mu(A_3)$ aligns with the definition of $\overline{\mu}_\alpha$.

Within a static sealing interface, the leakage channel is highly contingent upon the contact geometry, with the surface height distribution exerting a significant impact on the contact state. Concurrently, surface connectivity serves as a direct indicator of surface height distribution. Moreover, the magnitude of the connectivity class mirrors the resistance to leakage channels, analogous to electrical conductance. A reduced size of the connectivity class implies greater obstruction to lubricant flow. In contrast, a larger connectivity class size facilitates fluid passage through the channel, thereby increasing the likelihood of leakage channel formation. Consequently, the proposed connectivity function $\overline{\mu}_\alpha$ can be employed to anticipate the presence of leakage channels.

Correlation Parameters

As previously noted, the surface Z contains numerous connectivity classes. With the connectivity measure of an individual connectivity class established, it becomes essential to assess the interplay among various connectivity classes. To this end, three connectivity correlation parameters are introduced to quantify the relationship between any two arbitrary, independent connectivity classes. These parameters are the minimum connective distance mcd, the average connective distance acd, and the cross connectivity index θ. It should be noted that Euclidean distance is utilized in both mcd and acd. The specific definitions of these parameters are outlined below.

(1) Minimum connective distance

The minimum connective distance *mcd* characterizes the shortest distance separating two distinct connectivity classes. Let X_p and X_q represent two independent connectivity classes, with E^p and E^q being their respective edge matrices that meet the condition:

$$E^p = Ed(X_p) \tag{5.63}$$

$$E^q = Ed(X_q) \tag{5.64}$$

where *Ed* denotes a standard edge detector. The minimum connective distance mcd_{pq} between X_p and X_q is delineated as:

$$\begin{aligned} & mcd_{pq} = \min\left|E_i^p - E_j^q\right| = \sqrt{(x_i^p - x_j^q)^2 + (y_i^p - y_j^q)^2} \\ & (i = 1, 2, \ldots, mc_p, j = 1, 2, \ldots, mc_q) \end{aligned} \tag{5.65}$$

where mc_p is the number of E^p and mc_q is the number of E^q.

(2) Average connective distance

The average connective distance *acd* denotes the average separation between two distinct connectivity classes, determined by the centroids of these classes. The formula to compute the acd_{pq} between X_p and X_q is as follows:

$$acd_{pq} = \sqrt{(\overline{x}^p - \overline{x}^q)^2 + (\overline{y}^p - \overline{y}^q)^2} \tag{5.66}$$

$$\overline{x}^p = \sum_{i=1}^{mc_p} x_i^p, \overline{y}^p = \sum_{i=1}^{mc_p} y_i^p \tag{5.67}$$

$$\overline{x}^q = \sum_{j=1}^{mc_q} x_j^q, \overline{y}^q = \sum_{j=1}^{mc_q} y_j^q \tag{5.68}$$

where $(\overline{x}^p, \overline{y}^p)$ and $(\overline{x}^q, \overline{y}^q)$ are the center of X_p and X_q.

(3) Cross connectivity index

Cross connectivity index θ is introduced to quantify the degree of connection between two separate connectivity classes. This index is derived by independently dilating two distinct connectivity classes, X_p and X_q, using an identical structuring element until a new connectivity class BDX_{pq} emerges. This resulting class, BDX_{pq}, is referred to as the bi-dilational connectivity class, encompassing not only the original elements of X_p and X_q but also representing the initial connectivity class to form during the dilation process. Consequently, the expression for θ is outlined below:

$$BDX_{pq} = \delta_B^{\theta-1}(X_p + X_q) \tag{5.69}$$

$$\theta = \left\{\theta \in \mathbb{N}_+ : n(\delta_B^{\theta}(X_p + X_q)) = 2 \text{ and } n(\delta_B^{\theta-1}(X_p + X_q)) = 1, X_p, X_q \in P(E)\right\} \tag{5.70}$$

where $n(X)$ signifies the total count of connectivity classes within X. The cross connectivity index is capable of discerning the strength of correlation between various independent connectivity classes, which in turn has a direct bearing on the distribution of leakage channels and the sealing performance of the contact surface.

5.2.3.3 Leakage Channel Prediction

Drawing on the established leakage parameters, an innovative leakage channel prediction algorithm has been crafted. The central challenge of this prediction algorithm is to construct a leakage index Ψ that can explicitly identify the potential leakage channels on the surface. The framework of the algorithm is depicted in Fig. 5.31, with the detailed steps outlined below.

Step 1: Connectivity Class Identification

Upon generating the contact surface Z, the initial step involves identifying the connectivity class X_i of the surface based on the defined connectivity criteria. Concurrently, the total number of connectivity classes n is ascertained. Here, $i = 1, 2, \ldots, n$.

Step 2: Connectivity Parameters Computation

For each identified connectivity class, the connectivity parameters are computed. The average adjunctive multiscale connectivity functions $\overline{\mu}_\alpha(X_i)$ $(i = 1, 2, \ldots, n)$ are utilized to denote the connectivity size of the class X_i. For simplicity, $\overline{\mu}_{\alpha i} = \overline{\mu}_\alpha(X_i)$.

Step 3: Correlation Parameters Computation

Given the presence of multiple connectivity classes, the correlation parameters between each pair of classes are calculated. Let $Corr_{ij} = \left\{mcd_{ij}, acd_{ij}, \theta_{ij}\right\}$ represent the aggregate index of these correlation parameters, encompassing minimum connective distance, average connective distance, and cross connectivity index for connectivity classes X_i and X_j. Here, $i = 1, 2, \ldots, n, j = 1, 2, \ldots, n$ and $i \neq j$.

Step 4: Leakage Channel Identification

With both connectivity and correlation parameters determined, a leakage index Ψ is proposed to signify the potential leakage channel. This index, also known as the breakthrough index, is defined based on whether a connectivity class spans the entire surface. In engineering practice, the flow pressure is unidirectional, thus the traversal direction is considered in y direction of the entire surface. Assuming z_{ys} and z_{ye} are the minimum and maximum coordinate values of the surface Z, respectively, the breakthrough index for the connectivity class X_i is calculated as:

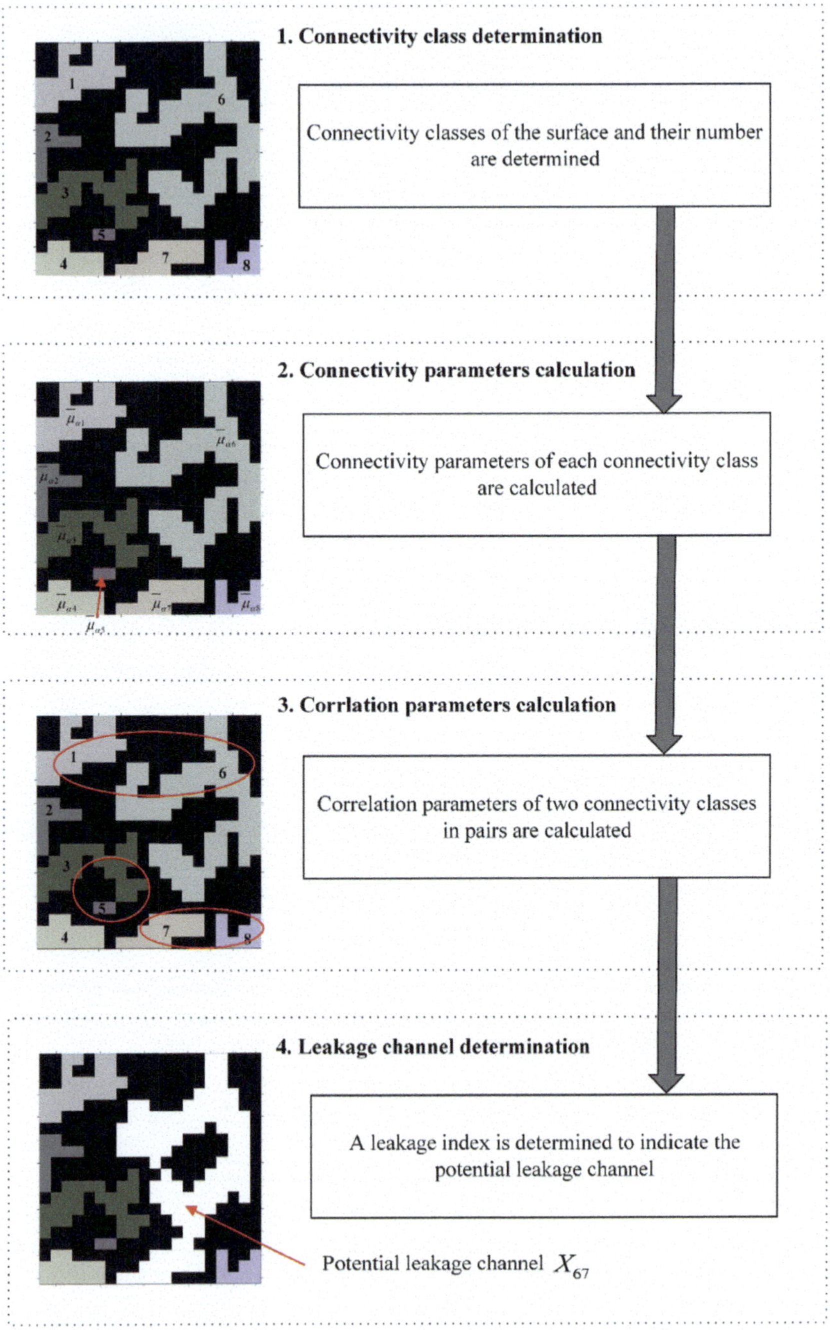

Fig. 5.31 The architecture of the leakage channel prediction algorithm

$$\Psi(X_i) = \begin{cases} 1, \ \min\{y_j^i\} = z_{ys} \text{ and } \max\{y_j^i\} = z_{ye} \\ 0, \ \text{else} \end{cases} \tag{5.71}$$

where $\min\{\cdot\}$ and $\max\{\cdot\}$ denote the operations of extracting the minimum and maximum values, respectively.

There are two scenarios Ψ_1 and Ψ_2 outlined in the definition of Ψ concerning the traversal of the entire surface. When $\Psi_1 = 1$ equals 1, the surface is fully traversed, whereas traversal is incomplete when $\Psi_2 = 0$.

Case 1:

In the scenario where $\Psi_1 = 1$, a significant potential leakage channel is present on the contact surface, and it must belong to one of the connectivity classes. Specifically, this can be articulated as:

- If $\Psi(X_i) = 1$, then the connectivity class X_i is identified as the potential leakage channel;
- If $\Psi(X_j) = 1$, then the connectivity class X_j is identified as the potential leakage channel.

Case 2:

When $\Psi_2 = 0$, there is no significant condition under which a single connectivity class can be deemed a potential leakage channel. In this case, the correlation and size of the connectivity classes are utilized to identify the potential leakage channel.

Initially, a concept termed the bi-closing connectivity class is introduced:

$$X_{pq} = \beta_B^h(X_p + X_q) \tag{5.72}$$

where $h = \left\{h \in \mathbb{N}_+ : n(\beta_B^h(X_p + X_q)) = 1\right\}$ represents the specific applicable scale. The bi-closing connectivity class is formed by closing any two independent connectivity classes, X_p and X_q, with the same structuring element until a new class, X_{pq}, is produced. Unlike the bi-dilational connectivity class X_{pq}, the bi-closing connectivity class X_{pq} does not surpass the boundary when the operation is repeatedly applied. It effectively preserves the characteristics of the original connectivity classes and reveals information regarding the potential leakage channel.

Secondly, the average adjunctive multiscale connectivity functions $\overline{\mu}_{\alpha i}$ for each connectivity class are determined, with the identification of the maximum $\overline{\mu}_\alpha(X_{k_0}) = \max\{\overline{\mu}_{\alpha i} | i = 1, 2, \ldots, n\}$ within the connectivity class X_{k_0}.

Following that, a pairwise comparison of all correlation parameters $Corr_{ij}$ between connectivity classes is conducted. The minimum values among three parameters—minimum connective distance, average connective distance, and cross connectivity index—are identified as follows:

$$mcd_{k_1k_2} = \min\left\{mcd_{ij} | i = 1, 2, \ldots n, j = 1, 2, \ldots n, i \neq j\right\} \tag{5.73}$$

$$acd_{k_3k_4} = \min\{acd_{ij}|i = 1, 2, \ldots n, j = 1, 2, \ldots n, i \neq j\} \quad (5.74)$$

$$\theta_{k_5k_6} = \min\{\theta_{ij}|i = 1, 2, \ldots n, j = 1, 2, \ldots n, i \neq j\} \quad (5.75)$$

Subsequently, the average adjunctive multiscale connectivity functions $\overline{\mu}_\alpha(X_{k_1k_2})$, $\overline{\mu}_\alpha(X_{k_3k_4})$, and $\overline{\mu}_\alpha(X_{k_5k_6})$ for the bi-closing connectivity classes $X_{k_1k_2}$, $X_{k_3k_4}$ and $X_{k_5k_6}$ are computed. Here, $k_0, k_1, k_2, k_3, k_4, k_5$ and $k_6 \in \{1, 2, \ldots, n\}$ represent the respective connectivity classes.

Finally, a comparison of the values among $\overline{\mu}_\alpha(X_{k_0})$, $\overline{\mu}_\alpha(X_{k_1k_2})$, $\overline{\mu}_\alpha(X_{k_3k_4})$ and $\overline{\mu}_\alpha(X_{k_5k_6})$ is performed, with the determination of the maximum $\overline{\mu}_\alpha(X_k)$ as defined as:

$$\overline{\mu}_\alpha(X_k) = \max\{\overline{\mu}_\alpha(X_{k_0}), \overline{\mu}_\alpha(X_{k_1k_2}), \overline{\mu}_\alpha(X_{k_3k_4}), \overline{\mu}_\alpha(X_{k_5k_6})\} \quad (5.76)$$

The subscript $k \in \{k_0, k_1k_2, k_3k_4, k_5k_6\}$ is used to denote the corresponding connectivity class. Consequently, the connectivity class or bi-closing connectivity class X_k is designated as the potential leakage channel.

Similarly, the leakage parameters for the entire contact surface are determined, with surface connectivity being influenced solely by the contact surface distribution. As previously stated, the sampling area size does not impact the distribution of the contact surface but rather the computation time required for the contact surface analysis. Consequently, the sampling area size does not alter surface connectivity and can be chosen based on engineering practices.

Example 1 Considering the computational efficiency, a 20 × 20 contact surface Z depicted in Fig. 5.32a. In Fig. 5.32b displays the same contact surface with connectivity classes distinguished by various colors, totaling $n = 8$ distinct classes. It is important to note that these colors do not represent height data but rather signify the separation of different connectivity classes.

Subsequently, connectivity parameters $\overline{\mu}_{\alpha i}$ $(i = 1, 2, \ldots, 8)$ and correlation parameters $Corr_{ij}$ $(i = 1, 2, \ldots, 8, j = 1, 2, \ldots, 8$ and $i \neq j)$ are calculated which are shown in Fig. 5.32c and d. The detailed results are shown in Tables 5.7 and 5.8.

Obviously, all the values of Ψ are equal to 0, it satisfies the case $\Psi_2 = 0$ and the results are obtained as follows:

$$\begin{cases} \overline{\mu}_{\alpha k_0} = \overline{\mu}_{\alpha 6} = 1.168 \\ mcd_{k_1k_2} = \{mcd_{17}, mcd_{18}, mcd_{25}, mcd_{35}, mcd_{46}, mcd_{78}\} = 1.41 \\ acd_{k_3k_4} = \{acd_{45}\} = 3.61 \\ \theta_{k_5k_6} = \{\theta_{12}, \theta_{16}, \theta_{23}, \theta_{34}, \theta_{35}, \theta_{36}, \theta_{45}, \theta_{47}, \theta_{67}, \theta_{78}\} = 1 \end{cases} \quad (5.77)$$

The maximum value of connectivity $\overline{\mu}_\alpha(X_k)$ are calculated as:

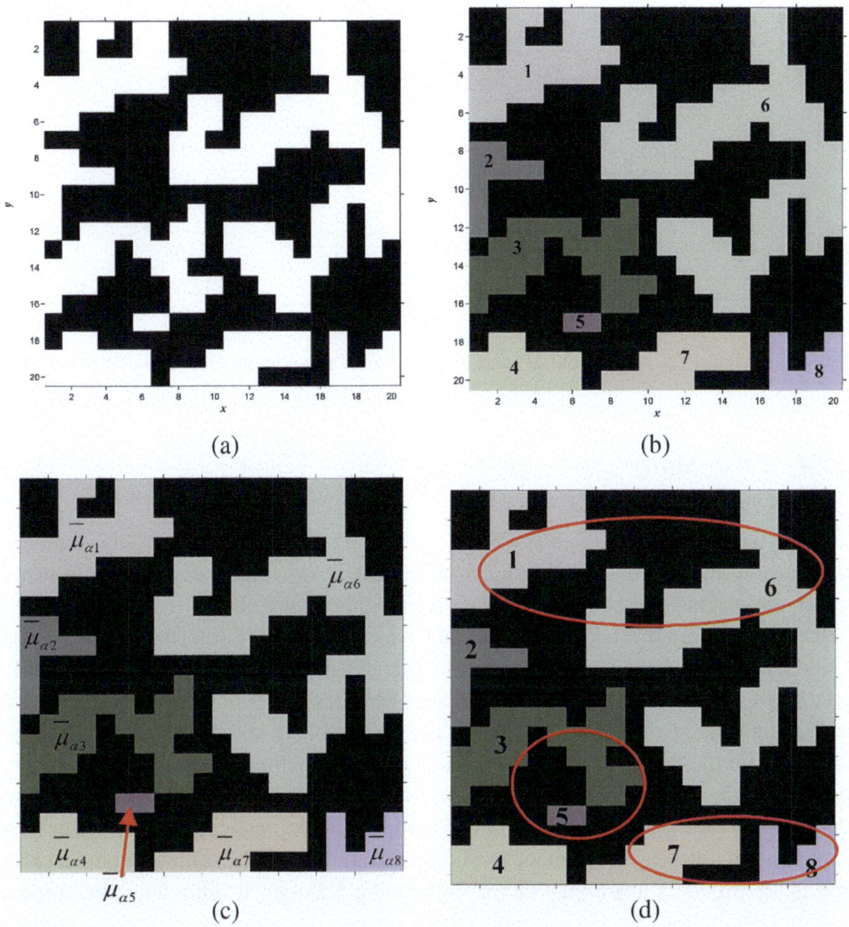

Fig. 5.32 **a** A 20×20 contact surface Z, **b** different connectivity classes, **c** connectivity parameters $\overline{\mu}_{\alpha i}$, **d** correlation parameters $Corr_{ij}$

$$\overline{\mu}_{\alpha}(X_k) = \max\left\{\begin{array}{c} \overline{\mu}_{\alpha}(X_6), \overline{\mu}_{\alpha}(X_{17}), \overline{\mu}_{\alpha}(X_{18}), \overline{\mu}_{\alpha}(X_{25}), \overline{\mu}_{\alpha}(X_{35}), \overline{\mu}_{\alpha}(X_{46}), \\ \overline{\mu}_{\alpha}(X_{78}), \overline{\mu}_{\alpha}(X_{45}), \overline{\mu}_{\alpha}(X_{12}), \overline{\mu}_{\alpha}(X_{16}), \overline{\mu}_{\alpha}(X_{23}), \overline{\mu}_{\alpha}(X_{34}), \\ \overline{\mu}_{\alpha}(X_{35}), \overline{\mu}_{\alpha}(X_{36}), \overline{\mu}_{\alpha}(X_{47}), \overline{\mu}_{\alpha}(X_{67}) \end{array}\right\}$$

$$= \overline{\mu}_{\alpha}(X_{67}) = 1.174 \tag{5.78}$$

Hence, the bi-closing connectivity class X_{67} is considered as the potential leakage channel which is found in Fig. 5.33.

Table 5.7 The results of connectivity parameters from X_1 to X_8

Connectivity class	s_{max}	$\overline{\mu}_\alpha$	Ψ
X_1	2	1.051	0
X_2	1	1.051	0
X_3	2	1.107	0
X_4	2	1.051	0
X_5	1	1.051	0
X_6	2	**1.168**	0
X_7	2	1.051	0
X_8	1	1.162	0

* The value in bold means the maximum value of the corresponding column

5.2.4 Case Study

Leakage within the vehicle powertrain is a critical concern, with particular implications for internal combustion engines where it can lead to compression loss, reduced power, and engine overheating. To illustrate the practical application of the proposed methodology, a case study was undertaken focusing on the interface between the cylinder block and head face of a vehicle engine. As shown in Fig. 5.34, a conformable gasket is typically utilized to ensure a seal between the block and head surfaces, preventing leakage during compression. The engine block and head surfaces in question are from the B12 series engine of an automotive manufacturer and are joined by tightening 10 bolts. Each bolt is tightened with a torque of 60 Nm, resulting in an approximate face pressure of 2 MPa between the block and head surfaces. Drawing from previous studies and contact mechanics references [9, 13, 14, 36], contact pressure is identified as a primary factor influencing leakage. It influences the contact surface distribution, which is connected to surface connectivity, hence the contact pressure is maintained at a consistent 2 MPa in the case study scenario.

The engine cylinder block surface, serving as a critical sealing surface, undergoes high-precision milling to ensure its finish. An example of a measured engine cylinder block surface is depicted in Fig. 5.35. Based on the practical experience of engineers, thirty distinct surface regions have been selected as representative areas where leakage is likely to occur. Each of these surface regions is divided into a 30×30 grid, with each grid measuring 0.2 mm, resulting in a sampling interval of 0.2 mm.

A surface region *SWvoid* is designated as a leakage surface region or an Out of Limit (OOL) leakage region when the leakage parameter *SWvc* exceeds a predetermined threshold. According to experimental data referenced in [51] the threshold *SWvc* is established at $SWvc \approx 0.300$ for the case study conditions. Consequently, the *SWvoid* values for each surface region are computed and documented in Table 5.9, with OOL leakage regions identified where the *SWvoid* value exceeds 0.3. For enhanced clarity, Fig. 5.36a presents a line chart of *SWvoid*, and the OOL leakage regions are both numbered and highlighted in red in Fig. 5.36b.

Table 5.8 The results of correlation parameters from (X_1, X_2) to (X_7, X_8)

Connectivity classes in pairs	*mcd*	*acd*	θ
(X_1, X_2)	2.00	6.71	**1**
(X_1, X_3)	8.00	10.05	3
(X_1, X_4)	13.00	16.03	6
(X_1, X_5)	7.00	14.14	6
(X_1, X_6)	12.00	11.18	**1**
(X_1, X_7)	**1.41**	17.46	8
(X_1, X_8)	**1.41**	21.26	12
(X_2, X_3)	2.00	5.66	**1**
(X_2, X_4)	7.00	10.20	3
(X_2, X_5)	**1.41**	9.43	5
(X_2, X_6)	6.08	13.04	2
(X_2, X_7)	5.00	14.14	7
(X_2, X_8)	3.16	19.72	11
(X_3, X_4)	3.00	6.32	**1**
(X_3, X_5)	**1.41**	4.12	**1**
(X_3, X_6)	2.24	10.30	**1**
(X_3, X_7)	10.77	8.49	2
(X_3, X_8)	9.22	14.32	5
(X_4, X_5)	6.08	**3.61**	**1**
(X_4, X_6)	**1.41**	15.56	5
(X_4, X_7)	15.00	8.00	**1**
(X_4, X_8)	14.00	15.00	5
(X_5, X_6)	5.00	12.04	4
(X_5, X_7)	9.49	5.39	2
(X_5, X_8)	8.06	12.17	5
(X_6, X_7)	14.14	11.40	**1**
(X_6, X_8)	13.00	11.70	2
(X_7, X_8)	**1.41**	7.00	**1**

* The value in bold means the minimum value of the corresponding column

The analysis reveals that thirteen surface regions, specifically Regions 2, 5, 7, 8, 10, 12, 16, 18, 20, 22, 23, 24, and 26, are identified as OOL leakage regions. Identifying these leakage regions is the first step; the subsequent step involves delineating the leakage channels within these areas. Furthermore, Region 12 exhibits the highest *SWvoid* value, while Region 11 shows the lowest *SWvoid* value. These two regions are selected to represent the leakage and no-leakage conditions, respectively, with their contact surfaces depicted in Fig. 5.37. It is evident that Region 11 is predominantly in contact, in contrast to Region 12, where the contact condition is the reverse

Fig. 5.33 The potential leakage channel X_{67}

Fig. 5.34 **a** Assembled engine cylinder head and block, **b** gasket and block

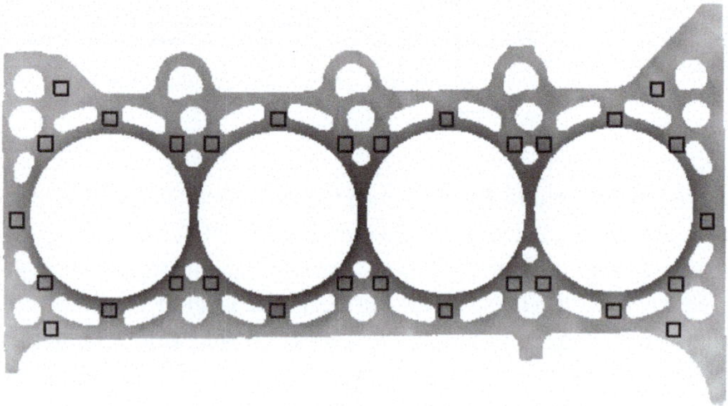

Fig. 5.35 Thirty surface regions

Table 5.9 The value of $SWvoid$ for each surface region

Surface region	$SWvoid$	Surface region	$SWvoid$	Surface region	$SWvoid$
Region 1	0.299	Region 11	0.229	Region 21	0.247
Region 2	**0.317**	Region 12	**0.384**	Region 22	**0.322**
Region 3	0.284	Region 13	0.274	Region 23	**0.353**
Region 4	0.253	Region 14	0.293	Region 24	**0.323**
Region 5	**0.303**	Region 15	0.280	Region 25	0.247
Region 6	0.286	Region 16	**0.367**	Region 26	**0.376**
Region 7	**0.362**	Region 17	0.260	Region 27	0.261
Region 8	**0.307**	Region 18	**0.310**	Region 28	0.282
Region 9	0.237	Region 19	0.298	Region 29	0.271
Region 10	**0.349**	Region 20	**0.314**	Region 30	0.288

* The values in bold denote that $SWvoid$ goes beyond $SWvc$

(with black indicating full contact and white indicating no contact). Therefore, further examination of Region 12 is warranted to determine the leakage channel, employing the methodology outlined in Sect. 3.

In Fig. 5.38a, the distinct connectivity classes within Region 12 are represented by different colors, totaling six classes, denoted as $n = 6$. The color scheme is not indicative of height variations but rather delineates separate connectivity classes. The necessary connectivity and correlation parameters for these classes are presented in Tables 5.10 and 5.11, respectively.

Since it doesn't exist $\Psi = 1$, the condition of case 2: $\Psi_2 = 0$ is considered. Then, the size of connectivity is calculated further.

The four indices are calculated as:

$$\begin{cases} \overline{\mu}_{\alpha k_0} = \overline{\mu}_{\alpha 4} = 1.201 \\ mcd_{k_1 k_2} = \{mcd_{15}, mcd_{16}, mcd_{24}\} = 1.41 \\ acd_{k_3 k_4} = \{acd_{56}\} = 7.62 \\ \theta_{k_5 k_6} = \{\theta_{23}, \theta_{34}, \theta_{35}, \theta_{36}, \theta_{46}\} = 1 \end{cases} \tag{5.79}$$

Finally, the maximum value of connectivity $\overline{\mu}_\alpha(X_k)$ are calculated as:

$$\begin{aligned} \overline{\mu}_\alpha(X_k) &= \max \begin{Bmatrix} \overline{\mu}_\alpha(X_4), \overline{\mu}_\alpha(X_{15}), \overline{\mu}_\alpha(X_{16}), \overline{\mu}_\alpha(X_{24}), \overline{\mu}_\alpha(X_{56}), \\ \overline{\mu}_\alpha(X_{23}), \overline{\mu}_\alpha(X_{34}), \overline{\mu}_\alpha(X_{35}), \overline{\mu}_\alpha(X_{36}), \overline{\mu}_\alpha(X_{46}) \end{Bmatrix} \\ &= \overline{\mu}_\alpha(X_{34}) = 1.272 \end{aligned} \tag{5.80}$$

So the bi-closing connectivity class X_{34} is considered as the potential leak channel of Region 12 which is shown in Fig. 5.38b.

For the additional twelve OOL leakage regions, namely Regions 2, 5, 7, 8, 10, 16, 18, 20, 22, 23, 24, and 26, the proposed methodology has been applied to each to identify the leakage channels. Figure 5.39 illustrates the leakage channels across these

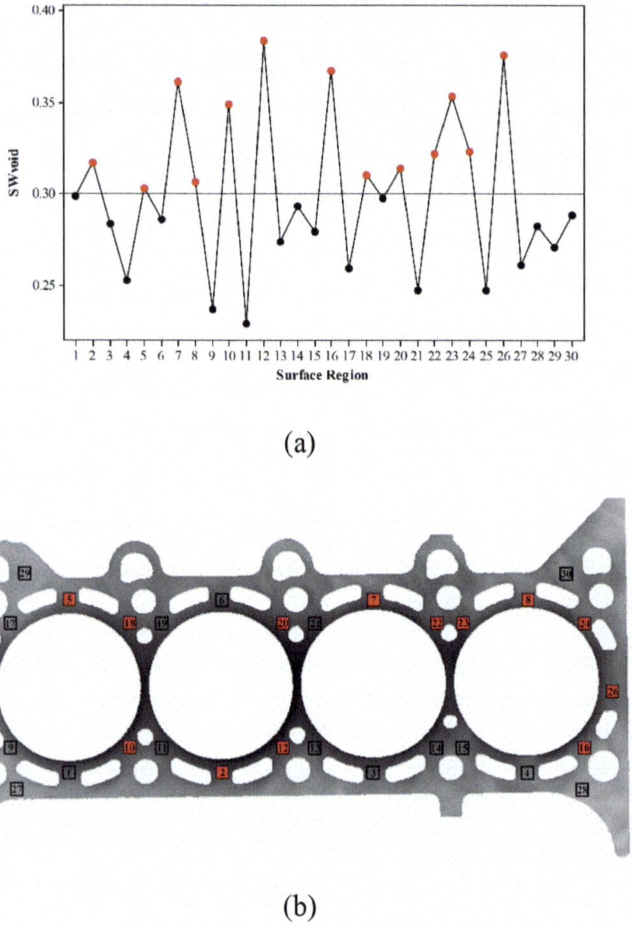

Fig. 5.36 **a** The line chart of *SWvoid*, **b** OOL leakage regions

regions. To enhance visibility, the leakage channels are marked with red curvilinear highlights.

The experimental outcomes substantiate the efficacy and precision of the proposed leakage channel prediction methodology, which offers significant insights into surface topography and the static sealing performance. Concurrently, the identification of potential leakage channels mirrors surface irregularities, facilitating the detection of machining defects. Conversely, the implementation of leakage-focused preventive measures prior to the product advancing to subsequent, more costly machining stages is crucial. Moreover, the accurate prediction of leakage channels ensures the quality of surface topography, thereby fulfilling the product's functional requirements and reducing the rate of defective parts.

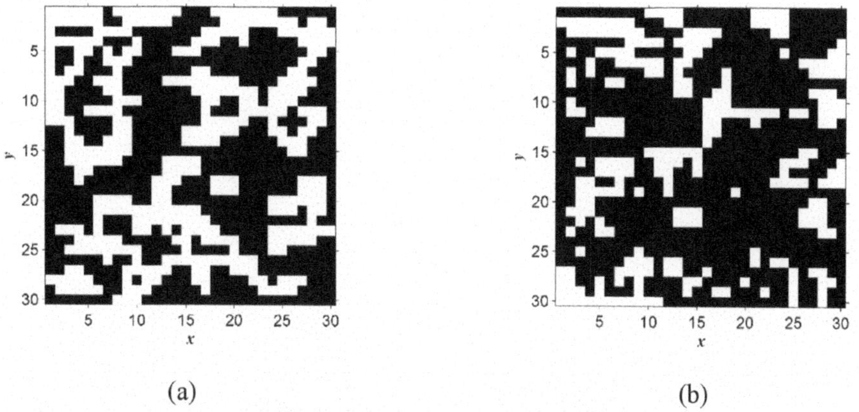

Fig. 5.37 **a** The contact surface of Region 12, **b** the contact surface of Region 11

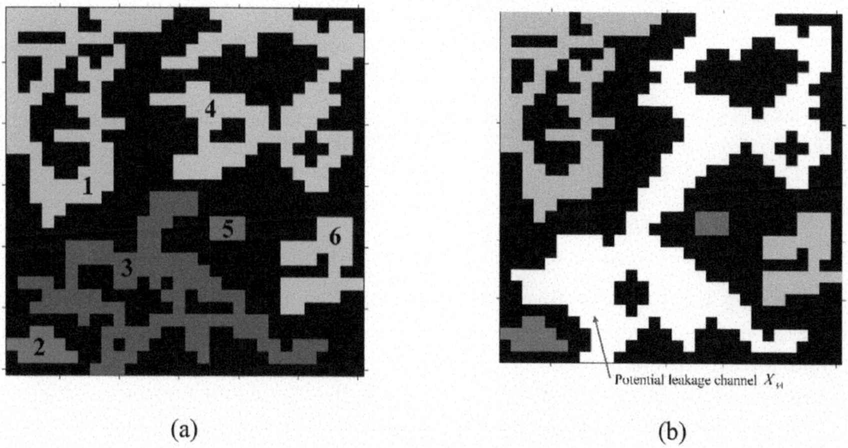

Fig. 5.38 **a** Different connectivity classes, **b** the leakage channel of Region 12

Table 5.10 The results of connectivity parameters from X_1 to X_6

Connectivity class	s_{max}	$\overline{\mu}_{\alpha}$	Ψ
X_1	2	1.168	0
X_2	2	1.078	0
X_3	2	1.107	0
X_4	2	**1.201**	0
X_5	1	1.051	0
X_6	2	1.107	0

* The value in bold means the maximum value of the corresponding column

Table 5.11 The results of correlation parameters from (X_1, X_2) to (X_5, X_6)

Connectivity classes in pairs	*mcd*	*acd*	θ
(X_1, X_2)	12.04	21.10	4
(X_1, X_3)	7.00	17.89	2
(X_1, X_4)	11.00	17.00	2
(X_1, X_5)	**1.41**	17.80	6
(X_1, X_6)	**1.41**	25.24	10
(X_2, X_3)	3.00	11.18	**1**
(X_2, X_4)	**1.41**	28.32	12
(X_2, X_5)	10.05	18.87	11
(X_2, X_6)	13.60	24.04	11
(X_3, X_4)	2.00	18.36	**1**
(X_3, X_5)	5.00	7.81	**1**
(X_3, X_6)	9.43	13.15	**1**
(X_4, X_5)	9.00	11.40	2
(X_4, X_6)	13.00	14.56	**1**
(X_5, X_6)	4.47	**7.62**	2

* The value in bold means the minimum value of the corresponding column

5.2.5 *Conclusions*

A pioneering methodology for predicting leakage channels has been introduced, grounded in the concept of surface connectivity. This methodology is structured around three core components: contact surface generation, definition of leakage parameters, and the prediction of leakage channels. The contact surface, which is indicative of potential leakage points, is established through the integration of a virtual gasket surface and a waviness surface. The methodology introduces two categories of leakage parameters—connectivity and correlation parameters—to characterize the attributes of the contact surface. An innovative algorithm for predicting leakage channels has been crafted, utilizing these parameters. A leakage index is constructed to explicitly identify the potential leakage channels on the surface. The methodology's accuracy and efficacy have been confirmed through experimental validation, offering substantial insights into surface topography and the performance of static seals.

Additionally, the area of surface topography and leakage in static sealing has not been extensively explored in previous studies, and the introduction of this approach is anticipated to address some of the existing research gaps. Building on the phases of leakage research, future investigations will delve into the calculation of leak rates based on 3D surface topography.

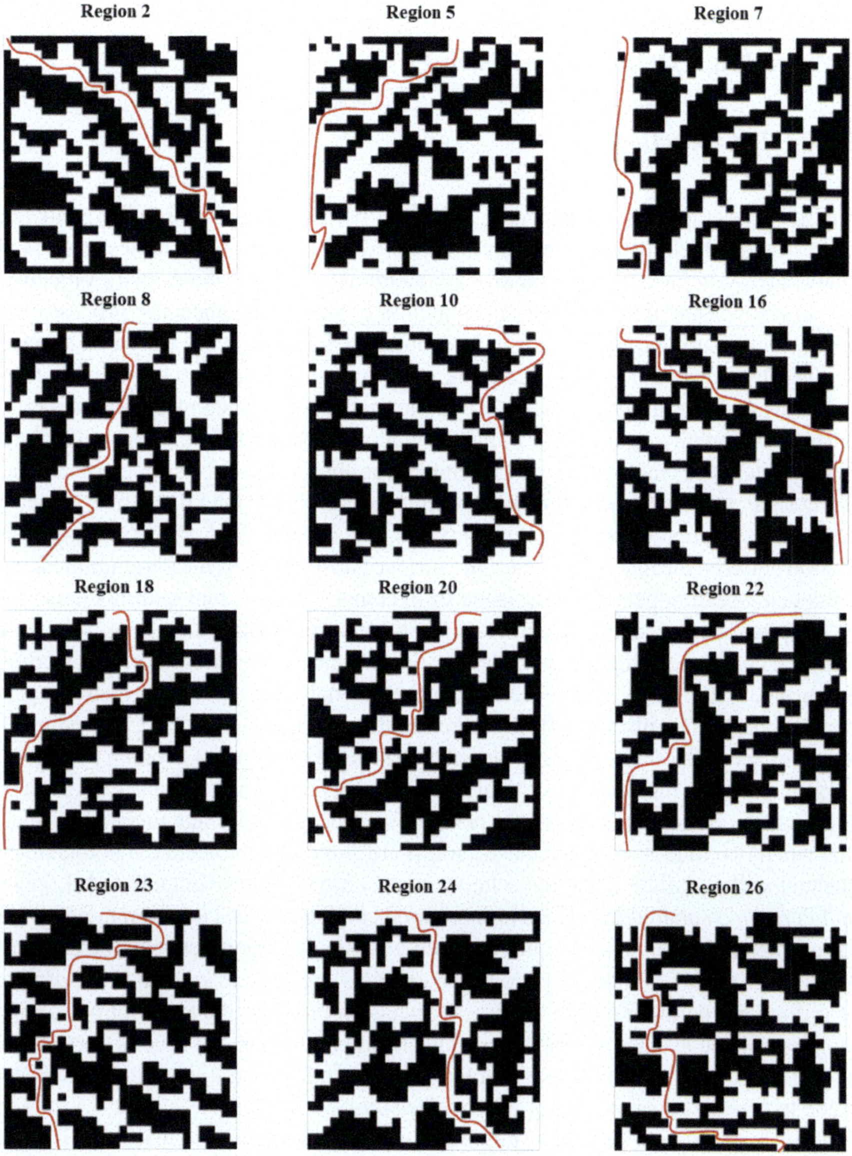

Fig. 5.39 Leakage channels of OOL surface regions

5.3 Evaluation of Surfaces Complementarity Based on High Definition Metrology

5.3.1 Introduction

Surface topography is intrinsically linked to both surface functionality and the processes by which surfaces are manufactured [67]. The majority of studies on surface characterization have focused on process monitoring [68], as the primary objective of surface measurement is to assist production engineers in enhancing process efficiency and ensuring quality control. In contrast to the extensive research on surface topography in relation to manufacturing processes, the correlation between surface functions and topography remains a complex issue for surface metrology experts. Surface topography significantly impacts functions such as friction, wear, and sealing [69]. Whitehouse employed two independent variables to describe surface functions: the normal separation between the two surfaces and their lateral relative velocity [70]. In the context of static sealing surfaces, where there is an absence of relative motion between the mating surfaces, the surface gap emerges as the paramount factor influencing sealing performance, aside from contact pressure and sealing elements.

Despite the significance of surface gaps in predicting surface functions, there is a practical challenge as these gaps cannot be directly measured. The majority of studies on surface sealing rely on theoretical surface models that do not account for gap measurements. For instance, Bottiglione developed a theoretical model to estimate fluid leakage in flat seals, utilizing percolation schemes and fractal geometry [38, 71]. Persson and Yang introduced a critical junction theory to forecast the leak rate of seals, drawing on contact mechanics and percolation theory [72]. Their theory indicates that when two randomly rough elastic solids are compressed together, a non-contact channel will percolate when the relative contact area is approximately 0.4. Lorenz and Persson compared the effective-medium theory and critical junction theory with experimental data on leakage and observed a good correlation between the theories and experimental findings [39]. However, engineering surfaces generated by turning or milling are neither random nor fractal, which contradicts the assumptions of the aforementioned leakage theories. In engineering applications, it is crucial to estimate leakage based on measured topography. Consequently, several attempts have been made to correlate surface topography with sealing performance.

French automotive sector has pioneered a profile motif method that establishes a correlation between motif parameters and surface functionality [73]. Extensive experimentation has demonstrated the impact of motif roughness and waviness parameters on the performance of static seals. F. Robbe-Valloire and M. Prat conducted an analysis of face-turned surface microgeometry, discovering that the variation in valley altitudes significantly affects sealing performance. Malburg introduced the concept of profile leakage parameters by employing a virtual gasket simulation on the waviness profile [46]. Given that profile data represents only a segment of the mating surfaces and is subject to change with different sampling positions, the reliability of leakage predictions based on profile data is inferior to those derived from surface topography data.

High definition metrology (HDM), a recently advanced technology, offers fresh prospects for assessing sealing performance through the topography data of entire surfaces. HDM, operating on the principle of laser holographic interference, is capable of measuring an engineering surface within 40 s and producing a vast array of data points. For workpieces that exceed the field of view, HDM can compile multiple views to create a comprehensive 3D height map of the surfaces. A depiction of the HDM device and an engine block as measured by HDM can be found in Fig. 5.1. Leveraging this innovative measurement platform, several studies have been conducted to link 3D surface topography with sealing performance [74–76]. Shao et al. expanded the virtual gasket concept to three dimensions using HDM data and introduced areal leakage parameters [74]. The incorporation of connectivity followed to identify leakage paths [75]. Ren et al. took into account the surface gap between mating surfaces and proposed a stochastic model for predicting leakage areas, drawing on leakage experiment data and employing a logistic regression method [76]. In Ren's model, the workpiece surfaces are assumed to have a regular rectangular shape, allowing for the calculation of the surface gap through straightforward algebraic operations.

For engineering surfaces with intricate geometries, such as those found on cylinder blocks and cylinder heads, virtual assembly technology is essential for aligning mating surfaces. Virtual assembly relies on high-precision models to simulate the spatial relationships between components. Jayaram et al. pioneered the analysis of assembly accuracy within a virtual assembly design environment [77]. Zhong et al. developed a meta-model for assembly tolerances using a description logic approach [78]. Gallegos-Nieto et al. introduced a haptic-enabled virtual assembly system capable of automatically generating and assessing assembly plans [79]. However, there has been minimal focus on the virtual assembly of point clouds that represent mating surfaces. To align the height maps of mating surfaces within a unified coordinate system, Zhang and Liu proposed an image-guided point cloud registration technique for the virtual measurement of car engine sealing states [80]. Zhang's method requires additional images and the selection of appropriate feature lines to facilitate the registration process. The introduction of automation into the registration process would enhance convenience in conducting such procedures within engineering practices.

Upon the registration of the two mating surfaces, the surface gap can be determined, and its complement, known as the sum surface, reflects the assembly conditions of the mating surfaces. The degree of complementarity between mating surfaces significantly influences the associated surface functions. There is a scarcity of research that quantifies surface complementarity. Furthermore, the correlation between the surface gap (or the sum surface) and the original mating surfaces is an area that has received minimal scholarly attention.

To address the aforementioned gaps, a formal quantitative definition of surface complementarity is introduced. A novel automatic virtual assembly algorithm has been crafted to align point clouds of engineering surfaces featuring complex geometries. Leveraging this automatic assembly algorithm, a comprehensive investigation into the relationships between the sum surface and the original mating surfaces is carried out through case studies.

5.3.2 *Methodology*

5.3.2.1 Definition of the Surfaces Complementarity

The compatibility of mating surfaces is predicated on the separation between two uneven surfaces. Prior to delving into the formal definition of surface compatibility, it is essential to understand the concept of surface separation and the corresponding complementary aggregate surface. As depicted in Fig. 5.40, the orange and blue profiles, z_1 and z_2, respectively, represent the initial state of contact between two mating surfaces. The normal separation between these surfaces is indicated by the red double-sided arrows in Fig. 5.40a. The green profile in Fig. 5.40b signifies the separation size between the surfaces, with the red single-sided arrows, which correspond in length to the red double-sided arrows in Fig. 5.40a, representing the separation size. The definition of the initial contact's separation or normal distance between the surfaces is encapsulated in Eq. (5.81). The black profile in Fig. 5.40c illustrates the profile of the aggregate surface, which complements the separation profile. The aggregate surface (*SS*) is defined by Eq. (5.82).

$$d = \max(z_1 + z_2) - (z_1 + z_2) \tag{5.81}$$

$$SS = \max(d) - d = \max(d) - \max(z_1 + z_2) + (z_1 + z_2) = C + z_1 + z_2 \tag{5.82}$$

In Eq. (5.82), the constant C signifies that the composite surface profile is shaped by the superimposition of the two mating surfaces, provided they are aligned. This additive relationship is also the basis for the term "sum surface." The reference plane for the sum surface is defined as the plane parallel to the mean line of the sum surface and passes through its highest point. The red arrows in Fig. 5.40c denote the normal distance from points on the sum surface to this reference plane. It is evident that $a_1 = b_1 = c_1$ and $a_2 = b_2 = c_2$ maintain their respective values. This correlation ensures that the sum surface maintains a normal distance to the reference plane equivalent to the initial normal distance of the original two rough surfaces in contact.

Significant variations in surface gap can arise when a single surface is paired with different counter-surfaces, as illustrated in Fig. 5.41. It is evident that profile B provides better compensation for profile A than profile C. Furthermore, the void area of the gap (VA_{gap}) between profiles A and B is considerably smaller compared to the gap area between profiles A and C. Consequently, the profile complementarity (*PC*) can be assessed by the void area of profiles (VA_p), as detailed in Eq. (5.83). Here, *sp*

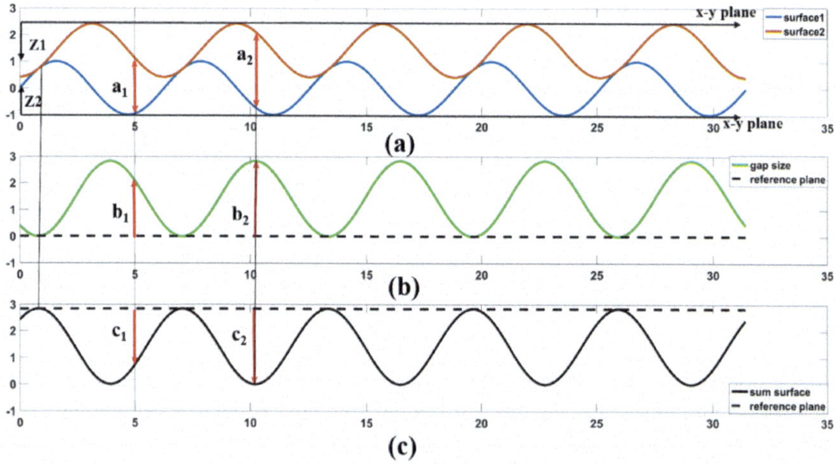

Fig. 5.40 Combination of two rough surfaces into one sum surface

stands for sum profile, which is analogous to the concept of the sum surface. Given that the sum profile is complementary to the profile gap, it follows that $VA_{gap} = VA_{sp}$.

$$PC = 1 - \frac{VA_{gap}}{VA_{p_1} + VA_{p_2}} = 1 - \frac{VA_{sp}}{VA_{p_1} + VA_{p_2}} \tag{5.83}$$

$$VA_p = \int_p \max(P) - P(x)dx \tag{5.84}$$

Drawing from the insights gained in Fig. 5.41, the complementarity index SC for the interaction between mating surfaces S_1 and S_2 can be delineated by Eq. (5.85). The term VV_s refers to the void volume of a surface, which is articulated in Eq. (5.86). Given that the sum surface is the complement to the surface gap, the void volume of the gap VV_{gap} is synonymous with the void volume of the sum surface VV_{ss}. In the scenario where two surfaces are perfectly complementary, the void volume of the surface gap is null, resulting in a complementary degree SC of 1. Conversely, when two surfaces exhibit perfectly mirrored topographies, the volume of the surface gap

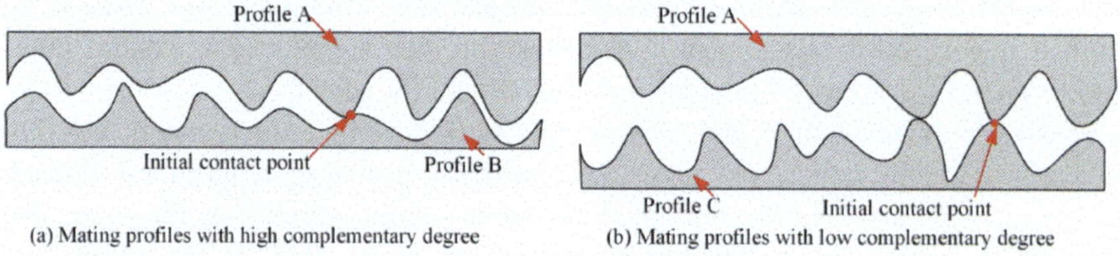

Fig. 5.41 Mating profiles with different complementary degrees

equates to the sum of the void volumes of each surface, yielding a complementary degree SC of 0. Thus, for engineering applications involving mating surfaces, their complementary degrees range between 0 and 1, with higher values of SC indicating superior compensation between the mating surfaces.

$$SC = 1 - \frac{VV_{gap}}{VV_{S_1} + VV_{S_2}} = 1 - \frac{VV_{SS}}{VV_{S_1} + VV_{S_2}} \tag{5.85}$$

$$VV_s = \iint_S (\max(S) - S(x, y))dxdy \tag{5.86}$$

In the context of actual engineering surfaces depicted in Fig. 5.1, discontinuities such as bores and holes are present. A modified concept of surface complementarity is introduced, utilizing surface masks denoted by B. The value $B(x, y) = 1$ signifies that the coordinate (x, y) is a valid point on the surface, while $\mathrm{B}(x, y) = 0$ signifies an invalid point. Given that engineering mating surfaces do not share identical shapes, a combined surface mask $B(x, y) = B_1(x, y)$ and $B_2(x, y)$ is established, assuming that B_1 and B_2 are correctly aligned. The surface gap and the sum surface are thus defined at positions where $B(x, y) = 1$, confirming that the coordinate (x, y) is valid on both mating surfaces. The void volume for the mating surfaces and the sum surface, incorporating surface masks, is articulated in Eq. (5.87), where the symbol "$\circ$" denotes a Hadamard product. With this adjusted definition of void volume, the concept of surface complementarity remains consistent.

$$\begin{aligned} VV_{S_1} &= \iint_{S_1} [\max(S_1) - S_1(x, y)] \circ B(x, y)dxdy \\ VV_{S_2} &= \iint_{S_2} [\max(S_2) - S_2(x, y)] \circ B(x, y)dxdy \\ VV_{SS} &= \iint_{SS} [\max(SS) - SS(x, y)] \circ B(x, y)dxdy \end{aligned} \tag{5.87}$$

For a clearer elucidation of the concept of surface complementarity, two illustrative examples are provided. Profiles of length L are generated using Eq. (5.88), with the function *randn* yielding values adhering to a standard normal distribution. Figure 5.42a presents the simulated profiles, each 1000 units in length, along with their corresponding sum profiles. These profiles are plotted at spatial intervals to prevent overlap. The length L is incrementally adjusted from 100 to 1000, increasing by 100 each time, and for each value of L, the simulation is conducted 1000 times. Consequently, the average profile complementarities are determined for each L. Figure 5.42b displays the simulated PC values across various vector sizes. The mean PC is calculated as 0.289, with a standard deviation of 0.0029. The linear trend line in Fig. 5.42b exhibits an exceedingly slight slope. This observation suggests that for profiles conforming to a normal distribution, their complementarity remains a constant, independent of the length of the profile. Fluctuations in the PC values are attributed to the inherent randomness of computer simulations.

$$z_1 = randn(L, 1),\ z_2 = randn(L, 1),\ SP = z_1 + z_2 \tag{5.88}$$

Surface simulations of analogous nature are executed through Eq. (5.89). Figure 5.43a illustrates a pair of simulated surfaces along with their aggregate surface. The fluctuation in L mirrors the profile simulation variations. The mean SC is 0.292, with a standard deviation of 0.0010. The linear trend line depicted in Fig. 5.43b possesses an insignificant slope, suggesting that SC is uncorrelated with matrix

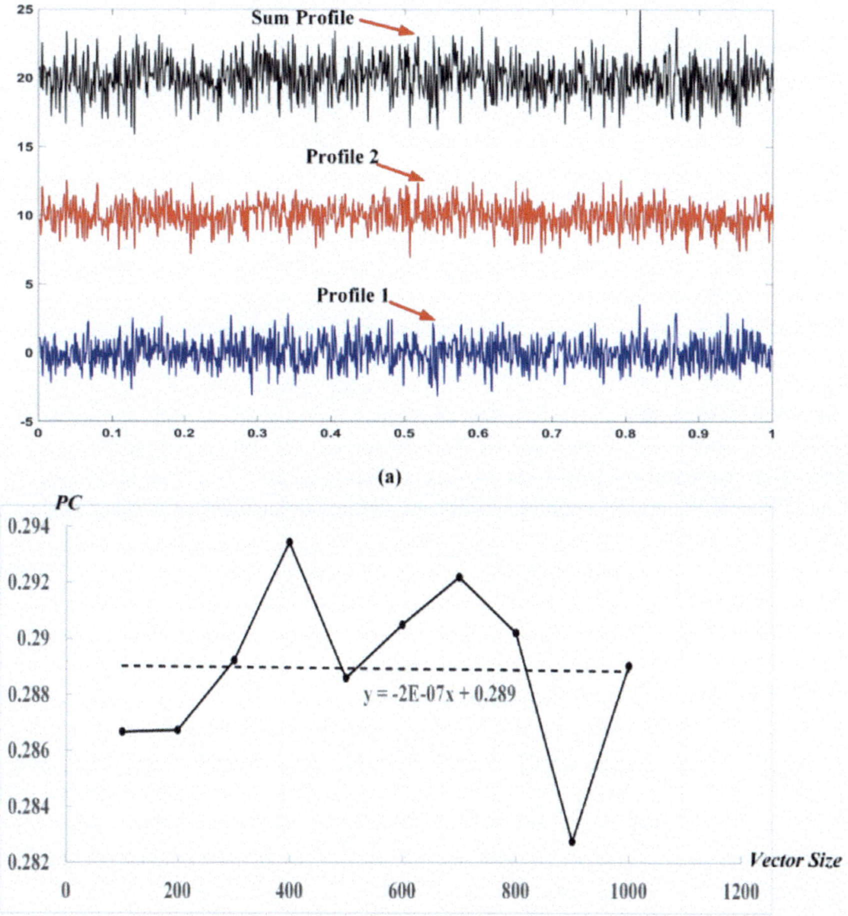

(b)

Fig. 5.42 Profile complementarities of simulated normal distributed profiles

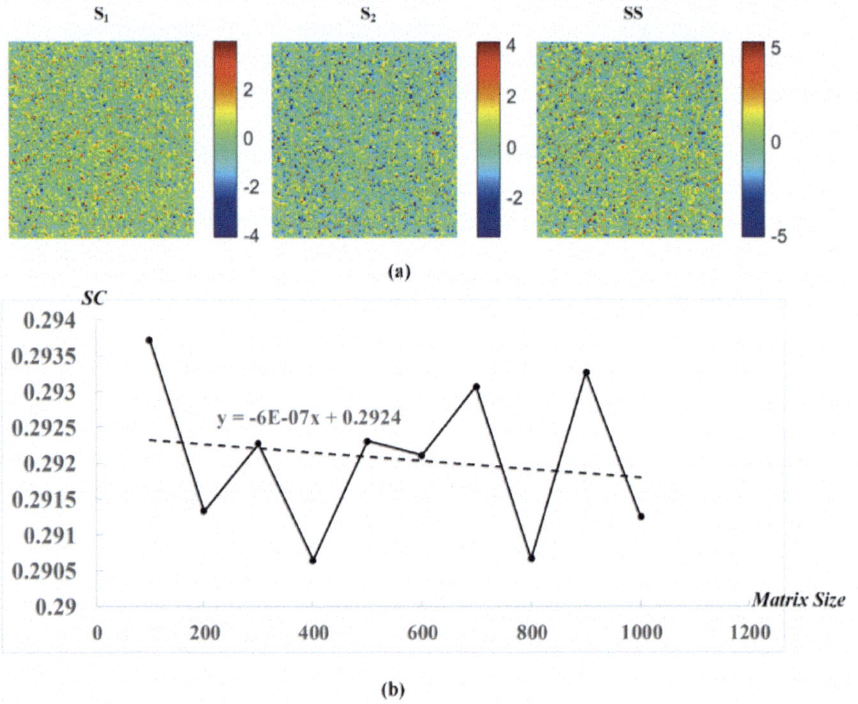

Fig. 5.43 Surface complementarities of simulated normal distributed surfaces

dimensions. Given that the normal distribution is a prevalent assumption in surface simulation studies, the characteristic that *PC* or *SC* remains constant irrespective of the dimensions of the profile or surface is advantageous. This trait facilitates the comparison of the complementarities across profiles or surfaces of varying sizes.

$$z_1 = randn(L, L),\ z_2 = randn(L, L),\ SS = z_1 + z_2 \tag{5.89}$$

5.3.2.2 Virtual Assembly Algorithm

Section 2.1 delineates the formal criteria for defining the sum surface and the complementarity of surfaces, assuming that the measurement data has been synchronized. However, synchronization in engineering applications is challenging due to the intricate geometry of actual surfaces. This section introduces an automated algorithm for aligning real engineering surfaces using HDM.

The HDM data of a planar surface can be transformed into a height-encoded grayscale image *G* and a binary image B (also known as the surface mask), following Wang's methodology [59]. $B(x, y) = 1$ if the point (x, y) is on the surface; otherwise,

$B(x, y) = 0$. Figure 5.44 displays an example of a cylinder block surface. When two mating surfaces are measured by HDM, the process for virtually assembling these surfaces is outlined in Fig. 5.45. Unlike conventional registration tasks that utilize images from the same scene, these registration surfaces originate from distinct workpieces. The topography of surfaces from different workpieces lacks correlation. Consequently, traditional registration methods that rely on intensity similarities are not directly applicable. The core of this virtual assembly algorithm involves registering the surface masks rather than the surface topography itself. Surface masks encompass crucial assembly features such as holes and bores, which are essential for aligning in physical assembly processes to fulfill mechanical functions. Registering surface masks simplifies the registration task of surface topography, and the transformation matrix derived from this registration can be directly applied to register surface topography. Thus, the pivotal step in the workflow is to obtain the transformation matrix through the registration of surface masks. With the transformation matrix in hand, the grayscale image depicting the topography of the mating surfaces can be aligned, and the sum surface can be computed.

Upon the precise alignment of the two surface masks, the masks overlap ratio (MOR) reaches its maximum. This characteristic transforms the task of registering surface masks into an optimization challenge. The optimization challenge can be articulated as follows:

$$\begin{aligned} &\max\ MOR(B_1, B_2(T(\Delta x, \Delta y, \theta))) \\ &s.t.\quad -n \le \Delta x \le n \\ &\qquad\quad -m \le \Delta y \le m \end{aligned}$$

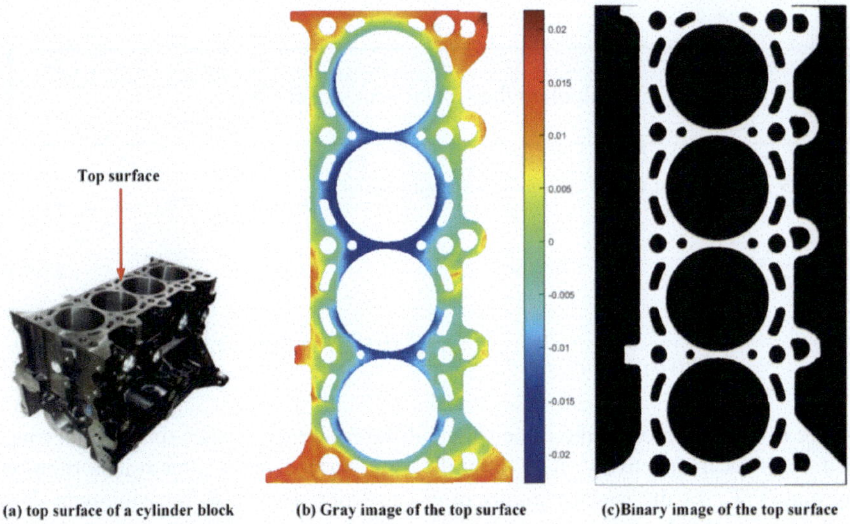

Fig. 5.44 Gray image and binary image of a cylinder block

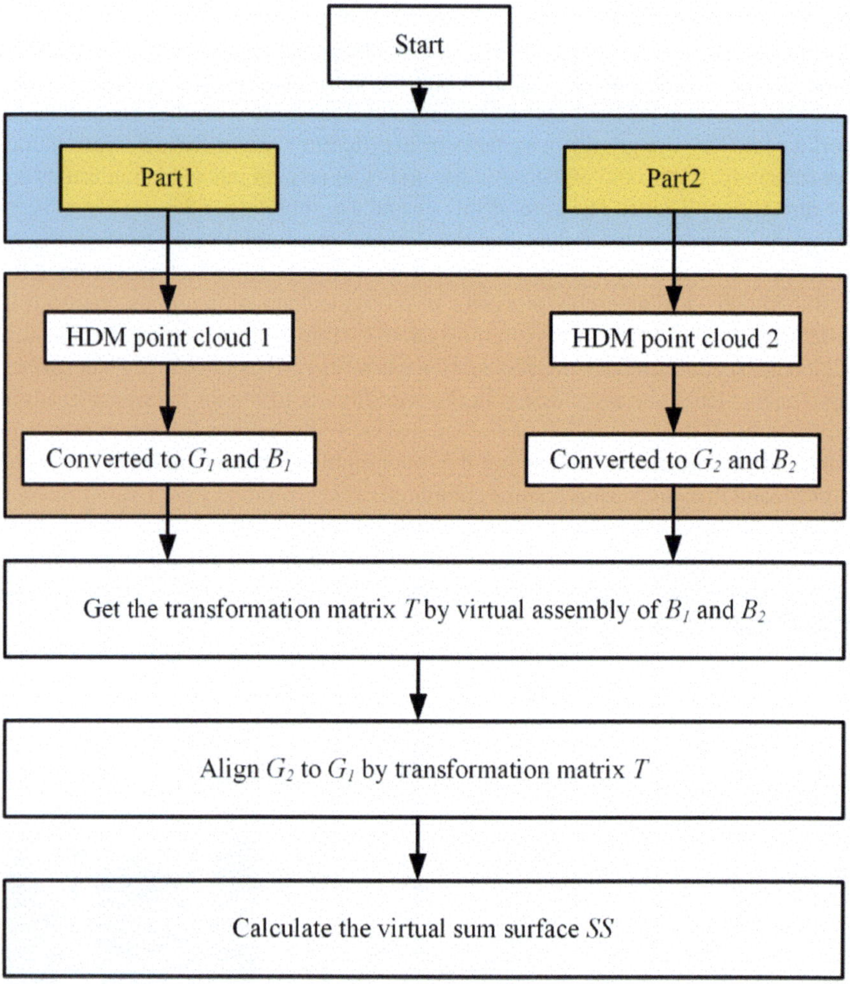

Fig. 5.45 Workflow of virtual assembling of mating surfaces

$$0 \leq \theta < 2\pi \tag{5.90}$$

$$MOR(B_1, B_2) = \frac{\sum_{i=1}^{m} \sum_{j=1}^{n} [if\ B_1(i,j) = B_2(i,j)\ 1\ else\ 0]}{mn} \tag{5.91}$$

where m and n represent the dimensions of the reference surface mask B_1 in rows and columns, respectively, $B_2(T)$ signifies the surface mask after transformation by matrix T. As the surface masks originate from real-world measurement data and the assembled components maintain consistent physical dimensions, scaling is excluded from the transformation parameters. Consequently, matrix T is constituted by two translational components, Δx and Δy, along with a rotational angle θ, as per Eq. (5.92).

$$T(\Delta x, \Delta y, \theta) = \begin{bmatrix} \cos\theta & \sin\theta & 0 \\ -\sin\theta & \cos\theta & 0 \\ \Delta x & \Delta y & 1 \end{bmatrix} \tag{5.92}$$

For addressing the optimization challenge outlined in Eq. (5.90), a particle swarm optimization (PSO) approach [81] is implemented to determine the optimal transformation parameters ($\Delta x, \Delta y, \theta$). The PSO is recognized for its efficient global search capabilities, swift search pace, and straightforward architecture. The sequence of steps for employing the PSO algorithm in the registration of surface masks is depicted in Fig. 5.46, with each phase elaborated as follows.

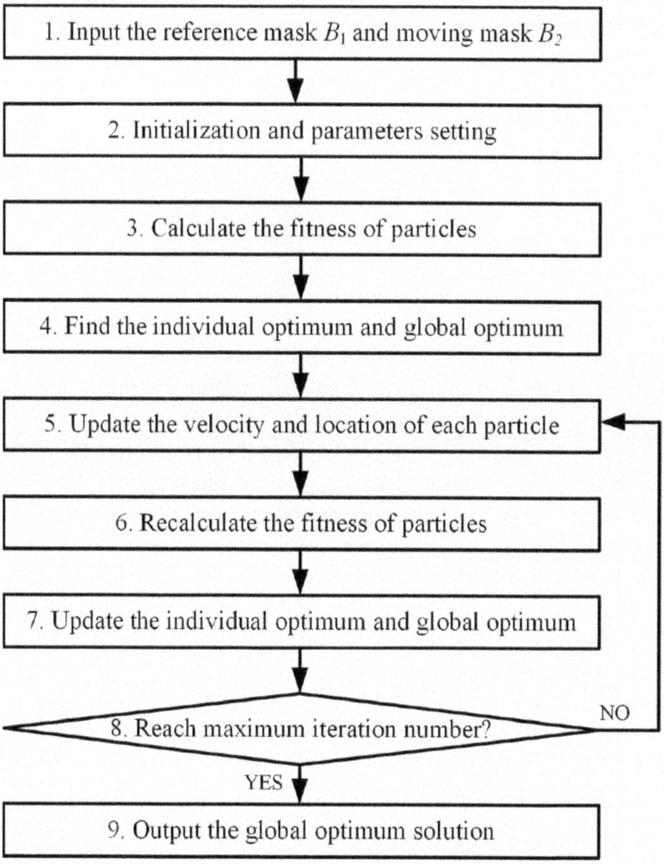

Fig. 5.46 Surface masks registration by the particle swarm optimization algorithm

Step 1: The reference image is set with the fixed surface mask B_1, while the moving surface mask B2, which is to be registered with B_1, is inputted. B_2 is flipped vertically to emulate the real assembly process's overturning operation.
Step 2: Initialize the acceleration coefficients, define the limits for particle positions and velocities, set the number of iterations and swarm size, and randomly assign the position x_k and velocity v_k for each particle.
Step 3: Assess the fitness of the particles, with the fitness being equivalent to the *MOR* value when B_2 is transformed in relation to B_1 (the transformation matrix is derived from the particle's position).
Step 4: In the initial generation, each particle's position is designated as the initial p_{best}, and the position exhibiting the highest fitness is designated as the initial g_{best}. Throughout the PSO evolution, p_{best} tracks the individual's best position, while g_{best} tracks the global best position.
Step 5: Each particle is updated, and they are directed towards the final global optimum using information from the current iteration's global optimum and their historical local optimum positions. The velocity and position update formulas for each particle are as follows:

$$v_{k+1} = \omega v_k + c_1 r_1 (p_{best} - x_k) + c_2 r_2 (g_{best} - x_k) \tag{5.93}$$

$$v_{k+1} = \begin{cases} v_{\max} & v_{k+1} > v_{\max} \\ -v_{\max} & v_{k+1} \leq -v_{\max} \end{cases} \tag{5.94}$$

$$x_{k+1} = x_k + v_k \tag{5.95}$$

where ω is the inertia weight, c_1 and c_2 are the acceleration coefficients, r_1 and r_2 are random numbers between [0,1], and x_{k+1} represents the current position of the particles, signifying the transformation parameters.
Step 6: Evaluate the fitness of the particles after their positions have been updated.
Step 7: Refresh the values of p_{best} and g_{best} accordingly.
Step 8: Determine if the predefined maximum number of iterations has been met. If the maximum number of iterations is reached, proceed to Step 9. If not, return to Step 5.
Step 9: Conclude the algorithm and present the global optimal solution, which consists of the transformation parameters that yield the maximum overlap ratio between the transformed B_2 and B_1.

Figure 5.47 illustrates an example of the evolution process within the automatic assembly algorithm. By maximizing the MOR value, the masks converge towards the correct assembly configurations.

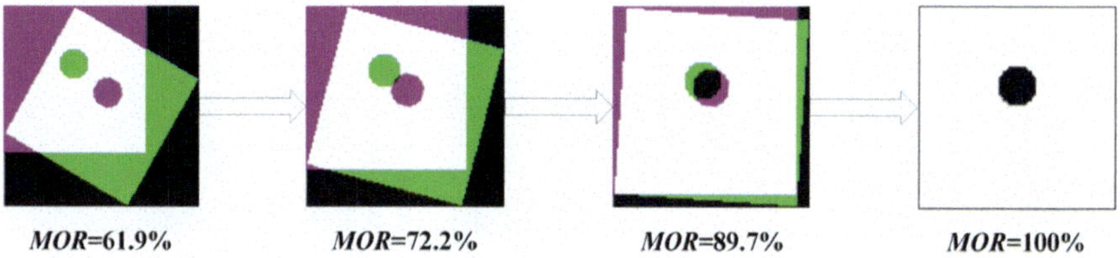

Fig. 5.47 The evolution processes of the automatic assembly algorithm

5.3.3 Case Study I

5.3.3.1 Virtual Assembly of a Cylinder Block Surface and a Cylinder Head Surface

Internal combustion engine leakage remains a critical concern, with the sealing effectiveness of the interface between the engine's cylinder block and cylinder head being directly influenced by their contact conditions. A compliant gasket is utilized at this interface to prevent leakage; however, due to form errors on both surfaces, the gasket may not always adequately fill the gap, leading to occasional leakage. Leakage can lead to a range of engine issues, including compression loss, reduced power, overheating, and potentially engine failure. To illustrate the applicability of the proposed methodology in engineering, a case study on this interface is undertaken.

The engine block and head under consideration are from an automotive manufacturer and are joined by securing 10 bolts, each with a torque of 60 Nm. The assembly and the mating surfaces are depicted in Fig. 5.48. The surfaces were scanned using the ShaPix3D® 3000 series, achieving a vertical resolution of 0.05 μm and a lateral resolution of 150 μm. The height measurement range is ± 5 mm, with a vertical accuracy of 1 μm. The field of view is 280 mm × 280 mm, and the maximum number of sampling points per view is 4 million. After data preprocessing, the resulting gray images, which represent the surfaces' height maps, are presented in Fig. 5.49. Utilizing the virtual assembly algorithm from Sect. 5.3.2.2, the virtual assembly outcome and sum surface are displayed in Fig. 5.50. In Fig. 5.50a, the magenta areas are on the block's surface but not on the head's surface, while the green areas indicate the reverse. The sum surface is determined from the white regions where both mating surfaces have valid data points, as shown in Fig. 5.50b. It is evident from Fig. 5.50b that the central gap is larger than the surrounding areas (lower heights on the sum surface corresponding to larger surface gaps at the same positions). Consequently, the connection part (encircled in red in Fig. 5.50b) between the second and third bores is identified as the area with the highest risk of gas leakage. Similar deductions can be made for other areas of interest.

(a)assembled engine cylinder head and block (b)Top surface of the cylinder block (c)Bottom surface of the cylinder head

Fig. 5.48 The assembled engine and the two mating surfaces

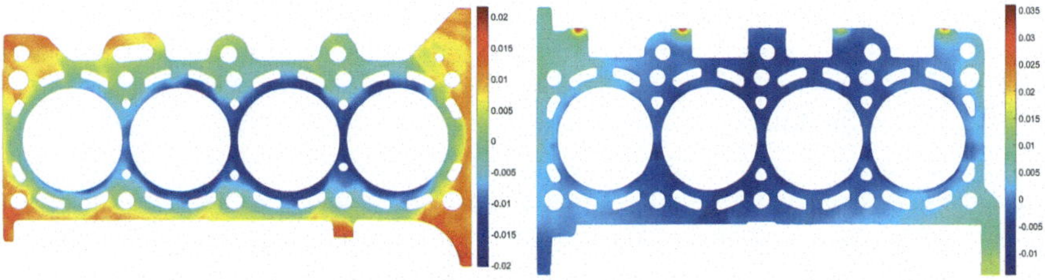

(a) The top surface of the cylinder block (b) The bottom surface of the cylinder head

Fig. 5.49 The height map of the mating surfaces

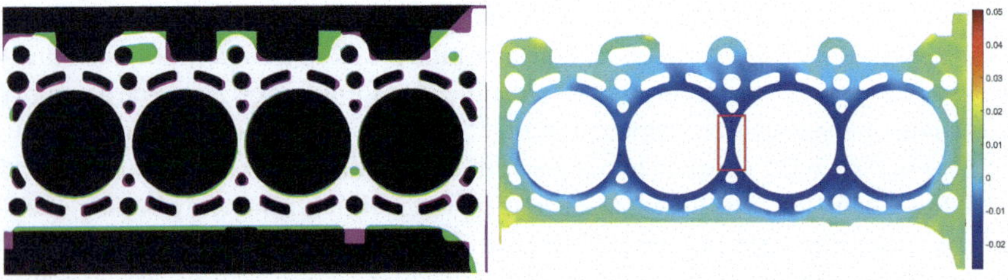

(a) Virtual assembly of the engine block and head (b) Sum surface of the mating surfaces

Fig. 5.50 Virtual assembly of the engine block and head

5.3.3.2 Comparison with GT Model

The "sum surface" concept has been established in tribology for some time. During the 1970s, to determine the contact characteristics between two rough surfaces, Greenwood and Tripp introduced the seminal GT model [82]. This model generates a sum surface through an equivalent relationship of asperity parameters, as depicted in Eq. (5.96). Here, σ^2 represents the variance of the height distribution of the asperities, R denotes the average radius of the asperities, and η signifies the distribution density of asperities.

$$\begin{cases} \sigma^2 = \sigma_1^2 + \sigma_2^2 \\ \frac{1}{R} = \frac{1}{R_1} + \frac{1}{R_2} \\ \eta = \eta_1 + \eta_2 \end{cases} \tag{5.96}$$

The widespread adoption of the sum surface concept in the GT model necessitates a comparison of the statistical asperity parameters with those of the traditional GT model. Employing the classical 9PP-3D definition of asperity-peaks [83], asperity parameters can be determined on 3D engineering surfaces. The radius R of an identified asperity-peak is derived by fitting a sphere to the 9 points of the peak using the least-squares method. To compare the asperity parameters of each mating surface and the sum surface, ten sampling regions (refer to Fig. 5.51) are selected.

The asperties parameters were determined for the mating surfaces, and their corresponding GT model parameters were derived using Eq. (5.96). These parameters were also computed for the sum surface and then contrasted with the GT model via Eq. (5.97). The disparities between the asperities' parameters as forecasted by the GT model and those calculated for the virtually assembled sum surface are detailed in Table 5.12. The final column presents the surface complementarity for the ten regions. The average discrepancies for σ^2, R and η are 24.8%, 30.4%, and 45.3%, respectively. These differences may stem from the phenomenon of surface complementarity. The average surface complementarity value for the ten regions is 16.5%, which is nearly half of the mean discrepancy of the three asperities' parameters.

$$\begin{cases} \mathrm{Diff}(\sigma^2) = \frac{|\sigma_{ss}^2 - \sigma_{GT}^2|}{\sigma_{GT}^2} \\ \mathrm{Diff}(R) = \frac{|R_{ss} - R_{GT}|}{R_{GT}} \\ \mathrm{Diff}(\eta) = \frac{|\eta_{ss} - \eta_{GT}|}{\eta_{GT}} \end{cases} \tag{5.97}$$

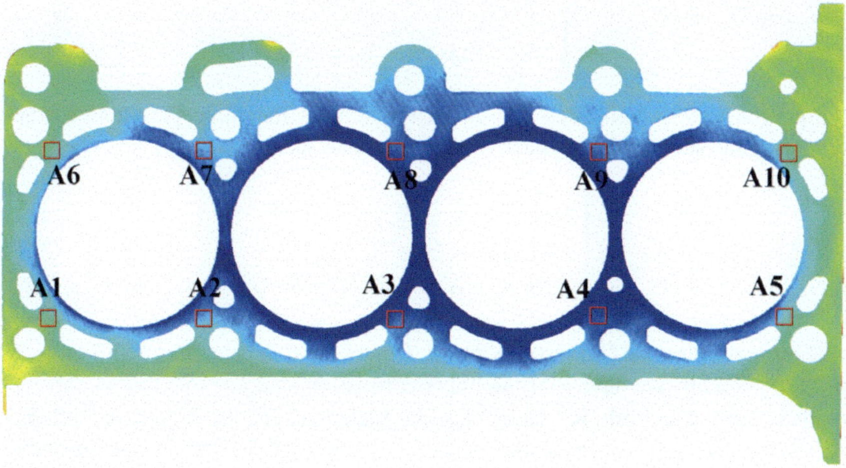

Fig. 5.51 Ten sampling regions of the mating surfaces

Table 5.12 Difference of asperities' parameters

Region	Diff (σ^2)	Diff (R)	Diff (η)	SC
1	0.575	0.486	0.387	0.172
2	0.312	0.331	0.547	0.096
3	0.176	0.222	0.520	0.116
4	0.226	0.344	0.391	0.296
5	0.060	0.092	0.520	0.185
6	0.258	0.434	0.411	0.034
7	0.101	0.185	0.439	0.031
8	0.162	0.382	0.493	0.19
9	0.046	0.236	0.396	0.319
10	0.559	0.327	0.421	0.206

5.3.3.3 Comparison of Areal Field Parameters

The sum surface is intimately connected to its mating surfaces, making it intriguing to examine the variations in certain traditional areal field parameters when comparing the sum surface to its mating counterparts. Within this section, the three classical areal field parameters—Sa, Sq, and St—are calculated across ten square samples extracted from the cylinder block (B), cylinder head (H), and the sum surface (SS). The definitions for these parameters are provided in Eq. (5.98). A tabulation of these parameter values can be found in Table 5.13.

$$\begin{cases} Sa = \frac{1}{A} \iint\limits_{A} |z(x, y)| dxdy \\ Sq = \sqrt{\frac{1}{A} \iint\limits_{A} z^2(x, y) dxdy} \\ St = \max(z) - \min(z) \end{cases} \tag{5.98}$$

Equation (5.99) delineates the disparities among the three parameters when comparing the sum surface to each of the individual surfaces.

$$\begin{cases} Diff\,(Sa) = \frac{Sa(B)+Sa(H)-Sa(SS)}{Sa(B)+Sa(H)} \\ Diff\,(Sq) = \frac{Sq(B)+Sq(H)-Sq(SS)}{Sq(B)+Sq(H)} \\ Diff\,(St) = \frac{St(B)+St(H)-St(SS)}{St(B)+St(H)} \end{cases} \tag{5.99}$$

The scatter plots of $Diff(Sa)$, $Diff(Sq)$, $Diff(St)$ and SC are shown in Fig. 5.52. $Diff(Sa)$ and $Diff(Sq)$ have a strong linear correlation, and $Diff(St)$ has an apparent positive correlation with SC. Generally, the higher of SC, the higher of $Diff(Sa)$, $Diff(Sq)$ and $Diff(St)$, and the lower of $Sa(SS)$, $Sq(SS)$ and $St(SS)$. The average value of $Diff(Sa)$, $Diff(Sq)$ and $Diff(St)$ are 0.152, 0.152 and 0.174 respectively, while the

Table 5.13 Areal field parameters of the ten square samples

Region	$Sa(10^{-3}$ mm)			$Sq(10^{-3}$ mm)			$St(10^{-3}$ mm)		
	B	*H*	*SS*	*B*	*H*	*SS*	*B*	*H*	*SS*
1	1.361	0.705	1.791	1.700	0.880	2.247	8.604	5.270	11.772
2	1.677	0.636	2.043	1.995	0.781	2.448	8.708	4.014	11.287
3	1.600	0.444	1.776	1.892	0.546	2.125	8.893	3.072	10.112
4	0.691	0.421	0.835	0.834	0.524	1.025	4.287	3.344	5.682
5	2.364	0.486	2.566	2.909	0.630	3.181	13.940	3.595	15.052
6	1.405	0.579	1.794	1.708	0.731	2.206	8.533	4.377	10.797
7	1.083	0.590	1.291	1.286	0.745	1.565	6.375	4.536	9.538
8	1.150	0.438	1.249	1.385	0.553	1.537	7.073	3.390	7.720
9	1.204	0.451	1.405	1.419	0.561	1.631	6.586	3.396	7.622
10	1.042	0.548	1.426	1.306	0.686	1.797	8.295	3.824	10.518

average of *SC* is 0.165, approximately the average value of *Diff*(*Sa*), *Diff*(*Sq*) and *Diff*(*St*).

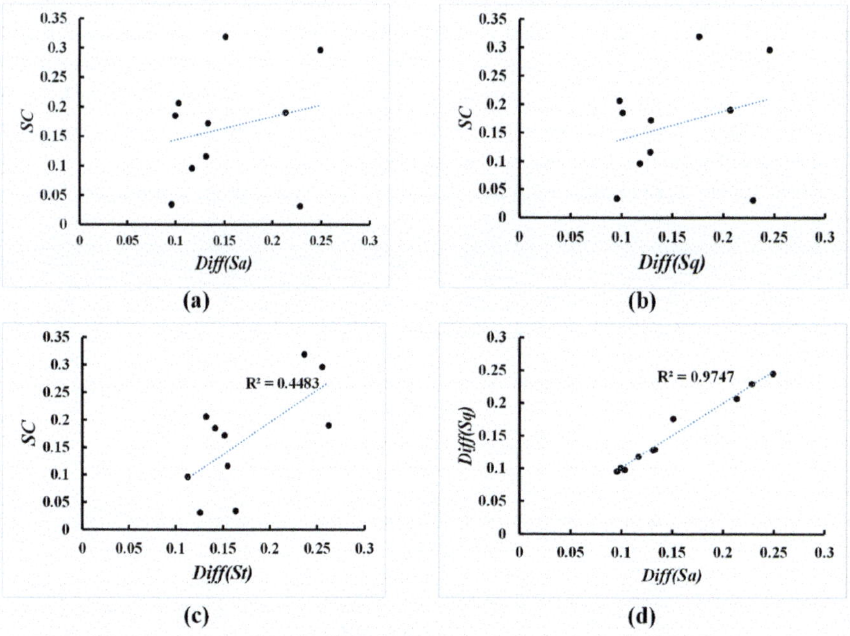

Fig. 5.52 Scatter plots of *Diff*(*Sa*), *Diff*(*Sq*), *Diff*(*St*) and *SC*

5.3.3.4 Comparison of Leakage Parameters

Leakage remains a critical issue at the interface between cylinder blocks and cylinder heads. Historically, flatness has been the primary quality control parameter for surface mating in manufacturing settings. However, even when both mating surfaces fall within flatness tolerances, leakage can still manifest upon mating. This implies that individual quality control of each surface does not ensure the sealing performance of the mating interface. It is valuable to investigate the functional prediction differences between individual surfaces and the sum surface, which can represent the actual mating conditions. For leakage prediction, Malburg introduced three profile parameters to assess the sealing capacity of profiles [10]. The original profile is filtered to extract the waviness component W, followed by the application of a morphological closing filter to simulate the virtual gasket C. Figure 5.53 provides an example of a waviness profile and its corresponding virtual gasket. In this context, the contact length percentage (*CLP*) and void area (*VA*) are utilized to assess the differences in sealing performance predictions between the two mating surfaces and the sum surface. The definitions of *CLP* and *VA* are reiterated in Eq. (5.100). Generally, a larger *VA* or a smaller *CLP* indicates poorer sealing performance of the profile.

The topography of surfaces surrounding the bores is essential for the sealing integrity of an engine's combustion chamber. To evaluate the sealing capacity, four enclosed circular regions around the bores have been selected for testing, as depicted in Fig. 5.54. The circular sampled profiles undergo filtration using a closed profile spline filter [84] to extract the waviness profiles *W*, with cut-off wavelengths set at 2.5 and 25 mm. Subsequently, these waviness profiles are subjected to morphological closing with a circle of radius 3000 mm. Figure 5.55 illustrates the waviness and virtual gasket of the L1 profile on the cylinder block, where the majority of the waviness profile is in contact with the virtual gasket. The leakage parameters for the

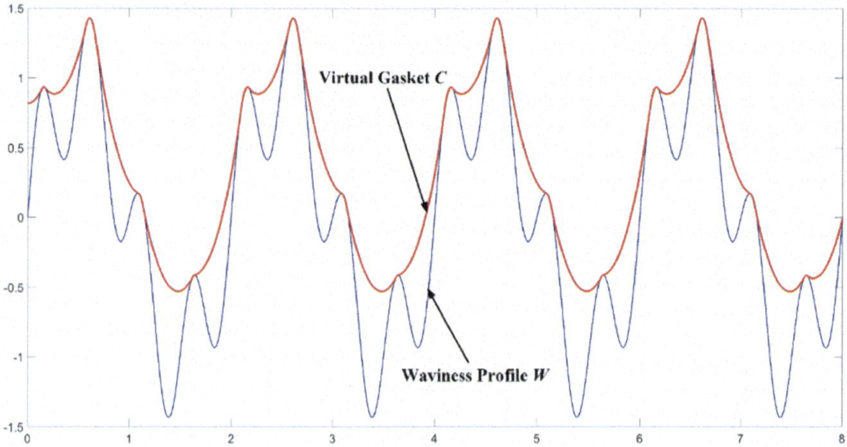

Fig. 5.53 Example of a waviness profile and its virtual gasket

four circular samples on the cylinder block (B), head (H), and sum surface (SS) are detailed in Table 5.14, with the final column indicating the complementarity of each pair of circular samples.

$$\begin{cases} CLP = \frac{\sum_n [if\,(C_i = W_i)1\; else\; 0]}{n} \cdot 100\% \\ VA = \int_0^l (C(x) - W(x))dx \end{cases} \tag{5.100}$$

Table 5.14 indicates that the CLP for the sum surface is comparable to that of the individual surfaces, as the virtual gasket largely encompasses the profiles, with the exception of the deepest valleys. The sum surface, relative to the individual mating surfaces, presents more pronounced peaks and more profound valleys. VA takes into account the depth variations of the valleys, and Table 5.14 demonstrates that $VA(SS) > VA(H), VA(B)$. $VA(SS) < VA(B) + VA(H)$ is due to the complementarity of the surfaces. The discrepancy in VA between the sum surface and the mating surfaces is quantified by Eq. (5.101). The mean difference in $Diff(VA) = 0.17$, and the he average $SC = 0.23$, which is nearly equivalent to the mean $Diff(VA)$.

$$Diff\,(VA) = \frac{VA(B) + VA(H) - VA(SS)}{VA(B) + VA(H)} \tag{5.101}$$

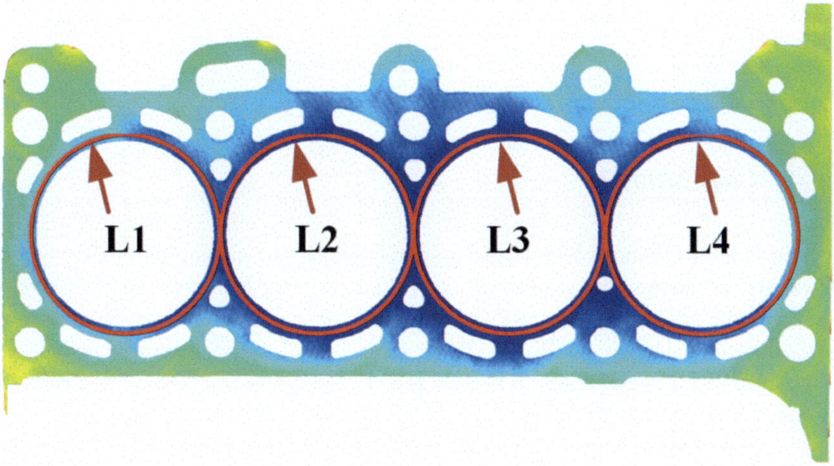

Fig. 5.54 Four line samples of the mating surfaces

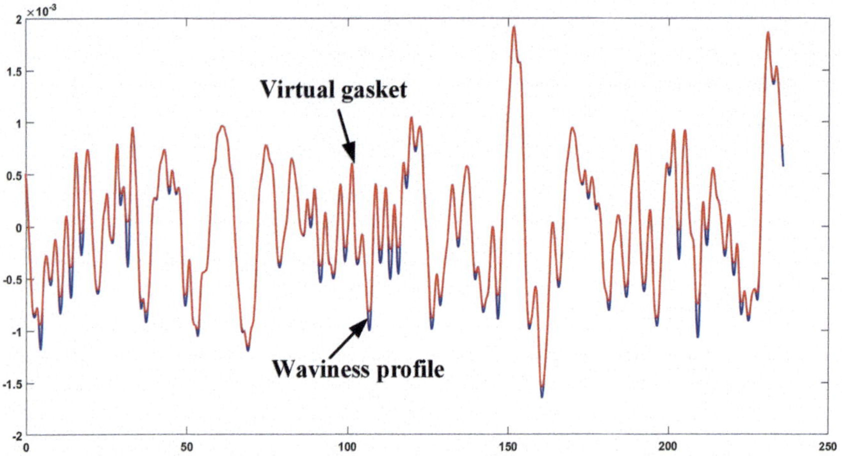

Fig. 5.55 Waviness profile and virtual gasket of L1 profile on the cylinder block

Table 5.14 Leakage parameters of four samples on three surfaces

Line	*CLP*			*VA* (10^{-3}mm^2)			*SC*
	B	*H*	*SS*	*B*	*H*	*SS*	
1	0.715	0.654	0.670	4.704	7.570	9.730	0.213
2	0.705	0.693	0.686	4.418	6.275	9.146	0.265
3	0.693	0.710	0.691	5.377	3.269	6.814	0.251
4	0.677	0.686	0.663	4.791	3.614	7.328	0.199

5.3.4 Case Study II

To ascertain the efficacy of the virtual assembly algorithm, an aluminum alloy block (referred to as the mini head) and a cast iron block (referred to as the mini block) were precision-machined in a laboratory setting. The surfaces of these two components were captured using the ShaPix3D® 1500 series, which offers a vertical resolution of 0.05 μm and a lateral resolution of 80 μm. The equipment's vertical precision is 1 μm, and it covers a field of view measuring 150 mm × 150 mm, with a maximum capacity of 4 million sampling points per view. Photographs of the mini head and mini block are depicted in Fig. 5.56a and b, respectively. The corresponding gray images, derived from the HDM data of the mating surfaces, are presented in Fig. 5.56c and d. Given that the mini block and mini head share the same shape upon a 90-degree rotation, four distinct assembly configurations are possible. Figure 5.57 illustrates the sum surfaces resulting from these four assembly configurations, utilizing the virtual assembly algorithm detailed in Sect. 5.3.2.2.

To assess the variations among the mating surfaces and the sum surfaces (which are four distinct sum surfaces resulting from different assembly orientations), three areal

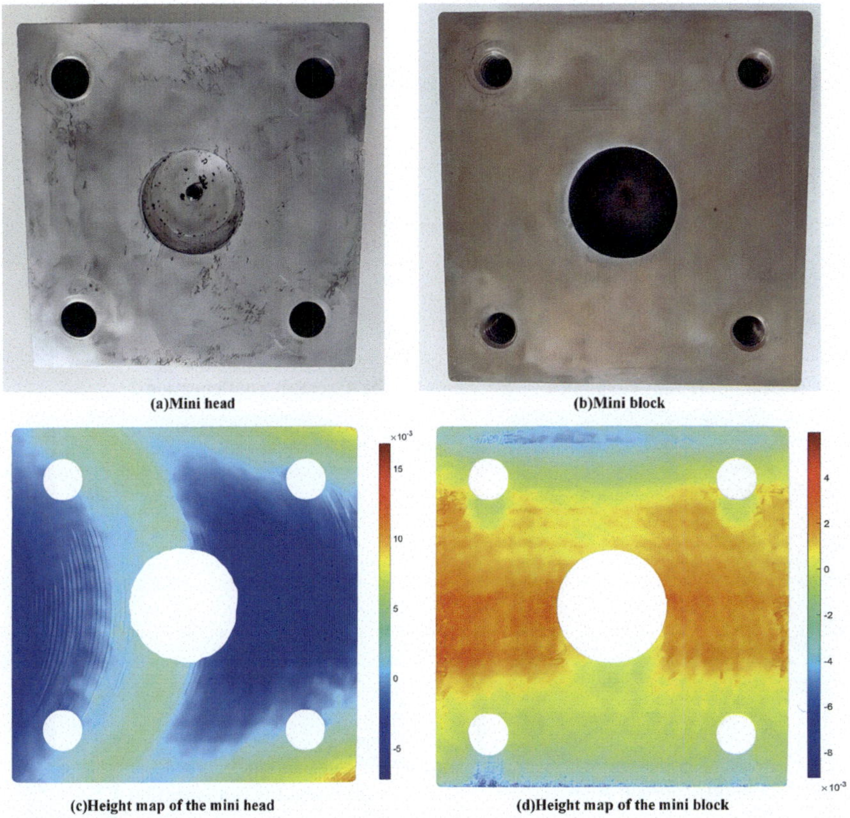

Fig. 5.56 The mini head and mini block and their height maps

field parameters and two leakage parameters were determined for these surfaces. The leakage parameters were derived from circular samples taken from the surfaces, with the sampling circle's position and dimensions detailed in Fig. 5.58. These circular samples were processed using a closed profile spline filter to extract the waviness profiles W, with cut-off wavelengths set at 2.5 and 25 mm. Subsequent to this, the waviness profiles underwent morphological closing with a circle of radius 4000 mm. The parameters for the individual surfaces and the sum surfaces are compiled in Tables 5.15 and 5.16, respectively. Table 5.16 indicates that identical surfaces can yield markedly different sum surfaces or surface gaps depending on the assembly orientation. Configuration 4 assembly exhibits the lowest surface complementarity and the greatest VA value, suggesting the poorest sealing performance. Conversely, configuration 1 assembly demonstrates the highest surface complementarity and the smallest VA value, indicating the best sealing performance. It is also observed that there is an inverse relationship between surface complementarity and VA values, suggesting that SC could serve as an effective indicator for sealing performance.

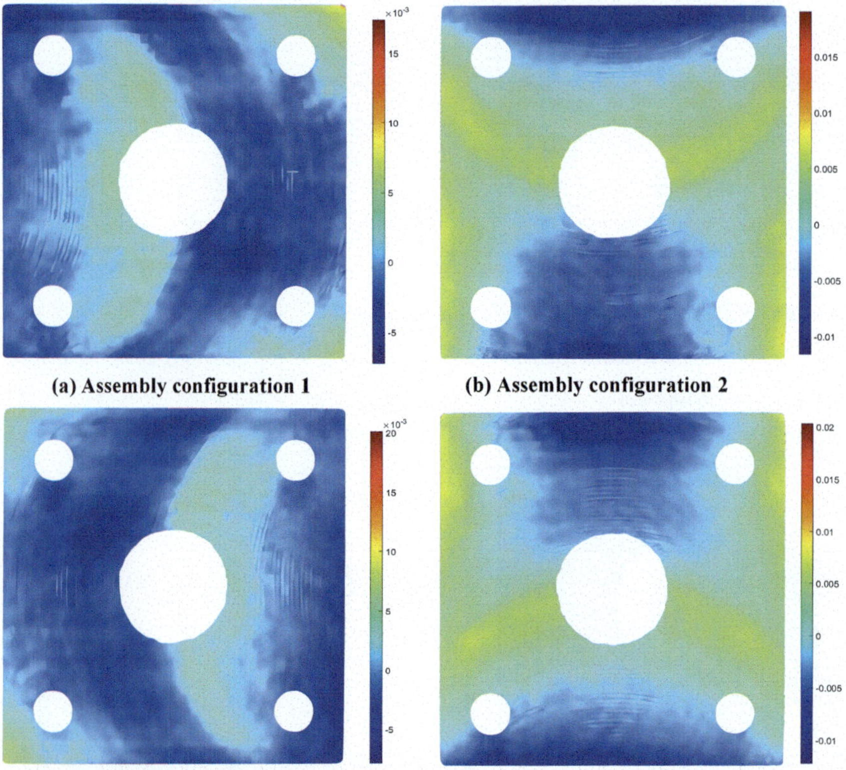

(a) Assembly configuration 1 **(b) Assembly configuration 2**

(c) Assembly configuration 3 **(d) Assembly configuration 4**

Fig. 5.57 Four types of assembly configuration

5.3.5 Conclusions

A new concept of surfaces complementarity (SC) which can measure how well two rough surfaces fit into each other is proposed. To make this new concept implementable in practice, a novel virtual assembly algorithm is proposed based on HDM. With this algorithm, the sum surface which is the complement of the surface gap could be constructed and the surfaces complementarity could be calculated. The first case of the interface between a cylinder block and a cylinder head shows the engineering application potential of the virtual assembly algorithm. The concept of sum surface is compared with traditional GT model by numerical asperity parameters, and the result shows the real sum surface has a certain deviation with GT model due to the phenomenon of surfaces complementarity. Areal field parameters comparison shows that the higher of the surfaces complementarity, the lower of the Sa, Sq and St value of the sum surface. Leakage parameters comparison shows that variation of VA has a closed relationship with surfaces complementarity. A case study of the square

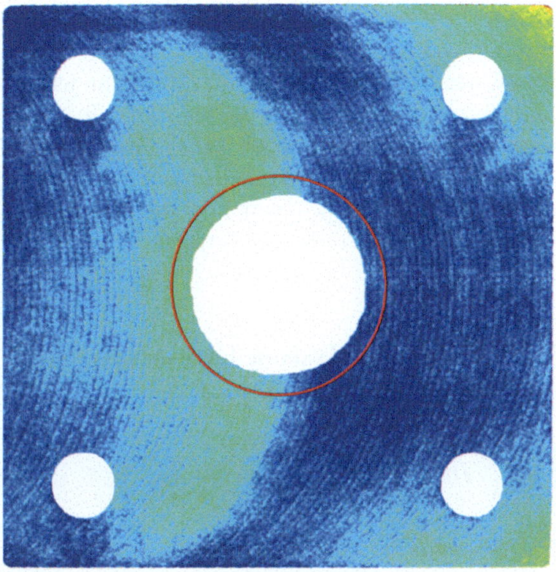

Fig. 5.58 Circle sampling of the mating surfaces

Table 5.15 Parameters of the mini block and mini head

Parameters	*Sa* (mm)	*Sq* (mm)	*St* (mm)	*CLP*	*VA* (10^{-3}mm^2)
Mini block	0.0015	0.0019	0.0135	0.580	3.322
Mini head	0.0022	0.0027	0.0236	0.586	5.451

Table 5.16 Parameters of the four sum surfaces

Assembly	*Sa* (mm)	*Sq* (mm)	*St* (mm)	*CLP*	*VA* (10^{-3}mm^2)	*SC*
1	0.0017	0.0021	0.0246	0.570	6.254	0.209
2	0.0031	0.0037	0.0306	0.562	6.430	0.134
3	0.0017	0.0021	0.0280	0.576	6.919	0.084
4	0.0031	0.0036	0.0326	0.562	7.261	0.075

surfaces shows the practical application potential of the virtual assembly algorithm to optimize the sealing performance of the mating surfaces.

Introducing the concept of Surfaces Complementarity (SC), a metric designed to quantify the degree of fit between two irregular surfaces, this section outlines a pioneering virtual assembly algorithm grounded in HDM. This algorithm enables the construction of a composite surface, representing the complement of the surface gap, and facilitates the computation of surface complementarity. The practical utility of the virtual assembly algorithm is demonstrated through an engineering case study involving the interface between a cylinder block and a cylinder head. A comparison of the composite surface concept with the traditional GT model using numerical asperity parameters reveals discrepancies between the actual composite surface and the GT model, attributed to the phenomenon of surface complementarity. An analysis of areal field parameters indicates that an increase in surface complementarity corresponds to a decrease in the Sa, Sq, and St values of the composite surface. Furthermore, a comparison of leakage parameters illustrates a strong correlation between the variation in VA and surface complementarity. A case study on square surfaces underscores the potential of the virtual assembly algorithm to enhance the sealing performance of mating surfaces.

5.4 Sealing Analysis of Face-Milled Surfaces Based on High Definition Metrology

5.4.1 Introduction

The prevention of leakage at the interface of the cylinder block and head is crucial, as it can lead to a decrease in engine power, excessive engine temperatures, and potential engine breakdowns. Figure 5.59 depicts a cylinder head affected by a leakage issue, with the affected area showing signs of carbon buildup and discoloration. The red-marked areas in Fig. 5.59 indicate the leakage spots, indicating that the sealing of this interface has failed.

The integrity of the seal at the interface between the cylinder block and the head is predominantly determined by the gasket, the topographical characteristics of the mating surfaces, and the torque applied to the bolts [85]. Elevated ignition pressures and temperatures exert significant stress on head gaskets. The multi-layer steel (MLS)

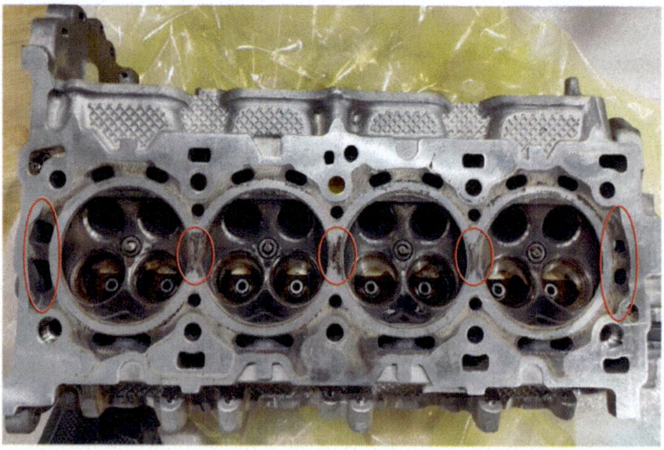

Fig. 5.59 A cylinder head with the leakage problem

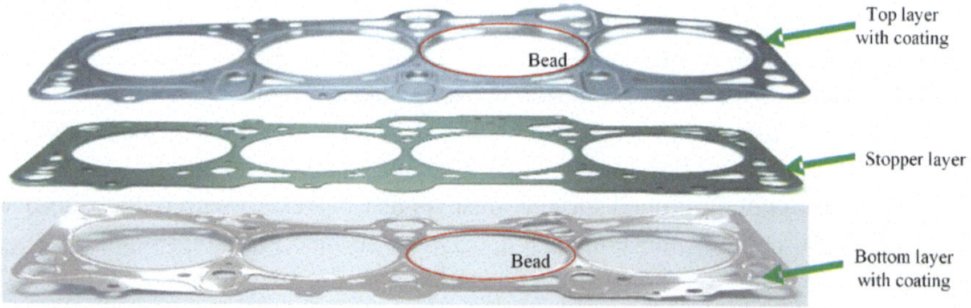

Fig. 5.60 A typical 3-layer MLS head gasket

gasket has been engineered to align with contemporary engine design trends, boasting enhanced sealing capabilities [86, 87]. Figure 5.60 illustrates a standard 3-layer MLS head gasket, comprising a top layer, a stopper layer, and a bottom layer. The presence of beads encircling combustion chambers and oil/coolant passages enhances localized sealing pressure (macro sealing), while full-surface or partial elastomer coatings augment sealing integrity (micro sealing) [88, 89].

Even with meticulous gasket design, leakage can manifest in actual operations owing to a variety of potential causes. Figure 5.61 details seven distinct types of media leakage that can affect MLS head gaskets. Among these, gas leakage between combustion chambers is the most detrimental and poses a challenging leakage issue to address. When high-temperature combustion gases bypass the gasket surface, the gasket is at risk of destruction and rapid loss of sealing efficacy. Consequently, the proposed approach concentrates primarily on sealing areas surrounding combustion chambers.

The efficacy of sealing at the interface between the cylinder block and the head is contingent upon both the gasket's integrity and the surface characteristics of the mating components. The finish quality of these surfaces is pivotal to the sealing outcome [90]. The surface topography of the mating surfaces, influenced by various process factors, encompasses a range of machining errors across different scales. To

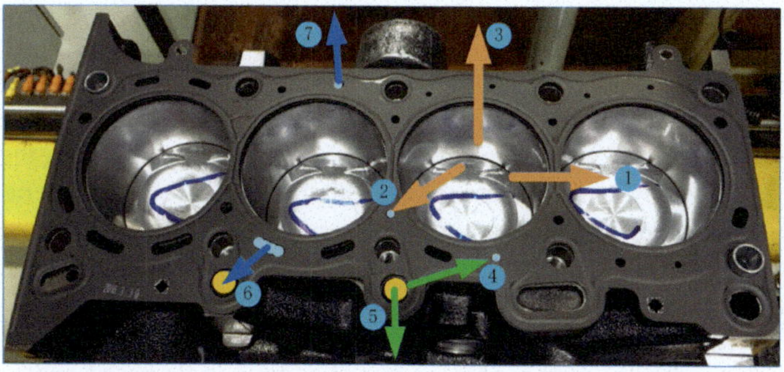

1.Gas leakage between combustion chambers
2.Gas leakage from combustion chamber to coolant circuit
3.Gas leakage to the environment
4.Oil leakage to the coolant circuit
5.Oil leakage to the environment
6.Water leakage to the lubricant circuit
7.Water leakage to the environment

Fig. 5.61 Seven types of media leakage on head gaskets

surpass the peak ignition pressure with sealing pressures around combustion chambers, the applied bolt torque in engineering practice is multiples of the theoretical values. However, excessive bolt torque not only elevates sealing pressures but also exacerbates cylinder bore distortion and diminishes the gasket's elastic recovery. The objective is to minimize bolt torque without compromising sealing efficacy.

Broadly speaking, improved surface finish quality correlates with reduced torque needed for sealing. Yet, the correlation between surface topography and sealing performance remains elusive. Gaining insights into the sealing characteristics of surface topographies is crucial for optimizing gasket design and minimizing the torque required for bolts. Considerable research has been conducted to elucidate the sealing capabilities of rough surface topographies.

Persson and colleagues introduced a critical junction theory to predict the leak rate of static seals, utilizing fractal geometry and percolation theory [34, 35, 91]. Bottiglione and team expanded this theory by incorporating multiple leakage channels to elucidate flat seal leakage [37, 38]. Persson et al. further advanced the effective-medium theory of leak rates to explain rubber seal leakage mechanisms, aligning well with experimental findings [39]. Pérez-Ràfols and colleagues developed a model examining metal-to-metal seal leakage, factoring in surface waviness and roughness [40]. Lorenz and Persson determined that seal leak rates are highly sensitive to skewness in surface height probability distributions, as confirmed by critical-junction theory and experimental validation [69]. Ledoux and team explored the influence of surface defect modal contents on flat seal efficiency, concluding that conventional roughness specifications inadequately ensure sealing performance [41]. Hao and colleagues assessed the impact of surface topography parameters on liquid film sealing, identifying waviness as the most influential factor [92]. Liu and team studied the effect of face-milled surface topography on adhesive sealing, discovering that surface roughness can increase adhesive contact area and strength [93]. Zhang and colleagues proposed a method to calculate leak channels and rates using surface topography measurements and finite element analysis [94]. Malburg focused on gasket adaptability, suggesting contact distribution simulation through morphological closings on waviness profiles [95]. Shao and team later extended this concept to three-dimensional scenarios, proposing a leakage monitoring method based on surface topography measurements [75, 96].

The majority of studies on leakage mechanisms presuppose isotropic surface topographies. However, surfaces generated by various machining processes exhibit distinct attributes. Face-turning and face-milling are prevalent techniques for creating smooth surfaces with anisotropic topographies, with the cutting tool paths being the primary source of anisotropy, a factor infrequently highlighted in existing literature. Deltombe and colleagues compared sealing rod surface finishes and determined that patterns perpendicular to the leak direction could boost sealing efficiency [97]. Yang and team explored the link between the percolation threshold and surface anisotropy, concluding that anisotropy significantly affects the threshold [98]. Liao and colleagues developed a leakage model for metallic static seals based on the micromorphology of face-turned surfaces [99]. Their experimental data and model indicated that radial leakage rates exceed circumferential rates under low contact pressure, shifting to circumferential leakage as contact pressure increases. Robbe-Valloire and Prat examined the sealing characteristics of face-turned surfaces using the waviness-motif method [100], emphasizing the importance of cutting tool paths in leakage estimation due to the disparity between radial and circumferential leakage. This analytical approach is also applicable to face-milled surfaces. The top surface of cylinder blocks and the bottom surface of cylinder heads, both critical sealing areas produced by face-milling, pose significant leakage risks. The proposed method aims to establish a sealing analysis model for face-milled surfaces and investigate the influence of tool path directions on sealing properties.

The foundation of sealing analysis for face-milled surfaces lies in capturing the full surface topography. In recent years, a non-contact measurement device known as high definition metrology (HDM) has been developed to meet the need for measuring large plane surfaces with complete resolution. HDM, operating on the principle of laser holographic interference, is capable of producing millions of data points for an engineering surface. Figure 5.62 displays the HDM equipment and an example of an engine head measured using HDM. Leveraging this innovative measurement platform, various studies on three-dimensional surface topography have been conducted, encompassing surface error assessment [16, 59], surface segmentation [101], and leakage path analysis [75, 102]. A comprehensive review of surface quality control and applications utilizing High Definition Metrology is detailed in a recently published volume [103].

Utilizing the recently developed measurement platform, the primary innovation of the proposed approach is the establishment of a sealing analysis framework for the topography of face-milled surfaces, which takes into account the trajectories

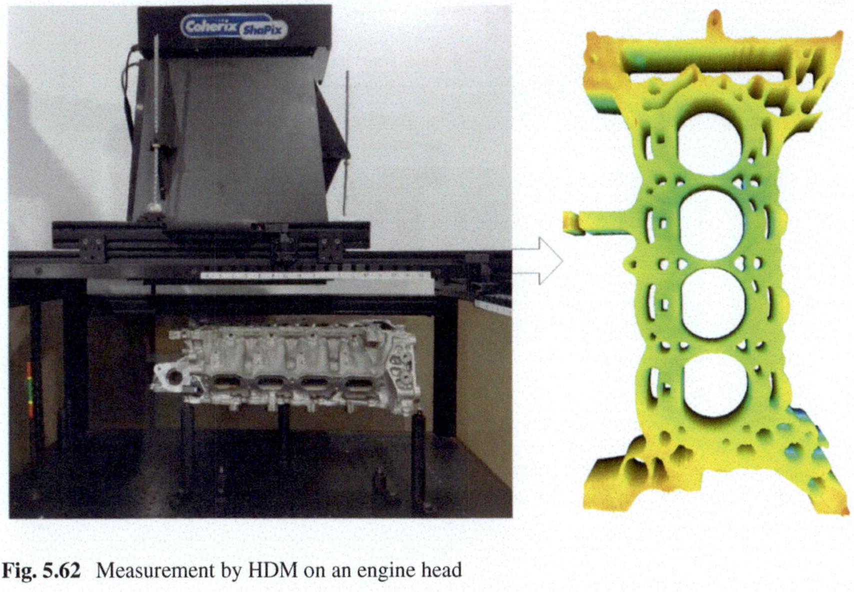

Fig. 5.62 Measurement by HDM on an engine head

of the cutting tools. Initial sampling of the topography surrounding the combustion chamber is conducted in accordance with the width of the gasket bead. The tool paths or channels are then reconstructed utilizing a novel surface segmentation technique [101]. The critical area encircling the combustion chamber is segmented into sub-regions based on the orientation of the channels, with each sub-region subjected to contact simulation and sealing analysis. The contact area percentage threshold (CAPT, defining the minimum contact area necessary to prevent leakage) is determined through a binary search algorithm.

5.4.2 *Methodology*

The sealing analysis method is depicted in Fig. 5.63 and is primarily composed of three key modules.

Module 1: The medium-scale topography of face-milled surfaces is characterized by periodic tool marks generated by tool trajectories, with the deep grooves of each path potentially serving as leakage channels. A surface segmentation method recently introduced by Yin et al. [101] is utilized to identify and reconstruct these potential leakage channels.

Module 2: Potential leakage channels are categorized into radial and circumferential based on their angular deviation from the radial direction. An angular threshold is established to differentiate between circumferential and radial channels. A region

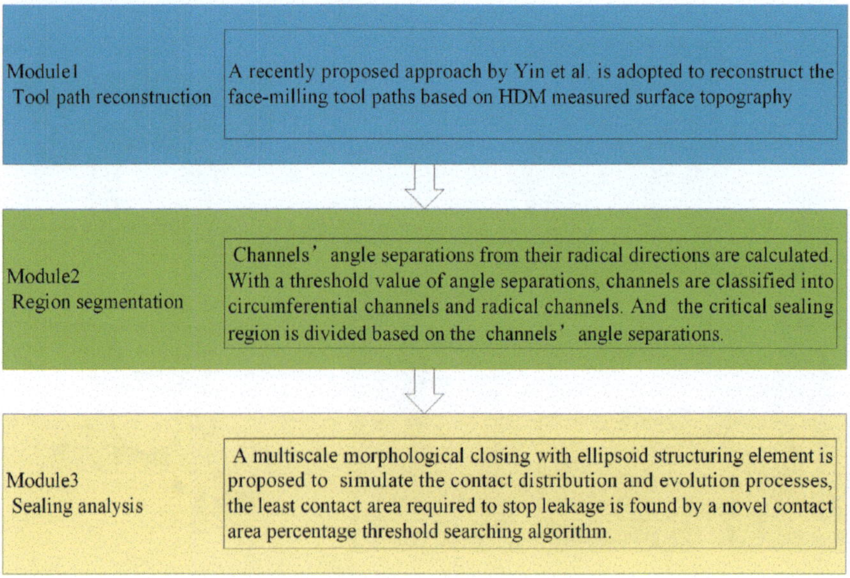

Fig. 5.63 Framework of the sealing analysis of face-milled surfaces

segmentation approach is proposed to partition the critical sealing topography into four distinct sub-regions, tailored to the varying orientations of the channels.

Module 3: For each sub-sealing region, a three-dimensional morphological closing operation is implemented to simulate the contact distribution and its progression at the interface between cylinder gaskets and face-milled surfaces. An innovative ellipsoid structuring element is introduced to expedite the closing operations and mitigate the end effects of morphological processing. To analyze the impact of channel directions on sealing properties, a contact area percentage threshold (CAPT) search algorithm is developed. This algorithm determines the minimum contact area necessary to prevent leakage in sealing regions with varying channel directions.

5.4.2.1 Tool Paths Reconstruction

Face-milled surfaces exhibit clear periodic tool marks, as illustrated in Fig. 5.64. In the proposed method, the local peaks of the milled surface topography are referred to as tool marks (the dotted lines in Fig. 5.64 delineate the concept of tool marks), while the local grooves are regarded as potential leakage channels. Owing to various processing errors, tool marks and grooves are affected by other machining imperfections, deviating from their ideal forms. To elucidate the correlation between the topography of face-milled surfaces and their sealing performance, the initial step involves reconstructing the tool paths from the measured surface topography. Local grooves and tool marks are complementary concepts. In an inverted face-milled surface topography, the tool marks transform into local grooves, and vice versa. Thus, the initial step concentrates solely on reconstructing tool marks, with local grooves being reconstructed similarly using the inverted topography. A recently introduced methodology for segmenting face-milled surface topography from prior research [101] is employed to reconstruct both tool marks and local grooves. A flowchart of the approach is depicted in Fig. 5.65, with the detailed procedures outlined below:

Step 1: Given that tool marks are intertwined with other machining errors, the milled surface topography requires filtering to extract the scale-limited surface, which is the filtered surface component within specific cutoff wavelengths that best highlights the tool marks. As tool marks are a result of periodic rotational motion, the scale-limited surface, which is predominantly composed of tool marks, should exhibit strong periodic characteristics. Consequently, a concept termed "periodic degree" is introduced to identify the scale-limited surface with the maximum periodic degree. The periodic degree (*PD*, as shown in Eq. (5.102)) is calculated as the ratio between the height of the secondary peak (H_{hp}) and the height of the highest peak (Hhp) on the areal auto-covariance function (*AACV*, as detailed in Eq. (5.103)) map. To filter out noisy spurious peaks, a "peak" is designated as any point that exceeds the values of forty-eight adjacent neighbors on the *AACV* map.

$$PD = \frac{H_{sp}}{H_{hp}} \tag{5.102}$$

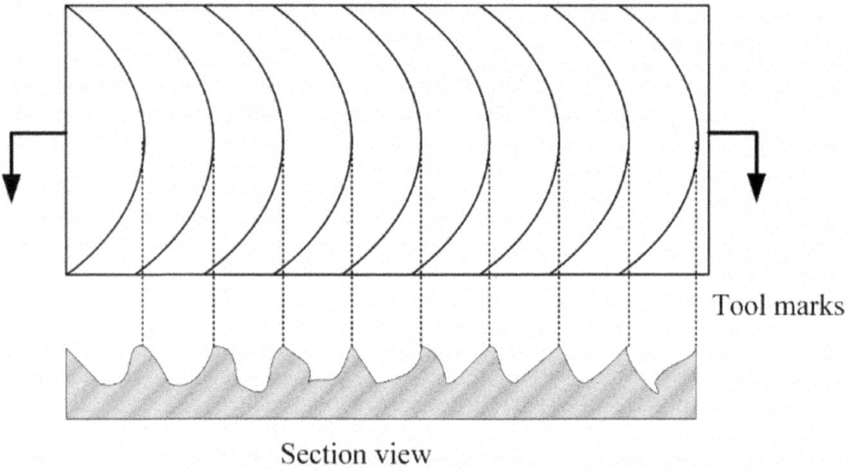

Fig. 5.64 Surface appearance of the face-milled workpiece

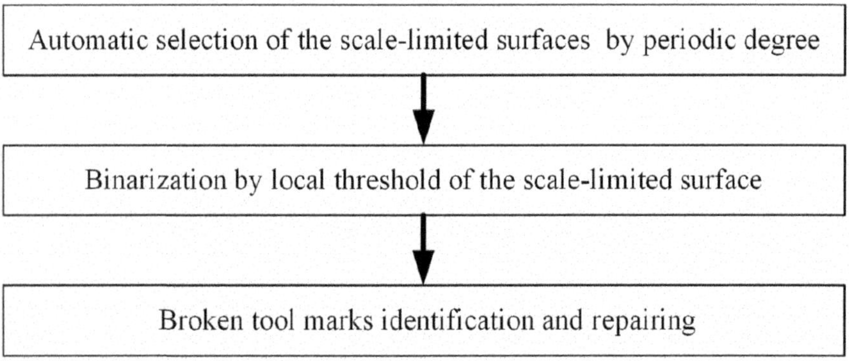

Fig. 5.65 The workflow to reconstruct tool paths

$$AACV(j',i') = \frac{1}{nx \times ny}\sum_{i=1}^{nx-i'}\sum_{j=1}^{ny-j'} z(j,i)z(j+j',i+i'), 0 \le j' \le ny-1, 0 \le i' \le nx-1 \quad (5.103)$$

A synthetic surface depicted in Fig. 5.66a, exhibiting form errors, periodic disturbances, and random noise, is analyzed through recursive biorthogonal wavelet decomposition, with $A_i = A_{i+1} + D_{i+1}$ defining $D_i = D_i^H + D_i^V + D_i^D$ and A_0 representing the pristine surface. The resultant scale-limited surfaces, $D_1 \sim D_5$, are visualized in Fig. 5.66b–f, while their *AACV* profiles are detailed in Fig. 5.66g–k.

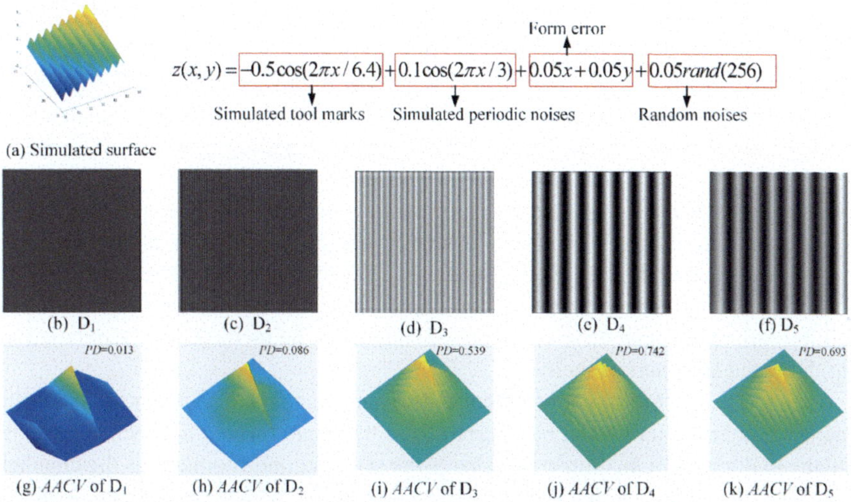

Fig. 5.66 Example of scale-limited surfaces selection by periodic degree

Employing periodicity assessment, surface D_4 is pinpointed for tool mark extraction, aligning with the anticipated outcome.

Step2: A local thresholding technique is employed on the chosen scale-limited surface to distinguish tool marks from the surrounding background. This method is advantageous for accommodating surface variations as it relies solely on the characteristics of a point's immediate vicinity. Let σ_{xy} and m_{xy} represent the standard deviation and mean height of the points in proximity to (x, y), respectively. The local threshold is formulated by Eq. (5.104), with a and b being nonnegative coefficients.

$$T_{xy} = a\sigma_{xy} + bm_{xy} \tag{5.104}$$

Step 3: Given the inherent irregularities of actual surfaces, tool marks may become disconnected post-thresholding. Typically, the size of these disconnections is smaller than the separation between neighboring tool marks. The inter-region distance (*IRD*) matrix can be utilized to detect fragmented tool marks. The matrix element $IRD(i, j)$ signifies the shortest distance between regions i and j, as delineated by Eq. (5.105). The maximum value of min(*IRD*) (the smallest value per column) approximates the gap between adjacent tool marks. Drawing from engineering expertise, the adaptive distance threshold DT can be established as a fraction of max(min(*IRD*)), as Eq. (5.106) illustrates.

$$IRD(i, j) = \min \sqrt{(x_p^i - x_q^j)^2 + (y_p^i - y_q^j)^2}$$
$$(p = 1, 2, \ldots, N_i, q = 1, 2, \ldots, N_j, i \neq j) \tag{5.105}$$

$$DT = C \times \max(\min(IRD)) \tag{5.106}$$

The fragmented tool marks can be restored using the specified repair algorithm (further details are available in reference [28]). The process for detecting and mending the discontinuous tool marks is depicted in Fig. 5.67a, while Fig. 5.67b–g provide a case study of the identification and repair procedures.

Deep grooves and tool paths serve as potential conduits for leakage. Radial channels are particularly prone to severe leakage, whereas circumferential channels can avert leakage due to their minimal contact area. To elucidate the inherent differential impact of channel orientations, an angle-based classification method is introduced to differentiate between these two types of channels. The methodology is detailed in Fig. 5.68 and summarized below:

Step 1: The configuration of potential leakage channels approximates a straight line. Channel orientation can be inferred from the angle formed between the x-axis and the major axis of an ellipse that shares the same second-moments as the channel area. The orientation is quantified by the acute angle, which varies from − 90 to 90 degrees. Figure 5.69 demonstrates an example of channel orientation. The left portion of the figure displays the channel area and its corresponding fitted ellipse, while the right portion presents the ellipse with solid blue lines indicating the axes; the orientation is the angle α between the horizontal black dotted line and the major axis. It should be noted that α is depicted as a negative value in Fig. 5.69.

Step2: The radial direction angle of channels is represented by the angle of the line connecting the center point of the bore and the centroid (average value of coordinates

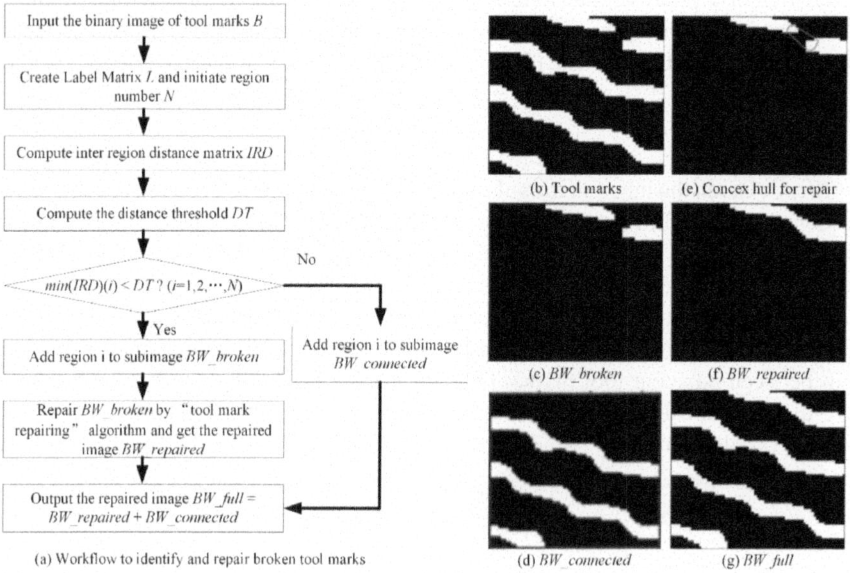

Fig. 5.67 Workflow and an example of broken tool marks identification and repairing

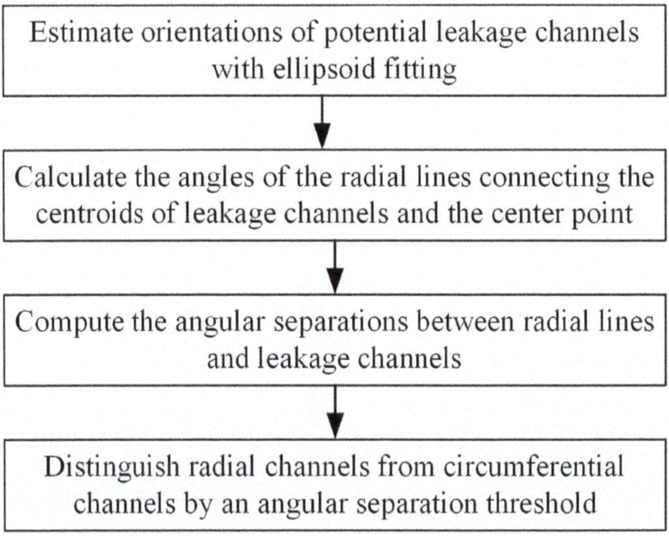

Fig. 5.68 An angle thresholding method to classify potential leakage channels

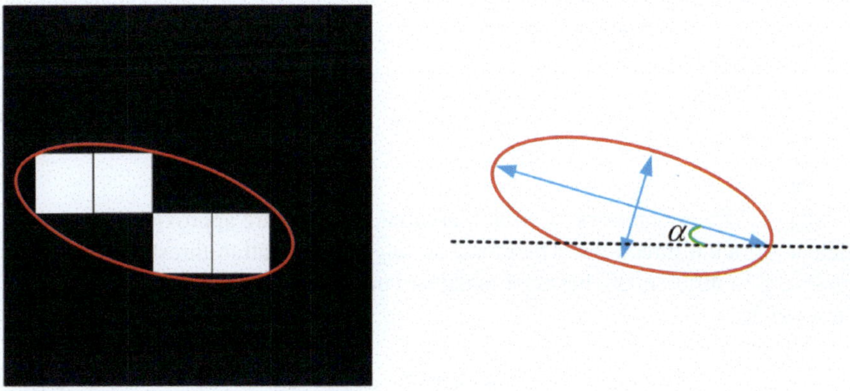

Fig. 5.69 Orientation estimation with ellipsoid fitting

in that region) of the channel regions. Suppose the centroid of a channel is (x, y) and the center point of the bore is (x_c, y_c), then the radial angle β of the channel could be calculated as Eq. (5.107). Figure 5.70a shows the radial angle with the x-axis. Note that β is also an acute angle ranging from -90 to 90 degrees.

$$\beta = \arctan \frac{y - y_c}{x_c - x} \tag{5.107}$$

Step 3: The angular separation γ, which is the angle between the radial line and the channel's orientation, indicates how much the channel deviates from the radial

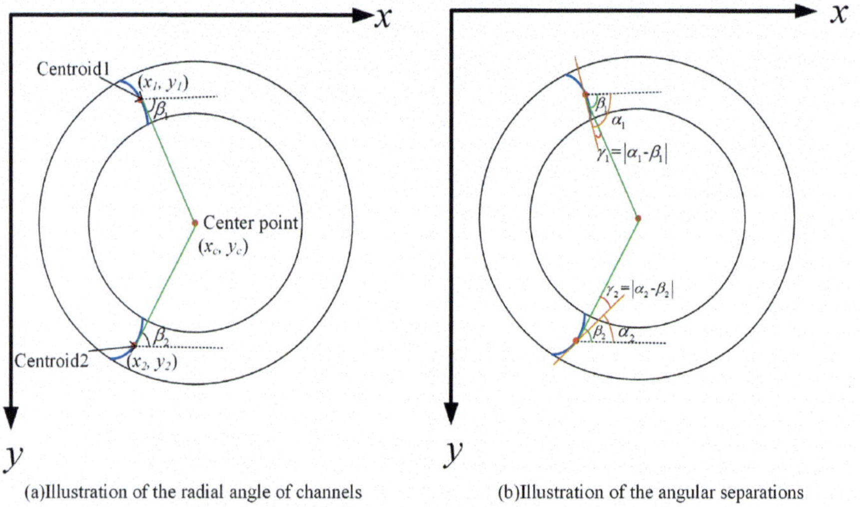

Fig. 5.70 Illustration of the radial angles and the angular separations

direction. To ensure that the angle values are neither negative nor obtuse, the angular separations are determined using Eq. (5.108). Figure 5.70b depicts the concept of angular separations.

$$\gamma = \begin{cases} |\alpha - \beta| & \textit{if } |\alpha - \beta| \le 90 \\ 180 - |\alpha - \beta| & \textit{if } |\alpha - \beta| > 90 \end{cases} \tag{5.108}$$

Step 4: The orientations of circumferential channels are approximately perpendicular to radial directions. Therefore, an angular separation threshold θ_T can be employed to differentiate between radial and circumferential channels, as shown in Eq. (5.109).

$$\begin{cases} \text{channel } i \in \text{radial channels} & \text{if } \gamma_i < \theta_T \\ \text{channel } i \in \text{circumferential channels} & \text{if } \gamma_i \ge \theta_T \end{cases} \tag{5.109}$$

For the purpose of elucidating the channel classification algorithm, a simulation of a face-milled tool path is conducted. The point cloud data for the face-milled surface is generated in accordance with Eq. (5.110). The function $H(R)$ delineates the tool's profile in the radial direction. A simulated point cloud representing the face-milled tool path is created based on the tool profile depicted in Fig. 5.71a, with milling parameters set at w = 136 rad/s, feed spacing = 8 mm/r, and the resulting point cloud is displayed in Fig. 5.71b.

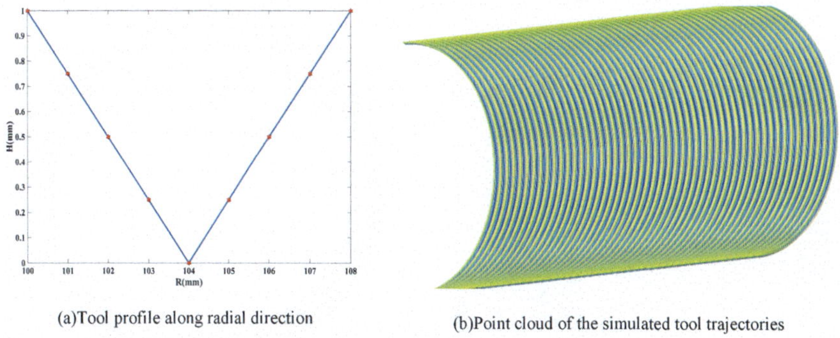

Fig. 5.71 The tool profile and the corresponding simulated tool trajectories

$$\begin{cases} x = R \sin wt + ft \\ y = R \cos wt \\ z = H(R) \end{cases} \tag{5.110}$$

The annular area surrounding the combustion chamber is of paramount importance for sealing purposes. An annular segment is extracted from the simulated topography, with an inner radius of 40 mm and an outer radius of 50 mm. Figure 5.72a displays the topography of this extracted area, where periodic tool marks and grooves are the predominant features. Grooves with heights below 0.1 mm are considered potential leakage paths, and these can be identified using a straightforward global thresholding technique. A region labeling method is then applied to tag these potential leakage paths, as depicted in Fig. 5.72b. Subsequently, employing the proposed angle-based channel classification algorithm, the angular separations between the channels' orientations and their radial directions are computed and presented in Fig. 5.73. It is evident from Fig. 5.73 that the angular separation for circumferential channels approaches 90 degrees, whereas for radial channels it is less than 70 degrees. Thus, an angular threshold θ_T of 80 degrees can be established to differentiate between circumferential and radial channels. The significance of this classification lies in the fact that radial channels, which link the combustion chamber to the outer space, pose a higher risk of leakage, whereas circumferential channels, connecting only the outer space, do not lead to gas leakage from the combustion chambers.

Based on the analysis above, radial channels face significantly different leakage risks from circumferential channels. It is essential to divide the ring-shaped sealing region into different parts according to the channels' directions to study their leakage characteristics. For an isotropic surface topography, the leakage direction is along the radial direction. So radial lines connecting the chamber center with endpoints of the two longest circumferential channels are adopted as segmentation lines. The principle of radial segmentation is outlined in Fig. 5.74. The segmentation result of the sampled ring-shaped region from simulated topography in Sect. 5.3.2.2 is shown in Fig. 5.75. Channels in sub-regions R1 and R4 are circumferential, while channels in sub-regions R2 and R3 are radial.

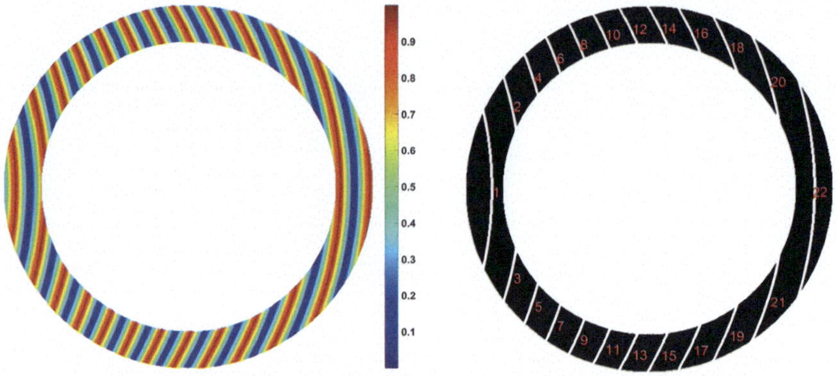

(a) Sampled ring region from the simulated topography

(b) Potential leakage channels and their labels

Fig. 5.72 A sampled ring-shaped topography and the potential leakage channels

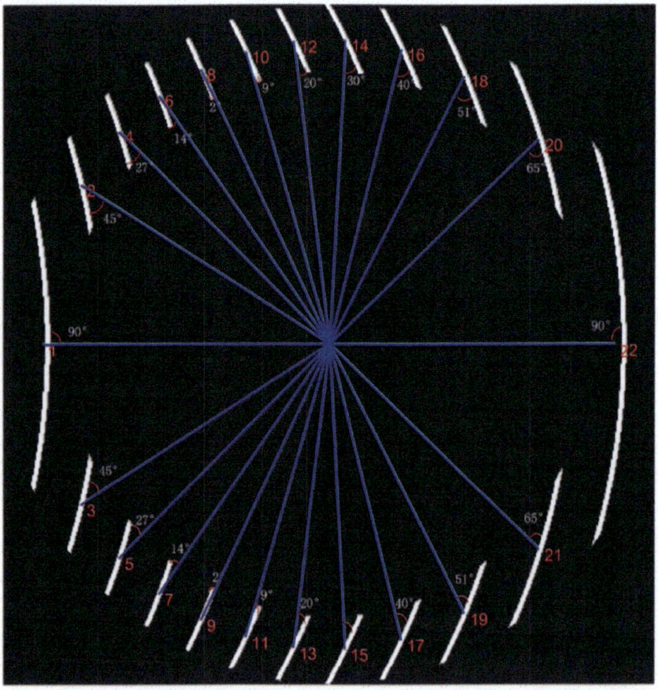

Fig. 5.73 Angular separations of the potential leakage channels with radial directions

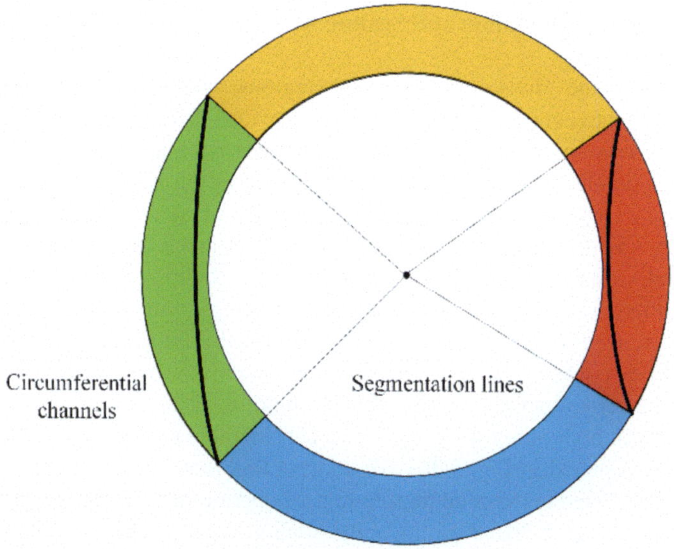

Fig. 5.74 The principle of sealing region segmentation

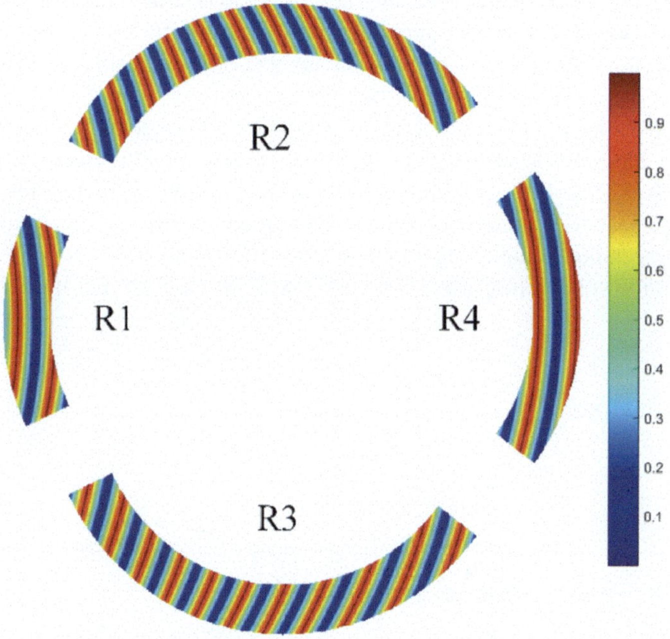

Fig. 5.75 The sealing region segmentation result of the simulated surface

5.4.2.2 Contact Simulation and Sealing Analysis

Subsequent to the division of sealing regions, the analysis of sealing is conducted through contact simulations. Gaskets with MLS coatings exhibit intricate compression-rebound properties, making the precise calculation of contact behavior at the interface challenging. Malburg [95] introduced an alternative approach for simulating the contact behavior of adaptable interfaces by employing morphological closings on undulating profiles. Shao et al. [96] further elaborated on the three-dimensional virtual gasket technique. However, the traditional circular structuring element utilized in profile filtering is excessively large (with a radius of 5000 mm as referenced in [95]), leading to significant end effects and elevated computational expenses. In the context of measured surface topography, the horizontal span typically surpasses the vertical span by a considerable margin. The circular element, having equal lengths in both dimensions, results in considerable redundancy in the vertical direction, as depicted in Fig. 5.76a. To address this issue, an elliptical structuring element is suggested to expedite morphological operations and mitigate end effects. As illustrated in Fig. 5.76a, a significantly smaller elliptical element could perform the same contact simulations as the large circle without the addition of superfluous points. The parameters of the ellipse are detailed in Fig. 5.76b, with R denoting the semi-major axis and H representing the semi-minor axis. This concept can be readily extended to three-dimensional scenarios, where the spherical element is replaced by a flattened ellipsoidal element, and the parameters of the ellipsoid are displayed in Fig. 5.77.

The distribution of contact areas at the interface between gaskets and the surfaces of cylinder blocks or heads is subject to variation due to bolt preload and combustion processes. A static structuring element is incapable of replicating the dynamic nature of contact distributions. Therefore, it is imperative to devise a dynamic contact simulation technique employing ellipsoid structuring elements of varying sizes. To investigate the contact properties associated with different structuring elements, a

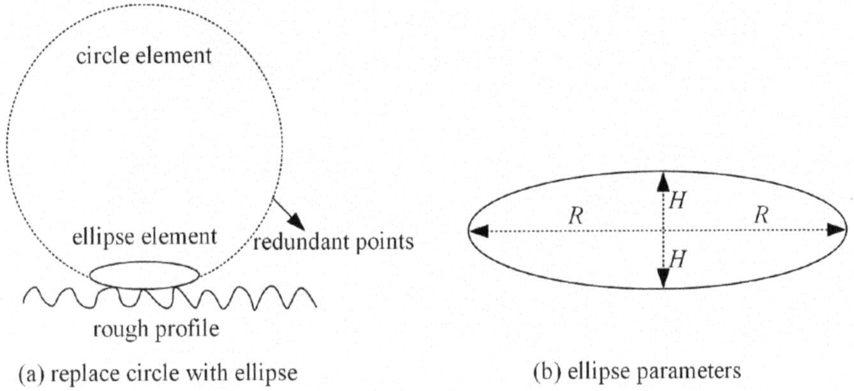

Fig. 5.76 The idea of replacing circle elements with ellipse elements

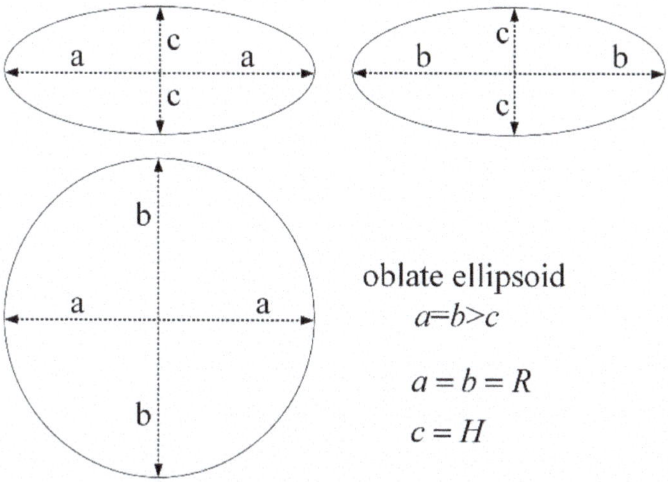

Fig. 5.77 Parameters of the oblate ellipsoid elements

simulated profile, as described by Eq. (5.111), is subjected to testing. Morphological closings are applied to this profile using eight distinct elliptical sizes, with the resulting closing profiles represented by red lines in Fig. 5.80, while the original profiles are indicated by blue lines. Observations from Fig. 5.78a–d reveal that, with the major axis length held constant, an increase in the minor axis length allows the closing profiles to penetrate deeper into the valleys. Similarly, from Fig. 5.78e–h, it is observed that with the minor axis length remaining fixed, a reduction in the major axis length enables the closing profiles to delve more deeply into the valleys. This finding is equally valid for three-dimensional morphological closings utilizing ellipsoid elements.

$$z = 0.01 \times (\sin(2\pi x/2) + 0.5\sin(2\pi x/0.5)) \tag{5.111}$$

Employing the recently introduced ellipsoid elements, the contact distribution is ascertained by the waviness component (*SW*) in conjunction with its morphological closing surface (*SC*). The contact surface, denoted as Contact, is conceptualized as a binary image, with its definition provided by Eq. (5.112). Concurrently, the contact area percentage (*CAP*) is delineated by Eq. (5.113).

$$Contact = \begin{cases} 0 \; SC(x,y) = SW(x,y) \text{ contact} \\ 1 \; SC(x,y) > SW(x,y) \text{ no contact} \end{cases} \tag{5.112}$$

$$CAP = \sum_{S} [if\,(SC(x,y) = SW(x,y))\ 1 \text{ else } 0]/S \tag{5.113}$$

A connectivity-based approach to leak detection has been introduced, utilizing the concept of surface connectivity, which is further elaborated in reference [75].

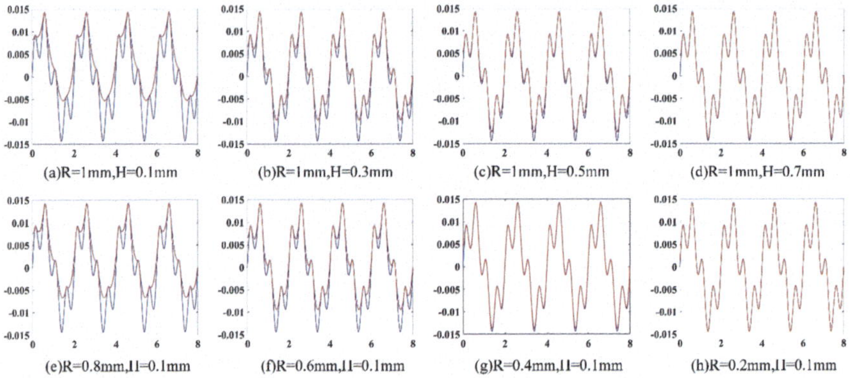

Fig. 5.78 Morphological closings using different sizes of ellipses

This method employs the 8-adjacency principle. In the event of a leakage channel within the sealed area under consideration, it would establish a connection between the outer space and the combustion chamber, as depicted in Fig. 5.79. Consequently, the area of the largest connectivity class would exceed the combined area of the outer space and the combustion chamber. The methodology for employing this concept to ascertain the presence or absence of a leak is detailed in Fig. 5.80, complete with pseudocode.

Utilizing the leak-detection algorithm enables dynamic sealing analysis. The critical parameter for ascertaining surface leakage is the contact area percentage (*CAP*), as defined by Eq. (5.113). Persson and Yang [91] noted that for two infinite elastic solids with randomly rough surfaces under compression, a non-contact channel

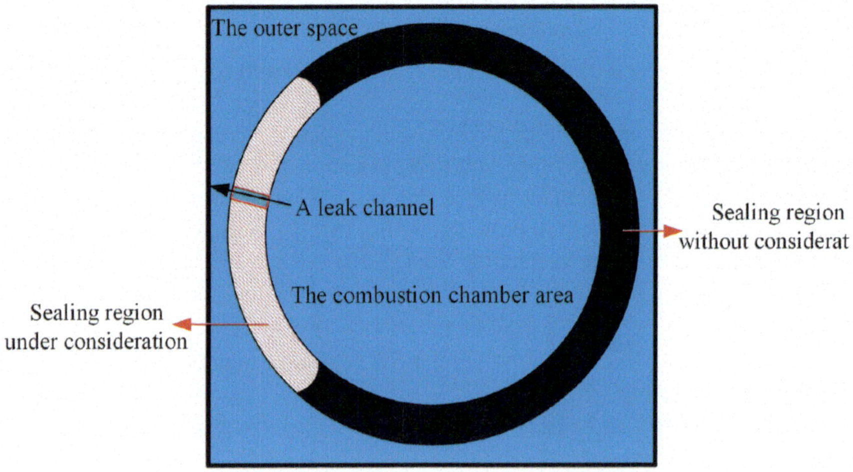

Fig. 5.79 The principle of detecting leakage by the area of connectivity classes

Algorithm : [*flag*, *CAP*] = RegionIsLeak(*W*,*M*,*RM*,*SE*)

Input: ring shaped surface waviness *W*, ring shaped mask *M*, regional mask *RM*, structuring element *SE*

Output: *flag*, 1 for leakage and 0 for no leakage, *CAP*, contact area percentage

```
C = imclose(W,SE); contact = (C-W)>0
[m,n]=size(M); ME = zeros(m+2,n+2); ME(2:m+1,2:n+1)=M;
AME = regionprops(~ME,'Area'); BaseA = AME(1).Area+AME(2).Area;
Leak =~M; Leak(RM)=contact(RM);
CAP=1-sum(contact(RM))/length(contact(RM));
LeakE = ones(m+2,n+2); LeakE(2:m+1,2:n+1)=Leak;
Larea = regionprops(LeakE,'Area');
If Larea(1).Area >BaseA
    flag =1
else
    flag = 0;
End if
```

Fig. 5.80 An algorithm to decide whether the considered region leaks or not

would form when the relative contact area approached 0.4, aligning with percolation theory. However, actual engineering surfaces are finite and exhibit distinct textures resulting from machining processes. Investigating the minimum *CAP* necessary for effective sealing on real engineering surfaces with a specified measurement resolution presents an intriguing research avenue.

Structuring elements of varying dimensions can be interpreted as representing different levels of observational magnification. The distribution of contact areas is influenced by the adjustment of these element sizes. A pivotal structuring element can be identified through the algorithm presented (as depicted in Fig. 5.81), with the associated *CAPT*(contact area percentage threshold) serving as the discriminating threshold to distinguish between leak and non-leak conditions.

Employing the *CAPT* searching algorithm on the simulated ring-shaped sealing region detailed in Sect. 5.3.2.2, one can discern the critical contact distribution necessary to prevent leakage, as illustrated in Fig. 5.82. The black regions represent areas of contact, whereas the green regions denote areas without contact.

Analysis of Fig. 5.82 reveals that minimal contact area is sufficient to prevent leakage in circumferential channel areas, whereas radial channel areas demand a significantly larger contact area for effective sealing. The CAPT values detailed in Table 5.17 corroborate the aforementioned finding.

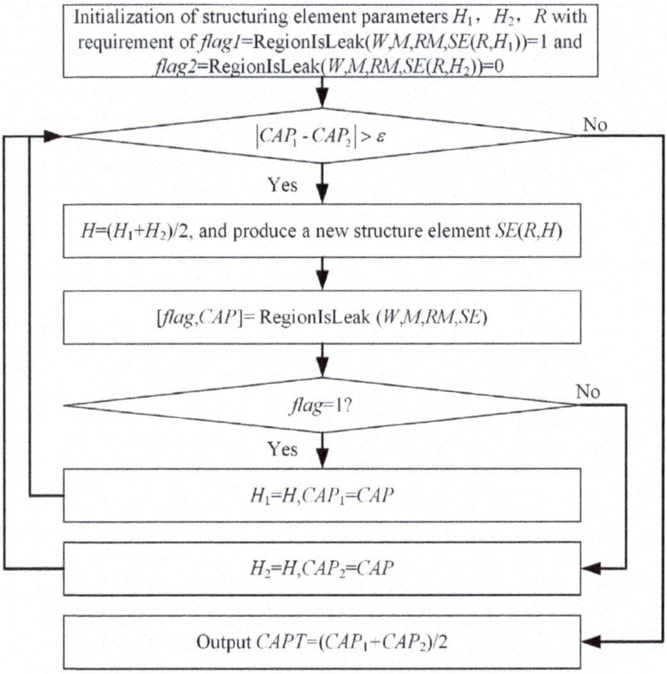

Fig. 5.81 The contact area percentage threshold searching algorithm

Fig. 5.82 The critical contact distribution of sub-regions on the simulated surface

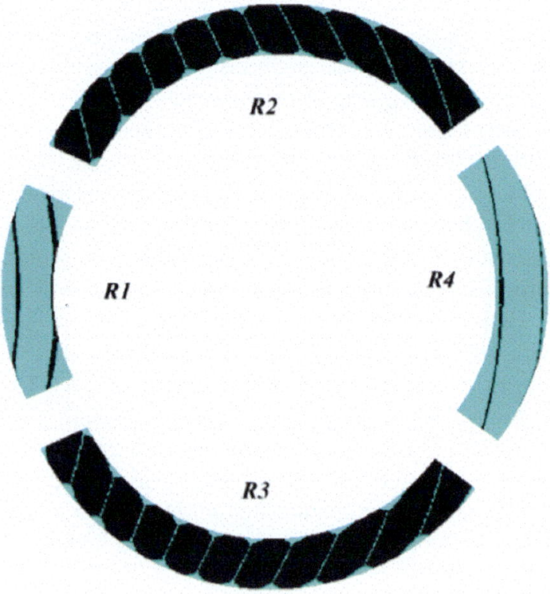

Table 5.17 *CAPT* values of the four sub-regions on the simulated surface

Region	R1	R2	R3	R4
CAPT	0.12	0.91	0.91	0.07

5.4.3 Case Study

The effectiveness of the proposed sealing analysis method is validated through the examination of a cylinder block from an engine plant. The top surface of the cylinder block is depicted in Fig. 5.83a, featuring three bores with surrounding material removed to lighten the block, resulting in a natural ring-shaped sealing area encircling the combustion chambers. The top surface was scanned using the ShaPix3D® 1500 series (HDM equipment, as referenced in Fig. 5.62), and the comprehensive surface topography is presented in Fig. 5.83b, with red circles indicating the bolt hole locations. Shapix3D has a lateral and vertical resolution of 80 μm and 0.05 μm, respectively, with a field of view measuring 150 mm × 150 mm and a capacity to capture up to 4 million points per view. For workpieces exceeding the field of view, Shapix3D can integrate point clouds from various angles to create a complete surface representation. The three ring-shaped sealing areas surrounding the combustion chambers are crucial for their sealing, as shown in the two-dimensional and three-dimensional views in Fig. 5.83c and d. The shared sealing area between adjacent combustion chambers is termed the "web area," highlighted in Fig. 5.83c with red boxes. The three bores are sequentially labeled 1, 2, and 3 from left to right in Fig. 5.83d.

The sealing area surrounding bore1 is investigated initially. To facilitate a detailed examination of the region's topography, an enlarged view of the ring-shaped topography is presented in Fig. 5.84, where tool paths are implicitly visible. Utilizing the tool paths reconstruction algorithm discussed in Sect. 2.1 on the inverted topography of this area, a binary image revealing potential leak channels is generated and displayed in Fig. 5.85.

Distinguishing between circumferential and radial channels involves calculating the angular deviation from the radial direction for each channel, as depicted in Fig. 5.86. The identification numbers and radial direction lines for each channel are illustrated in Fig. 5.86a. An angular threshold of 80 degrees effectively differentiates circumferential channels from radial ones, as demonstrated in Fig. 5.86b.

Employing the region segmentation technique outlined in Sect. 5.3.2.2, the outcomes of segmentation on the measured surface topography and on the filtered waviness—utilizing areal spline filters with cut-off wavelengths of 0.8 mm and 2.5 mm—are depicted in Fig. 5.87a and b, respectively.

Dynamic sealing analysis is conducted on the four waviness sub-regions using an ellipsoid structuring element $SE(R, H)$. Here, R is set at 0.4 mm, and H is varied to emulate contact distribution and determine the *CAPT* value. The transition processes (from leakage to non-leakage) in the contact distribution for sub-regions R1 and R4, which are circumferential channels, are illustrated in Fig. 5.88. White areas represent

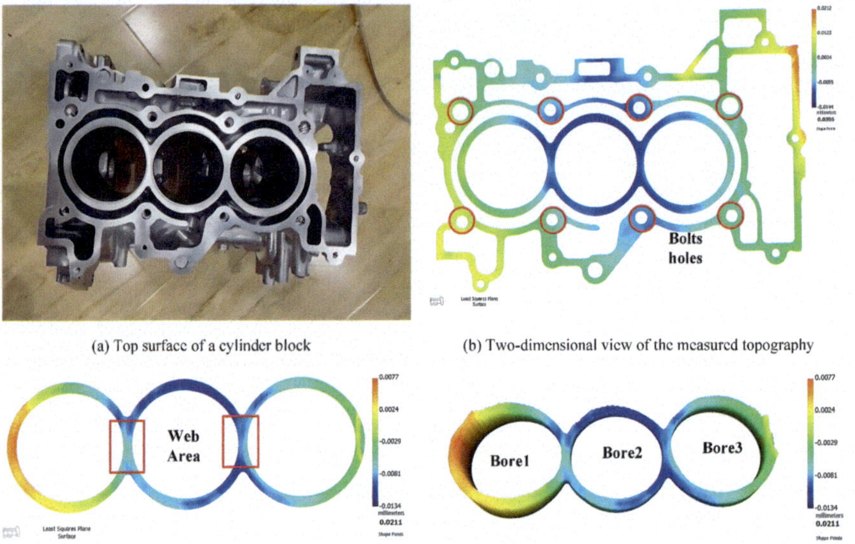

(a) Top surface of a cylinder block (b) Two-dimensional view of the measured topography

(c) Two-dimensional view of surface topography around bores (d) Three-dimensional view of surface topography around bores

Fig. 5.83 The cylinder block and its topography measurement results

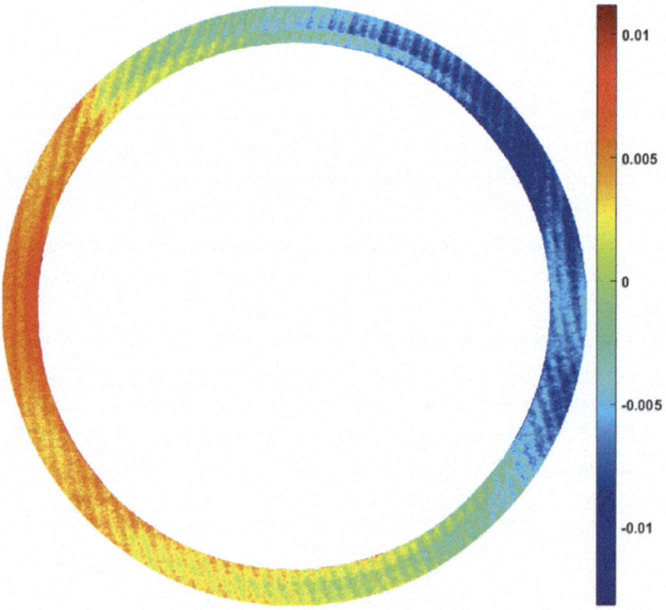

Fig. 5.84 A close view of the first ring-shaped sealing region

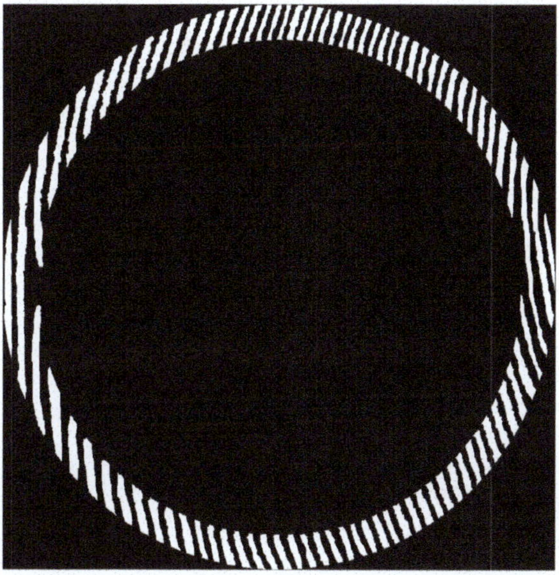

Fig. 5.85 A binary image indicating the potential leak channels

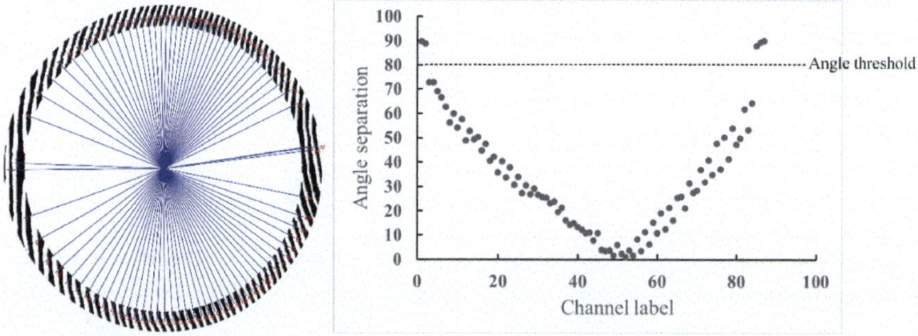

(a) Labels of the potential leakage channels (b) Angle separations distribution and their threshold

Fig. 5.86 The angle separations distribution of channels and their threshold

non-contact regions, and black areas denote contact points. The final column displays the critical contact distribution, signifying the minimum contact points necessary for preventing leakage.

The transition of contact distribution from leakage to non-leakage for radial channel regions R2 and R3 is presented in Fig. 5.89. The bottom row illustrates the critical contact distribution, where all potential leakage paths are sealed by contact points.

The CAPT value is derived from the critical contact distribution, with the CAPT values for the four sub-regions detailed in Table 5.18. The average CAPT for circumferential channel regions is approximately 0.56, whereas for radial channel regions it is around 0.70, which is higher than that of the circumferential channels. In contrast

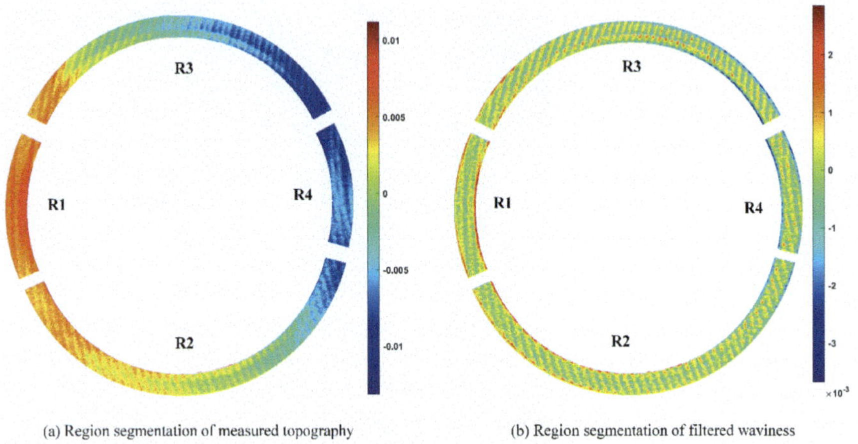

Fig. 5.87 Region segmentation on the measured topography and waviness

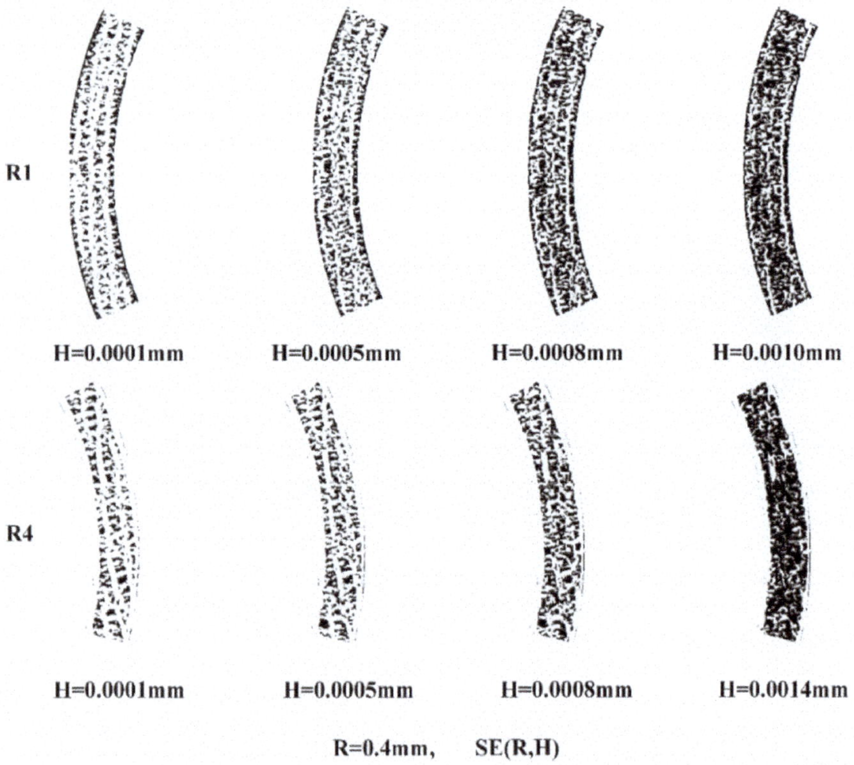

Fig. 5.88 The evolution processes of contact distribution of sub-regions R1 and R4

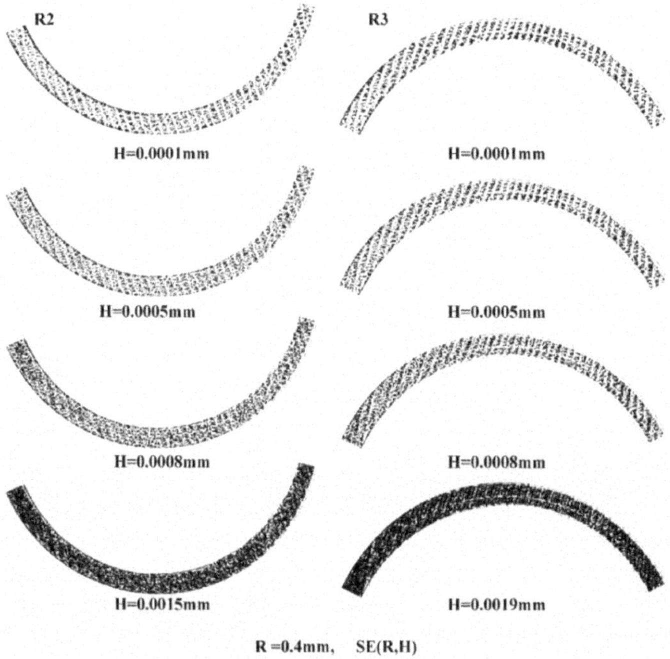

Fig. 5.89 The evolution processes of contact distributions of sub-regions R2 and R3

to the simulation in Sect. 5.3.2.3, the CAPT for circumferential channels in actual engineering surfaces is significantly higher. This is attributed to the irregularities of real surface topographies, where circumferential tool marks are fragmented, creating more microchannels that connect the outer space to the inner space via circumferential channels. For radial channels, the CAPT values on actual engineering surfaces are lower than those on simulated surfaces, as radial channels are more susceptible to blockage due to tool wear and machining errors. Compared to Persson's theory [91], the critical contact area percentage of face-milled surfaces exceeds the theoretical threshold of 0.4. This discrepancy arises because face-milled surface topography does not conform to the assumptions of fractal and isotropic surface micro-geometry posited in Persson's theory. The directions of texture or tool paths significantly influence the sealing properties of a surface. Generally, circumferential textures enhance surface sealing, while radial textures are more prone to leakage. Fortunately, the sealing regions of circumferential channels largely coincide with the web area between adjacent combustion chambers, as shown in Fig. 5.83c. In essence, the web area can achieve effective sealing with reduced contact pressure compared to other radial channel regions. This finding aligns with the distribution of bolts, which, as depicted in Fig. 5.83b, are positioned such that the web areas receive less torque pressure than other regions. Consequently, the circumferential channels in the web area are more likely to form effective seals under lower contact pressures.

Table 5.18 *CAPT* values of the four sealing sub-regions

Region	R1	R2	R3	R4
CAPT	0.55	0.67	0.73	0.56

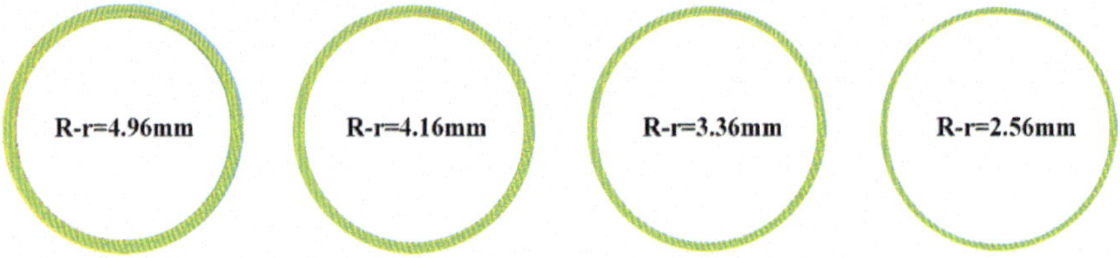

Fig. 5.90 Width reduced ring-shaped sealing regions

The CAPT value of sealing regions is also influenced by the width of the ring-shaped area defined by the outer radius R and the inner radius r. In Fig. 5.84, the width R-r is 5.76 mm. Figure 5.90 displays the waviness of four ring-shaped sealing regions with reduced widths. Figure 5.91 plots the CAPT values calculated using the same segmentation lines (four sealing sub-regions R1 ~ R4) across ring-shaped sealing regions of varying widths. As the ring width diminishes, the CAPT value rises, indicating that a higher percentage of contact area is required for effective sealing in narrower ring-shaped regions. The limiting case is line sealing, where a 100% contact area percentage is necessary to prevent leakage. In gasket design, the width of the beads encircling the combustion chambers is a critical variable. Analysis suggests that the bead design width should extend sufficiently to span at least one complete circumferential channel, which can prevent gas leakage between adjacent bores with a comparatively low contact area percentage.

To ascertain the effect of channel orientation on CAPT values in sealing areas, 12 ring-shaped sealing regions were sampled from the top surfaces of three cylinder blocks and the bottom surface of a cylinder head. The topography of these four surfaces and the 12 ring-shaped sealing regions (labeled Ring1 to Ring12) are depicted in Fig. 5.92. Following the same analytical procedures, each sealing region was partitioned into two circumferential channel regions (R1 and R4) and two radial channel regions (R2 and R3), with the respective CAPT values computed and documented in Table 5.19. The CAPT value for region Ri is indicated as CAPTi. Figure 5.93 presents a bar chart of the CAPT values for the 12 ring-shaped regions, revealing that min(CAPT2, CAPT3) exceeds max(CAPT1, CAPT4) for each region. On average, circumferential channel regions necessitate 54% contact area for effective sealing, whereas radial channel regions demand 72% contact area to prevent leakage. This finding robustly corroborates the conclusion that circumferential channels or paths are beneficial for sealing surfaces, whereas radial channels are more susceptible to leakage.

The analysis of the real surfaces was conducted using *MATLAB* R2020a on a desktop computer equipped with 8 GB of RAM and a 3.60 GHz i7-7700 CPU. The

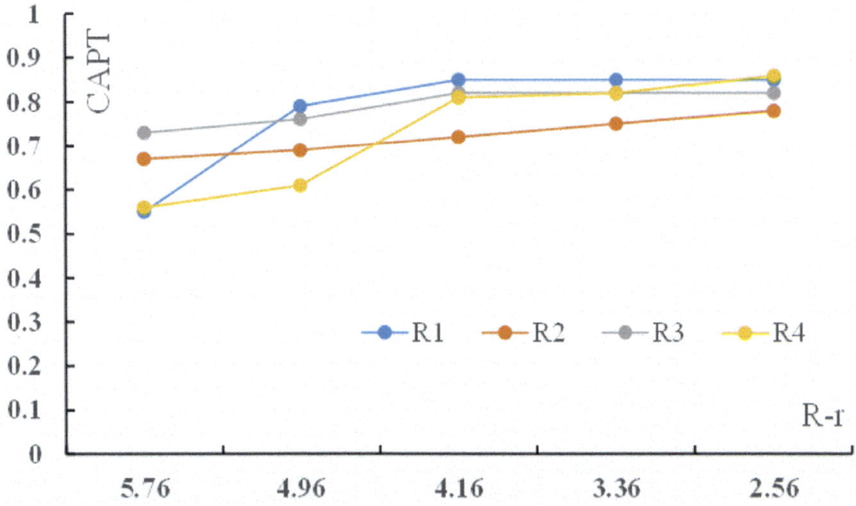

Fig. 5.91 *CAPT* values vary with the width of the ring-shaped region

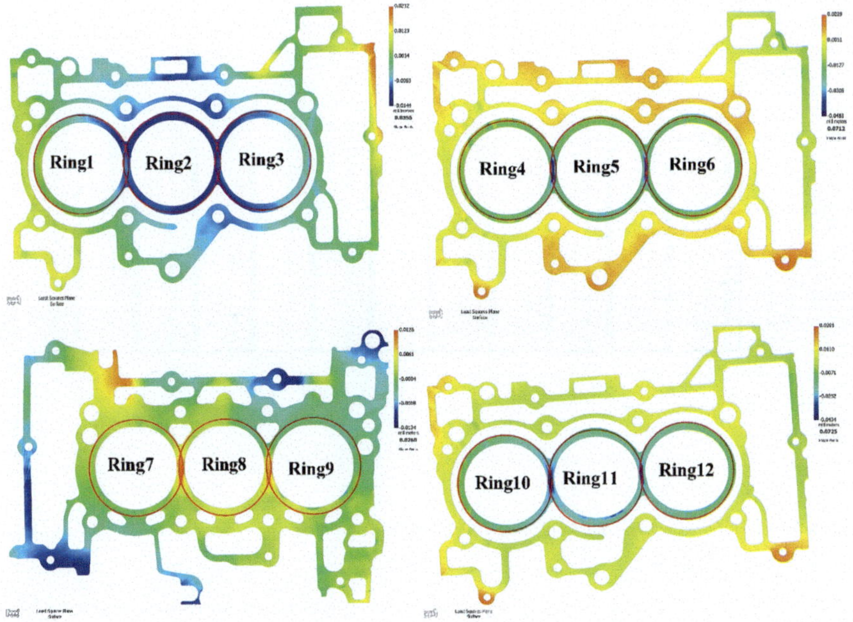

Fig. 5.92 12 ring-shaped sealing regions

Table 5.19 *CAPT* values of 12 ring-shaped sealing regions

Sample	*CAPT*1	*CAPT*2	*CAPT*3	*CAPT*4
Ring1	0.55	0.67	0.73	0.56
Ring2	0.54	0.59	0.68	0.58
Ring3	0.59	0.76	0.76	0.51
Ring4	0.52	0.67	0.72	0.47
Ring5	0.40	0.67	0.70	0.57
Ring6	0.52	0.67	0.71	0.47
Ring7	0.54	0.85	0.93	0.69
Ring8	0.58	0.76	0.84	0.58
Ring9	0.61	0.70	0.75	0.63
Ring10	0.62	0.69	0.71	0.54
Ring11	0.47	0.65	0.62	0.49
Ring12	0.47	0.69	0.71	0.51

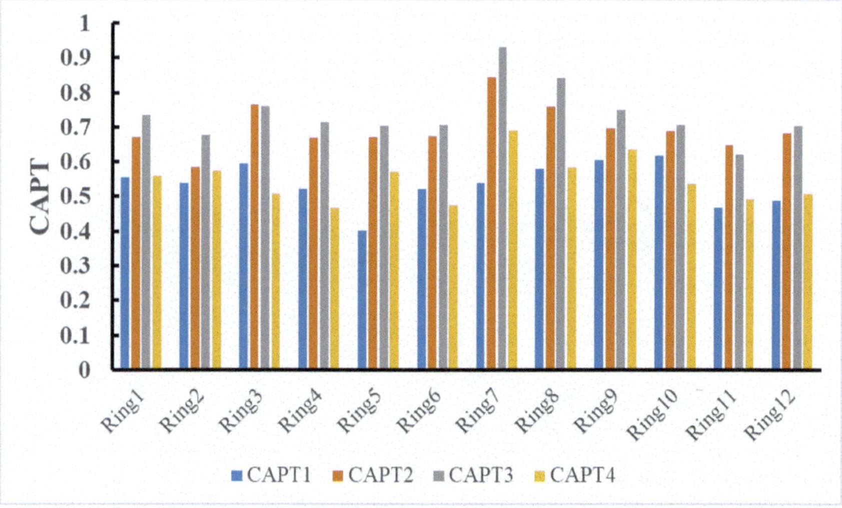

Fig. 5.93 *CAPT* values of each ring-shaped sealing region

mean execution time for analyzing a single ring-shaped sealing region was 44.67 s. The methodology presented is suitable for all engineering surfaces produced by face-milling with disc milling cutters. Notably, the contact simulation and sealing analysis technique (Module 3) is applicable to a variety of engineering surfaces that utilize conformable gaskets for sealing.

5.4.4 Conclusions

A methodology for analyzing the sealing performance of face-milled surfaces has been introduced. Initially, tool paths are delineated through a novel surface segmentation approach. Subsequently, the ring-shaped sealing area is segmented into four sub-regions based on the orientation of the channels. A method for calculating angular separation is introduced to establish a threshold that differentiates circumferential from radial channels. An ellipsoid structuring element is proposed to model the contact interaction between rough surfaces and gaskets via morphological closing operations. Additionally, a CAPT search algorithm is developed to determine the minimum contact area percentage necessary to seal leakage paths.

Real-case studies reveal that the direction of channels significantly influences the sealing characteristics of surfaces. Circumferential channels or paths are beneficial for sealing, whereas radial channels are more susceptible to leakage. The width of the ring-shaped sealing region also affects its CAPT value, with narrower regions demanding a higher contact area percentage for effective sealing. Future research will focus on optimizing cylinder head gasket design based on the analysis of surface topographies. Moreover, experimental validation of predicted contact states and enhancement of the sealing efficiency of MLS gaskets will be pursued in subsequent studies.

References

1. Ramasamy SK, Raja J (2013) Performance evaluation of multi-scale data fusion methods for surface metrology domain. J Manuf Syst 32:514–522
2. ISO 4287 (1997) Geometrical product specifications (GPS)-surface texture: profile method: terms, definitions and surface texture parameters
3. Bhuiyan MSH, Choudhury IA, Dahari M (2014) Monitoring the tool wear, surface roughness and chip formation occurrences using multiple sensors in turning. J Manuf Syst 33:476–487
4. Haruyama S, Nurhadiyanto D, Choiron MA, Kaminishi K (2013) Influence of surface roughness on leakage of new metal gasket. Int J Press Vessels Pip 111–112:146–154
5. Ledoux Y, Lasseux D, Favreliere H, Samper S, Grandjean J (2011) On the dependence of static flat seal efficiency to surface defects. Int J Press Vessels Pip 88:518–529
6. Malburg MC (2003) Surface profile analysis for conformable interfaces. J Manuf Sci Eng 125:624
7. Bottiglione F, Carbone G, Mangialardi L, Mantriota G (2009) Leakage mechanism in flat seals. J Appl Phys 106:104902
8. Bottiglione F, Carbone G, Mantriota G (2009) Fluid leakage in seals: An approach based on percolation theory. Tribol Int 42:731–737
9. Hanada H, Saito T, Hasegawa M, Yanagi K (2008) Sophisticated filtration technique for 3D surface topography data of rectangular area. Wear 264:422–427
10. Tomonori G, Kazuhisa Y (2009) An optimal discrete operator for the two-dimensional spline filter. Meas Sci Technol 20:125105
11. Fu S, Muralikrishnan B, Raja J (2003) Engineering surface analysis with different wavelet bases. J Manuf Sci Eng 125:844
12. Jiang XQ, Blunt L, Stout KJ (2000) Development of a lifting wavelet representation for surface characterization. Proc R Soc A Math Phys Eng Sci 456:2283–2313

13. Raja J, Muralikrishnan B, Fu S (2002) Recent advances in separation of roughness, waviness and form. Precis Eng 26:222–235
14. Huang Z, Shih AJ, Ni J (2006) Laser interferometry hologram registration for three-dimensional precision measurements. J Manuf Sci Eng 128:1006
15. Wang M, Xi L, Du S (2014) 3D surface form error evaluation using high definition metrology. Precis Eng 38:230–236
16. Du S, Liu C, Huang D (2015) A shearlet-based separation method of 3D engineering surface using high definition metrology. Precis Eng 40:55–73
17. Du S, Liu T, Huang D, Li G (2018) A fast and adaptive bi-dimensional empirical mode decomposition approach for filtering of workpiece surfaces using high definition metrology. J Manuf Syst 46:247–263
18. Du S, Huang D, Wang H (2015) An adaptive support vector machine-based workpiece surface classification system using high-definition metrology. IEEE Trans Instrum Meas 64(10):2590–2604
19. Du S, Liu C, Xi L (2014) A selective multiclass support vector machine ensemble classifier for engineering surface classification using high definition metrology. J Manuf Sci Eng 137:011003
20. Shao Y, Du S, Xi L (2017) 3D machined surface topography forecasting with space-time multioutput support vector regression using high definition metrology. In: Proceedings of the ASME 2017 international design engineering technical conferences & computers and information in engineering conference
21. Wang M, Ken T, Du S, Xi L (2015) Tool wear monitoring of wiper inserts in multi-insert face milling using three-dimensional surface form indicators. J Manuf Sci Eng 137:031006
22. Nguyen HT, Wang H, Hu SJ (2013) Characterization of cutting force induced surface shape variation in face milling using high-definition metrology1. J Manuf Sci Eng 135:041014
23. Nguyen HT, Wang H, Tai BL, Ren J, Jack HuS, Shih A (2015) High-definition metrology enabled surface variation control by cutting load balancing1. J Manuf Sci Eng 138:021010
24. Shao Y, Yin Y, Du S, Xia T, Xi L (2018) Leakage monitoring in static sealing interface based on three dimensional surface topography indicator. J Manuf Sci Eng 140:101003–101012
25. Zhang M, Levina E, Djurdjanovic D, Ni J (2008) Estimating distributions of surface parameters for classification purposes. J Manuf Sci Eng 130:031010
26. Liao Y, Stephenson DA, Ni J (2012) Multiple-scale wavelet decomposition, 3D surface feature exaction and applications. J Manuf Sci Eng 134:011005
27. Krommweh J (2010) Tetrolet transform: a new adaptive Haar wavelet algorithm for sparse image representation. J Vis Commun Image Represent 21:364–374
28. Krommweh J, Ma J (2010) Tetrolet shrinkage with anisotropic total variation minimization for image approximation. Signal Process 90:2529–2539
29. Golomb SW (1994) Polyominoes. Princeton University Press, Princeton
30. ISO 16610–29 (2015) Geometrical product specifications (GPS)-filtration part 29: linear profile filters: spline wavelets
31. Flitney RK (2011) Seals and sealing handbook. Elsevier, Oxford
32. Lebeck AO (1981) Hydrodynamic lubrication in wavy contacting face seals—a two dimensional model. J Lubr Technol 103(4):578–586
33. Robbe-Valloire F, Prat M (2008) A model for face-turned surface microgeometry. Application to the analysis of metallic static seals. Wear 264(11–12):980–989
34. Aharony A, Stauffer D (2003) Introduction to percolation theory. Taylor & Francis, London
35. Persson BNJ, Albohr O, Creton C, Peveri V (2004) Contact area between a viscoelastic solid and a hard, randomly rough, substrate. J Chem Phys 120(18):8779–8793
36. Persson BNJ, Yang C (2008) Theory of the leak-rate of seals. J Phys Condens Matter 20(31):315011
37. Bottiglione F, Carbone G, Mangialardi L, Mantriota G (2009) Leakage mechanism in flat seals. J Appl Phys 106(10):104902
38. Bottiglione F, Carbone G, Mantriota G (2009) Fluid leakage in seals: an approach based on percolation theory. Tribol Int 42(5):731–737

39. Lorenz B, Persson BNJ (2010) Leak rate of seals: effective-medium theory and comparison with experiment. Eur Phys J E 31(2):159–167
40. Pérez-Ràfols F, Larsson R, Almqvist A (2016) Modelling of leakage on metal-to-metal seals. Tribol Int 94:421–427
41. Ledoux Y, Lasseux D, Favreliere H, Samper S, Grandjean J (2011) On the dependence of static flat seal efficiency to surface defects. Int J Press Vessels Pip 88(11–12):518–529
42. Marie C, Lasseux D (2007) Experimental leak-rate measurement through a static metal seal. J Fluids Eng 129(6):799–805
43. Lorenz B, Persson BNJ (2009) Leak rate of seals: comparison of theory with experiment. EPL (Europhysics Letters) 86(4):44006
44. Lorenz B, Persson BNJ (2010) On the dependence of the leak rate of seals on the skewness of the surface height probability distribution. EPL (Europhysics Letters) 90(3):38002
45. Vlădescu S-C, Putignano C, Marx N, Keppens T, Reddyhoff T, Dini D (2018) The percolation of liquid through a compliant seal—an experimental and theoretical study. J Fluids Eng 141(3):031101–031112
46. Malburg MC (2003) Surface profile analysis for conformable interfaces. J Manuf Sci Eng 125(3):624–627
47. Liao Y, Stephenson DA, Ni J (2012) Multiple-scale wavelet decomposition, 3D surface feature exaction and applications. J Manuf Sci Eng 134(1):011005
48. Jianjun S, Chenbo M, Jianhua L, Qiuping Y (2018) A leakage channel model for sealing interface of mechanical face seals based on percolation theory. Tribol Int 118:108–119
49. Zhang F, Liu J, Ding X, Yang Z (2017) An approach to calculate leak channels and leak rates between metallic sealing surfaces. J Tribol 139(1):011708
50. Ren J, Park C, Wang H (2018) Stochastic modeling and diagnosis of leak areas for surface assembly. J Manuf Sci Eng 140(4):041011
51. Shao Y, Yin Y, Du S, Xia T, Xi L (2018) Leakage monitoring in static sealing interface based on three dimensional surface topography indicator. J Manuf Sci Eng 140(10):101003–101012
52. Liu X, Liu K, Wang W, Gui C (2009) Connectivity characterization of three-dimensional surface topography based on mathematical morphology. Proc Inst Mech Eng Part J-J Eng Tribol 223(J7):941–947
53. Huang Z, Shih AJ, Ni J (2006) Laser interferometry hologram registration for three-dimensional precision measurements. J Manuf Sci Eng 128(4):1006–1013
54. Du S, Fei L (2015) Co-kriging method for form error estimation incorporating condition variable measurements. J Manuf Sci Eng 138(4):041003
55. Huang D, Du S, Li G, Wu Z (2017) A systematic approach for online minimizing volume difference of multiple chambers in machining processes based on high-definition metrology. J Manuf Sci Eng 139(8):081003
56. Shao Y, Du S, Xi L (2017) 3D machined surface topography forecasting with space-time multioutput support vector regression using high definition metrology. In: 37th Computers and information in engineering conference, vol 1, p V001T002A069
57. Shao Y, Wang K, Du S, Xi L (2018) High definition metrology enabled three dimensional discontinuous surface filtering by extended tetrolet transform. J Manuf Syst 49:75–92
58. Wang M, Shao Y, Du S, Xi L (2015) A diffusion filter for discontinuous surface measured by high definition metrology. Int J Precis Eng Manuf 16(10):2057–2062
59. Wang M, Xi L, Du S (2014) 3D surface form error evaluation using high definition metrology. Precis Eng 38(1):230–236
60. Tzafestas CS, Maragos P (2002) Shape connectivity: multiscale analysis and application to generalized granulometries. J Math Imaging Vis 17(2):109–129
61. Serra J (2000) Connections for sets and functions. Fund Inform 41(1):147–186
62. ISO (2015) Geometrical product specifications (GPS)-filtration part 1: overview and basic concepts. International Organization for Standardization, Geneva, Switzerland, Standard No. ISO 16610-1:2015
63. ISO (2015) Geometrical product specifications (GPS)-filtration part 22: linear profile filters: spline filter. International Organization for Standardization, Geneva, Switzerland, Standard No. ISO 16610-22:2015

64. ISO (2015) Geometrical product specifications (GPS)-filtration part 40: morphological profile filters: basic concepts. International Organization for Standardization, Geneva, Switzerland, Standard No. ISO 16610-40:2015
65. Johnson KL (1985) Contact mechanics. Cambridge University Press, New York
66. Lou S, Jiang XQ, Bills PJ, Scott PJ (2013) Defining true tribological contact through application of the morphological method to surface topography. Tribol Lett 50(2):185–193
67. Whitehouse D (2002) Handbook of surface and nanometrology, 2nd edn. CRC Press, University of Warwick Coventry
68. Cabuil N, Le Gouil A, Doclot O, Dickson B, Lagha A, Aminpur M, Chaton C, Royer JC (2007) Process monitoring and surface characterization with in-line XPS metrology. Solid State Technol 50:48–51
69. Lorenz B, Persson BNJ (2010) On the dependence of the leak rate of seals on the skewness of the surface height probability distribution. EPL 90(3)
70. Whitehouse DJ (2001) Function maps and the role of surfaces. Int J Mach Tools Manuf 41(13):1847–1861
71. Bottiglione F, Carbone G, Mangialardi L, Mantriota G (2009) Leakage mechanism in flat seals. J Appl Phys 106(10)
72. Persson BNJ, Yang C (2008) Theory of the leak-rate of seals. J Phys-Condens Matter 20(31)
73. ISO (1996) Geometrical product specifications (GPS)-surface texture: profile method-motif parameters. International Organization for Standardization, Geneva, Switzerland, Standard No. ISO 12085:1996
74. Shao YP, Yin YX, Du SC, Xia TB, Xi LF (2018) Leakage monitoring in static sealing interface based on three dimensional surface topography indicator. ASME Trans Manuf Sci Eng 140:101003
75. Shao YP, Yin YX, Du SC, Xi LF (2019) A surface connectivity based approach for leakage channel prediction in static sealing interface. J Tribol 141(6):062201
76. Ren J, Park C, Wang H (2018) Stochastic modeling and diagnosis of leak areas for surface assembly. J Manuf Sci Eng
77. Jayaram S, Yong W, Jayaram U, Lyons K, Hart P, A virtual assembly design environment, proceedings IEEE virtual reality (Cat. No. 99CB36316), pp 172–179
78. Zhong Y, Qin Y, Huang M, Lu W, Chang L (2014) Constructing a meta-model for assembly tolerance types with a description logic based approach. Comput Aided Des 48:1–16
79. Gallegos-Nieto E, Medellin-Castillo Hugo I, Xiu-Tian Y, Corney J (2020) Haptic-enabled virtual planning and assessment of product assembly. Assem Autom 40(4):641–654
80. Zhang X, Liu H (2017) Image guided as-assembled car engine sealing surface modeling. In: International conference on robotics and automation engineering, pp 317–321
81. Eberhart R, Kennedy J (1995) A new optimizer using particle swarm theory, MHS'95. In: Proceedings of the sixth international symposium on micro machine and human science, pp 39–43
82. Greenwood JA, Tripp JH (1970) The contact of two nominally flat rough surfaces. Proc Inst Mech Eng 185(1):625–633
83. Kalin M, Pogačnik A (2013) Criteria and properties of the asperity peaks on 3D engineering surfaces. Wear 308(1):95–104
84. Krystek M (1996) Form filtering by splines. Measurement 18(1):9–15
85. Flitney R (2007) Chapter 2—Static seals. In: Flitney R (ed) Seals and sealing handbook, 5th edn. Elsevier Science, Oxford, pp 7–104
86. Cierocki K, Ermert T (2003) Topographic stopper for cylinder head gaskets. Mtz Worldwide 64(1):9–12
87. Kestly M, Unseld G, Weiss A, Ludwig J (2003) MLS cylinder head gasket wave stopper, SAE technical paper 2003-01-0474
88. Katsurai T, Adachi K, Nishiyama T, Murakami O (2009) Cylinder head gasket for high combustion pressure diesel engines. SAE Int J Engines 2(1):941–947
89. Capretta R, Ohigashi H (1995) Design methodology for automotive multi-layer steel cylinder head gaskets. SAE technical paper-950322

90. Ma Z, Yu T, Xing Y, Liu L, Hu M (2020) A method for studying the sealing performance of cylinder head matched surface. In: RICAI 2020: 2020 2nd international conference on robotics, intelligent control and artificial intelligence
91. Persson BNJ, Yang C (2008) Theory of the leak-rate of seals. J Phys-Condens Matter 20(31):315011
92. Hao MM, Wang YL, Li ZT, Sun XH (2018) Effects of surface topography on hydrodynamic performance of liquid film seals considering cavitation. Ind Lubr Tribol 70(6):984–992
93. Liu S, Jin S, Zhang X, Wang L, Mei B, Hu B (2017) Controlling topography of machined surface for adhesive-sealing. In: ASME 2017 12th international manufacturing science and engineering conference collocated with the JSME/ASME 2017 6th international conference on materials and processing
94. Zhang F, Liu J, Ding X, Yang Z (2016) An approach to calculate leak channels and leak rates between metallic sealing surfaces. J Tribol 139(1):011708
95. Malburg MC (2003) Surface profile analysis for conformable interfaces. J Manuf Sci E T ASME 125(3):624–627
96. Shao YP, Yin YX, Du SC, Xia TB, Xi LF (2018) Leakage monitoring in static sealing interface based on three dimensional surface topography indicator. J Manuf Sci Eng Trans ASME 140:101003
97. Deltombe R, Bigerelle M, Jourani A (2015) Analysis of the effects of different machining processes on sealing using multiscale topography. Surf Topogr Metrol Prop 4(1):015003
98. Yang Z, Liu J, Ding X, Zhang F (2018) The effect of anisotropy on the percolation threshold of sealing surfaces. J Tribol 141(2):022203
99. Liao C, Xu X, Fang H, Wang H, Man M (2015) A leakage model of metallic static seals based on micromorphology characteristics of turning flange surface. Ind Lubr Tribol 67(6):572–581
100. Robbe-Valloire F, Prat M (2008) A model for face-turned surface microgeometry application to the analysis of metallic static seals. Wear 264(11–12):980–989
101. Yin Y, Shao Y, Wang K, Du S, Xi L (2020) Segmentation of workpiece surfaces with tool marks based on high definition metrology. J Manuf Process 57:268–287
102. Ren J, Park C, Wang H (2018) Stochastic modeling and diagnosis of leak areas for surface assembly. J Manuf Sci Eng Trans ASME 140(4):041011
103. Du S, Xi L (2019) High definition metrology based surface quality control and applications. Springer, Singapore

Chapter 6
Modeling and Management of Variation Propagation

6.1 State Space Modeling of Variation Propagation in Multistage Turning Processes

6.1.1 Introduction

Minimizing process variation to enhance quality is crucial for the success of manufacturing enterprises in the competitive global market [1]. While statistical process control (SPC) is extensively employed to monitor process variation across various applications [2–4], it is not designed to analyze the propagation of variation within manufacturing processes. The production of complex workpieces involves numerous machining stages, and as workpieces progress through these stages, variations in their characteristics are introduced. If characteristics machined in earlier stages serve as the reference for later stages, the variations will propagate to newly created characteristics and accumulate in the final product. Consequently, variations can be categorized into two types: those generated in the current stage and those inherited from prior stages. The final product's variations are the sum of variations from all stages involved in the process.

Research into variation propagation modeling and monitoring for multistage processes has been a focal point of intensive study over the past few decades [5, 6]. Engineering models that directly link the engineering knowledge of process variation sources with key product characteristic (KPC) measurements have garnered significant interest. Among these engineering models, the state-space based model stands out as an effective framework, which can be broadly categorized into two types: the multistage assembly process (MAP) model and the multistage milling process (MMP) model. The state space model for variation propagation was initially introduced for two-dimensional assembly processes [7] and subsequently explored and applied in three-dimensional assembly processes [8, 9]. In assembly processes, the primary drivers of dimension variation propagation are fixture re-orientation

S. Du and L. Xi, *Quality Control for Variable Stiffness Structured Workpiece*,
https://doi.org/10.1007/978-981-96-5376-8_6

errors, workpiece fabrication errors, and workpiece joint errors, such as those found in lap, butt, and mixed joints. In contrast, in machining processes, the combination of locating datum errors and fixture errors complicates variation propagation, making them the predominant factors in the propagation of variations.

In the context of MMP, several researchers have delved into the three-dimensional propagation of variations at a systemic level using state space models. Huang et al. [10] expanded the variation propagation model through an approximate linearization technique. Zhou et al. [11] crafted a variation propagation model and derived explicit expressions for datum and fixture errors. Huang and Shi [12] advanced the state space modeling approach to encompass serial-parallel hybrid MMPs with multiple process routes. Loose et al. [13] formulated a variation propagation model applicable to general fixture setup schemes. Loose et al. [14] incorporated geometric dimensioning and tolerancing (GD&T) into dimensional variation models for MMP. Cao et al. [15] put forth a performance evaluation and enhancement methodology for MMPs that involve rework loops. Abellan-Nebot et al. [17] extended the state space model for MMP to account for machining-induced errors. Comprehensive accounts of existing research on variation propagation modeling and its applications can be found in a dedicated monograph [18] and a survey [19].

However, the existing state space models have primarily focused on variation propagation in MMPs for the fabrication of cubic workpieces. Given the differences in the principles of variation propagation between MMPs for cubic workpieces and MTPs for rotary workpieces, these MMP models cannot be directly applied to MTPs [20]. Figure 6.1 illustrates the distinction in locating schemes between MMP and MTP. The left portion of Fig. 6.1 depicts the locating scheme for a typical cubic workpiece, such as that of an engine cylinder head, while the right portion shows the locating scheme for a rotary workpiece.

MTPs are extensively utilized in the manufacturing industry for the production of rotary workpieces. Because of the distinct causes of variation propagation and the scarcity of research in this area, there is a need to analyze and model the variation propagation specific to MTPs. Hence, the aim of this section is to introduce a generic framework for three-dimensional variation propagation modeling for MTPs of rotary workpieces. This involves developing explicit three-dimensional expressions for datum error and fixture error, thereby advancing the understanding and control of variation in the manufacturing of such components.

6.1.2 Workpiece Random Geometric Error Representation

6.1.2.1 Coordinate System Definition

To derive the expressions for variations, we start by defining four types of coordinate systems:

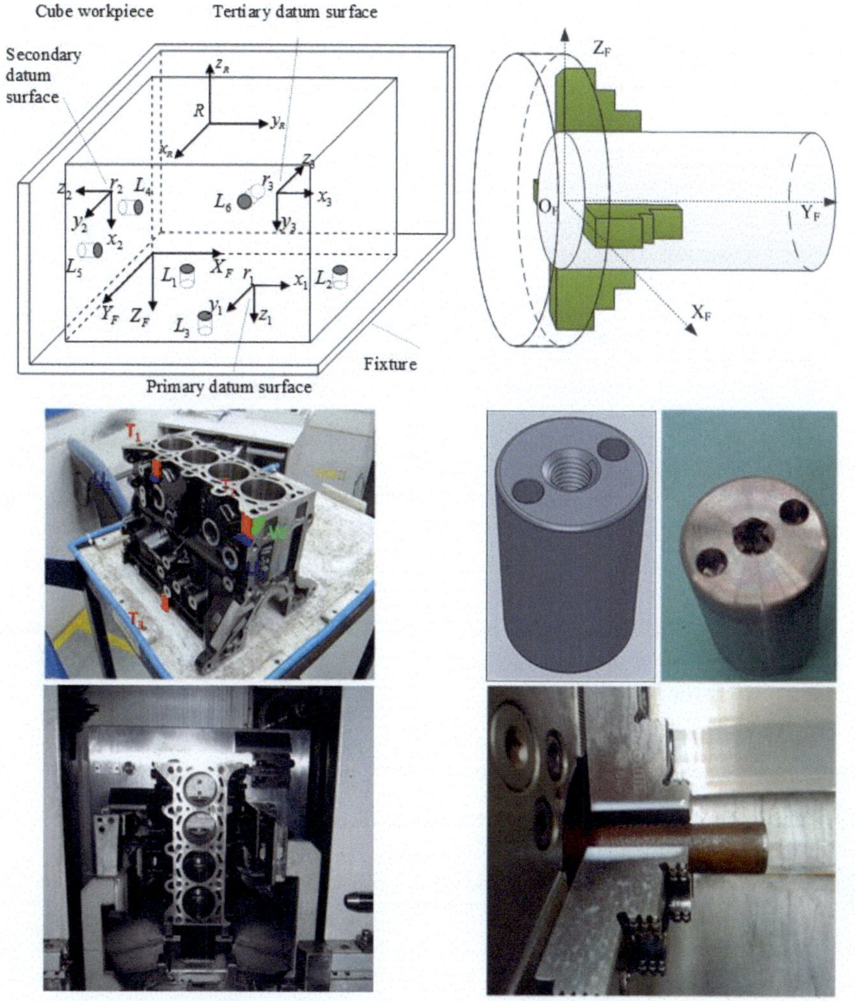

(a) "3-2-1" locating scheme for MMP of cube workpiece

(b) jaw chuck locating scheme for MTP of rotary workpiece

Fig. 6.1 Difference of locating scheme between MMP and MTP

1. The Reference Coordinate System (RCS): This system is rigidly linked to a specific workpiece and serves to represent its location and orientation. It is also referred to as the workpiece coordinate system.
2. The Fixture Coordinate System (FCS): This system delineates the actual fixture setup that secures the workpiece and is determined by the actual positions of the locators.

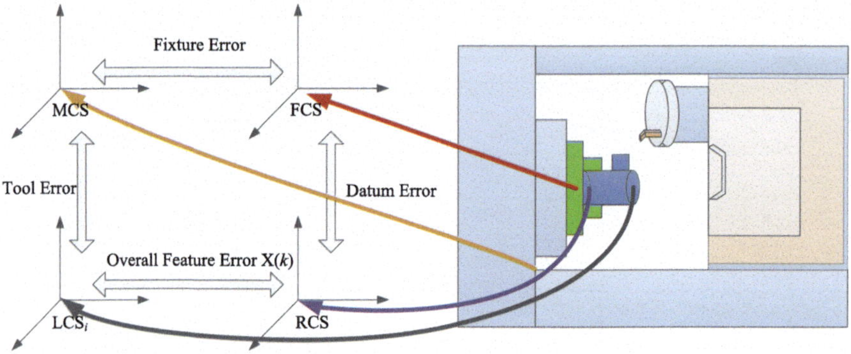

Fig. 6.2 Illustration of coordinate system and error

3. The Nominal Fixture Coordinate System (^{0}FCS): This system defines the theoretical or ideal fixture setup, based on the nominal positions of each locator. The notation with a left superscript 0 is used to denote the nominal condition. As the 0FCS is fixed to the machine, it can also be regarded as the Machine Coordinate System (MCS).
4. The Local Coordinate System (LCS): This system is attached to specific features on the workpiece and represents the actual location and orientation of these features. Error Definition

Four types of errors, derived from the relationships between the aforementioned coordinate systems, are outlined below:

1. Tool Error: This error represents the deviation of the LCS with respect to another reference coordinate system.
2. Fixture Error: This error is the deviation of the actual FCS with respect to the nominal 0FCS.
3. Datum Error: This error refers to the positional and angular discrepancy of the actual RCS with respect to the FCS, typically arising from inaccurate positioning datums.
4. Overall Characteristic Error: This error is the deviation of the LCS with respect to the RCS, encapsulating the overall position and orientation deviation of the workpiece's characteristics.

A visual representation of all the coordinate systems and their interrelationships is provided in Fig. 6.2.

6.1.2.2 Characteristic and Variation Representation

A workpiece's characteristic is defined by its position, orientation, and dimensions. In this section, we use differential motion vectors (DMV) as the state vector to represent the variations in position and orientation of a characteristic. As detailed in

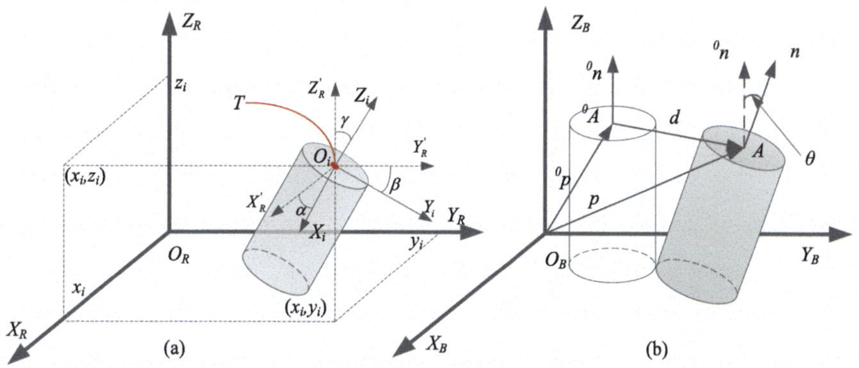

Fig. 6.3 Feature and deviation representation

Sect. 6.1.2.1, the relationship between the FCS and RCS can be leveraged to describe the position and orientation of a workpiece's characteristic. In Fig. 6.3, a cylindrical characteristic is depicted, with its position indicated by the coordinates of a specific point O_i, its orientation defined by the rotation between coordinate systems $X_i|X_R$, $Y_i|Y_R$ and $Z_i|Z_R$, and its size specified by the cylinder's diameter and depth. The characteristic is thus represented by its position and orientation: $(x_i, y_i, z_i, \alpha, \beta, \gamma)$, where (x_i, y_i, z_i) denotes the coordinates of point T in $O_RX_RY_RZ_R$, and (α, β, γ) represents the rotational degree required to align coordinate system $O_iX_iY_iZ_i$ with coordinate system $O_RX_RY_RZ_R$.

The Vectorial Dimensioning and Tolerancing (VD&T) representation encompasses both the positional and orientational data of a characteristic [14], capturing its six degrees of freedom. The variation of a characteristic during the turning process is manifested through alterations in its location and orientation across these six degrees of freedom, denoted as:

$$x_A^B = [(d_A^B)^T \quad (\theta_A^B)^T]^T \tag{6.1}$$

where $d_A^B = [d_{Ax}^B\ d_{Ay}^B\ d_{Az}^B]^T$ is the positional error, and $\theta_A^B = [\theta_{Ax}^B\ \theta_{Ay}^B\ \theta_{Az}^B]^T$ is rotational error.

6.1.3 Variation Coupling Analysis

Direct application of DMVs to fundamental arithmetic operations like addition and subtraction is not feasible due to the presence of Euler angles. To address this, a novel set of rules for DMV computation has been formulated in this section, enabling the analysis of variation coupling.

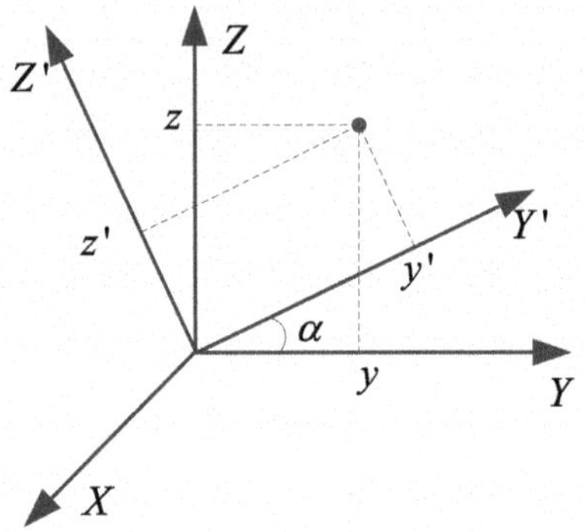

Fig. 6.4 Illustration of rotation matrix

6.1.3.1 Rotation and Translation Matrix

Prior to delving into the transformation matrix, it is essential to first introduce its two fundamental components:

Rotation Matrix

Let $Rot(k, \theta)$ denote the rotation matrix for the operation of rotating an object around the k-axis by an angle θ. Suppose a point possesses coordinates (x, y, z) in the initial coordinate system, which indicate the position and orientation of the object prior to rotation, and coordinates (x', y', z') in the subsequent coordinate system, reflecting the position and orientation post-rotation, as depicted in Fig. 6.4.

A correlation between the two sets of coordinate values is defined as $(x, y, z, 1)^T = Rot(k, \theta)(x', y', z', 1)^T$, with the fourth element '1' included to ensure uniformity in matrix operations. If a coordinate system undergoes a rotation around the X-axis by an angle α, and a point p has a coordinate value (x, y, z) in the original coordinate system and a coordinate (x', y', z') in the new system, then the following two equations are valid:

$$y = y' \cos\alpha - z' \sin\alpha,\ z = y \sin\alpha + z' \cos\alpha \tag{6.2}$$

Written in matrix form, the Eq. (6.26) turns into

$$\begin{bmatrix} x \\ y \\ z \\ 1 \end{bmatrix} = \begin{bmatrix} 1 & 0 & 0 & 0 \\ 0 & \cos\alpha & -\sin\alpha & 0 \\ 0 & \sin\alpha & \cos\alpha & 0 \\ 0 & 0 & 0 & 1 \end{bmatrix} \begin{bmatrix} x' \\ y' \\ z' \\ 1 \end{bmatrix} \tag{6.3}$$

Noted as:

$$[M] = Rot(X, \alpha)[M'] \tag{6.4}$$

Translation Matrix

Denote $Trans(x, y, z)Trans(x, y, z)$ as the translation matrix for translating an object by displacements (t_x, t_y, t_z) along the X\Y\Z axes, respectively. A point p with coordinates (x, y, z) in the original coordinate system is transformed to coordinates (x', y', z') in the new system. The relationship between these coordinates is established as: $(x, y, z, 1)^T = Trans(t_x, t_y, t_z)(x', y', z', 1)^T$, with the addition of a fourth '1' element to ensure matrix operation uniformity. The translation matrix is thus structured as:

$$Trans(t_x, t_y, t_z) = \begin{bmatrix} 1\,0\,0\,t_x \\ 0\,1\,0\,t_y \\ 0\,0\,1\,t_z \\ 0\,0\,0\,1 \end{bmatrix} \tag{6.5}$$

6.1.3.2 Transformation Matrix

The movement of an object involves translation and rotation of that object, and the corresponding transformation matrix for the movement of an object can be represented through the translation and rotation matrix. Note H_i^R as the transformation matrix for the operation of moving an object. A point p has coordinate value (x, y, z) in the original coordinate system, and a coordinate (x', y', z') in the new coordinate system, Then a relationship between the two coordinate values are: $(x, y, z, 1)^T = H_i^R(x', y', z', 1)^T$.

The displacement of an object encompasses both translational and rotational components, and the corresponding transformation matrix for such movement can be constructed using both translation and rotation matrices. Let H_i^R represent the transformation matrix for the movement of an object. If a point p has coordinate values (x, y, z) in the original coordinate system and (x', y', z') in the new coordinate system, then the relationship between these two sets of coordinates is given by: $(x, y, z, 1)^T = H_i^R(x', y', z', 1)^T$.

The motion of a rigid body can be broken down into a sequence of rotations followed by a translation. For example, one might first rotate about the X-axis by an angle α, then about the Y-axis by an angle β, and finally about the Z-axis by an angle γ, concluding with a translation by the vector (t_x, t_y, t_z). The resultant transformation matrix for this sequence of operations is constructed as follows:

$$H_{\mathrm{i}}^R = Trans(t_x, t_y, t_z) \cdot Rot(z, \gamma) \cdot Rot(y, \beta) \cdot Rot(x, \alpha) \tag{6.6}$$

where $Rot(x, \alpha)$ and $Trans(t_x, t_y, t_z)$ are listed in Eqs. (6.4) and (6.5), $Rot(y, \beta)$,$Rot(z, \gamma)$ and H_i^R are:

$$Rot(y,\beta)=\begin{bmatrix}\cos\beta & 0 & \sin\beta & 0\\ 0 & 1 & 0 & 0\\ -\sin\beta & 0 & \cos\beta & 0\\ 0 & 0 & 0 & 1\end{bmatrix},\ Rot(z,\gamma)=\begin{bmatrix}\cos\gamma & -\sin\gamma & 0 & 0\\ \sin\gamma & \cos\gamma & 0 & 0\\ 0 & 0 & 1 & 0\\ 0 & 0 & 0 & 1\end{bmatrix} \tag{6.7}$$

$$H_i^R = \begin{bmatrix}\cos\beta\cos\gamma & \sin\alpha\sin\beta\cos\gamma-\cos\alpha\sin\gamma & \cos\alpha\sin\beta\cos\gamma+\sin\alpha\sin\gamma & t_x\\ \cos\beta\sin\gamma & \sin\alpha\sin\beta\sin\gamma+\cos\alpha\cos\gamma & \cos\alpha\sin\beta\sin\gamma-\sin\alpha\cos\gamma & t_y\\ -\sin\beta & \sin\alpha\cos\beta & \cos\alpha\cos\beta & t_z\\ 0 & 0 & 0 & 1\end{bmatrix} \tag{6.8}$$

6.1.3.3 Variation Coupling

In this subsection, a novel set of calculation principles has been developed to analyze the relationships between various sources of variation and their interactive effects. The homogenous transformation matrix (HTM) is utilized to represent the transformation of coordinate systems. Let H_{i}^R denote the HTM for the actual position and orientation of the characteristic, ${}^0H_{\mathrm{i}}^R$ denote the HTM for the nominal position and orientation of the ith characteristic, and the adjustment of the HTM be noted as δH_i^R. The HTM between the RCS and the actual characteristic is considered as a two-step transformation process: (i) from the actual LCS_i to ${}^0LCS_i$, and (ii) from ${}^0LCS_i$ to the RCS.

$$H_{\mathrm{i}}^R={}^0H_{\mathrm{i}}^R\cdot\delta H_i^R \tag{6.9}$$

where $\delta H_i^R = \begin{bmatrix}1 & -\theta_{iz}^R & \theta_{iy}^R & d_{ix}^R\\ \theta_{iz}^R & 1 & -\theta_{ix}^B & d_{iy}^R\\ -\theta_{iy}^R & \theta_{ix}^R & 1 & d_{iz}^R\\ 0 & 0 & 0 & 1\end{bmatrix} = I_{4\times4}+\Delta_i^R,\ \Delta_i^R=\begin{bmatrix}0 & -\theta_{iz}^R & \theta_{iy}^R & d_{ix}^R\\ \theta_{iz}^R & 0 & -\theta_{ix}^B & d_{iy}^R\\ -\theta_{iy}^R & \theta_{ix}^R & 0 & d_{iz}^R\\ 0 & 0 & 0 & 0\end{bmatrix}$
is the differential transformation matrix (DTM) associating to the DMV x_i^R, $x_i^R = [\,(d_i^R)^T\ (\theta_i^R)^T\,]^T$. Define the function $f(x_i^R)\rightarrow\Delta_i^R$ as the relationship between the elements of the DMV and the DTM. Thus:

$$H_{\mathrm{i}}^R={}^0H_{\mathrm{i}}^R\cdot\delta H_i^R={}^0H_{\mathrm{i}}^R\cdot\left(I_{4\times4}+\Delta_i^R\right)={}^0H_{\mathrm{i}}^R+{}^0H_{\mathrm{i}}^R\cdot\Delta_i^R \tag{6.10}$$

Assume there are two characteristics on the rotary workpiece, noted as LCS_i, $i = 1, 2$, and the HTM from LCS_1 to RCS, and form LCS_2 to LCS_1 are known, then

$$H_1^R={}^0H_1^R+{}^0H_1^R\cdot\Delta_1^R,\ \text{and}\ H_2^1={}^0H_2^1+{}^0H_2^1\cdot\Delta_2^1 \tag{6.11}$$

and,

$$H_2^R = H_1^R \cdot H_2^1 = ({}^0H_1^R + {}^0H_1^R \cdot \Delta_1^R) \cdot ({}^0H_2^1 + {}^0H_2^1 \cdot \Delta_2^1) \tag{6.12}$$

Extend H_2^R, and neglect the second-order small values, and the Eq. (6.13) can be obtained.

$$H_2^R = {}^0H_2^R\left(I + \left({}^0H_2^1\right)^{-1} \cdot \Delta_1^R \cdot {}^0H_2^1 + \Delta_2^1\right) \tag{6.13}$$

Since

$$H_2^R = {}^0H_2^R(I + \Delta_2^R) \tag{6.14}$$

Combining the Eqs. (6.13) and (6.14),

$$\Delta_2^R = \left({}^0H_2^1\right)^{-1} \cdot \Delta_1^R \cdot {}^0H_2^1 + \Delta_2^1 \tag{6.15}$$

According to the work of Craig [21], $\left({}^0H_2^1\right)^{-1} \cdot \Delta_1^R \cdot {}^0H_2^1 = {}^T\Delta_1^R$, where ${}^T\Delta_1^R = f\left(\begin{bmatrix} \left({}^0R_2^1\right)^T & -\left({}^0R_2^1\right)^T \cdot {}^0\hat{t}_2^1 \\ \mathbf{0} & \left({}^0R_2^1\right)^T \end{bmatrix} x_1^R\right)$, that is $\Delta_2^R = {}^T\Delta_1^R + \Delta_2^1$.
$\Delta_2^R = {}^T\Delta_1^R + \Delta_2^1$ is written as:

$$f\left(x_2^R\right) = f\left(\begin{bmatrix} \left({}^0R_2^1\right)^T & -\left({}^0R_2^1\right)^T \cdot {}^0\hat{t}_2^1 \\ \mathbf{0} & \left({}^0R_2^1\right)^T \end{bmatrix} x_1^R\right) + f\left(x_2^1\right) \tag{6.16}$$

Thus, according to the one-by-one mapping, the Eq. (6.16) is rewritten as:

$$x_2^R = Q_2^1 x_1^R + x_2^1 \tag{6.17}$$

where $Q_2^1 = \begin{bmatrix} \left({}^0R_2^1\right)^T & -\left({}^0R_2^1\right)^T \cdot {}^0\hat{t}_2^1 \\ \mathbf{0} & \left({}^0R_2^1\right)^T \end{bmatrix}$.

It is clear that the ${}^0FCS$ and MCS are fixed relative to each other in terms of position and orientation, hence there exists a nominal *HTM* transformation matrix between the two coordinate systems, and ${}^0FCS$ is used to represent MCS to simplify the computation for this fixed relationship. The imperfection in the fixture results in a misalignment between ${}^0FCS$ and FCS, which is denoted as $x_{FCS}^{{}^0FCS}$.

When characteristics with flaws from previous machining stages are utilized as the datum, the misalignment between the datum characteristic and the fixture—referred to as datum error—is introduced into the process. This misalignment is denoted as $x_{FCS}^{RCS} x_{FCS}^{RCS}$, or more succinctly, as $x_F^{R(k)}$ at stage k.

Given that the characteristics machined are directly shaped by the tool path, which is typically programmed based on ${}^{0}FCS$, the imperfections in the machined characteristic at the current stage are reflected in the actual path taken. Therefore, the tool error is considered as the deviation of the actual tool path from its intended, programmed path, denoted as $x_{i(k)}^{{}^{0}F}$, where $i(k)$ is the ith characteristic machined at stage k.

The characteristic error is also influenced by the imperfections in the fixture setup, meaning the characteristic error at the current stage is a combined result of both tool error and fixture error. As per Eq. (6.17), the characteristic error contributed by the current stage is given by the combined effect of these errors.

$$x_{i(k)}^{F(k)} = Q_{i(k)}^{{}^{0}F(k)} \cdot x_{{}^{0}F(k)}^{F(k)} + x_{i(k)}^{{}^{0}F(k)} \tag{6.18}$$

Given that the sources of characteristic error can generally be divided into two categories—those from the current stage and those from previous stages (the flawed datum)—the characteristic error also includes a contribution from the datum error $x_F^{R(k)}$. The combined effects of these two sources are represented by Eq. (6.19).

$$x_{i(k)}^{R(k)} = Q_{i(k)}^{F(k)} x_F^{R(k)} + x_{i(k)}^{F(k)} \tag{6.19}$$

The relationships between different coordinate systems and their corresponding DMVs, as discussed, can be encapsulated in Fig. 6.5. The fundamental concept of Fig. 6.5 is that each type of error can be depicted by a nominal HTM and a DMV (color-coded for clarity). For instance, fixture error represents the deviation of MCS with respect to FCS as shown in Fig. 6.2. This can be further analyzed by breaking it down into two steps: (1) the nominal HTM from MCS to ^{0}FCS, and (2) the corresponding DMV x_F^M. (Note that x_F^M and $x_F^{{}^{0}FCS}$ are equivalent through the nominal HTM from MCS to ^{0}FCS. We use the notation x_F^M instead of $x_F^{{}^{0}FCS}$ for two reasons. First, to maintain consistency within the section. The DMVs are assessed within the actual coordinate system rather than the nominal one. Second, to distinguish DMVs based on different coordinate systems, such as $x_{LCS_d}^{FCS}$ and $x_{LCS_d}^{RCS}$ in Fig. 6.5). Both steps are indicated with bold black lines. Another example is provided to illustrate tool error, marked with light pink lines at the bottom of Fig. 6.5. The tool error can be inferred through (1) the HTM from MCS to ^{0}LCSi, and (2) from 0LCSi to LCSi (which is the DMV $x_{LCS_i}^M$).

6.1.4 Derivation of the State Space Model for MTPs

6.1.4.1 Framework

In a MTP consisting of N stages, there are two primary types of variations at any given stage k (as depicted in Fig. 6.6): (i) variations propagated from previous stages

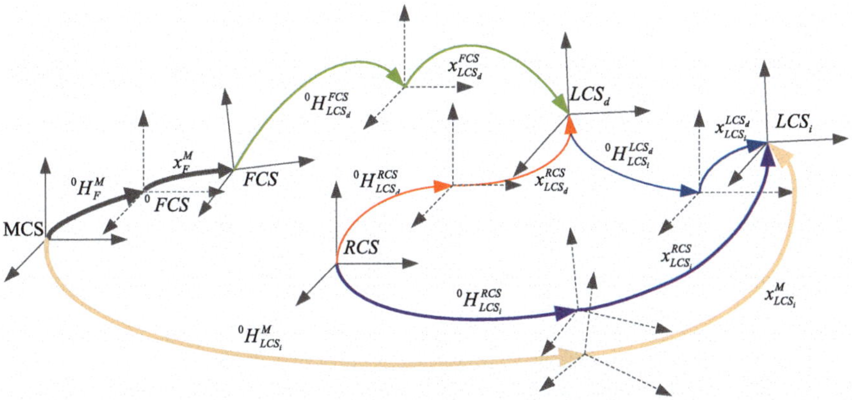

Fig. 6.5 Differential motion representation of variation accumulation of rotary workpiece

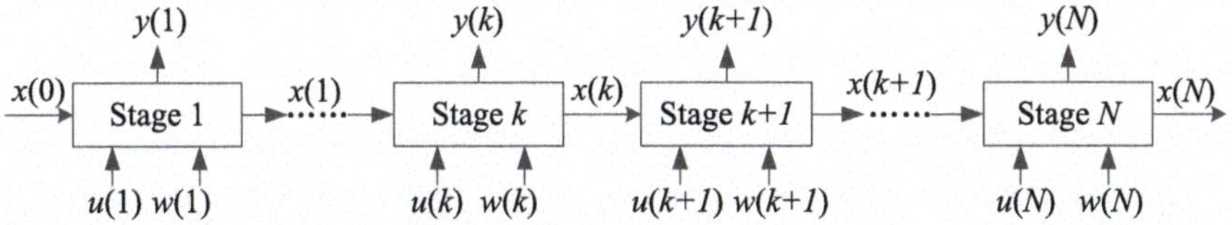

Fig. 6.6 Variation propagation of MTP

due to imperfect datum characteristics, and (ii) variations generated in the current stage, which include fixture error, tool error, and other related errors.

The property of stage-wise variation accumulation justifies the use of the state space model, and the accumulation of variations for MTP can be represented as follows:

$$\begin{aligned}\mathbf{x}(k) &= \mathbf{A}(k)\mathbf{x}(k-1) + \mathbf{B}(k)\mathbf{u}(k) + \mathbf{v}(k) \\ \mathbf{y}(k) &= \mathbf{C}(k)\mathbf{x}(k) + \mathbf{w}(k)\end{aligned} \tag{6.20}$$

where k indicates the stage index, $\mathbf{x}(k)$ denotes the product variation, and $\mathbf{y}(k)$ is the corresponding measurement relative to $\mathbf{x}(k)$. The vector $\mathbf{u}(k)$ accounts for the errors introduced at the local stage, which includes both the tool error and fixture error as detailed in this section. The matrix $\mathbf{A}(k)$ represents the dynamics of repositioning during the local stage, $\mathbf{B}(k)$ is the matrix that describes the transformation of variations due to local operations, $\mathbf{C}(k)$ is the observation matrix, and $\mathbf{v}(k)$ and $\mathbf{w}(k)$ are the noise terms associated with the model's state and observation equations, respectively.

The complete computational sequence is outlined in Fig. 6.7. At the commencement of each stage within the MTPs, the rotary workpiece undergoes relocation from stage $k-1$ to stage k, effecting a transition in the RCS from one stage to the next (refer to S1). Broadly, the errors engendered by the fixturing process can be categorized

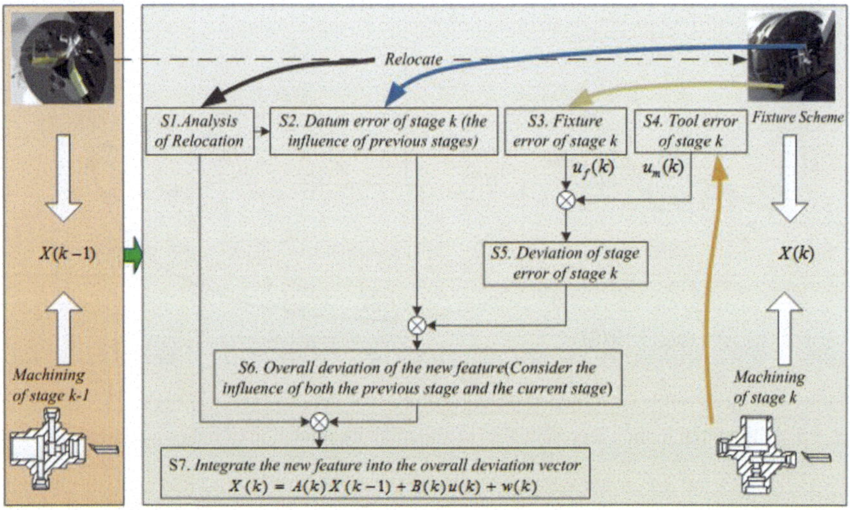

Fig. 6.7 Dimensional error computation procedure

into two segments: (i) the datum error stemming from the positioning datum (refer to S2), inherited from the preceding stages; (ii) the fixture error attributable to the defective fixture setup and degraded locators (refer to S3). As the turning process is executed under these suboptimal fixturing and positioning conditions, the variation is subject to the influence of both types of errors (refer to S4 and S5). The alteration in the workpiece characteristic represents an aggregate outcome influenced by the cumulative effects from the antecedent stages as well as the current stage k (refer to S6). Ultimately, the state space model is formulated (S7).

6.1.4.2 Derivation of Model

S1: Relocation Error Analysis

When a workpiece transitions from stage $k-1$ to stage k, the RCS shifts from $R(k-1)$ to $R(k)$, and the variation of characteristics is consequently altered due to this change in RCS. If the HTM from $R(k)$ to each FCS is known, then, based on Eq. (6.17), the DMV of a characteristic's variation within the coordinate system $R(k)$ is given by:

$$x_i^{R(k)} = x_i^{R(k-1)} - Q_i^{R(k)} \cdot x_R^{R(k-1)} \tag{6.21}$$

Let $\mathbf{x}_1(k)$ denote the state vector that encapsulates the variation of all characteristics; it is a compilation of the DMVs, $x_i^{R(k)}$, of each characteristic at the onset of stage k. Similarly, let $\mathbf{x}(k-1)$ represent a stack of $x_i^{R(k-1)}$ after all operations at stage $k-1$

have been completed. In the event that there are M characteristics to be machined, then:

$$\mathbf{x}_1(k) = \mathbf{A}_1(k)\mathbf{x}(k-1) \tag{6.22}$$

where, $\mathbf{x}_1(k) = \begin{bmatrix} x_1^{R(k)} \\ \vdots \\ x_i^{R(k)} \\ \vdots \\ x_R^{R(k)} \\ \vdots \\ x_M^{R(k)} \end{bmatrix}$, $\mathbf{A}_1(k) = \begin{bmatrix} \mathbf{I} \cdots 0 \cdots & 0 & \cdots 0 \\ \vdots \; \vdots \; \vdots \; \vdots & \vdots & \vdots \; \vdots \\ 0 \cdots \mathbf{I} \cdots & -\mathbf{Q}_i^{R(k)} & \cdots 0 \\ \vdots \; \vdots \; \vdots \; \vdots & \vdots & \vdots \; \vdots \\ 0 \cdots 0 \cdots & 0 & \cdots 0 \\ \vdots \; \vdots \; \vdots \; \vdots & \vdots & \vdots \; \vdots \\ 0 \cdots 0 \cdots & 0 & \cdots \mathbf{I} \end{bmatrix}$, $\mathbf{x}(k-1) = \begin{bmatrix} x_1^{R(k-1)} \\ \vdots \\ x_i^{R(k-1)} \\ \vdots \\ x_r^{R(k-1)} \\ \vdots \\ x_M^{R(k-1)} \end{bmatrix}$.

If the RCS is not changed when relocating a workpiece from stage $k-1$ to stage k, then $\mathbf{x}_1(k) = \mathbf{x}(k-1)$.

S2: Datum Error Analysis

Datum error arises from imperfections in the datum characteristics machined in previous stages, leading to a deviation between the RCS and the FCS. For instance, in a turning process, the presence of datum error can result in coaxiality variation (refer to Fig. 6.8). This is illustrated by the datum error, where the upper line represents the nominal position of the datum characteristic, and the lower line indicates the actual position of the datum characteristic.

This section examines the datum error associated with the four-jaw chuck fixturing scheme, a principal strategy in turning processes. As illustrated in Fig. 6.9a, a 4-jaw chuck clamping system establishes a FCS with the Z axis perpendicular to the turning axis, the Y axis aligned with the turning axis, and the X axis perpendicular to the plane of $Y_F O_F Z_F$. Figure 6.9b demonstrates that the jaws restrict movement along the X and Z axes, as well as rotation around these axes; the end of the jaw limits movement along the Y-axis, and with adequate clamping force, rotation around the Y-axis is also restricted. Under these conditions, a virtual locator is positioned on the tangent plane where locators L_1 and L_2 are placed, constraining the rotational

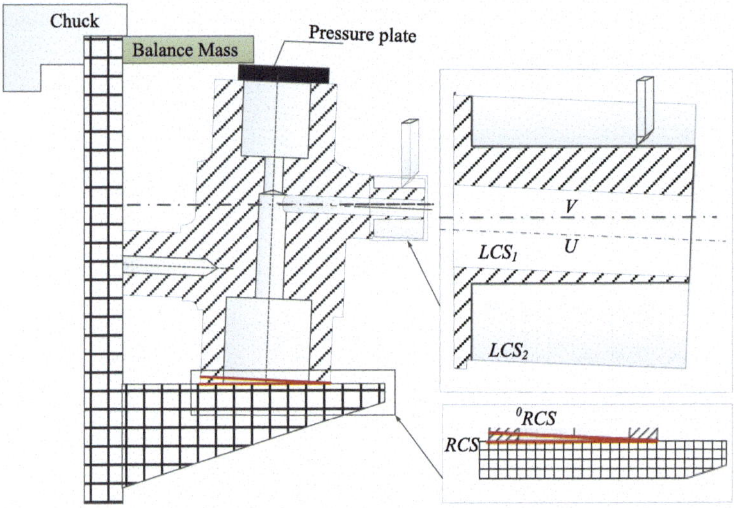

Fig. 6.8 Effects illustration of Datum error in rotary workpieces turning process

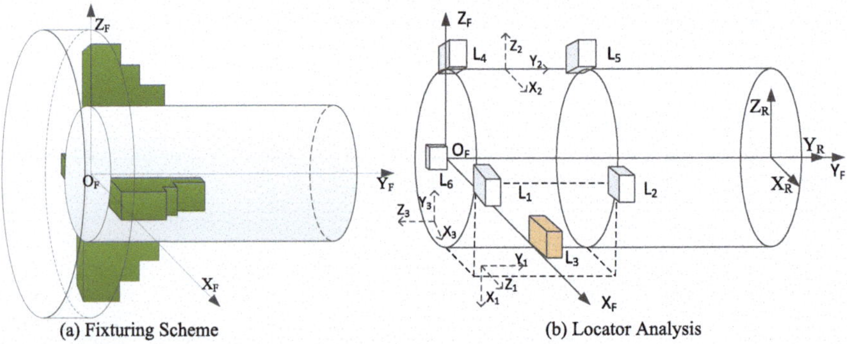

Fig. 6.9 Illustration of 4-jaws chuck fixturing scheme in MTP

motion around the Y-axis. Consequently, the locators $L_1 \sim L_6$ (the latter being on the end plane) restrict all six degrees of freedom of the workpiece.

The plane that includes locator $L_1 \sim L_3$ is designated as the primary datum plane ($O_1X_1Y_1Z_1$), which limits one translational degree of freedom and two rotational degrees of freedom. The secondary datum characteristic plane, where locators L_4 and L_5 are placed, is the tangential plane ($O_2X_2Y_2Z_2$), and the tertiary datum plane is the end plane that contains locator L_6 ($O_3X_3Y_3Z_3$).

The datum error resulting from the flawed datum at stage k is recognized as the difference between $R(k)$ and $FCS(k)$, denoted as $x_{FCS(k)}^{R(k)}$. The primary datum deviation with respect to $R(k)$ is denoted as $x_1^{R(k)}$, the secondary datum deviation with respect to $R(k)$ is denoted as $x_2^{R(k)}$, and the tertiary datum deviation with respect to $R(k)$ is denoted as $x_3^{R(k)}$. Accordingly, $x_{FCS(k)}^{R(k)}$ is expressed as a linear combination of $x_1^{R(k)}$, $x_2^{R(k)}$ and $x_3^{R(k)}$. Let $T_1(k)$, $T_2(k)$ and be the coefficient matrices, and $x_{FCS(k)}^{R(k)}$ is expressed as:

$$x_{FCS(k)}^{R(k)} = T_1(k)x_1^{R(k)} + T_2(k)x_2^{R(k)} + T_3(k)x_3^{R(k)} \tag{6.23}$$

$p_1 \sim p_6$ are the points on the workpiece that directly touching the six locators, $L_1 \sim L_6$, on the fixture, with coordinate value $p_1^F \sim p_6^F$ with respect to FCS respectively. Under the direct touching condition, the Z value of each locator is zero. That is:

$$\begin{cases} [H_R^1 \cdot H_F^R \cdot \tilde{p}_1^F]_3 = 0 \\ [H_R^1 \cdot H_F^R \cdot \tilde{p}_2^F]_3 = 0 \\ [H_R^1 \cdot H_F^R \cdot \tilde{p}_3^F]_3 = 0 \\ [H_R^2 \cdot H_F^R \cdot \tilde{p}_4^F]_3 = 0 \\ [H_R^2 \cdot H_F^R \cdot \tilde{p}_5^F]_3 = 0 \\ [H_R^3 \cdot H_F^R \cdot \tilde{p}_6^F]_3 = 0 \end{cases} \tag{6.24}$$

Take p_1 for example, given that $H_R^1 = (\overline{\Delta}_1^R + I) \cdot {}^0H_R^1$ and $H_F^R = {}^0H_F^R \cdot (\Delta_F^R + I)$.

$$[(\overline{\Delta}_1^R \cdot {}^0H_F^1 + {}^0H_F^1 \cdot \Delta_F^R + {}^0H_F^1) \cdot \tilde{p}_1^F]_3 = 0 \tag{6.25}$$

Under nominal fixture locating condition, $[{}^0H_F^1 \cdot \tilde{p}_1^F]_3 = 0$, the Eq. (6.25) is rewritten as:

$$[\Delta_1^R \cdot {}^0H_F^1 \cdot \tilde{p}_1^F]_3 = [{}^0H_F^1 \cdot \Delta_F^R \cdot \tilde{p}_1^F]_3 \tag{6.26}$$

Note $\Delta_1^R = \begin{bmatrix} \hat{\theta}_2^R & d_2^R \\ 0 & 1 \end{bmatrix}$, ${}^0H_F^1 = \begin{bmatrix} {}^0n_F^1 & {}^0o_F^1 & {}^0a_F^1 & {}^0t_F^1 \\ 0 & 0 & 0 & 1 \end{bmatrix}$, then the two sides of the Eq. (6.25) is converted to:

$$[\Delta_1^R \cdot {}^0H_F^1 \cdot \tilde{p}_1^F]_3 = [\ [\theta_1^R \times {}^0n_F^1]_3\ [\theta_1^R \times {}^0o_F^1]_3\ [\theta_1^R \times {}^0a_F^1]_3\ [\theta_1^R \times {}^0t_F^1 + d_1^R]_3\ ] \cdot \tilde{p}_1^F$$

$$[{}^0H_F^1 \cdot \Delta_F^R \cdot \tilde{p}_1^F]_3 = \left[[{}^0a_1^F]^T\ [p_1^F \times {}^0a_1^F]^T \right] \cdot x_F^R$$

Thus,

$$\begin{aligned} &\left[[{}^0a_1^F]^T [p_1^F \times {}^0a_1^F]^T \right] \cdot x_F^R \\ &= [[\theta_1^R \times {}^0n_F^1]_3 [\theta_1^R \times {}^0o_F^1]_3 [\theta_1^R \times {}^0a_F^1]_3 [\theta_1^R \times {}^0t_F^1 + d_1^R]_3] \cdot \tilde{p}_1^F \end{aligned} \tag{6.27}$$

By applying the same procedure from Eqs. (6.25) to (6.27), the latter five equations in equation set (6.24) for the touching condition of $p_2 \sim p_6$ can be obtained,

$$\begin{bmatrix} [{}^0a_1^F]^T & [p_1^F \times {}^0a_1^F]^T \\ [{}^0a_1^F]^T & [p_2^F \times {}^0a_1^F]^T \\ [{}^0a_1^F]^T & [p_3^F \times {}^0a_1^F]^T \\ [{}^0a_2^F]^T & [p_4^F \times {}^0a_2^F]^T \\ [{}^0a_2^F]^T & [p_5^F \times {}^0a_2^F]^T \\ [{}^0a_3^F]^T & [p_6^F \times {}^0a_3^F]^T \end{bmatrix} x_F^R$$

$$= \begin{bmatrix} [\,[\theta_1^R \times {}^0n_F^1]_3\ [\theta_1^R \times {}^0o_F^1]_3\ [\theta_1^R \times {}^0a_F^1]_3\ [\theta_1^R \times {}^0t_F^1 + d_1^R]_3\,] \cdot \tilde{p}_1^F \\ [\,[\theta_1^R \times {}^0n_F^1]_3\ [\theta_1^R \times {}^0o_F^1]_3\ [\theta_1^R \times {}^0a_F^1]_3\ [\theta_1^R \times {}^0t_F^1 + d_1^R]_3\,] \cdot \tilde{p}_2^F \\ [\,[\theta_1^R \times {}^0n_F^1]_3\ [\theta_1^R \times {}^0o_F^1]_3\ [\theta_1^R \times {}^0a_F^1]_3\ [\theta_1^R \times {}^0t_F^1 + d_1^R]_3\,] \cdot \tilde{p}_3^F \\ [\,[\theta_2^R \times {}^0n_F^2]_3\ [\theta_2^R \times {}^0o_F^2]_3\ [\theta_2^R \times {}^0a_F^2]_3\ [\theta_2^R \times {}^0t_F^2 + d_2^R]_3\,] \cdot \tilde{p}_4^F \\ [\,[\theta_2^R \times {}^0n_F^2]_3\ [\theta_2^R \times {}^0o_F^2]_3\ [\theta_2^R \times {}^0a_F^2]_3\ [\theta_2^R \times {}^0t_F^2 + d_2^R]_3\,] \cdot \tilde{p}_5^F \\ [\,[\theta_3^R \times {}^0n_F^3]_3\ [\theta_3^R \times {}^0o_F^2]_3\ [\theta_3^R \times {}^0a_F^3]_3\ [\theta_3^R \times {}^0t_F^3 + d_3^R]_3\,] \cdot \tilde{p}_6^F \end{bmatrix} \tag{6.28}$$

The computation of Eq. (6.28) is simplified by the orthogonality of $O_1X_1Y_1Z_1$, $O_2X_2Y_2Z_2$ and $O_3X_3Y_3Z_3$. Given the nominal position of the three datum characteristics and the position of the six locators with respect to FCS as:

$$p_1^F = [r\ p_{1y}^F\ p_{1z}^F]^T, p_2^F = [r\ p_{2y}^F\ p_{2z}^F]^T, p_3^F = [r\ p_{3y}^F\ p_{3z}^F]^T, p_4^F = [p_{4x}^F\ p_{4y}^F\ r]^T$$
$$p_5^F = [p_{5x}^F\ p_{5y}^F\ r]^T, p_6^F = [p_{6x}^F\ 0\ p_{6z}^F]^T.$$

$${}^0H_F^1 = \begin{bmatrix} 0 & 0 & -1 & r \\ 0 & 1 & 0 & {}^0t_{Fy}^1 \\ 1 & 0 & 0 & 0 \\ 0 & 0 & 0 & 1 \end{bmatrix} {}^0H_F^2 = \begin{bmatrix} 1 & 0 & 0 & 0 \\ 0 & 1 & 0 & {}^0t_{Fy}^2 \\ 0 & 0 & 1 & r \\ 0 & 0 & 0 & 1 \end{bmatrix} {}^0H_F^3 = \begin{bmatrix} 1 & 0 & 0 & {}^0t_{Fx}^3 \\ 0 & 0 & 1 & {}^0t_{Fy}^3 \\ 0 & -1 & 0 & 0 \\ 0 & 0 & 0 & 1 \end{bmatrix}$$

where r is the radius of the rotary workpiece.

The equation set (6.28) is simplified to the Eq. (6.28).

$$\begin{bmatrix} 1 & 0 & 0 & 0 & p_{1z}^F & -p_{1y}^F \\ 1 & 0 & 0 & 0 & p_{2z}^F & -p_{2y}^F \\ 1 & 0 & 0 & 0 & p_{3z}^F & p_{3y}^F \\ 0 & 0 & 1 & p_{4y}^F & -p_{4x}^F & 0 \\ 0 & 0 & 1 & p_{5y}^F & -p_{5x}^F & 0 \\ 0 & -1 & 0 & p_{6z}^F & 0 & -p_{6x}^F \end{bmatrix} \cdot \begin{bmatrix} d_{Fx}^R \\ d_{Fy}^R \\ d_{Fz}^R \\ \theta_{Fx}^R \\ \theta_{Fy}^R \\ \theta_{Fz}^R \end{bmatrix}$$

$$= \begin{bmatrix} (p_{1y}^F + {}^0t_{Fy}^1) \cdot \theta_{1\alpha}^R + (p_{1z}^F - R_0) \cdot \theta_{1\beta}^R + d_{1z}^R \\ (p_{2y}^F + {}^0t_{Fy}^1) \cdot \theta_{1\alpha}^R + (p_{2z}^F - R_0) \cdot \theta_{1\beta}^R + d_{1z}^R \\ (p_{3y}^F + {}^0t_{Fy}^1) \cdot \theta_{1\alpha}^R + (p_{3z}^F - R_0) \cdot \theta_{1\beta}^R + d_{1z}^R \\ (p_{4y}^F + {}^0t_{Fy}^2) \cdot \theta_{2\alpha}^R - p_{4x}^F \cdot \theta_{2\beta}^R + d_{2z}^R \\ (p_{5y}^F + {}^0t_{Fy}^2) \cdot \theta_{2\alpha}^R - p_{5x}^F \cdot \theta_{2\beta}^R + d_{2z}^R \\ (p_{6z}^F + {}^0t_{Fy}^3) \cdot \theta_{3\alpha}^R - (p_{6x}^F + {}^0t_{Fx}^3) \cdot \theta_{3\beta}^R + d_{3z}^R \end{bmatrix} \tag{6.29}$$

There are six variables and six equations in equation set (6.29), thus the equation set (6.29) can be solved by arranging the result in the form of the Eq. (6.23). Then T_1, T_2 and T_3 is obtained:

$$T_1 = \begin{bmatrix} 0\,0\,1 & {}^0t^1_{Fy} & -r & 0 \\ 0\,0\,0 & -p^F_{6x} & \frac{-p^F_{6z}(p^F_{4x}-p^F_{5x})}{p^F_{4y}-p^F_{5y}} & 0 \\ 0\,0\,0 & 0 & \frac{-(p^F_{4x}\cdot p^F_{5y}-p^F_{5x}\cdot p^F_{4y})}{p^F_{4y}-p^F_{5y}} & 0 \\ 0\,0\,0 & 0 & \frac{p^F_{4x}-p^F_{5x}}{p^F_{4y}-p^F_{5y}} & 0 \\ 0\,0\,0 & 0 & 1 & 0 \\ 0\,0\,0 & -1 & 0 & 0 \end{bmatrix},$$

$$T_2 = \begin{bmatrix} 0\,0\,0 & 0 & 0 & 0 \\ 0\,0\,0 & -p^F_{6z} & \frac{-p^F_{6z}(p^F_{4x}-p^F_{5x})}{p^F_{4y}-p^F_{5y}} & 0 \\ 0\,0\,0 & {}^0t^2_{Fy} & \frac{p^F_{4x}\cdot p^F_{5y}-p^F_{5x}\cdot p^F_{4y}}{p^F_{4y}-p^F_{5y}} & 0 \\ 0\,0\,0 & 0 & \frac{-(p^F_{4x}-p^F_{5x})}{p^F_{4y}-p^F_{5y}} & 0 \\ 0\,0\,0 & 0 & 0 & 0 \\ 0\,0\,0 & 0 & 0 & 0 \end{bmatrix}, T_3 = \begin{bmatrix} 0\,0\,0 & 0 & 0 & 0 \\ 0\,0\,1 & -p^F_{6z} - {}^0t^3_{Fy} & p^F_{6x} + {}^0t^3_{Fx} & 0 \\ 0\,0\,0 & 0 & 0 & 0 \\ 0\,0\,0 & 0 & 0 & 0 \\ 0\,0\,0 & 0 & 0 & 0 \\ 0\,0\,0 & 0 & 0 & 0 \end{bmatrix}$$

Hence, the datum error occurred at stage k is:

$$X^{R(k)}_{FCS(k)} = A_2(k)X_1(k) \tag{6.30}$$

where $A_2(k) = [0 \ldots T_1(k) \ldots T_2(k) \ldots T_3(k) \ldots 0]_{6\times 6M}$

S3: Fixture Error Analysis

Fixture error represents the actual positional and orientational deviation of the FCS with respect to ${}^0FCS$. The presence of fixture error during the machining process can also lead to coaxiality error. As depicted in Fig. 6.10, if a fixture error is present, it results in the coaxiality error of the machined characteristic.

As shown in Fig. 6.9b, a workpiece is located by six locators, $L_1 \sim L_6$. The error of $L_1 \sim L_6$ is represented by $u_f(k) = [\Delta L_{1z}, \Delta L_{2z}, \Delta L_{3z}, \Delta L_{4z}, \Delta L_{5z}, \Delta L_{6z}]^T$. The fixture error is obtained according to the work of Cai, Hu, and Yuan [22]:

$$x^{FCS(k)}_{{}^0FCS(k)} = A_3(k)u_f(k) = -T_4(k)u_f(k) \tag{6.31}$$

where

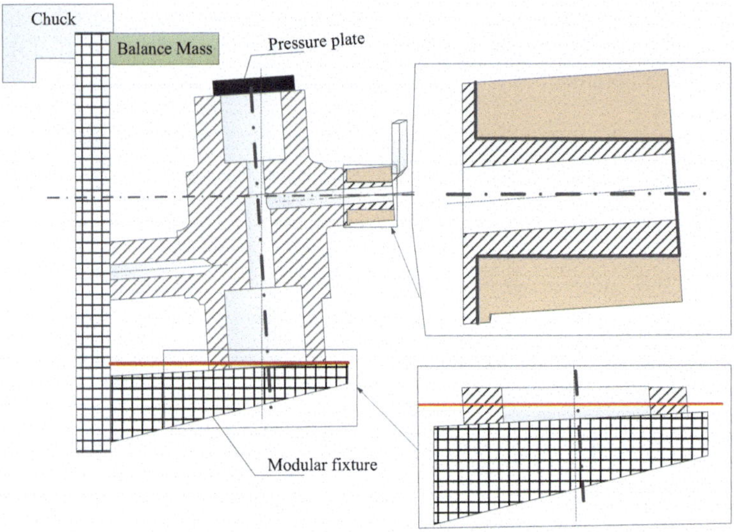

Fig. 6.10 Effects illustration of fixture error in rotary parts machining process

$$T_4 = \begin{bmatrix} 0 & 0 & 0 & 1 & 0 & 0 \\ \frac{L_{6x}}{L} & \frac{-L_{6x}}{L} & 0 & \frac{-L_{6z}}{L} & \frac{L_{6z}}{L} & 1 \\ 1 & 0 & 0 & 0 & 0 & 0 \\ \frac{1}{L} & \frac{-1}{L} & 0 & 0 & 0 & 0 \\ \frac{L_{3y}-L}{L \cdot L_{3z}} & \frac{-L_{3y}}{L \cdot L_{3z}} & \frac{1}{L_{3z}} & 0 & 0 & 0 \\ 0 & 0 & 0 & \frac{-1}{L} & \frac{1}{L} & 0 \end{bmatrix}$$

S4: Turning Tool Error Analysis

Given that the tool path is programmed in ${}^{0}FCS$, the tool error is considered the deviation of the actual cutting tool path from the programmed (nominal) tool path within ${}^{0}FCS$. The tool error for a characteristic machined at stage k is represented as such:

$$x_{i(k)}^{{}^{0}FCS} = [\Delta x, \Delta y, \Delta z, \Delta \psi, \Delta \theta, \Delta \varphi] \tag{6.32}$$

As shown in Fig. 6.11, the tool is programmed to follow a path $M_0 \rightarrow {}^{0}M_1$, but the actual tool path is $M_0 \rightarrow M_1$. The discrepancy results in an angular deviation around the X-axis, leading to a perpendicularity error.

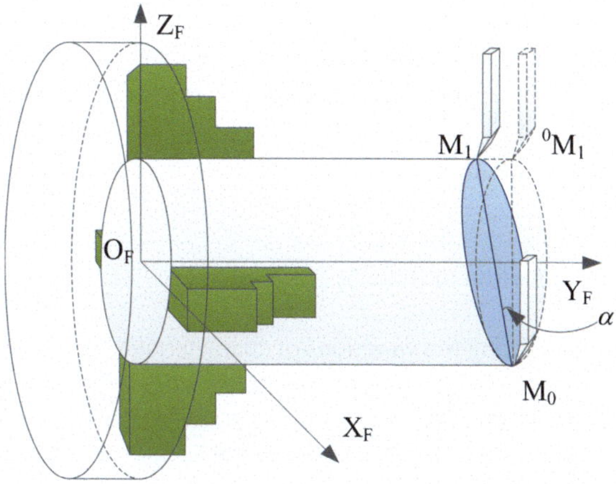

Fig. 6.11 Illustration of tool error

S5: Stage Error Analysis

If there are $l(k)$ newly generated characteristics at stage k, then according to the Eq. (6.18), the stage error of stage k is

$$u_s(k) = A_4(k)x_{{}^0FCS_k}^{FCS_k} + u_t(k) \tag{6.33}$$

where $A_4(k) = [G_1(k)^T \cdots G_i(k)^T \cdots G_l(k)^T]^T$, $G_i(k) = \begin{bmatrix} \left({}^0R_{i(k)}^{{}^0FCS(k)}\right)^T & -\left({}^0R_{i(k)}^{{}^0FCS(k)}\right)^T \cdot \left({}^0\hat{\mathrm{t}}_{i(k)}^{{}^0FCS(k)}\right) \\ 0 & \left({}^0R_{i(k)}^{{}^0FCS(k)}\right)^T \end{bmatrix}$, $u_t(k)$ is a stack of $x_{i(k)}^{{}^0FCS}$, corresponding to tool errors for all characteristics machined at stage k, and $u_s(k)$ is the error contributed by current stage, which is a coupling error of fixture error and tool error, both $u_t(k)$ and $u_s(k)$ have a dimension of $6l(k) \times 1$, and $A_4(k)$ has a dimension of $6l(k) \times 6$.

S6: Variation of Characteristics Generated at Stage K

The variation of the newly generated characteristic is represented by $X^{new}(k) = \left[\left(x_{1(k)}^{R(k)}\right)^T \cdots \left(x_{i(k)}^{R(k)}\right)^T \cdots \left(x_{l(k)}^{R(k)}\right)^T\right]^T$. According to the Eq. (6.43), it is calculated:

$$X^{new}(k) = A_5(k) \cdot x_{FCS(k)}^{R(k)} + u_s(k) \tag{6.34}$$

where $u_s(k)$ is the error induced by stage k, $x_{FCS(k)}^{R(k)}$ is datum error, $A_5(k) = [D_1(k)^T \cdots D_i(k)^T \cdots D_l(k)^T]^T$ is coefficient matrix, in which $D_i(k)^T = \begin{bmatrix} \left({}^0R_{i(k)}^{FCS_k}\right)^T & -\left({}^0R_{i(k)}^{FCS_k}\right)^T \cdot \left({}^0\hat{t}_{i(k)}^{FCS_k}\right) \\ 0 & \left({}^0R_{i(k)}^{FCS_k}\right)^T \end{bmatrix}$.

S7: Deduction of State Space Model

By integrating the variations of characteristics generated at stage k and the variations of other characteristics from upstream stages, the state vector is derived by combining the variation of characteristics generated before stage $k(\mathbf{x}_1(k)$ (as shown in Eq. (6.22)) with the variation generated at stage k.

$$x(k) = x_1(k) + A_6(k)x_{new}(k) \tag{6.35}$$

where $A_6(k)$ is a selective matrix with all the identity elements corresponding to the newly generated dimensional error at stage k (see Eq. (6.36)).

$$A_6(k) = \begin{bmatrix} 0 & \cdots & 0 & \cdots & 0 \\ \vdots & \vdots & \vdots & \vdots & \vdots \\ I_{6\times6} & \cdots & 0 & \cdots & 0 \\ \vdots & \vdots & \vdots & \vdots & \vdots \\ 0 & \cdots & I_{6\times6} & \cdots & 0 \\ \vdots & \vdots & \vdots & \vdots & \vdots \\ 0 & \cdots & 0 & \cdots & I_{6\times6} \\ \vdots & \vdots & \vdots & \vdots & \vdots \\ 0 & \cdots & 0 & \cdots & 0 \end{bmatrix} \tag{6.36}$$

By substituting Eqs. (6.30) to (6.34) into Eq. (6.35), the state space model for MTPs is formulated.

$$\begin{aligned} X(k) &= X_1(k) + A_6(k)X^{new}(k) \\ &= [A_1(k) + A_6(k)A_5(k)A_2(k)A_1(k)]X(k-1) \\ &\quad + A_6(k)\left[A_4(k)A_3(k)\ I_{6\times6}\right]\begin{bmatrix} u_f(k) \\ u_t(k) \end{bmatrix} \\ &= A(k)X(k-1) + B(k)u(k) \end{aligned} \tag{6.37}$$

So far, all of the coefficients of state space model (6.20) for MTP are obtained.

$$A(k) = [A_1(k) + A_6(k)A_5(k)A_2(k)A_1(k)]$$

$$B(k) = \left[A_6(k)A_4(k)A_3(k)\ A_6(k) \right] \tag{6.38}$$

The coefficient for the measurement can be inferred using a similar methodology. To avoid excessive length, the details are omitted in this section.

6.1.5 Case Study

6.1.5.1 Experiment Setup

The proposed state space model for MTPs is validated using a specific type of rotary workpiece, manufactured by a domestic machining company that primarily produces valve shells. Figure 6.12a shows a raw part of the valve shell; Figure 6.12b presents a 3-D model of the finished product. Figure 6.12c displays a cross-sectional view of the valve shell along axes A, B, C, and D after the MTP. Owing to confidentiality requirements, the tolerances in the part drawing have been removed. Figure 6.13 illustrates the actual tuning process of the valve shell.

The turning process is carried out using a Computer Numerical Control (CNC) machining center (CTX420). The entire process is executed under two distinct fixture schemes. One of these schemes is depicted in Fig. 6.14, while the other involves a

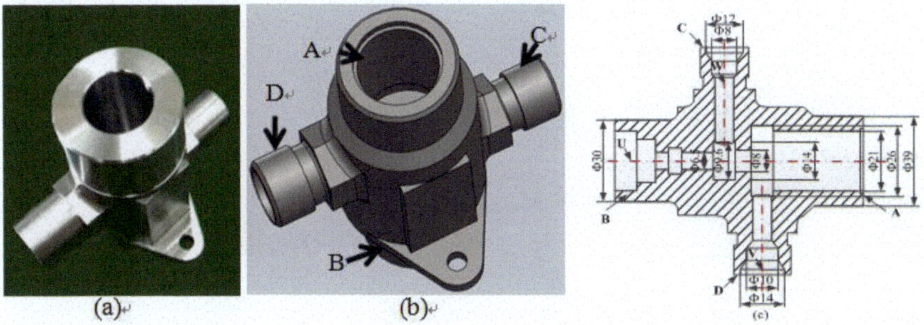

Fig. 6.12 Valve shell part model

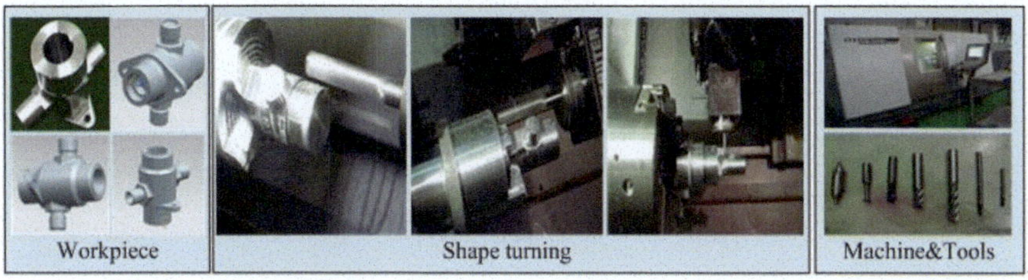

Fig. 6.13 Workpiece turning process

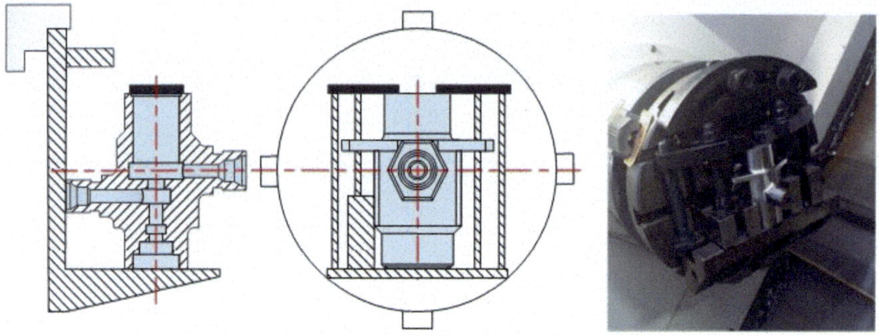

Fig. 6.14 Illustration of the first fixture scheme

chuck jaws setup as shown in Fig. 6.15. The turning process is primarily divided into five operations, designated as OP10 to OP50 (refer to Fig. 6.16). The tolerances for the workpiece are detailed in Table 6.1.

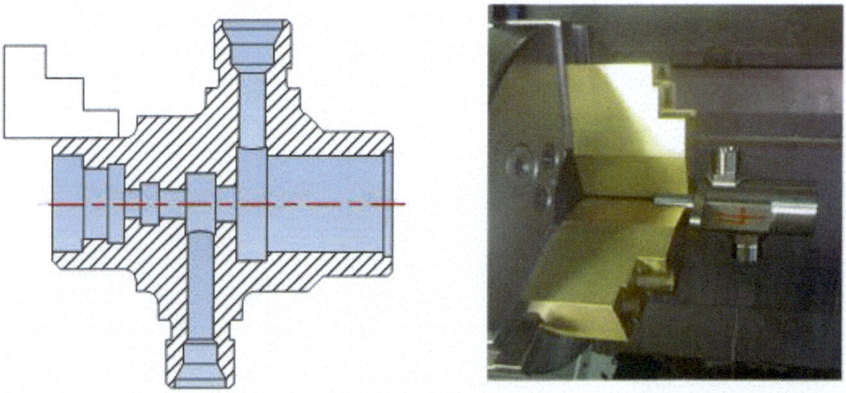

Fig. 6.15 Illustration of the second fixture scheme

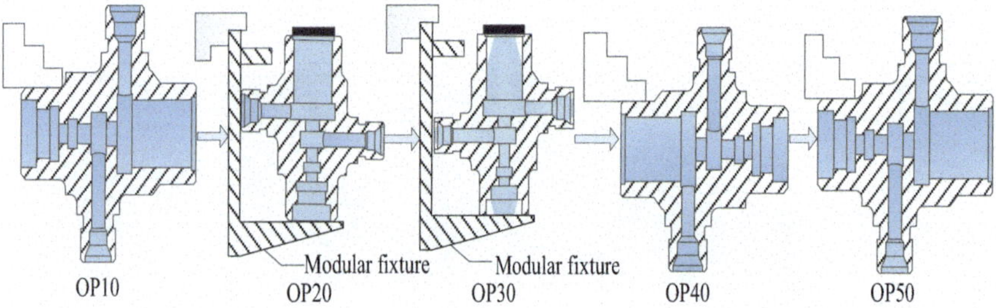

Fig. 6.16 Illustration of valve shell MTP

Table 6.1 Nominal positions and orientations of KPCs

No.	Fixture scheme	KPCs	Designed tolerances	Nominal position t_n^R / mm	Nominal angular ω_n^R/ rad
OP10	B, 4-jaws Chuck	Hole Φ21	Coaxiality0.1|U|	[0, 0, − 29]	[0, 0, 0]
		ExcircleΦ39	Not a KPC	[0, 0, 0]	[0, 0, 0]
OP20	B downward, modular fixture (Process end D)	Hole Φ14	Cylindricity0.02|V|	[− 42,0, − 32.5]	[π/2, 0, π/2]
		Hole Φ10	Cylindricity0.02|V|	[− 32,0, − 32.5]	[π/2, 0, π/2]
OP30	B downward, modular fixture (Process end C)	Hole Φ12	Cylindricity0.02|W|	[− 42,0, − 44.5]	[− π/2,0, − π/2]
		Hole Φ8	Cylindricity0.02|W|	[− 32,0, − 32.5]	[− π/2,0, − π/2]
OP40	A, 4-jaws Chuck	Slot Φ9.6	Coaxiality0.05|U|	[0, 0, − 66.5]	[0, π, 0]
		Hole Φ6.5	Coaxiality0.05|U|	[0, 0, − 62.5]	[0, π, 0]
		ExcircleΦ30	Not a KPC	[0, 0, − 79.5]	[0, π, 0]
OP50	B, 4-jaws Chuck	Slot Φ26	Depth 36 ± 0.1|A|	[0, 0, − 36]	[0, 0, 0]
		Slot Φ14	Depth48 ± 0.1|A|	[0, 0, − 48]	[0, 0, 0]
		Hole Φ8	Coaxialityjn0.05|U|	[0, 0, − 35.5]	[0, 0, 0]

End planes A and B are critical datum features whose deviations, when introduced as datum errors, impact the machining precision of other characteristics, including holes and slots. The circle machined at operation OP10 serves as the datum for operation OP40, hence any error from OP10 will accumulate in the variation of characteristics machined at OP40. Likewise, the variation of characteristics machined at OP40 is a significant source of variation for OP50. The turning process of such a valve shell exemplifies a typical MTP where variations are propagated, transmitted, and compounded through all stages.

6.1.5.2 Results and Discussion

For each stage of the process, the variation of the workpiece characteristics can be automatically computed once the fixture error and tool error data are input into the program. This three-dimensional variation propagation model for rotary workpieces is employed for the prediction of KPC variations. Table 6.2 presents the predicted values of KPC variations, assuming a fixture error at each stage of [0, 0.01, 0, 0, 0, 0].

The actual measurements for all the characteristics are detailed in Table 6.3. A comparison between Tables 6.2 and 6.3 reveals that the overall differences between the predicted values and the actual measurements are acceptably small. For instance,

Table 6.2 Predicted values of KPCs' variations

KPCs	Designed tolerances	*x*	*y*	*z*	*i*	*j*	*k*
Hole Φ21	Coaxiality0.1|U|	0	−0.0193	0	0.0007	0	0
ExcircleΦ39	Not a KPC	0	−0.0387	0	0	0	0
Hole Φ14	Cylindricity0.02|V|	0.0120	0.0280	0	0	−0.0002	−0.0008
Hole Φ10	Cylindricity0.02|V|	0.0120	0.0213	0	0	−0.0002	−0.0008
Hole Φ12	Cylindricity0.02|W|	−0.0040	0.0280	0	0	−0.0002	−0.0008
Hole Φ8	Cylindricity0.02|W|	−0.0040	0.0213	0	0	−0.0002	−0.0008
Slot Φ9.6	Coaxiality0.05|U|	0	−0.0027	0	0.0013	0	0
Hole Φ6.5	Coaxiality0.05|U|	0	0.0017	0	0.0007	0	0
ExcircleΦ30	Not a KPC	0	−0.0193	0	0.0013	0	0
Slot Φ26	Depth 36 ± 0.1|A|	0	0	0	0.0007	0	0
Slot Φ14	Depth48 ± 0.1|A|	0	0	0	0.0007	0	0
Hole Φ8	Coaxiality ± 0.05|U|	0	0	0	0.0007	0	0

regarding the characteristic Ø8, the coaxiality requirement is 0.02 mm with respect to axis W. The prediction error is 0.004 mm, indicating a minor deviation between the model's prediction and the actual values. Similar prediction errors are observed for other characteristics when comparing the predicted values with the actual measurements. One primary reason for these discrepancies may be attributed to the tool error, which was assumed to be zero in the prediction model during the experiment. However, in reality, the tool error during the turning process has nonzero values. These minor differences between the predicted values and the actual measurements confirm the validity of the proposed model.

Furthermore, the proposed model not only predicts the variation values but also illustrates the variation patterns. Consider holes Φ14 and Φ10 from operation OP20 as examples; both are machined under the same modular fixture scheme. They exhibit a distinctly similar variation pattern because they share the same fixture setup and tool cutting conditions. Any differences in their patterns can be attributed to their positional relationships relative to the datum, as well as other factors such as thermal variations and changes in cutting forces during the machining process.

Table 6.3 Measurement values of KPCs' variations

KPCs	Designed tolerances	x	y	z	i	j	k
Hole Φ21	Coaxiality0.1|U|	0	– 0.020	0	0.001	0	0
ExcircleΦ39	Not a KPC	0	– 0.040	0.001	0	0	0
Hole Φ14	Cylindricity0.02|V|	0.013	0.027	0	0	0	– 0.001
Hole Φ10	Cylindricity0.02|V|	0.013	0.023	0	0	0	– 0.001
Hole Φ12	Cylindricity0.02|W|	– 0.004	0.027	0	0	0	– 0.001
Hole Φ8	Cylindricity0.02|W|	– 0.004	0.022	0.001	0	0	– 0.001
Slot Φ9.6	Coaxiality0.05|U|	0	– 0.002	0	0.001	0	0
Hole Φ6.5	Coaxiality0.05|U|	0	0.002	0	0.001	0	0
ExcircleΦ30	Not a KPC	0	– 0.020	0.000	0.001	0	0
Slot Φ26	Depth 36 ± 0.1|A|	0	0	0.001	0.001	0	0
Slot Φ14	Depth48 ± 0.1|A|	0	0	0	0.001	0	0
Hole Φ8	Coaxiality ± 0.05|U|	0	0	0.001	0.001	0	0

6.1.6 Conclusions

Drawing upon the definitions of the four types of coordinate systems and the four types of errors, this section conducts an analysis of fixture error, datum error, and their interactive effects within the context of a jaws chuck fixturing scheme. A state space model for three-dimensional variation propagation in MTP is developed based on an analysis of the turning procedure and the coupling analysis, effectively capturing the transmission and accumulation of product characteristic variations across different stages. A case study involving the turning of a valve shell is used to validate the proposed model, and the results demonstrate that the model achieves satisfactory predictive accuracy. This proposed model can be extended to applications such as tolerance allocation, process control, fault diagnosis, and process improvement.

6.2 Three-Dimensional Tolerance Modeling and Analysis of Variation Propagation

6.2.1 Introduction

Achieving precise dimensional and geometric accuracy is a critical engineering goal in the machining of workpieces. However, imperfections in fixture locators and variations induced by machining can cause deviations of quality characteristics from their intended target values. Moreover, these variations are not solely dependent on the current machining process; they are also influenced by upstream errors when the corresponding features serve as locating datums in the current stage [18]. Therefore, studying the expression, propagation, and accumulation of variations in multi-stage machining processes is essential for controlling process capability, enhancing product quality, reducing manufacturing costs, and shortening the ramp-up time.

The Stream of Variation (SoV) theory stands as a pivotal approach in the analysis of variation propagation within manufacturing systems, encapsulating both product and process knowledge and elucidating the correlation between Key Control Characteristics (KCC) and Key Product Characteristics (KPC) [18]. Originating with Hu [23] in the context of automotive assembly, the SoV model for multi-stage assembly processes (MAPs) has garnered significant interest [24–26]. Concurrently, the application of SoV theory in multi-stage machining processes (MMPs) has become a focal point over the past two decades. Djurdjanovic and Ni [27] utilized Taylor series expansion to derive a linear state space model, while Huang and Shi [28] introduced an implicit nonlinear model to describe the propagation and accumulation of variations in MMPs. Zhou et al. [16] investigated the vectorial representation of variations, formulating detailed mathematical expressions for datum-induced and fixture-induced errors. The differential motion vector (DMV)-based model marked an innovative approach for variation analysis in MMPs. This model has since been extended to encompass machining-induced variations [29–32], fixture configurations [13, 33], machine tool variations [34, 35], expanded application scopes [20, 36], measurement variations [37–40], and the integration of Geometric Dimensioning and Tolerancing (GD&T) [14, 41]. Yang et al. [42] synthesized the three predominant variation propagation models in use. Comprehensive accounts of existing research and applications in variation propagation modeling are detailed in a dedicated monograph [14] and a review article [19]. Collectively, the evolution of SoV model research exhibits a pattern characterized by a shift from assembly to machining processes, a transition from vague system descriptions to explicit variation source representations, and an expansion of target products from simple box parts to more intricate components, such as rotating parts.

Despite significant advancements in the variation modeling of MMPs over the past two decades, certain limitations persist, constraining their industrial applicability. The universality and precision of Stream of Variation (SoV) models are in need of further investigation. These primary limitations are outlined below.

As demonstrated in Table 6.4, which uses automobile engine manufacturing as an example, workpieces of various shapes and their respective variation propagation research methods differ. Box-type and revolving-type parts involve distinct machining techniques, locating schemes, and predominant error sources [36]. The engine block, for instance, features six independent locating cast datum planes for the rough machining process, which cannot be managed by a simple 3-2-1 locating scheme [33]. Thus, amidst product diversification, there is a growing interest in variation modeling and quality control for workpieces of general shapes. Among existing SoV models, variations are typically expressed using DMVs and propagated via Homogeneous Transformation Matrices (HTMs), which is a relatively versatile approach. It has been feasible to model the variation propagation for both box-type and revolving-type parts individually. However, due to the differences highlighted in Table 6.4, the necessity to choose a specific model for each unique workpiece and each distinct machining stage remains a prerequisite for SoV model application, posing significant inconvenience in industrial settings. Therefore, developing a unified model for workpieces of diverse shapes to facilitate variation modeling is a worthwhile area of research.

The precision of SoV models is another challenge. Conventional variation propagation models predominantly focus on modeling fixture errors and datum-induced errors, often treating factors like cutting-tool wear as noise. Abellan-Nebot et al. [29] expanded the model to include machining-induced factors, but the complexity increases with the addition of numerous coordinate systems, leading to potential omissions of error terms in each extended system. Consequently, a model that succinctly encompasses all critical elements of the manufacturing system and clearly identifies the main error sources would be beneficial for both the accuracy and simplicity of studying variation propagation.

Tolerance analysis in three dimensions is adept at capturing subtle feature variations within specified tolerances and their sequential transfers within a dimensional chain, a technique frequently utilized in assembly processes [43]. Chen et al. [44] reviewed four predominant methods for this form of analysis: the three-dimensional direct linearization method (3D-DLM), tolerance-map (T-Map), matrix model, and the Jacobian-Torsor model. Of these, the Jacobian-Torsor model [45] stands out for its structured clarity, algorithmic brevity, and depth of engineering meaning. It merges the strengths of the small displacement torsor (SDT) model for representing variations with the Jacobian matrix's proficiency in propagating them. Consequently, employing three-dimensional tolerance analysis, notably the Jacobian-Torsor method, is a laudable strategy for crafting variation propagation models in the context of machining processes [46, 47]. However, the existing studies are sparse and confined either to single-stage scenarios [48] or to box-type workpieces processed using a rudimentary 3-2-1 locating scheme [48, 49]. There is a clear need for further research that extends the applicability of these models to multi-stage processes and workpieces of diverse shapes.

Table 6.4 Typical workpieces and related variation researches

	Box-type parts	Revolving-type parts
Typical workpieces	Engine head, Engine block	Crankshaft, Valve shell
Machining methods	Milling	Turning
Locating schemes	Simple 3-2-1 locating Plane-hole locating Six independent datums locating	Three-jaws chuck locating Four-jaws chuck locating
Dominant errors	1. Perpendicularity error 2. Flatness error 3. Locators wear 4. Cutting tool error	1. Cylindricity error 2. Coaxiality error 3. Jaws wear 4. Centering error 5. Cutting tool error
Typical researches	Zhou [16], Loose [13] and Yang [33]	Du [20, 36]

Addressing the aforementioned limitations holistically, a novel variation propagation model for multi-stage machining processes of workpieces with general shapes is developed, utilizing an enhanced three-dimensional tolerance analysis approach. This model establishes an "assembly chain" that encompasses all critical components of the manufacturing system, including the workpiece, fixture, cutting tool, and machine tool, thereby pinpointing the primary sources of error. The model achieves the equivalent transformation of the workpiece-fixture interface, enabling the modeling of workpieces of various shapes irrespective of their machining techniques and locating schemes. The Jacobian-Torsor model is employed to articulate and disseminate variations within the defined assembly chain. Leveraging these methodologies, the approach not only broadens the applicability of variation propagation models to workpieces of general shapes but also offers an innovative assembly-centric view of the variation stream and incorporates additional error sources, thereby enhancing the precision of predictive modeling.

6.2.2 *Workpiece Random Variation Representation*

With the assumption of a rigid workpiece, the variation propagation model in the manufacturing system is characterized by considering five key elements: the machine tool, cutting tool, fixture locators, workpiece datum features, and machining features.

6.2.2.1 Coordinate Systems Definition

For ease of reference, this section establishes definitions for five coordinate systems central to the derivation process (notations with a left superscript "0" indicate the nominal condition):

GCS (Global Coordinate System): Serves as the master reference for all coordinate systems.
RCS (Reference Coordinate System): Tied to a specific workpiece, it captures the workpiece's position and orientation. It is also known as PCS (Part Coordinate System).
FCS (Fixture Coordinate System): Outlines the actual configuration of the fixture, based on the true positions of the locators.
${}^{0}FCS$ (Nominal Fixture Coordinate System): Depicts the ideal setup of the fixture, determined by the intended positions of the locators.
LCS (Local Coordinate System): Associated with a particular feature on the workpiece, it signifies the position and orientation of the machining feature.

6.2.2.2 Error Source and Error Definition

In the conventional SoV model, the interactions between the coordinate systems detailed in Sect. 6.2.2.1 result in four distinct error types. Figure 6.17 illustrates the relationships between these systems and their associated errors: The datum error, originating from prior stages, is the discrepancy between FCS and RCS. The fixture error, a consequence of locator imperfections, is the deviation between FCS and the Nominal ^{0}FCS. Machining error is the divergence of the cutting tool from its intended path with respect to ^{0}FCS, which is equivalent to the deviation of LCS relative to ^{0}FCS. The overall feature error is the deviation of the LCS in relation to the RCS, encapsulating the total positional and orientational variations of a workpiece feature. The reference [16] provides detailed mathematical formulations for the various errors within the traditional SoV model.

Concurrently, in MMPs, the sources of error and their propagation can be distinctly portrayed through the assembly-like assembly of various elements within the manufacturing system. As depicted in Fig. 6.18, a typical vertical milling setup can be broken down into three distinct stages.

Setup Operation 1: Mounting the cutting tool onto the machine tool spindle: In addition to the errors resulting from the alignment of the spindle and the cutting tool holder, the wear of the cutting tool from previous operations also contributes to the cutting tool error.

Setup Operation 2: Mounting the fixture onto the worktable: The fixture itself may exhibit locator errors due to manufacturing precision, wear, deformation, and other factors accumulated from prior uses. Furthermore, the alignment of the fixture with the worktable introduces additional errors.

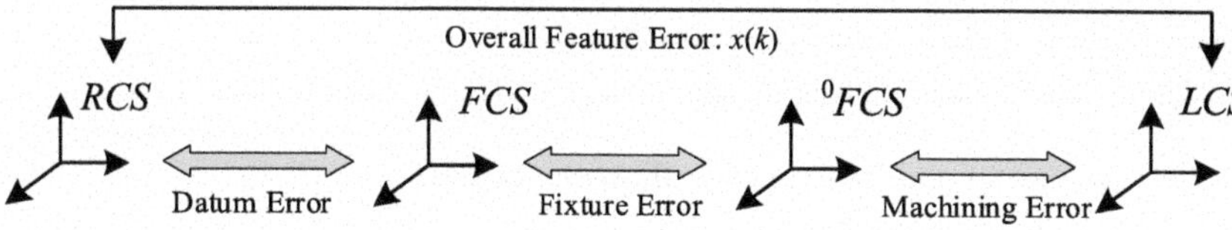

Fig. 6.17 Error source in traditional SoV model

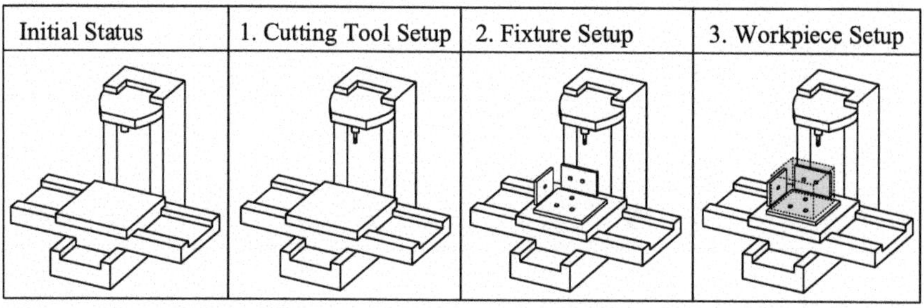

Fig. 6.18 Installation and error introduction process in assembly perspective

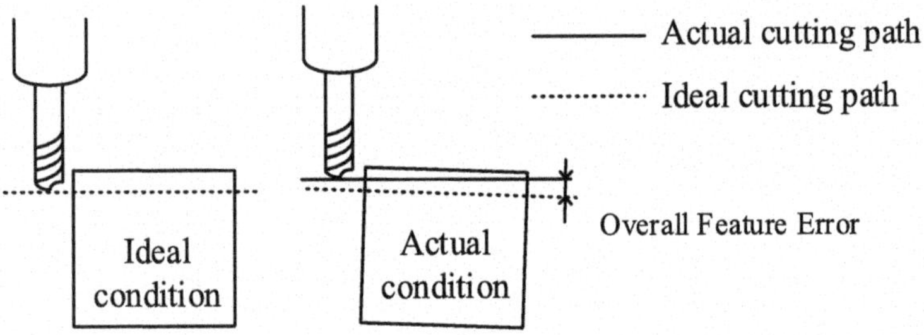

Fig. 6.19 The overall variation of machining process

Setup Operation 3: Mounting the workpiece onto the fixture: By utilizing specific features of the workpiece as datums for positioning, the errors associated with these datum features are transmitted to the machining feature, resulting in datum-induced error.

Ultimately, these errors accumulate and manifest on the machining feature. The overall variation is represented by the divergence of the actual cutting path of the tool from its ideal state relative to the RCS, as illustrated in Fig. 6.19.

Following the sequence of installation and the introduction of errors within the manufacturing system, the error contributions in the traditional SoV model are restructured to align with the assembly pattern. The total variation of the machining feature in MMPs encompasses four components: machine tool-worktable error, cutting tool-spindle error, fixture-induced error, and datum-induced error. This reorganization, in contrast to the traditional SoV model that primarily accounts for the last two types of errors, allows for a more comprehensive consideration of factors through a concise representation. Likewise, the turning process for revolving parts can be reconceptualized using the same approach.

6.2.2.3 Variation and Propagation Representation

The SDT model characterizes the dimensional and geometric deviations between an ideal feature N_{i} and its actual counterpart A_{i}, denoted as **T** [50]. This model comprises a rotation deviation vector $\boldsymbol{\rho}_{A,N} = \left[\alpha, \beta, \gamma\right]^T$, which signifies the rotation about the x, y, and z axes, and a translation deviation vector $\boldsymbol{\varepsilon}_{A,N} = [u, v, w]^T$, which signifies the displacement along the x, y, and z axes relative to the coordinate origin. The SDT for a general feature is articulated as follows:

$$\mathbf{T} = \left[\, \boldsymbol{\rho}_{A,N} \ \boldsymbol{\varepsilon}_{A,N} \,\right] = \begin{bmatrix} \alpha & u \\ \beta & v \\ \gamma & w \end{bmatrix} \tag{6.39}$$

The SDT model has been established in the literature to define the variations for various feature types [51]. Two of the most frequently utilized are illustrated

Table 6.5 SDTs according to the type of surface features

Plane	Cylinder
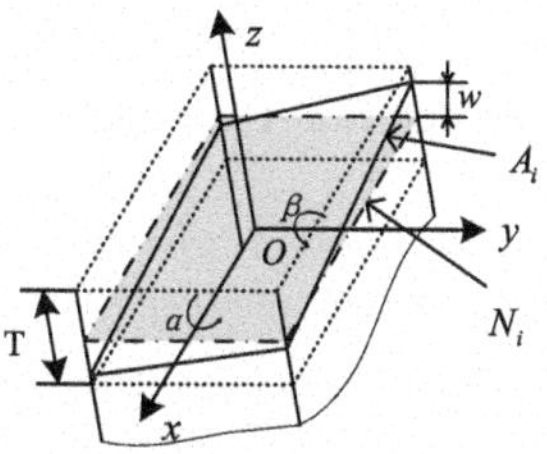	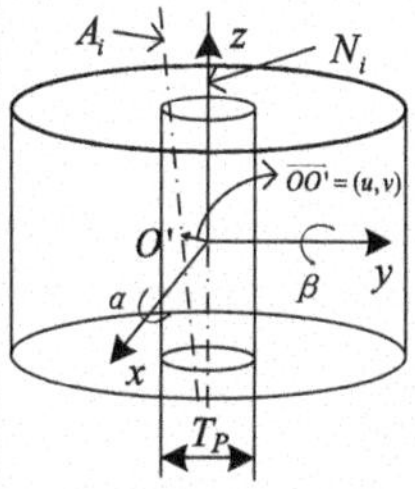
$\mathbf{T}_{\text{plane}} = \begin{Bmatrix} \alpha & U \\ \beta & U \\ U & w \end{Bmatrix}$	$\mathbf{T}_{\text{cylinder}} = \begin{Bmatrix} \alpha & u \\ \beta & v \\ U & U \end{Bmatrix}$

in Table 6.5. For example, if the normal vector of a planar feature is aligned with the z direction, the plane exhibits rotational variations around the x and y axes and translational variations along the z axis. Other variations—such as rotations around the z axis and translations in the x and y directions—do not alter the surface's characteristics; hence, these deviations are considered negligible (denoted as U), and they can be assumed to be zero to streamline the calculation process.

The SDT model offers an intuitive representation of variations, but it is less convenient for the propagation of these variations. Consequently, the Jacobian matrix is introduced to facilitate the propagation of variations. The Jacobian matrix for the ith feature functional elements (FEs) can be articulated as follows:

$$[\mathbf{J}]_{FE_i} = \begin{bmatrix} \left[\mathbf{R}_0^i\right]_{3\times 3} & \left[\mathbf{W}_i^n\right]_{3\times 3} \cdot \left[\mathbf{R}_0^i\right]_{3\times 3} \\ [\mathbf{0}]_{3\times 3} & \left[\mathbf{R}_0^i\right]_{3\times 3} \end{bmatrix}_{6\times 6} \tag{6.40}$$

where $\mathbf{R}_0^i$ represents the rotation matrix of the ith feature w.r.t. GCS; $\mathbf{W}_i^n$ is a skew-symmetric matrix associated with a vector $\mathbf{d}_i^n$, defined as Eq. 7.3.

$$\left[\mathbf{W}_i^n\right]_{3\times 3} = \begin{bmatrix} 0 & dz_i^n & -dy_i^n \\ -dz_i^n & 0 & dx_i^n \\ dy_i^n & -dx_i^n & 0 \end{bmatrix} \tag{6.41}$$

where $\mathbf{d}_i^n = \left[dx_i^n, dy_i^n, dz_i^n\right]^T$ is the translation vector defining the position change of machining surface (nth feature) relative to ith feature in the direction of three coordinate axes, which can be obtained by $dx_i^n = dx_n - dx_i$, $dy_i^n = dy_n - dy_i$, $dz_i^n = dz_n - dz_i$.

6.2.3 The New Variation Propagation Model in Machining Processes

6.2.3.1 The Construction of Variation Propagation Chain

In three-dimensional tolerance analysis, there are two categories of tolerances: functional requirement (FR) and functional element (FE). The FR represents the design's target value and is the goal of the tolerance analysis; it is typically a closed-loop system. FE refers to the tolerance of a feature involved in the tolerance transmission process and can be further divided into internal functional element (IFE) and contact functional element (CEF).

When applying the three-dimensional tolerance analysis method to variation propagation in MMPs, it is essential to first identify FR and FEs based on the error sources and their propagation paths. This method primarily considers tolerances and their coordination in the same direction as the FR. The sequential connection relationships of the manufacturing system components along the z-direction are depicted in Fig. 6.20a. The deviation of the actual cutting path from the ideal path with respect to RCS is considered the FR, while the remaining error factors that constitute the total variation are classified as FEs. The constructed error propagation chain is IFE_1-CFE_1-IFE_2-CFE_2-IFE_3-CFE_3-IFE_4-FR-IFE_5-CFE_4-IFE_6.

In engineering practice, the errors involved in the connection between the fixture and the workpiece extend beyond the z-axis direction. For instance, in a simple 3-2-1 locating scheme, the normal vectors of the secondary and tertiary datum planes are aligned with the x-axis and y-axis, respectively. Consequently, the directions of fixture errors in the locators do not align with the direction of the FR [16]. These two primary error factors significantly impact the FR, necessitating an expansion

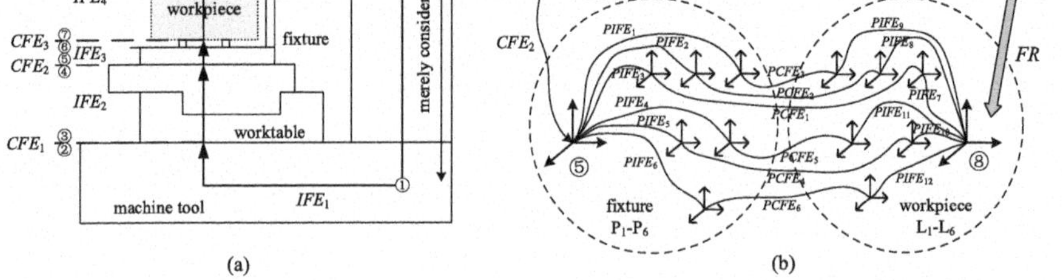

Fig. 6.20 Assembly connection and variation propagation chain

of the connection relationship between the fixture and the workpiece as shown in Fig. 6.20a.

For workpieces of general shapes, constraining all six degrees of freedom is essential for proper machining. By incorporating connections oriented towards the x-axis and y-axis, the variation propagation chain that was initially direct from IFE_3-CFE_3-IFE_4 is expanded into a partial parallel connection, as depicted in Fig. 6.20b. Here, $P_1 - P_6$ denotes the six fixture locators, and $L_1 - L_6$ denotes the six contacting datums of the workpiece. The Parallel Functional Element (PFE) is a type of FE that forms parallel connections in assemblies, participating in variation propagation and significantly affecting the FR. The PFE facilitates an equivalent connection between the workpiece and the fixture, akin to the connection between two sets of points capable of restricting the six degrees of freedom of the workpiece. This approach enables the modeling of general shape workpieces, irrespective of their specific machining methods and locating schemes.

To address this model, this paper transforms the mixed structure, which comprises both partial parallel and global series connections, into a purely series structure. As illustrated in Fig. 6.21, the connection between the two dotted coordinate systems encompasses both fixture error and datum-induced error across all directions. A new concept, the Virtual Contact Functional Element (VCFE), is introduced to represent the CFE between virtual coordinate systems. Section 6.2.3.2 will detail the conversion method and outline the calculation of the torsor for the VCFE.

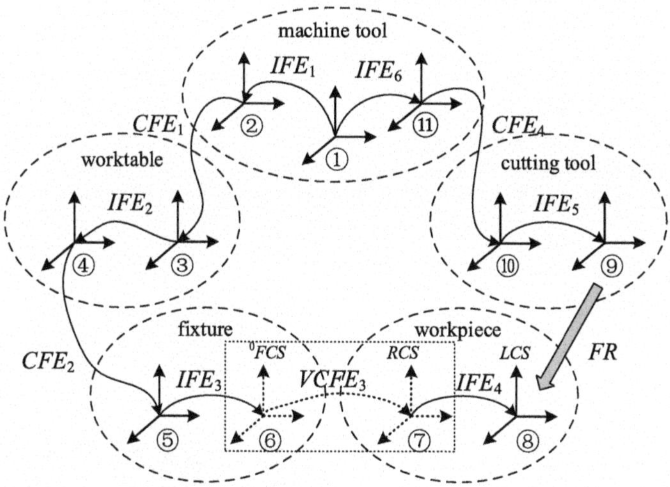

Fig. 6.21 The conversion of variation propagation chain

6.2.3.2 The Solution of VCFE Torsor

For simplicity and to maintain general applicability, Fig. 6.22 displays a segment of a general-shaped workpiece in contact with a single fixture locator. The determination of the VCFE torsor is outlined in four sequential steps.

Constraint Construction

In proper operating condition, the contact point on the workpiece and the corresponding locator of fixture should be guaranteed touching and these two points should be in the same tangent plane. Therefore, the constraint equation for *i*th locator is:

$$\mathbf{n}_i'^T \cdot \boldsymbol{\varepsilon}_i' = \mathbf{n}_i'^T \cdot \mathbf{t}_i' \tag{6.42}$$

where $\mathbf{n}_i'^T$ represents the outgoing normal vector of *i*th contact point of the workpiece, $\mathbf{t}_i'$ is the position of *i*th contact point in *RCS*. $\boldsymbol{\varepsilon}_i'$ is the position of *i*th fixture locator in *RCS*, but the position can only be delivered in ^0FCS directly. Denoting $\mathbf{t}_i$ as the position of *i*th fixture locator in ^0FCS, $\mathbf{R}_{RCS}$ and $\mathbf{t}_{RCS}$ as the rotational matrix and translational vector from *RCS* to ^0FCS, $\boldsymbol{\varepsilon}_i'$ can be written as $\boldsymbol{\varepsilon}_i' = \mathbf{R}_{RCS}^T \cdot (\mathbf{t}_i - \mathbf{t}_{RCS})$.

Rewriting Eq. 7.4, the *i*th contact point meets the following condition:

$$\mathbf{f}_i = \mathbf{n}_i'^T \cdot \boldsymbol{\varepsilon}_i' - \mathbf{n}_i'^T \cdot \mathbf{t}_i' = \mathbf{n}_i'^T \cdot \mathbf{R}_{RCS}^T \cdot (\mathbf{t}_i - \mathbf{t}_{RCS}) - \mathbf{n}_i'^T \cdot \mathbf{t}_i' = \mathbf{0} \tag{6.43}$$

To ensure that all six degrees of freedom of the workpiece are adequately constrained, *i* can range from 1 to 6. By considering all contact points collectively, the constraint equation for the entire fixture-workpiece connection system can be formulated as follows:

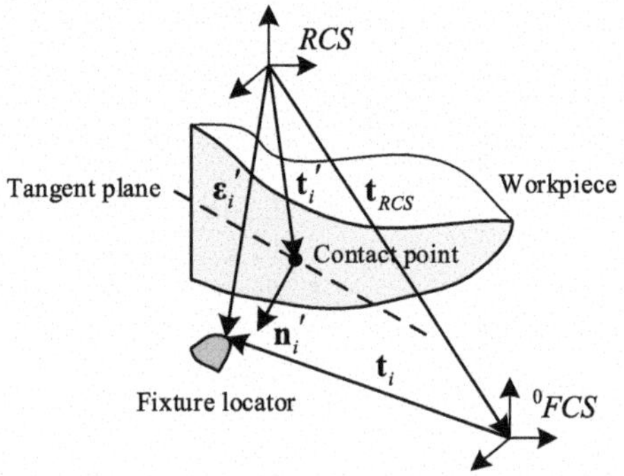

Fig. 6.22 Contact condition of a general shape workpiece with one fixture locator

$$\Phi = \begin{bmatrix} \mathbf{f}_1 \\ \mathbf{f}_2 \\ \mathbf{f}_3 \\ \mathbf{f}_4 \\ \mathbf{f}_5 \\ \mathbf{f}_6 \end{bmatrix} = \begin{bmatrix} \mathbf{n}_1'^T \cdot \mathbf{R}_{RCS}^T \cdot (\mathbf{t}_1 - \mathbf{t}_{RCS}) - \mathbf{n}_1'^T \cdot \mathbf{t}_1' \\ \mathbf{n}_2'^T \cdot \mathbf{R}_{RCS}^T \cdot (\mathbf{t}_2 - \mathbf{t}_{RCS}) - \mathbf{n}_2'^T \cdot \mathbf{t}_2' \\ \mathbf{n}_3'^T \cdot \mathbf{R}_{RCS}^T \cdot (\mathbf{t}_3 - \mathbf{t}_{RCS}) - \mathbf{n}_3'^T \cdot \mathbf{t}_3' \\ \mathbf{n}_4'^T \cdot \mathbf{R}_{RCS}^T \cdot (\mathbf{t}_4 - \mathbf{t}_{RCS}) - \mathbf{n}_4'^T \cdot \mathbf{t}_4' \\ \mathbf{n}_5'^T \cdot \mathbf{R}_{RCS}^T \cdot (\mathbf{t}_5 - \mathbf{t}_{RCS}) - \mathbf{n}_5'^T \cdot \mathbf{t}_5' \\ \mathbf{n}_6'^T \cdot \mathbf{R}_{RCS}^T \cdot (\mathbf{t}_6 - \mathbf{t}_{RCS}) - \mathbf{n}_6'^T \cdot \mathbf{t}_6' \end{bmatrix} = \begin{bmatrix} \mathbf{0} \\ \mathbf{0} \\ \mathbf{0} \\ \mathbf{0} \\ \mathbf{0} \\ \mathbf{0} \end{bmatrix} = \mathbf{0} \tag{6.44}$$

Defining $\mathbf{N}' = \begin{bmatrix} \mathbf{n}_1'^T & & \mathbf{0} \\ & \ddots & \\ \mathbf{0} & & \mathbf{n}_6'^T \end{bmatrix}$, $\mathbf{P}' = \left[\mathbf{t}_1'^T\,\mathbf{t}_2'^T \ldots \mathbf{t}_6'^T\right]^T$ as the orientation and position collection of contact points under *RCS* which contains the information of datum features, $\mathbf{L} = \left[\mathbf{t}_1^T\,\mathbf{t}_2^T \cdots \mathbf{t}_6^T\right]^T$ as the position collection of fixture locators in ${}^0FCS$ which contains the information of fixture. $\mathbf{x} = \left[\mathbf{t}_{RCS}^T\,\boldsymbol{\theta}_{RCS}^T\right]^T$ contains the position and orientation information of the transformation from *RCS* to ${}^0FCS$, $\boldsymbol{\theta}_{RCS}$ is three Euler angles of $\mathbf{R}_{RCS}$. Equation 7.6 can be expressed as:

$$\boldsymbol{\Phi}(\mathbf{N}', \mathbf{P}', \mathbf{L}, \mathbf{x}) = \mathbf{0} \tag{6.45}$$

Error Factors Extraction

Taking into account the deviation in the process, the deviation in datum ($\delta\mathbf{N}'$ and $\delta\mathbf{P}'$) and the error in fixture locators ($\delta\mathbf{L}$) cause the deviation of $\mathbf{x}$ jointly. The deviation of $\mathbf{x}$ ($\delta\mathbf{x}$) is the torsor of *VCFE* (denoted as $\mathbf{T}_{VCFE}$). Taking differential operation to Eq. 6.46.

$$\frac{\partial\boldsymbol{\Phi}}{\partial\mathbf{x}}\delta\mathbf{x} = -\frac{\partial\boldsymbol{\Phi}}{\partial\mathbf{N}'}\delta\mathbf{N}' - \frac{\partial\boldsymbol{\Phi}}{\partial\mathbf{P}'}\delta\mathbf{P}' - \frac{\partial\boldsymbol{\Phi}}{\partial\mathbf{L}}\delta\mathbf{L} \tag{6.46}$$

$\frac{\partial\boldsymbol{\Phi}}{\partial\mathbf{N}'}$ is always equal to zero because the outgoing normal vector is perpendicular to the tangent plane. $-\frac{\partial\boldsymbol{\Phi}}{\partial\mathbf{P}'}$ is equal to $\mathbf{N}'$ according to geometry. $\delta\mathbf{L}$ represents the fixture error and it is commonly expressed by $\mathbf{U} = \left[\mathbf{u}_1^T\,\mathbf{u}_2^T \cdots \mathbf{u}_6^T\right]^T$.

To establish the state space equation, the Eq. 6.46 can be rewritten as Eq. 6.47 after extracting the collection of datum error torsors $\mathbf{T}_C$:

$$\mathbf{C} \cdot \mathbf{T}_{VCFE} = \mathbf{A} \cdot \mathbf{T}_C + \mathbf{B} \cdot \mathbf{U} \tag{6.47}$$

where **A**, **B** and **C** are coefficient matrices, $\mathbf{A} \cdot \mathbf{T}_C = -\frac{\partial\boldsymbol{\Phi}}{\partial\mathbf{P}'}\delta\mathbf{P}'$, $\mathbf{B} \cdot \mathbf{U} = -\frac{\partial\boldsymbol{\Phi}}{\partial\mathbf{L}}\delta\mathbf{L}$, $\mathbf{C} = \frac{\partial\boldsymbol{\Phi}}{\partial\mathbf{x}}$, collective torsors $\mathbf{T}_C = \left[\mathbf{T}'_1{}^T\,\mathbf{T}'_2{}^T \cdots \mathbf{T}'_6{}^T\right]^T$, $\mathbf{T}'_i$ is the datum error torsor of *i*th contact point.

The Expression of Coefficient Matrices

(a) Calculation of coefficient matrix **A**

$\delta \mathbf{t}'_i = \mathbf{t}'_i - {}^0\mathbf{t}'_i$ contains the information about the position error of the ith contact point under *RCS*. $\delta \mathbf{P}' = \left[\delta \mathbf{t}_1'^T \ \delta \mathbf{t}_2'^T \ldots \delta \mathbf{t}_6'^T\right]^T$ corresponds to the collective vector of $\delta \mathbf{t}'_i$. To obtain coefficient matrix **A**, the linear equation $\mathbf{N}' \cdot \delta \mathbf{P}' = \mathrm{A} \cdot \mathbf{T}_C$ from (6.46) and (6.47) can be used, which establishes the relation between $\delta \mathbf{t}'_i$ and $\mathbf{T}'_i$.

Defining $\mathbf{t}''_i$ as the position of ith contact point under *LCS*, the value of $\mathbf{t}'_i$ can be determined using homogeneous transformation matrices from *RCS* to *LCS* as:

$$\begin{bmatrix} \mathbf{t}'_i \\ 1 \end{bmatrix} = ({}^0\mathbf{H}'_i) \cdot \mathbf{H}(\mathbf{T}'_i) \cdot \begin{bmatrix} \mathbf{t}''_i \\ 1 \end{bmatrix} \tag{6.48}$$

$$\begin{bmatrix} {}^0\mathbf{t}'_i \\ 1 \end{bmatrix} = ({}^0\mathbf{H}'_i) \cdot \begin{bmatrix} {}^0\mathbf{t}''_i \\ 1 \end{bmatrix} \tag{6.49}$$

Under the small deviation assumption, the contact position in the LCS can be considered constant (i.e., $\mathbf{t}''_i = {}^0\mathbf{t}''_i$). By rewriting Eq. (6.29) and substituting it into Eq. (6.48):

$$\begin{bmatrix} \mathbf{t}'_i \\ 1 \end{bmatrix} = ({}^0\mathbf{H}'_i) \cdot \mathbf{H}(\mathbf{T}'_i) \cdot ({}^0\mathbf{H}'_i)^{-1} \cdot \begin{bmatrix} {}^0\mathbf{t}'_i \\ 1 \end{bmatrix} \tag{6.50}$$

The relation between $\delta \mathbf{t}'_i$ and $\mathbf{T}'_i$ can be obtained as Eq. 6.51. Therefore, according to the linear equation $\mathbf{N}' \cdot \delta \mathbf{P}' = \mathbf{A} \cdot \mathbf{T}_C$ and Eq. 6.52, general expression of coefficient matrix **A** can be solved as follows.

$$\begin{bmatrix} \delta \mathbf{t}'_i \\ 1 \end{bmatrix} = ({}^0\mathbf{H}'_i) \cdot \left[\mathbf{H}(\mathbf{T}'_i) - \mathbf{I}_{4\times 4}\right] \cdot ({}^0\mathbf{H}'_i)^{-1} \cdot \begin{bmatrix} {}^0\mathbf{t}'_i \\ 1 \end{bmatrix} \tag{6.51}$$

Defining the following matrix operators to extract specific parts from the matrix: $\boldsymbol{\sigma}_1 = \begin{bmatrix} 1 & 0 & 0 & 0 \\ 0 & 1 & 0 & 0 \\ 0 & 0 & 1 & 0 \end{bmatrix}$, $\boldsymbol{\sigma}_2 = \begin{bmatrix} 0 & 0 & 0 & 0 \\ 0 & 0 & -1 & 0 \\ 0 & 1 & 0 & 0 \\ 0 & 0 & 0 & 0 \end{bmatrix}$, $\boldsymbol{\sigma}_3 = \begin{bmatrix} 0 & 0 & 1 & 0 \\ 0 & 0 & 0 & 0 \\ -1 & 0 & 0 & 0 \\ 0 & 0 & 0 & 0 \end{bmatrix}$, $\boldsymbol{\sigma}_4 = \begin{bmatrix} 0 & -1 & 0 & 0 \\ 1 & 0 & 0 & 0 \\ 0 & 0 & 0 & 0 \\ 0 & 0 & 0 & 0 \end{bmatrix}$, operators $\boldsymbol{\lambda}_i^m$ is adopted to obtain **A**. Based on Eq. 6.51, $\boldsymbol{\lambda}_i^m$ is expressed as follows:

$$\begin{cases} \boldsymbol{\lambda}_i^m = \mathbf{n}_i'^T \cdot \boldsymbol{\sigma}_m \cdot {}^0\mathbf{H}_i' & \text{(when } m = 1\text{)} \\ \boldsymbol{\lambda}_i^m = \mathbf{n}_i'^T \cdot \begin{bmatrix} {}^0\mathbf{a}_i' \cdot \boldsymbol{\sigma}_m \cdot ({}^0\mathbf{H}_i')^{-1} \\ {}^0\mathbf{b}_i' \cdot \boldsymbol{\sigma}_m \cdot ({}^0\mathbf{H}_i')^{-1} \\ {}^0\mathbf{c}_i' \cdot \boldsymbol{\sigma}_m \cdot ({}^0\mathbf{H}_i')^{-1} \end{bmatrix} \cdot \begin{bmatrix} {}^0\mathbf{t}_i' \\ 1 \end{bmatrix} & \text{(when } m = 2, 3, 4\text{)} \end{cases} \tag{6.52}$$

where ${}^0\mathbf{a}_i'$, ${}^0\mathbf{b}_i'$, ${}^0\mathbf{c}_i'$ are the components of ${}^0\mathbf{H}_i'$, and are organised as ${}^0\mathbf{H}_i' = \left[{}^0\mathbf{a}_i'^T\ {}^0\mathbf{b}_i'^T\ {}^0\mathbf{c}_i'^T\ {}^0\mathbf{d}_i'^T\right]^T$.

Thus, the coefficient matrix **A** can be depicted as the subsequent block matrix:

$$\mathbf{A}\ (i,\ 6(i-1)+1:6i) = \left[\left[\lambda_i^1\right]_{(1)}\ \left[\lambda_i^1\right]_{(2)}\ \left[\lambda_i^1\right]_{(3)}\ \lambda_i^2\ \lambda_i^3\ \lambda_i^4\right] (i = 1, 2, \ldots, 6)$$

$$\mathbf{A} = \begin{bmatrix} \left[\lambda_1^1\right]_{(1)}\ \left[\lambda_1^1\right]_{(2)}\ \left[\lambda_1^1\right]_{(3)}\ \lambda_1^2\ \lambda_1^3\ \lambda_1^4 & & \mathbf{0} \\ & \ddots & \\ \mathbf{0} & & \left[\lambda_6^1\right]_{(1)}\ \left[\lambda_6^1\right]_{(2)}\ \left[\lambda_6^1\right]_{(3)}\ \lambda_6^2\ \lambda_6^3\ \lambda_6^4 \end{bmatrix}$$

(b) Calculation of coefficient matrix **B**

From Eq. 6.47, $\mathbf{B} = -\frac{\partial \Phi}{\partial \mathbf{L}}$. Coefficient matrix **B** can be obtained based on the geometry directly as follows:

$$\mathbf{B} = -\frac{\partial \Phi}{\partial \mathbf{L}} = -\mathbf{N}' \cdot \begin{bmatrix} \mathbf{R}_{RCS}^T & & \\ & \ddots & \\ & & \mathbf{R}_{RCS}^T \end{bmatrix}$$

(c) Calculation of coefficient matrix **C**

$\mathbf{x} = \left[\mathbf{t}_{RCS}^T\ \boldsymbol{\theta}_{RCS}^T\right]^T = \left[u\ v\ w\ \alpha\ \beta\ \gamma\right]^T$ contains the position and orientation information of the transformation from *RCS* to ${}^0FCS$. From Eq. 6.47, $\mathbf{C} = \frac{\partial \Phi}{\partial \mathbf{x}}$ is the Jacobian of $\Phi(\mathbf{N}', \mathbf{P}', \mathbf{L}, \mathbf{x}) = \mathbf{0}$ about **x**, and it can be expressed as $\mathbf{C} = [\mathbf{C}_1^T\ \mathbf{C}_1^T \ldots \mathbf{C}_6^T]^T$, where $\mathbf{C}_i$ is a 1 by 6 vector as follows:

$$\mathbf{C}_i = \left[-\mathbf{n}_i'^T \cdot \mathbf{R}_{RCS}^T \cdot \frac{\partial \mathbf{t}_{RCS}}{\partial u}\ \ -\mathbf{n}_i'^T \cdot \mathbf{R}_{RCS}^T \cdot \frac{\partial \mathbf{t}_{RCS}}{\partial v}\ \ -\mathbf{n}_i'^T \cdot \mathbf{R}_{RCS}^T \cdot \frac{\partial \mathbf{t}_{RCS}}{\partial w}\right.$$
$$\left.\mathbf{n}_i'^T \cdot \frac{\partial \mathbf{R}_{RCS}^T}{\partial \alpha} \cdot \mathbf{t}_i'\ \mathbf{n}_i'^T \cdot \frac{\partial \mathbf{R}_{RCS}^T}{\partial \beta} \cdot \mathbf{t}_i' \mathbf{n}_i'^T \cdot \frac{\partial \mathbf{R}_{RCS}^T}{\partial \gamma} \cdot \mathbf{t}_i'\right] = [-\mathbf{n}_i'^T\ \ \mathbf{n}_i' \times \mathbf{t}_i']$$

The Acquisition of Parameters

From the first three steps, the torsor of *VCFE* can be calculated by the following equation:

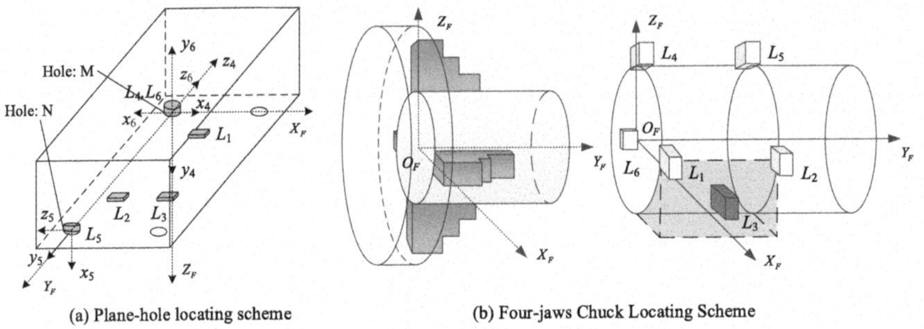

Fig. 6.23 Equivalent conversion of different locating scheme

$$\mathbf{T}_{VCFE} = \mathbf{C}^{-1} \cdot (\mathbf{A} \cdot \mathbf{T}_C + \mathbf{B} \cdot \mathbf{U}) \tag{6.53}$$

Beyond the fixture error **U**, determining the coordinates and orientation of the workpiece's contact points within *RCS* ($\mathbf{N}'$ and $\mathbf{P}'$) and the coordinates of the fixture locators within ${}^0FCS$ (**L**) are crucial for the computation of $\mathbf{T}_{VCFE}$. Extracting these vector components from the actual machining process is therefore essential.

As detailed in Sect. 6.2.1, workpieces of varying geometries demand distinct machining techniques and positioning strategies. In the case of a basic 3-2-1 locating scheme or a scheme utilizing six independent datums, identifying six contact points and six fixture locators is a direct process for vector acquisition. However, for more complex locating schemes, such as the plane-hole or four-jaw chuck methods, obtaining six distinct points is less clear-cut and necessitates an equivalent transformation.

As depicted in Fig. 6.23a, a typical plane-hole locating scheme is presented. Locators L_1, L_2, L_3 make contact with the back surface, thereby constraining three degrees of freedom (α, β, w). A round pin is equivalent to two virtual locators (L_4, L_6) positioned at the same place, engaging with Hole M and restricting two degrees of freedom (u, v). An edge cutting pin is equivalent to L5, which interacts with Hole N and constrains one degree of freedom (γ). Through this conversion, the two pins are transformed into three distinct independent fixture locators with defined coordinates and normal vectors. This implies that all vectors for $\mathbf{N}'$, $\mathbf{P}'$ and **L** can be directly extracted.

Similarly, for the four-jaw chuck locating scheme as shown in Fig. 6.23b, the chuck's jaws restrict movement along the *x*/*z*-axis (u, w) and rotation around the x/z-axis(α, γ). The end plane of the jaw limits movement along the y-axis (v). With sufficient clamping force, rotation around the y-axis (β) is also restricted. To achieve an equivalent conversion, a virtual locator, L_3, is positioned on the tangent plane of locators L_1 and L_2. It is postulated that this virtual plane is part of the workpiece; thus, locators L_1, L_2, and L_3 interact with this virtual plane, constraining three degrees of freedom (β, γ, u). Locators L_4 and L_5 together restrict two degrees of freedom (α, w). Finally, locator L_6 constrains one degree of freedom (v).

6.2.3.3 Jacobian-Torsor Model to Solve FR

As illustrated in Fig. 6.21, the logical relationships between the functional requirement (FR) and the functional elements (FEs) within the variation propagation chain have been established. Given that all key elements in this chain have been aligned in the same direction through the construction of the Virtual Contact Functional Element (VCFE) in Sect. 6.2.3.2, the Jacobian-Torsor model serves as an effective method to establish the mathematical relationships necessary to determine the FR at stage k. The arithmetic formulation is expressed as follows [28]:

$$\mathbf{T}_{FR}(k) = \begin{bmatrix} u \\ v \\ w \\ \alpha \\ \beta \\ \gamma \end{bmatrix}_{FR} = \left[[\mathbf{J}]_{FE_1} \cdots [\mathbf{J}]_{FE_n} \right] \cdot \left[\begin{bmatrix} u \\ v \\ w \\ \alpha \\ \beta \\ \gamma \end{bmatrix}_{FE_1}^T \cdots \begin{bmatrix} u \\ v \\ w \\ \alpha \\ \beta \\ \gamma \end{bmatrix}_{FE_n}^T \right]^T + o(k) \tag{6.54}$$

where $\mathbf{T}_{FR}(k)$ is the torsor associated with the functional requirement at stage k, $\mathbf{J}_{FE_i}$ is the Jacobian matrix that illustrates the geometric relationship of the ith feature functional element (FE) with respect to the machining feature FR, and $\mathbf{T}_{FE_i}$ is the individual torsor of each feature in the chain. i ranges from 1 to n, with n denoting the total number of functional elements. $o(k)$ represents the unmodeled system noise.

Figure 6.24 outlines the steps for employing the variation propagation model, with k representing the stage index. The vector $\mathbf{T}(k)$ is a compilation of deviations for all critical features on the workpiece following the kth stage. Each feature's deviation is depicted by a torsor, as shown in Table 6.2. For features that have not been created by the end of the kth stage, the corresponding components in $\mathbf{T}(k)$ are set to zero; otherwise, the zero components are substituted with the deviation torsors. The detailed procedure for each step is described below:

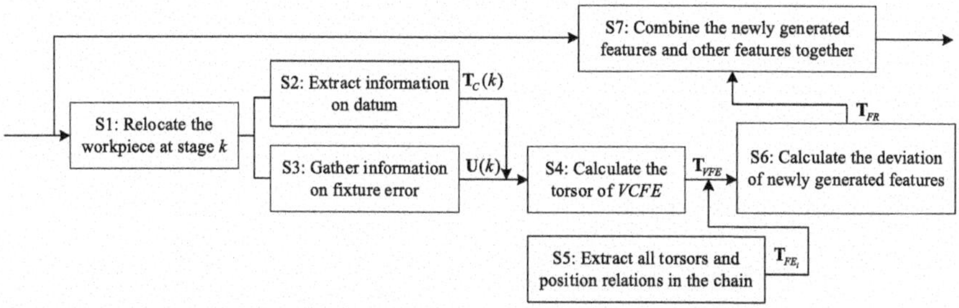

Fig. 6.24 Steps of the derivation of variation propagation model

Step 1: Post each stage, the workpiece may require repositioning based on the machining feature and the method used. This step involves constructing the variation propagation chain and establishing the actual positional relationships between each feature at this stage.
Step 2: Only the errors in the datum features of the workpiece affect the deviation of the newly generated feature. Thus, we extract $\mathbf{T}_C(k)$ from $\mathbf{T}(k)$ by applying a selection matrix. This matrix includes the deviations of all datum features at this stage. Additionally, the position and orientation of the contact points ($\mathbf{N}'(k)$ and $\mathbf{P}'(k)$) must be determined.
Step 3: Compile the fixture error $\mathbf{U}(k) = \left[\mathbf{u}_1^T \ \mathbf{u}_2^T \ \cdots \ \mathbf{u}_6^T\right]^T$ and the positions of the locators $L(k)$.
Step 4: Compute $\mathbf{T}_{VCFE}$ using the method described in Sect. 6.2.3.2.
Step 5: Aggregate all error torsors within the variation propagation chain, which includes errors from the machine tool and worktable, cutting tool and spindle, and $\mathbf{T}_{VCFE}$ encompassing fixture-induced and datum-induced errors. Their positional relationships must be articulated using Jacobian matrices.
Step 6: Calculate the deviation of the newly generated feature using the unified Jacobian-Torsor model.
Step 7: Replace the components in $\mathbf{T}(k)$ corresponding to the newly generated feature with its deviation $\mathbf{T}_{FR}$ to obtain the vector $\mathbf{T}(k+1)$ (i.e., integrate the newly generated features with the existing features on the workpiece).

This sequence of steps will be iterated for each stage, culminating in a chain-like state space model. The subsequent section will present practical examples to demonstrate the efficacy of this model.

6.2.4 *Case Study*

To validate the versatility of the proposed variation propagation model for workpieces of diverse shapes, two distinct types of workpieces have been selected for study. Figure 6.25a displays an automotive engine block of box-type, which is processed using the EX-CELL-O machining center as shown in Fig. 6.25b. The valve shell, a revolving-type component, along with its clamping arrangements, are depicted in Fig. 6.25c–e.

6.2.4.1 Experimental Setup

Engine Block Machining

The engine block machining process comprises five stages. It incorporates a six independent datums locating scheme for the initial three operations and a plane-hole locating scheme for the last two operations. The first two operations involve

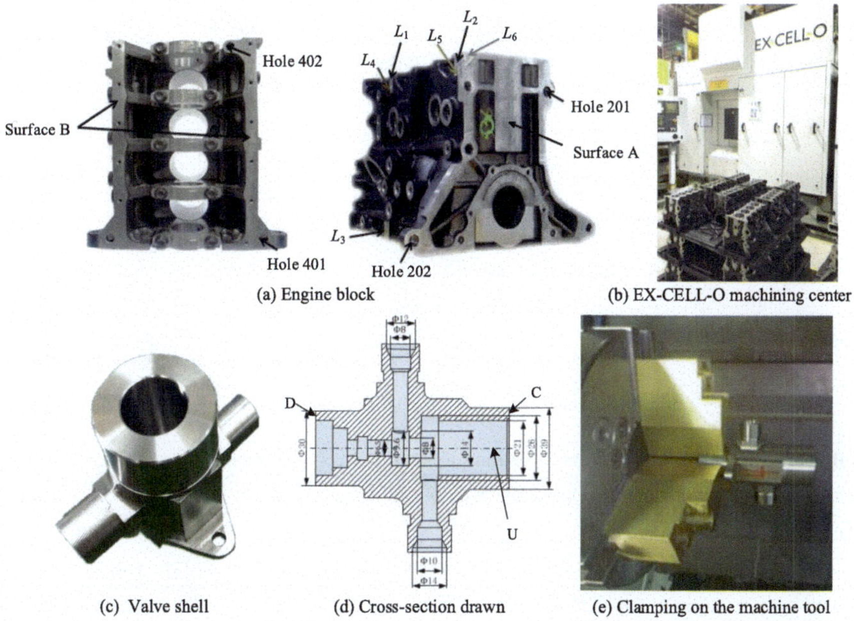

Fig. 6.25 The workpieces in the case study

milling Surface A and semi-finish-milling Surface B, succeeded by spot drilling Holes 401 and 402. Features L_1 through L_6 serve as the rough datum features for operations OP10 to OP30. Notably, this setup deviates from the conventional 3-2-1 locating scheme, as no two of the six datums are coplanar. The deviation of Surface B machined at OP20 and the deviations of the two holes drilled at OP30, considered as datum-induced errors, are significant contributors to the variations in the machining features at OP40. Likewise, the errors in features generated from OP10 and OP40 accumulate deviations at OP50. A detailed description of the operations and the nominal locations of key features relative to RCS is presented in Table 6.6.

Valve Shell Turning

The process in question is a five-stage turning operation performed on a CNC machining center (Gildemeister CTX420). A description of the operations and the nominal locations of key features relative to RCS is provided in Table 6.7. This process vividly illustrates the propagation and accumulation of variations. For example, end plane C and the excircle turned at operation OP10 are used as the datums for operation OP40, thereby conveying the machining error from the earlier stage to the subsequent one. Similarly, the excircle machined in OP40 serves as the datum for OP50, which in turn impacts the precision of the machining features at OP50.

Table 6.6 Operations and nominal locations of key features (engine block)

No.	Locating datum	Operation descriptions	${}^{0}\boldsymbol{\omega}_{0}^{R}$	${}^{0}\mathbf{t}_{0}^{R}$
OP10	$L_1, L_2, L_3, L_4, L_5, L_6$	Mill Surface A	[0, 0, 0]	[− 170.5, 138, 0]
OP20		Semi-finish-mill surface B	[0, π/2, π/2]	[18, 248, − 30.5]
OP30		Spot drill Hole 401	[π/2, 0, 0]	[18, 73, − 312.5]
		Spot drill Hole 402	[0, 0, π/2]	[18, 27, − 17]
OP40	Surface B, Hole 401, 402	Spot drill Hole 201	[− π/2, 0, 0]	[− 176.5, 201, 0]
		Spot drill Hole 202	[0, π/2, 0]	[0, 0, 0]
OP50	Surface A, Hole 201, 202	Finish-mill Surface B	[0, π/2, π/2]	[17.5, 248, − 30.5]

Table 6.7 Operations and nominal locations of key features (valve shell)

No.	Locating datum	Operation descriptions	${}^{0}\boldsymbol{\omega}_{0}^{R}$	${}^{0}\mathbf{t}_{0}^{R}$
OP10	D	Drill Hole Φ 21	[0, 0, 0]	[0, 0, − 29]
		Lathe Excircle Φ 39	[0, 0, 0]	[0, 0, 0]
OP20	D downward with modular fixture	Drill Hole Φ 14	[π/2, 0, π/2]	[− 42, 0, − 32.5]
		Drill Hole Φ 10	[π/2, 0, π/2]	[− 32, 0, − 32.5]
OP30		Drill Hole Φ 12	[− π/2, 0, − π/2]	[− 42, 0, − 44.5]
		Drill Hole Φ 8	[− π/2, 0, − π/2]	[− 32, 0, − 32.5]
OP40	C	Lathe Slot Φ 9.6	[0, π, 0]	[0, 0, − 66.5]
		Drill Hole Φ 6.5	[0, π, 0]	[0, 0, − 62.5]
		Lathe Excircle Φ 30	[0, π, 0]	[0, 0, − 79.5]
OP50	D	Lathe Slot Φ 26	[0, 0, 0]	[0, 0, − 36]
		Lathe Slot Φ 14	[0, 0, 0]	[0, 0, − 48]
		Drill Hole Φ 8	[0, 0, 0]	[0, 0, − 35.5]

6.2.4.2 Parameters Acquisition

To establish the variation propagation chain as depicted in Fig. 6.21, it is essential to take into account the entire manufacturing system and to determine the unknown parameters in Eq. (6.54). Using operation OP50 of the engine block machining as an example, Table 6.8 Manufacturing system components at OP50 details FE components and their positional relationships based on the actual EX-CELL-O machining center. The z-axis direction is aligned with the feed direction. Torsors can be either calculated or directly measured within the manufacturing system. Jacobian matrices can be derived from Eq. (6.40), utilizing the last two columns of Table 6.8 for this purpose.

To confirm the accuracy of the model, intentional fixture errors are introduced at each stage of the engine block machining process. The inputs to the model that

Table 6.8 Manufacturing system components at OP50

No.		*FE* Components	Torsor parameters	${}^{0}\boldsymbol{\omega}_{0}^{G}$	${}^{0}\mathbf{t}_{0}^{n}$
Line I	1	Machine tool IFE_1	w, α, β	[0, 0, 0]	[0, 0, – 370]
	2	Machine tool to worktable CFE_1	w, α, β	[0, 0, 0]	[0, 0, – 370]
	3	Worktable upper plane IFE_2	w, α, β	[0, 0, 0]	[0, 0, – 283]
	4	Worktable to fixture CFE_2	w, α, β	[0, 0, 0]	[0, 0, – 283]
	5	Fixture Coordinate System IFE_3	$u, v, w, \alpha, \beta, \gamma$	[0, – π/2, 0]	[248, 30.5, – 17.5]
	6	Fixture to workpiece $VCFE_3$	$u, v, w, \alpha, \beta, \gamma$	[0, – π/2, 0]	[248, 30.5, – 17.5]
	7	Local Coordinate System IFE_4	$u, v, w, \alpha, \beta, \gamma$	[0, 0, 0]	[0, 0, 0]
Line II	1	Machine tool spindle IFE_6	u, v, w, α, β	[0, π, 0]	[0, 0, 128]
	2	Spindle to cutting tool CFE_4	u, v, α, β	[0, π, 0]	[0, 0, 128]
	3	Cutting tool tip IFE_5	w	[0, π, 0]	[0, 0, 0]

correspond to these fixture errors are as follows:

$$\mathbf{U}(1) = [0\ \ 0\ \ 0\ \ 0\ \ 0\ \ 0\ \ 0\ \ 0\ \ 0.018\ \ 0\ \ 0\ \ 0.023\ \ 0\ \ 0\ \ 0\ \ 0\ \ 0\ \ 0]^T$$
$$\mathbf{U}(2) = [0\ \ 0\ \ 0\ \ 0\ \ 0\ \ 0\ \ 0\ \ 0\ \ 0.018\ \ 0\ \ 0\ \ 0.023\ \ 0\ \ 0\ \ 0\ \ 0\ \ 0\ \ 0]^T$$
$$\mathbf{U}(3) = [0\ \ 0\ \ 0\ \ 0\ \ 0\ \ 0.009\ \ 0\ \ 0\ \ 0.006\ \ 0\ \ 0\ \ 0.008\ \ 0\ \ 0\ \ 0\ \ 0\ \ 0\ \ 0]^T$$
$$\mathbf{U}(4) = [0\ \ 0\ \ 0\ \ 0\ \ 0\ \ 0\ \ 0\ \ 0\ \ 0.016\ \ 0\ \ 0\ \ 0\ \ 0\ \ 0\ \ 0\ \ 0\ \ 0\ \ 0]^T$$
$$\mathbf{U}(5) = [0\ \ 0\ \ 0\ \ 0\ \ 0\ \ 0.019\ \ 0\ \ 0\ \ 0.019\ \ 0\ \ 0\ \ 0\ \ 0\ \ 0\ \ 0\ \ 0\ \ 0\ \ 0]^T.$$

The rough datum adopted in first three operations is the input of the model as well. The error is the collection of datum features' torsors, $\mathbf{T}_{\mathrm{C}}(1) = [0; 0; 0.007; 0; 0; 0; 0; 0; 0.006; 0; 0; 0; 0; 0; 0; 0; 0; 0; -0.012; 0; 0; 0; 0; 0; 0.006; 0; 0; 0; 0; 0; 0; 0.004; 0; 0; 0; 0]$. All the other error torsors in manufacturing system are set as zeros.

6.2.4.3 Results and Discussion

Adhering to the procedure outlined in Sect. 6.2.3, the variation propagation model has been implemented using MATLAB® 2016b. All vectors and matrices are computed using the relevant equations. The model is designed to automatically output the variation of key features for each stage once the error torsors and positional relationships are input into the program.

In the context of engine block machining, operations OP10 and OP50 are critical for machining key features, which are commonly machined in engineering practice. Consequently, Surface A and Surface B are measured using a coordinate measurement machine (CMM) post-machining. Given the complexity of the engine block and the variety of locating methods, relying solely on a single traditional Stream of Variation (SoV) model is insufficient. Currently, a composite approach that selects specific models for specific operations is often employed in such scenarios. However, the proposed variation propagation model has the capability to manage all operations and all types of locating schemes. The key data in the torsor is the translation deviation vector, which is perpendicular to the plane. Table 6.9 presents the measurement results and a comparison between the traditional method, the current method, and the proposed model.

For the valve shell turning process, the majority of the features produced across the five stages are considered key features in engineering practice. Using Hole Φ6.5 after OP40 and Hole Φ8 after OP50 as examples, the measurement outcomes and a comparison among the traditional method, the current method, and the proposed model are detailed in Table 6.10. In this scenario, the fixture error at each stage is assumed to be [0; 0; 0; 0; 0; 0.01; 0; 0; 0; 0; 0; 0; 0; 0; 0; 0; 0; 0].

The advantages of the proposed model are examined in two dimensions: its alignment with actual experimental data and its contrast with current methodologies.

In the context of the former comparison, the aggregate discrepancies between the projected values and the empirical data for all pivotal features are acceptably minor, thereby attesting to the model's efficacy. A case in point is the valve shell turning outcome for Hole Ø6.5 post-OP40, with actual measurements of [0; 0.002; 0; 0.001; 0; 0] juxtaposed against the model's forecast of [0; 0.0022; 0; 0.0012; 0; 0]. These minor deviations are primarily ascribed to certain probabilistic elements that are inherently challenging to encapsulate within the model, including the effects of vibration or temperature fluctuations on the machining processes and subsequent

Table 6.9 Variation results comparison for the engine block machining

Key feature		Surface A after OP10 (mm) w.r.t. the rough datum L_6			Surface B after OP50 (mm) w.r.t. the datum Hole 202		
		u	v	w	u	v	w
Traditional SoV model		–	–	×	×	–	–
Composed method	By [14]	–	–	– 0.0087	×	–	–
	By [8]	–	–	×	0.0254	–	–
Proposed model		–	–	– 0.0082	0.0271	–	–
Measurement		–	–	– 0.006	0.031	–	–

Table 6.10 Variation results comparison for the valve shell turning

Key feature	Hole Φ6.5 after OP40 (mm) w.r.t. the axis U						Hole Φ8 after OP50 (mm) w.r.t. the axis U					
	u	v	w	α	β	γ	u	v	w	α	β	γ
Traditional SoV model	–	×	–	×	–	–	–	–	×	×	–	–
Modified method [18]	–	0.0017	–	0.0007	–	–	–	–	0.0007	0.0007	–	–
Proposed model	–	0.0022	–	0.0012	–	–	–	–	0.0012	0.0012	–	–
Measurement	–	0.002	–	0.001	–	–	–	–	0.001	0.001	–	–

measurements. These uncertainties and probabilistic elements adhere to a normal distribution and are delineated by $o(k)$ within the framework of the proposed model.

When juxtaposing the proposed model with existing methodologies, its superiority manifests in universality and precision. The proposed model surpasses current approaches in terms of universality. Traditional SoV models are incapable of managing six independent datums locating schemes or revolving-type workpieces, rendering them unsuitable for certain scenarios. Modified models often necessitate a tailored approach for complex workpieces, with selections based on specific machining processes, lacking universality across various workpieces and stages. The proposed model, however, adeptly handles complex box parts and accurately predicts machining precision for revolving workpieces, achieving universality in variation propagation modeling for general shape workpieces, irrespective of the machining method and locating scheme. In terms of accuracy, the proposed model outperforms existing methods when compared with real experimental data. While existing SoV models concentrate on modeling fixture and datum-induced errors, other factors such as cutting-tool wear, spindle-thermal variations, and geometric and kinematic variations are considered stochastic noises, denoted by $w(k)$ [8, 14, 18]. Within the proposed model, additional sources of variability, including machine tool error and cutting-tool wear, are explicitly incorporated into the calculations through the torsor $\mathbf{T}_{FE_i}$, rather than being lumped together as unmodeled noise. Consequently, the model's variance and central tendency, represented by $o(k)$, are more tightly controlled than $w(k)$ in machining processes, indicating a narrower confidence interval under the same significance level, thus enhancing the accuracy of the prediction results.

6.2.5 Conclusion

A novel variation propagation model for multi-stage machining of workpieces with diverse geometries has been introduced. This model offers a visual representation of the variation propagation chain, thereby broadening the applicability of existing

SoV models. By reassessing error contributions from an assembly perspective within machining processes, the model encompasses all critical elements of the manufacturing system, enabling the consideration of a wider range of error sources and improving the precision of predictive outcomes. The model, which is applicable to workpieces of various shapes, is resolved using an enhanced three-dimensional tolerance analysis approach that integrates the strengths of the SDT model and the Jacobian matrix. Case studies of the multi-stage machining processes for an automotive engine block and a valve shell validate the reliability and accuracy of the proposed methodology, taking into account the complete manufacturing system.

For subsequent research, the incorporation of multi-scale variations into the variation propagation model is suggested. The current study is confined to dimensional errors; however, multi-scale variations, including geometric error, waviness, and roughness, also contribute to error propagation and accumulation, impacting the overall variation of the machined feature. Additionally, the current research assumes workpieces to be rigid. Future studies should address the variability in machining due to workpiece deformations induced by cutting tool forces or clamping forces. Furthermore, the application of this model holds significant engineering value, such as in process control and enhancement, tolerance allocation, and fault diagnosis for workpieces of general shapes.

6.3 Variation Management of Key Control Characteristics in Multistage Machining Processes Considering Quality-Cost Equilibrium

6.3.1 Introduction

Multistage machining processes (MMPs) are extensively employed in the manufacturing industry to produce high-quality products by material removal from casting blanks. Key Product Characteristics (KPCs) are essential features that define a workpiece's design to ensure it meets specific functional requirements. Consequently, ensuring that the quality specifications of KPCs are met within MMPs is a critical concern [18].

In MMPs, tooling elements serve to position, secure, or machine the unfinished workpieces. During a particular stage, the actual machined features may deviate from their intended design values due to the imperfections in fixture locators or other tooling elements. Should any of these off-target features be employed as datum features in subsequent stages, their deviations will propagate and accumulate, ultimately impacting the machining precision of KPCs. Key Control Characteristics (KCCs), which encompass process-related knowledge, are the tooling elements whose variations are the primary source of deviations in KPCs [52]. Hence, to sustain a high level of precision in KPCs, it is essential and beneficial to manage variations in KCCs.

Capturing the introduction and propagation of variations from KCCs to KPCs forms the foundation of variation management. The Stream of Variation (SoV) theory stands as one of the most potent methodologies for uncovering the mapping relationship between KCCs and KPCs [18]. Zhou et al. [16] delved into vector deviation representation and introduced a Differential Motion Vector (DMV)-based state space model to articulate the SoV. This research provided detailed mathematical formulations for fixture and datum variations, marking a pioneering approach to modeling variation propagation in MMPs. Following this, the model was extended to encompass fixture layouts [13, 33, 53], variations induced by machining [29, 54], a broadened range of application objects [20, 36, 55], and the integration of Geometric Dimensioning and Tolerancing (GD&T) [14]. With these advancements, the rules governing variation propagation from KCCs to KPCs in MMPs have become increasingly clear.

Nevertheless, KCCs possess inherent imperfections, both static and dynamic, which pose significant challenges for variation management [56]. Static imperfection pertains to the manufacturing deviations inherent to the KCC itself, whereas dynamic imperfection relates to the degradation of KCCs due to wear or other factors within MMPs. Tolerancing and maintenance are the two principal strategies to combat these imperfections. For MMPs, tolerancing establishes the initial tolerance conditions for tooling elements, that is, it allocates permissible variation ranges. However, the allocated tolerances can only ensure product quality in the early stages of production if the degradation of KCCs over time is not taken into account. Maintenance periodically restores deteriorated KCCs to their initial state to prevent the production of out-of-tolerance workpieces caused by dynamic imperfections. Consequently, a novel concept of variation management emerges, which integrates tolerancing and maintenance.

Tolerancing, a common practice in multistage assembly processes (MAPs), distributes final quality specifications across the tolerances of individual components, aiming to minimize manufacturing costs while ensuring product functionality is maintained [57–59]. Traditional product-oriented tolerancing concentrates on assigning tolerances to product variables, essentially allocating the assembly's quality specifications to the KPCs of each component. The majority of related studies construct cost-tolerance functions with the goal of minimizing costs and solve them using optimization techniques [60–62]. Although product-oriented tolerancing is crucial for product optimization design, it often neglects process knowledge. Given that variations in KCCs are the primary source of product quality degradation, the optimal assignment of tolerances to KCCs throughout the manufacturing process has become a focal point. Ding et al. [52] initially established a process-oriented tolerancing approach for automotive body MAPs. However, this method arbitrarily set the replacement cycle of fixture elements to half a year, not fully leveraging the dynamic degradation patterns of KCCs. Chen et al. [56] expanded on Ding's research by integrating process-oriented tolerancing with maintenance, comprehensively optimizing the tolerances and maintenance strategies for assembly fixtures, which led to both quality enhancement and cost reduction.

These foundational works have established the groundwork for KCC variation management. However, their focus was primarily on MAPs, where the variation of assembly fixtures was the main concern for KCC variation management. In MMPs, there is a broader range of KCCs, each with distinct cost functions and degradation models, such as cutting tools and various fixture elements within different fixture schemes. Consequently, KCC variation management in MMPs is more complex and less researched. Huang and Shi [63] were the first to develop a tolerance stackup model for MMPs, assigning tolerances to fixture locators by maximizing the variances of KCCs within the constraints of part quality specifications. At the time, due to the immaturity of SoV research, this work utilized a two-dimensional variation propagation model and considered only three point-to-surface fixture locators for all KCCs, assuming independence and normal distribution among them. Building on the three-dimensional SoV model, Liu et al. [64] selected the optimal setup planning for MMPs using a process-oriented tolerancing approach. Chen et al. [65] and Zhang et al. [66] optimized KCC design through tolerance synthesis to enhance quality and reduce costs. Since these models did not account for the dynamic degradation of KCCs, maintenance planning could not be integrated, and optimizations were achieved by altering the layout or process of the manufacturing system. For mature manufacturing systems, such as automotive engine cylinder block machining, an optimal KCC variation management scheme that combines process-oriented tolerancing with maintenance planning is often the preferred choice for most manufacturing enterprises [67–69]. Abellán-Nebot et al. [70] introduced a new process-oriented tolerancing approach, considering two additional KCCs—cutting tools and spindle temperature. Degradation models for locators and cutting tools were defined to facilitate maintenance integration. This method has shown effective performance in MMPs with point-to-surface locating schemes.

Despite certain achievements in the concurrent optimization of tolerancing and maintenance planning for KCCs within MMPs, three primary constraints are evident in the existing studies: (i) The scope of application is confined. The research is restricted to the traditional 3-2-1 positioning scheme, focusing solely on point-to-surface locators. Alternative setups, such as the pin-hole positioning scheme, prevalent in MMPs, necessitate the introduction of distinct KCC types. (ii) Quality constraints are limited to dimensional tolerances, neglecting geometric tolerances, which are equally vital aspects of quality specifications that warrant consideration. (iii) Balancing quality and cost within a single-objective optimization framework is challenging when quality is treated as a constraint and cost minimization as the objective. In the revised approach, where the objective function is defined as a combination of quality loss and variation management expenses, the two elements are not directly comparable, thus lacking a scientific basis for assigning appropriate weights.

A novel approach to variation management for KCCs within MMPs has been introduced, addressing the quality-cost balance. This paper's principal contributions address the aforementioned challenges in three key areas: (i) For the first time, the variation management of KCCs in MMPs is extended to include the pin-hole positioning scheme. The paper introduces fixture pins as an additional KCC type in MMPs, with a comprehensive examination of their tolerance-variation relationship, degradation patterns, and cost functions. (ii) The workpiece's quality specifications now encompass both traditional dimensional tolerances and geometric tolerances, thereby broadening the scope of quality control. (iii) A dual-objective optimization framework is presented to achieve a balance between quality and cost, utilizing a modified Chebyshev goal programming (MCGP) method to resolve the model.

6.3.2 *The Proposed Method*

6.3.2.1 Overview and Framework

Figure 6.26 illustrates the logical structure of the proposed methodology, highlighting the connections between quality and cost within MMPs. The overarching framework is organized into three distinct tiers: the operational level, the decision-making level, and the objective level.

The operational layer captures the dynamic process of variation propagation from KCC to KPC. As outlined in Sect. 6.3.1, the deviations in tooling elements are represented by the process variables of KCCs. The variation propagation mechanism inherent to MMPs facilitates the transfer of these KCC variations to KPCs, which results in final product deviations and a decline in product quality. Considering an MMP with n tooling elements—comprising p locators/pins and q cutting tools—distributed across N stages, and m KPCs present in the workpiece, the variations of the n KCCs are denoted by $\mathbf{U} = [U_1\, U_2 \cdots U_r \cdots U_n]^T$, and the deviations of the m KPCs are represented by $\mathbf{Y} = [Y_1\, Y_2 \cdots Y_h \cdots Y_m]^T$. According to the Theory of Variations (SoV), the deviations in KPCs, Y, can be expressed as a function of U.

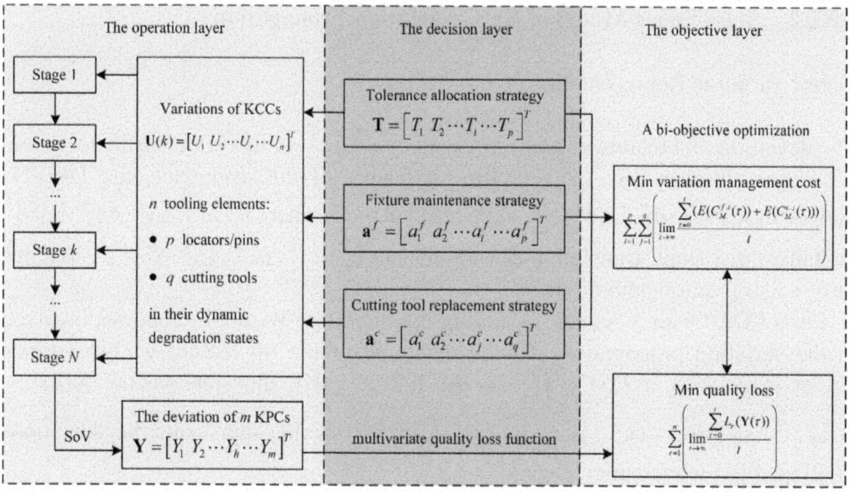

Fig. 6.26 The logic framework of the proposed method

Section 6.3.2.2 will further develop this relationship and present the detailed variation propagation model.

The decision layer represents the outcome of the KCC variation management strategy. As referenced in Sect. 6.3.1, the imperfections of KCCs are characterized by their tolerance and degradation, which are static and dynamic attributes, respectively. The variable $\mathbf{T} = \left[T_1\, T_2 \cdots T_i \cdots T_p\right]^T$ represents the tolerance allocated to locators/pins within KCCs, establishing the initial permissible variation range for U. Throughout the operation of MMPs, U dynamically changes due to the ongoing degradation of KCCs, potentially leading to a decline in product quality and eventual failure when KPC deviations surpass the specified quality limits at some point. Scheduled maintenance can prevent such failures by restoring degraded KCCs to their original state. In this paper, a preventive maintenance policy is adopted, with $\mathbf{a}^f = \left[a_1^f\, a_2^f \cdots a_i^f \cdots a_p^f\right]^T$ denoting the maintenance interval for locators/pins and $\mathbf{a}^c = \left[a_1^c\, a_2^c \cdots a_j^c \cdots a_q^c\right]^T$ representing the replacement interval for cutting tools. Both T and a are pivotal decision variables that influence the condition of KCCs within MMPs, and their specific effects on KCC variations will be further elaborated in Sect. 6.3.2.3.

The objective layer outlines the optimization objectives of this section, which are to minimize the variation management cost and to minimize quality loss. Typically, stricter tolerances or more frequent maintenance leads to increased variation management costs, while reducing quality loss. These two objectives are inherently conflicting. Consequently, a bi-objective optimization approach is introduced to identify the equilibrium between quality and cost. The formulation of the objective functions, constraints, and the optimization model will be detailed in Sect. 6.3.2.4.

6.3.2.2 State Space Model of KCC Variation Propagation

Vector Variation Representation

The deviation of a feature can be represented by DMV of its own coordinate system (CS) w.r.t. another CS [16]. For the ith feature of the workpiece, the DMV is defined by $\mathbf{x}_i^R = \left[\, (\mathbf{d}_i^R)^T\ (\boldsymbol{\theta}_i^R)^T \,\right]^T$ w.r.t. reference CS, where $\mathbf{d}_i^R = \left[\, \Delta x_i^R\ \Delta y_i^R\ \Delta z_i^R \,\right]^T$ contains three small translation deviations and $\boldsymbol{\theta}_i^R = \left[\, \Delta\alpha_i^R\ \Delta\beta_i^R\ \Delta\gamma_i^R \,\right]^T$ contains three small rotation deviations.

For a MMP with N stages, assuming that there are W quality features involved in the variation propagation and the deviation of the ith feature w.r.t. reference CS at stage k is a DMV $\mathbf{x}_{k,i}^R$ $(i = 1, 2, \ldots, W)$, the state vector $\mathbf{X}(k) = \left[(\mathbf{x}_{k,1}^R)^T\ (\mathbf{x}_{k,2}^R)^T \cdots (\mathbf{x}_{k,W}^R)^T\right]^T$ is a stack of DMVs that represents the deviations of all quality features after stage k.

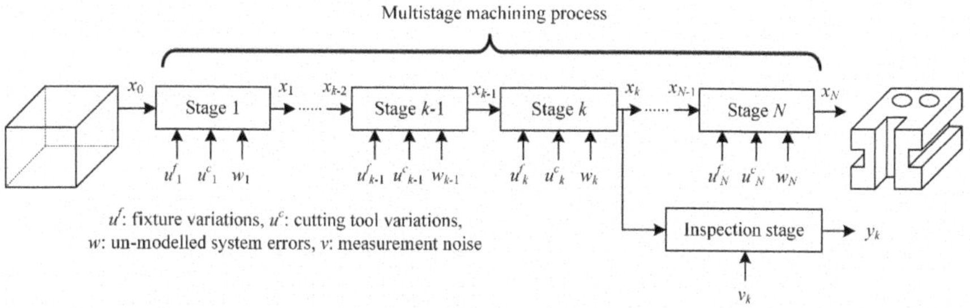

Fig. 6.27 Variation propagation by state space representation

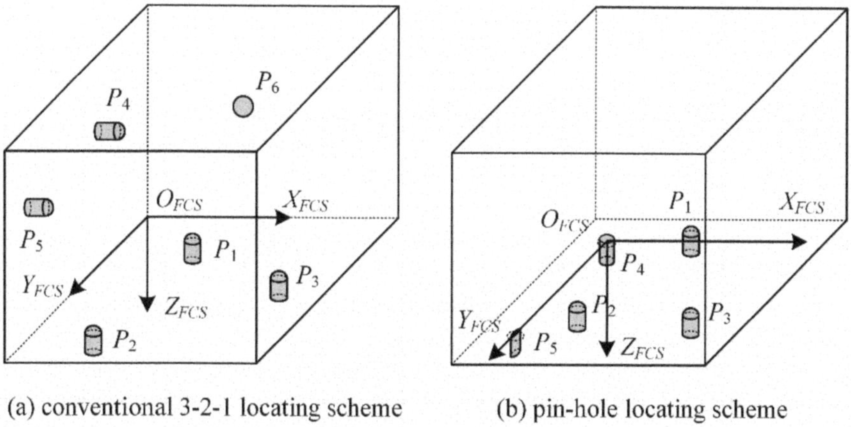

Fig. 6.28 The two most representative locating schemes in MMPs

Variation Contributors and Propagation Model

A thorough comprehension of the variation propagation mechanism is essential for managing variations in KCCs. This paper identifies three principal sources of variation within MMPs: fixture variations, cutting tool variations, and datum variations. The propagation of these variations and their contributors can be conceptualized through state space representation as depicted in Fig. 6.27.

Figure 6.28 illustrates the two predominant positioning schemes within MMPs: the traditional 3-2-1 positioning arrangement and the pin-hole positioning arrangement. Consequently, the fixture comprises two categories of KCC: locators and pins. These elements introduce variations into MMPs distinctively due to disparities in their contact forms, tolerance-variation relationships, and degradation models.

Cutting tools undergo gradual wear throughout their operational lifecycle. In contrast to fixtures, it is presupposed that cutting tools are flawless at the outset, implying that variations in cutting tools stem exclusively from degradation. As a result, maintenance planning for cutting tools is essential, while tolerancing is not required.

Datum variations arise when features manufactured in earlier stages serve as reference datums. Although these datum features do not directly engender KCC, they significantly influence the propagation of KCC variations.

A state space model is formulated to characterize the variation propagation within MMPs, encapsulated within a state equation and a measurement equation:

$$\mathbf{X}(k) = \mathbf{A}(k-1) \cdot \mathbf{X}(k-1) + \mathbf{B}(k) \cdot \mathbf{U}(k) + \mathbf{w}(k) \tag{6.55}$$

$$\mathbf{Y}(k) = \mathbf{C}(k) \cdot \mathbf{X}(k) + \mathbf{v}(k) \tag{6.56}$$

where $\mathbf{U}(k)$ represents the random variable of KCC variations introduced at stage k, encompassing both fixture and cutting tool variations;$\mathbf{Y}(k)$ denotes the deviation vector of the KPCs at stage k; $\mathbf{w}(k)$ and $\mathbf{v}(k)$ correspond to the unmodeled system errors and measurement noise, respectively. The coefficient matrices A and B are derived from process knowledge, including the datum transformation between stages and the layout of KCCs on each stage, while C is determined by the positioning of KPCs and the measurement of datum features. The specific expressions for these matrices can be found in reference [16].

Consequently, within the state equation, $\mathbf{A}(k-1) \cdot \mathbf{X}(k-1)$ signifies the datum variations inherited from preceding stages, and $\mathbf{B}(k) \cdot \mathbf{U}(k)$ denotes the deviations introduced at the current stage attributed to fixtures and cutting tools. In the measurement equation, $\mathbf{C}(k) \cdot \mathbf{X}(k)$ indicates the KPC deviations with respect to the measurement datum.

In general, the assessment of product quality is based on the KPC deviations at the ultimate stage, specifically when $k = N$ within the state space model. By substituting Eq. (6.55) into (6.56), the process can be understood.

$$\mathbf{Y}(N) = \sum_{k=1}^{N} \mathbf{C}(N)\Phi(N,k)\mathbf{B}(k)\mathbf{U}(k) + \mathbf{C}(N)\Phi(N,0)\mathbf{X}(0) + \varepsilon \tag{6.57}$$

where the state transition matrix $\Phi(N,k) = \mathbf{A}(N-1)\mathbf{A}(N-2)\cdots\mathbf{A}(k)$ and $\Phi(k,k) = \mathbf{I}$.

With initial conditions set to zero, i.e. $\mathbf{X}(0) = \mathbf{0}$ and considering the uncertainty term ε to be negligible due to the minimal contribution of un-modelled noise to the additional variations in MMP [71] , the covariance calculation for both sides of Eq. (6.57) post-simplification allow for the approximation of variation propagation as follows:

$$\boldsymbol{\Sigma}_{\mathrm{Y}} = \sum_{k=1}^{N} \boldsymbol{\gamma}(k)\boldsymbol{\Sigma}_{\mathrm{U}}(k)\boldsymbol{\gamma}^{T}(k) = \boldsymbol{\Gamma}\boldsymbol{\Sigma}_{\mathrm{U}}\boldsymbol{\Gamma}^{T} \tag{6.58}$$

where $\boldsymbol{\Sigma}_{\mathrm{Y}}$ and $\boldsymbol{\Sigma}_{\mathrm{U}}(k)$ represent the covariance matrices of $\mathbf{Y}(N)$ and $\mathbf{U}(k)$, $\boldsymbol{\gamma}(k) = \mathbf{C}(N)\boldsymbol{\Phi}(N,k)\mathbf{B}(k)$, $\boldsymbol{\Gamma} = [\gamma((1)\cdots\gamma(N)]$, and $\boldsymbol{\Sigma}_{\mathrm{U}} = diag(\boldsymbol{\Sigma}_{\mathrm{U}}(1),\cdots,\boldsymbol{\Sigma}_{\mathrm{U}}(N))$.

6.3.2.3 Distribution of KCC Variations Considering Process Degradation

KCC Variations in Fixture

This section will deduce the distribution of KCC variations for two distinct locating schemes within fixtures, highlighting the differences between locators and pins in terms of contact form, tolerance-variation relationship, and degradation model. The pin-hole locating scheme introduces the pin as a novel object for variation management in MMPs.

(a) Conventional 3-2-1 locating scheme

The locating pair of conventional 3-2-1 locating scheme is in the form of point-to-surface contact in single direction as shown in Table 6.11. T_i is the tolerance of ith locator, which limits the initial variation range of this KCC. Tighter tolerance means higher manufacturing requirement and cost, so tolerancing tends to allocate tighter tolerances to locators that have a greater impact on KPCs. The variation motion of locator is only in one direction. Here, z direction is taken as the example. The deviation δ_i is between the actual and ideal position of the locator, which is initially bounded by $[-T_i/2,\ T_i/2]$ and follows a normal distribution of $N\left(0,\ (T_i/6)^2\right)$. Since $E[\cdot]$ is the expectation operator, the statistics regarding ΔZ before operation are.

In the conventional 3-2-1 locating scheme, the locating pair is characterized by a point-to-surface contact in a single direction, as illustrated in Table 6.8. The tolerance T_i of the ith locator restricts the initial variation range of this KCC. A tighter tolerance implies a higher manufacturing requirement and cost, thus tolerancing typically assigns tighter tolerances to locators with a more significant influence on KPCs. The variation motion of the locator is unidirectional, with the z direction serving as an example. The deviation δ_i represents the difference between the actual and ideal positions of the locator, initially constrained by $[-T_i/2,\ T_i/2]$ and following a normal distribution $N\left(0,\ (T_i/6)^2\right)$. Given that $E[\cdot]$ denotes the expectation operator, the statistical properties of ΔZ before operation are denoted as

$$E[\Delta Z] = E[\delta] = 0 \tag{6.59}$$

$$\sigma_Z^2 = E[\Delta Z^2] = E[\delta^2] = E^2[\delta] + D[\delta] = \frac{T^2}{36} \tag{6.60}$$

The wear of locators leads to process degradation. According to maintenance handbooks [72] , the material wear rate of locators is generally constant throughout the process. Considering the impact of the coating on the locator surface, the material

Table 6.11 Different types of locating pair and its variation

Locating pair	Point-to-surface locating pair	Four-way pin-hole locating pair	Two-way pin-hole locating pair
Diagram	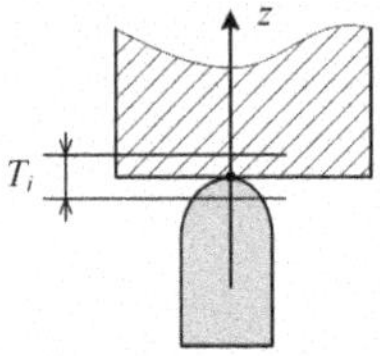	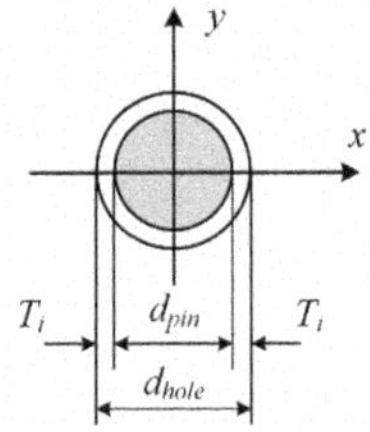	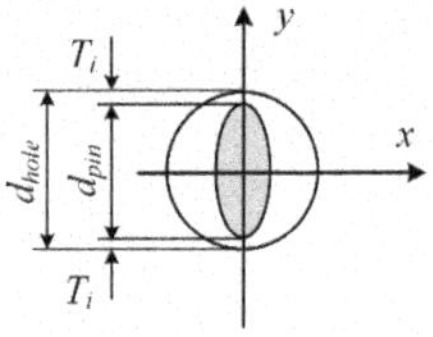
Deviation	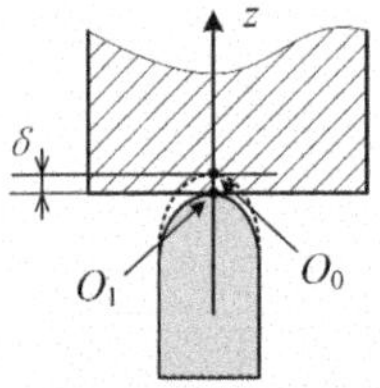	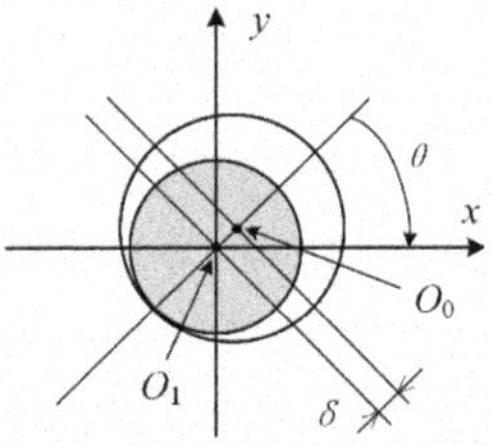	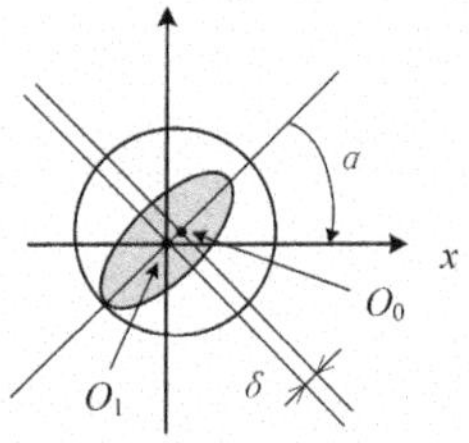

[a] O_0 is the ideal position and O_1 is the actual position

wear rate is initially low until the coating degrades. Thus, for simplicity, the locator degradation can be modeled by a quadratic curve with respect to operation time [70]. The deviation of the locating pair, accounting for degradation, can be updated as follows:

$$d_i(\tau) = \delta_i + \Delta_i(\tau) \tag{6.61}$$

where δ_i is the initial variation determined by the tolerance, $\Delta_i(\tau)$ is the aggregated wear at age τ which is defined by $\Delta_i(\tau) = G_k \cdot \tau^2$ for locator, G_k is a constant indicates the wear rate of locators at stage k. Equation (6.61) means that the locating variation changes with the wear of locator.

Consequently, by inserting Eq. (6.61) into Eq. (6.60), one can determine the variance of the locator variation that incorporates process degradation. This variance U_i is formulated as:

$$\begin{aligned} Var(U_i) &= \sigma_Z^2(\tau) = E[(\delta + \Delta(\tau))^2] = E[\delta^2 + 2\delta\Delta(\tau) + \Delta^2(\tau)] \\ &= E[\delta^2] + 2E[\delta]\mu_{\Delta\tau} + (\mu_{\Delta\tau})^2 + D[\Delta(\tau)] = \frac{T^2}{36} + (\mu_{\Delta\tau})^2 + \tau \cdot \sigma_\Delta^2 \end{aligned} \tag{6.62}$$

This formula will be iteratively applied to each point-to-surface locating pair throughout the stages of the MMPs, enabling the expression of the fixture part of $\mathbf{\Sigma}_{\mathrm{U}}$ in Eq. (6.58) in terms of the tolerance and operation time of the respective locator.

(b) Pin-hole locating scheme

In addition to the point-to-surface locating pair discussed previously, the pin-hole locating scheme, as depicted in Table 6.11, must account for two types of locating pairs: the four-way pin-hole and the two-way pin-hole. The interaction between the pin and the datum hole is cylindrical in nature. The diameter or major axis of the pin is denoted by d_{pin}, and the diameter of the datum hole is d_{hole}. The tolerance T_i of the ith pin represents the maximum allowable clearance δ_i. The clearance, which is the deviation of the hole center from the pin center, is initially confined within $[0,\ T_i]$ and is distributed normally as $N\left(T_i/2,\ (T_i/6)^2\right)$.

In the context of the four-way pin-hole locating pair, let θ represent the contact orientation. The deviations in the x and y directions are as follows:

$$\Delta X = \delta\cos\theta,\ \Delta Y = \delta\sin\theta \tag{6.63}$$

Given that the clearance for a four-way locating pair is isotropic, the orientation angle θ follows a uniform distribution $U(0, 2\pi)$. Thus, the variation statistics for a four-way locating pair can be formulated as:

$$E[\Delta X] = E[\delta\cos\theta] = E[\delta]\cdot E[\cos\theta] = 0 \tag{6.64}$$

$$E[\Delta Y] = E[\delta\sin\theta] = E[\delta]\cdot E[\sin\theta] = 0 \tag{6.65}$$

$$\sigma^2_{X,4\text{-way}} = E[\Delta X^2] = E[\delta^2]\cdot E[\cos^2\theta] = (E^2[\delta] + D[\delta])\cdot\frac{1}{2} = \frac{5T^2}{36} \tag{6.66}$$

$$\sigma^2_{Y,4\text{-way}} = E[\Delta Y^2] = E[\delta^2]\cdot E[\sin^2\theta] = (E^2[\delta] + D[\delta])\cdot\frac{1}{2} = \frac{5T^2}{36} \tag{6.67}$$

$$\mathrm{Cov}(\Delta X, \Delta Y) = E[\Delta X\,\Delta Y] = E[\delta^2\sin\theta\cos\theta] = E[\delta^2]\cdot E[\sin\theta]\cdot E[\cos\theta] = 0 \tag{6.68}$$

In the case of the two-way pin-hole locating pair, the variation occurs solely in the direction defined by α, with the deviations in the x and y directions being:

$$\Delta X = \kappa\cdot\delta\sin\alpha,\ \Delta Y = \kappa\cdot\delta\cos\alpha \tag{6.69}$$

where κ is a sign variable, which determines the sign of variation according to which side of the datum hole touches the pin. α is a fixed value according to the orientation of the two-way pin. Therefore, the statistics regarding ΔX and ΔY can be given as

$$E[\Delta X] = E[\kappa \cdot \delta \sin\alpha] = E[\kappa] \cdot E[\delta] \cdot E[\sin\alpha] = 0 \tag{6.70}$$

$$E[\Delta Y] = E[\kappa \cdot \delta \cos\alpha] = E[\kappa] \cdot E[\delta] \cdot E[\cos\alpha] = 0 \tag{6.71}$$

$$\sigma^2_{X,2\text{-way}} = E[\kappa^2 \cdot \delta^2 \sin^2\alpha] = E[\delta^2] \cdot E[\sin^2\alpha] = \frac{5T^2}{18}\sin^2\alpha \tag{6.72}$$

$$\sigma^2_{Y,2\text{-way}} = E[\kappa^2 \cdot \delta^2 \cos^2\alpha] = E[\delta^2] \cdot E[\cos^2\alpha] = \frac{5T^2}{18}\cos^2\alpha \tag{6.73}$$

$$\begin{aligned}\mathrm{Cov}(\Delta X, \Delta Y) &= E[\kappa^2 \cdot \delta^2 \sin\alpha\cos\alpha] = E[\delta^2] \cdot E[\sin\alpha] \cdot E[\cos\alpha] \\ &= \frac{5T^2}{18}\sin\alpha\cos\alpha \end{aligned} \tag{6.74}$$

Equation (6.61) provides the formula for locating variation considering process degradation, thus identifying an appropriate degradation model for the pin-hole locating pair is crucial for determining $\Delta_i(\tau)$ in Eq. (6.61). Archard [73] initially developed a sliding wear model to represent the physical wear mechanism of tools. Building upon this, Jin and Chen [74] introduced a stochastic degradation model. The cumulative wear after time τ is denoted as

$$\Delta_i(\tau) = \Delta_i(\tau - 1) + \Delta_i^r(\tau) \tag{6.75}$$

where $\Delta_i^r(\tau)$ is the wear increment at time τ, and it is characterized by a lognormal distribution Lognorm$(\mu_\Delta, \sigma_\Delta^2)$. The mean wear rate μ_Δ over time τ is described by

$$\mu_\Delta(\tau) = \mu_0 + \mu_1 e^{-\beta\tau} \tag{6.76}$$

which includes two elements: μ_0 represents the constant wear rate, while $\mu_1 e^{-\beta\tau}$ accounts for the exponential decay from the initial wear rate $\mu_0 + \mu_1$ to μ_0.

Consequently, the deviations for both types of locating pairs are revised. By substituting Eq. (6.61) into Eqs. (6.66), (6.67), (6.72) and (6.73), the variances in the two directions can be formulated as:

$$\sigma^2_{X,4\text{-way}}(\tau) = E[(\delta + \Delta(\tau))^2 \cdot \cos^2\theta] = \frac{1}{2}E[(\delta + \Delta(\tau))^2] \tag{6.77}$$

$$\sigma^2_{Y,4\text{-way}}(\tau) = E[(\delta + \Delta(\tau))^2 \cdot \sin^2\theta] = \frac{1}{2}E[(\delta + \Delta(\tau))^2] \tag{6.78}$$

$$\sigma^2_{X,2\text{-way}}(\tau) = E[\kappa^2 \cdot (\delta + \Delta(\tau))^2 \cdot \sin^2\alpha] = \sin^2\alpha \cdot E[(\delta + \Delta(\tau))^2] \tag{6.79}$$

$$\sigma^2_{Y,2\text{-way}}(\tau) = E[\kappa^2 \cdot (\delta + \Delta(\tau))^2 \cdot \cos^2\alpha] = \cos^2\alpha \cdot E[(\delta + \Delta(\tau))^2] \tag{6.80}$$

Since U_i includes deviations in both directions, the variance of the pin variation considering process degradation can be expressed as

$$\mathrm{Var}(U_{i,4-\mathrm{way}}) = \sqrt{(\sigma^2_{X,4-\mathrm{way}}(\tau))^2 + (\sigma^2_{Y,4-\mathrm{way}}(\tau))^2} = \frac{\sqrt{2}}{2}E[(\delta + \Delta(\tau))^2] \quad (6.81)$$

$$\mathrm{Var}(U_{i,2-\mathrm{way}}) = \sqrt{(\sigma^2_{X,2-\mathrm{way}}(\tau))^2 + (\sigma^2_{Y,2-\mathrm{way}}(\tau))^2} = E[(\delta + \Delta(\tau))^2] \quad (6.82)$$

where $E[(\delta+\Delta(\tau))^2] = E[\delta^2+2\delta\Delta(\tau)+\Delta^2(\tau)] = \frac{5T^2}{18}+T\cdot\mu_{\Delta\tau}+(\mu_{\Delta\tau})^2+\tau\cdot\sigma^2_{\Delta}$.

These formulas will be applied iteratively to each pin-hole locating pair throughout the stages of MMPs within the pin-hole locating scheme. In conjunction with the variance determined by Eq. (6.62) for the three locators, the fixture component of $\mathbf{\Sigma}_{\mathrm{U}}$ in Eq. (6.58) can be articulated in terms of the tolerance and operational duration of the respective fixture elements.

KCC Variations of Cutting Tool

The cutting tool constitutes a significant KCC within MMPs. It is typically presupposed that cutting tools are acquired without any initial variation. Throughout the machining process, the cutting tool undergoes wear, which induces variation. Consequently, the emphasis of cutting tool variation management lies in maintenance planning, specifically the strategy for tool replacement.

In standard machining operations, the wear of cutting tools is often represented as a third-order polynomial function of operational time in existing studies, whereas in high-speed machining, it is more commonly modeled as a second-order polynomial function [75]. For the sake of generality, the wear variable of cutting tools in MMPs is hypothesized to adhere to a quadratic curve, expressed as

$$w_j(\tau) = E_k \cdot \tau + F_k \cdot \tau^2 \quad (6.83)$$

where $w_j(\tau)$ is the wear of the jth cutting tool after time τ, E_k and F_k are wear rate coefficients at stage k. Thus, the cutting tool variation, denoted as $w_j(\tau)$, is a stochastic variable with its probability density function derived as follows.

For the jth cutting tool, a^c_j is defined as the decision variable for the replacement cycle. Consequently, $\tau \in [0, a^c_j]$ and $w^{\max}_j$ represent the admissible maximum wear of the cutting tool at time $\tau = a^c_j$. By discretizing the operational time, the variable τ can be treated as a uniform distribution $U(0, a^c_j)$. Therefore, the probability density function of the jth cutting tool wear can be derived from Eq. (6.84). For a function $y = r(x)$, if r is differentiable and the probability density function of x is known as $f_X(x)$, the probability density function of y can be calculated using the following formula:

$$g_Y(y) = f_X(r^{-1}(y)) \cdot \left| \frac{dr^{-1}(y)}{dy} \right| \tag{6.84}$$

Therefore, the probability density function of cutting tool variation is

$$g(w_j) = \frac{1}{a_j^c} \cdot \left| \frac{1}{\sqrt{(E_k)^2 + 4 \cdot F_k \cdot w_j}} \right| \tag{6.85}$$

Following extensive numerical simulations, the variance of the cutting tool variation can be determined as a function of operational time. This variance will be iteratively applied to each stage of the MMPs, allowing the cutting tool component of $\mathbf{\Sigma}_{\mathrm{U}}$ in Eq. (6.52) to be expressed.

6.3.2.4 Optimization Model Construction

Traditional variation management models focus on single-objective optimization [63], or they combine the objectives by summing quality loss and variation management costs [70]. A bi-objective optimization model, resolved through the MCGP method, can identify the balance point between quality and cost, circumventing the challenge of assigning weights to these two components [74] . Furthermore, beyond dimensional tolerance constraints, geometric tolerances must also be incorporated into the constraints of the optimization model. This section details the definition of objective functions and constraints, as well as the resolution of the optimization model.

Objective Functions Definition

In the context of this bi-objective optimization model, the objective functions are designed to achieve two goals: the minimization of variation management costs and the reduction of quality loss.

(a) The objective function of variation management cost

Taking into account the two types of KCC in MMPs, the variation management cost encompasses the following:

(1) *Fixture cost*: The tolerance-cost function for each locator/pin is inversely proportional in this section, expressed as $C_T^i = w_i/T_i$, $i = 1, ..., p$, where T_i is the allocated tolerance for the ith locator/pin and w_i is the corresponding weight coefficient. Besides the tolerance cost, scheduled maintenance incurs an additional fixed cost c_{0i}^f(attributable to labor and management) for fixture elements. Thus, the fixture cost comprises both tolerance cost and fixed cost. The cost incurred at time t can be defined as:

$$C_M^f(t) = \sum_{i \in S_t} (C_T^i + c_{0i}^f) = \sum_{i \in S_t} \left(\frac{w_i}{T_i} + c_{0i}^f \right), \; i = 1, \ldots, p \tag{6.86}$$

where S_t is the index set of locators/pins that are subject to scheduled maintenance at time t, which is determined by the maintenance strategy $\mathbf{a}^f$. The notation $i \in S_t$ signifies that the ith locator/pin is scheduled for maintenance at time t according to the plan. Consequently, $C_M^f(t)$ is contingent on both decision parameters and can be formulated as $C_M^f(t;\ \mathbf{T}, \mathbf{a}^f)$. A tighter tolerance or a more frequent maintenance schedule will result in increased costs.

(2) *Cutting tool cost*: Similar to the fixture cost, cutting tool cost induced at time t can be defined as

$$C_M^c(t) = \sum_{j \in S_t} (c_j^c + c_{0j}^c), \; j = 1, \ldots, q \tag{6.87}$$

where c_j^c is the cost of cutting tool itself, c_{0j}^c is the fixed cost of jth cutting tool replacement, $j \in S_t$ means that the jth cutting tool needs to replace at time t according to the schedule. Hence, $C_M^c(t)$ depends on the decision parameter of $\mathbf{a}^c$ and can be described as $C_M^c(t;\ \mathbf{a}^c)$. A high replacement frequency increases the cost.

(b) The objective function of quality loss

The variation propagation model introduced in Sect. 6.3.2.2 can effectively forecast the variations of KPCs, represented as $\mathbf{Y}$. Adhering to the Taguchi quality loss principle of "the smaller the better," the quality loss function is quadratic in nature. Pignatiello [77] expanded the quality loss function to accommodate multi-dimensional variables. In this section, Eq. (7.50) is utilized to represent the multivariate quality loss function:

$$L(\mathbf{Y}) = \mathbf{Y}^T \mathbf{S} \mathbf{Y} \tag{6.88}$$

where $\mathbf{S} = q\mathbf{I}$ is a symmetric positive-definite matrix defining the contribution of each KPC to product quality loss.

Given that Y is contingent on the variations of KCCs, which in turn are determined by time t and decision parameters T and a, the quality loss function $L(\mathbf{Y})$ is influenced by all three variables and can be articulated as $L(\mathbf{Y}(t);\ \mathbf{T}, \mathbf{a})$.

Constraints Definition

The constraints of the optimization model are primarily categorized into two types: product quality constraints and decision parameter limit constraints. The former is the primary focus and will be detailed in this section, while the latter will be presented directly within the optimization model in the subsequent section.

Product tolerances establish the design specifications for the workpiece. To maintain product quality throughout the MMP, both dimensional and geometric tolerances must be met simultaneously. This means that both types of tolerances should be mathematically incorporated into the constraint equations. In this section, it marks the inaugural expansion of variation management in MMP to include the geometric dimension.

As detailed in Sect. 6.3.2.2, the KPC deviation vector can be denoted by $\mathbf{Y} = \left[(\mathbf{y}_1^R)^T \cdots (\mathbf{y}_m^R)^T\right]^T$, and the deviation of an individual KPC can be expressed in the form of DMV as $\mathbf{y}_h^R = \left[\Delta x_h^R\ \Delta y_h^R\ \Delta z_h^R\ \Delta\alpha_h^R\ \Delta\beta_h^R\ \Delta\gamma_h^R\right]^T$. In accordance with the principle of invariance, tolerances only apply to the non-invariant elements of DMV [78]. Planar and cylindrical KPCs constitute the majority in MMPs. A planar feature has three invariants: two translations parallel to the plane and one rotation around the normal vector of the plane, whereas a cylindrical feature has two invariants: translation along its axis and rotation around its axis. Therefore, the constraint tolerance for planar KPCs can be simplified to $[0\,0\,w\,\alpha\,\beta\,0]^T$ and for cylindrical KPCs to $[u\,v\,0\,\alpha\,\beta\,0]^T$, where $\mathbf{T}_h^{KPC} = \left[u, v, w, \alpha, \beta, \gamma\right]$ encompasses six constraint elements in the form of DMV for general features.

Based on the variation propagation model given by Eq. (7.58), the variability of a KPC, indexed as h, can be computed as:

$$\boldsymbol{\Sigma}_{\mathbf{y}_h} = \boldsymbol{\Gamma}_h \boldsymbol{\Sigma}_{\mathbf{U}} \boldsymbol{\Gamma}_h^T, \quad h = 1, ..., m \tag{6.89}$$

where $\boldsymbol{\Sigma}_{\mathbf{y}_h}$ is the 6×6 covariance matrix indicating the deviation of hth KPC, whose six diagonal elements are the variances of six elements in $\mathbf{y}_h = \left[\Delta x\ \Delta y\ \Delta z\ \Delta\alpha\ \Delta\beta\ \Delta\gamma\right]^T$, i.e., $\boldsymbol{\Sigma}_{\mathbf{y}_h}[\rho, \rho] = \text{Var}(\mathbf{y}_h[\rho])$, $\rho = 1, 2, \ldots, 6$. To ensure the quality, it should be restricted by the tolerance $\mathbf{T}_h^{KPC}$, which can be denoted by $\boldsymbol{\Sigma}_{\mathbf{y}_h}[\rho, \rho] \leq \text{Var}(\mathbf{T}_h^{KPC}[\rho])$.

For managing dimensional and geometric tolerances, Loose et al. [14] pioneered the integration of Geometric Dimensioning and Tolerancing (GD&T) and constrained deviations based on boundary points. However, the number of constraint equations is contingent upon the number of boundary points, with each KPC necessitating at least four boundary points. This implies that a multitude of nonlinear constraints would demand a complex and time-consuming optimization algorithm. Consequently, a novel constraint method is required for $\boldsymbol{\Sigma}_{\mathbf{y}_h}$ to address these challenges.

Table 6.12 outlines the general deviations in both dimensional and geometric scales for two types of features and establishes the constraint relationships between DMV elements based on various tolerances. For each non-invariant element, the permissible range is clearly defined. The planar KPC is subject to constraints by parallelism tolerance ε in addition to dimensional tolerance T_d. Roy and Li [79] introduced the variation constraint inequalities for planar features. For other geometric tolerances in planar features, such as angularity and perpendicularity, the approach is to multiply the deviation DMV by a homogeneous transformation matrix. For the cylindrical KPC, the centerline is constrained. The case in Table 6.12 is denoted with the tolerance $\phi\omega$ with respect to the GD&T datum features A and B.

Table 6.12 Constraint relations considering geometric tolerances

	The planar KPC	The cylindrical KPC
Tolerance condition	Dimensional tolerance T_d parallelism tolerance ε	Position tolerance $\phi\omega$
Schematic diagram	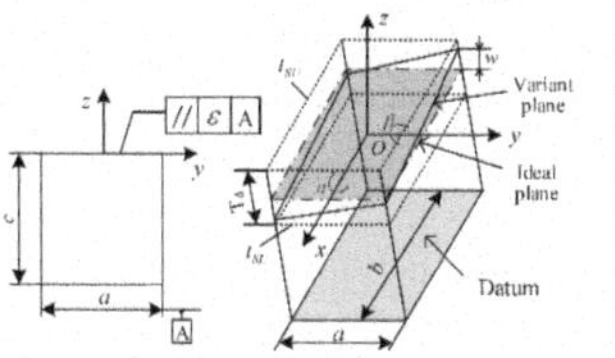	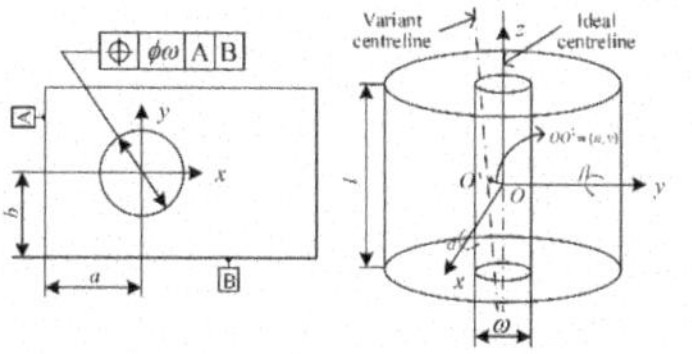
Constraint relations	$-\varepsilon/2 \leq (\alpha \cdot y + \beta \cdot x) \leq \varepsilon/2$ $-T_d/2 \leq (w+\alpha \cdot y+\beta \cdot x) \leq T_d/2$	$(u + \beta \cdot z)^2 + (v + \alpha \cdot z)^2 \leq (\omega/2)^2$
Allowable range of DMV non-invariant elements	$-\varepsilon/b \leq \alpha \leq \varepsilon/b$ $-\varepsilon/a \leq \beta \leq \varepsilon/a$ $-T_d/2 \leq w \leq T_d/2$	$-\omega/2l \leq \alpha \leq \omega/2l$ $-\omega/2l \leq \beta \leq \omega/2l$ $-\omega/2 \leq u \leq \omega/2$ $-\omega/2 \leq v \leq \omega/2$

The variance of the elements of interest in $\mathbf{T}_h^{KPC}$ is simulated using the Monte Carlo method. For example, if w is the focus of the planar feature, it has a variance of 0.0126 when $T_d = 0.1$ mm mm, $\varepsilon = 0.05$ mm, $a = 300$ mm, $b = 100$ mm, and the number of simulations is 10000. This means the quality constraint for this KPC is $\mathbf{\Sigma}_{\mathbf{Y}_h}[3, 3] \leq 0.0126$.

Optimization Model

With the objective functions and constraints defined, this section will develop the optimization model for managing KCC variations within MMPs.

The total variation management cost at time t is computed as follows:

$$C(t) = C_M^f(t;\ \mathbf{T}, \mathbf{a}^f) + C_M^c(t;\ \mathbf{a}^c) \tag{6.90}$$

The long-run expected variation management cost per unit time is

$$\begin{aligned}\Phi(\mathbf{T}, \mathbf{a}) \equiv \lim_{t\to\infty} \frac{\sum_{\tau=0}^{t} E(C(\tau))}{t} &= \lim_{t\to\infty} \frac{\sum_{\tau=0}^{t} E\Big(\sum_{i\in S_t} (\frac{w_i}{T_i} + c_{0i}^f) + \sum_{j\in S_t} (c_j^c + c_{0j}^c)\Big)}{t} \\ &= \sum_{i=1}^{p}\sum_{j=1}^{q}\left(\lim_{t\to\infty} \frac{\sum_{\tau=0}^{t} (E(C_M^{f,i}(\tau)) + E(C_M^{c,j}(\tau)))}{t}\right)\end{aligned} \tag{6.91}$$

where $C_M^{f,i}(\tau) \equiv \begin{cases} w_i/T_i + c_{0i}^f, & \text{if } i \in S_t \\ 0, & \text{otherwise} \end{cases}$ and $C_M^{c,j}(\tau) \equiv \begin{cases} c_j^c + c_{0j}^c, & \text{if } j \in S_t \\ 0, & \text{otherwise} \end{cases}$.

For the quality loss $L(\mathbf{Y}(t);\ \mathbf{T},\mathbf{a})$, the expected value of quality loss is

$$E(L(\mathbf{Y})) = E\left(\sum_{i=1}^{m}\sum_{j=1}^{m} s_{ij}Y_iY_j\right) = \sum_{i=1}^{m}\sum_{j=1}^{m} s_{ij}Cov(Y_i, Y_j)$$
$$= \sum_{r=1}^{n}\left(\mathbf{\Gamma}_{(:,r)}^T\mathbf{S}\mathbf{\Gamma}_{(:,r)}\right)\mathrm{Var}(U_r) = \sum_{r=1}^{n}\rho_r\mathrm{Var}(U_r) \tag{6.92}$$

where s_{ij} is the (i, j)th element of matrix $\mathbf{S}$, $\rho_r \equiv \mathbf{\Gamma}_{(:,r)}^T\mathbf{S}\Gamma_{(:,r)}$, and $\rho_r\mathrm{Var}(U_r)$ denotes the contribution of rth KCC to the quality loss.

Therefore, the long-run expected cost per unit time can be obtained by

$$\Psi(\mathbf{T}, \mathbf{a}) \equiv \lim_{t\to\infty}\frac{\sum_{\tau=0}^{t} E(L(\mathbf{Y}(\tau)))}{t}$$
$$= \lim_{t\to\infty}\frac{\sum_{\tau=0}^{t}\left(\sum_{r=1}^{n}\rho_r\mathrm{Var}(U_r)\right)}{t} = \sum_{r=1}^{n}\left(\lim_{t\to\infty}\frac{\sum_{\tau=0}^{t} L_r(\mathbf{Y}(\tau))}{t}\right) \tag{6.93}$$

The aim of the model is to reduce both the variation management cost and the quality loss, thereby determining the optimal decision parameters for variant management in MMPs that will balance quality and cost. The optimization model is structured as:

$$\min\begin{cases} \Phi(\mathbf{T}, \mathbf{a}) = \sum_{i=1}^{p}\sum_{j=1}^{q}\left(\lim_{t\to\infty}\frac{\sum_{\tau=0}^{t}\left(E(C_M^{f,i}(\tau)) + E(C_M^{c,j}(\tau))\right)}{t}\right) \\ \Psi(\mathbf{T}, \mathbf{a}) = \sum_{r=1}^{n}\left(\lim_{t\to\infty}\frac{\sum_{\tau=0}^{t} L_r(\mathbf{Y}(\tau))}{t}\right) \end{cases} \tag{6.94}$$

subjected to $\mathbf{\Sigma}_{\mathbf{y}_h}[\rho, \rho] \le \mathrm{Var}(\mathbf{T}_h^{KPC}[\rho])$, $0 \le T_i \le T_i^{\max}$, $a_i^f \ge 0$, $a_j^c \ge 0$, $\forall i, j, r, h, \rho$.

6.3.3 *Case Study*

6.3.3.1 Problem Description

The machining of an automotive engine cylinder block, as depicted in Fig. 6.29, exemplifies a complex MMP that encompasses over twenty stages. The top surface of the engine cylinder block acts as the mating surface with the engine cylinder head

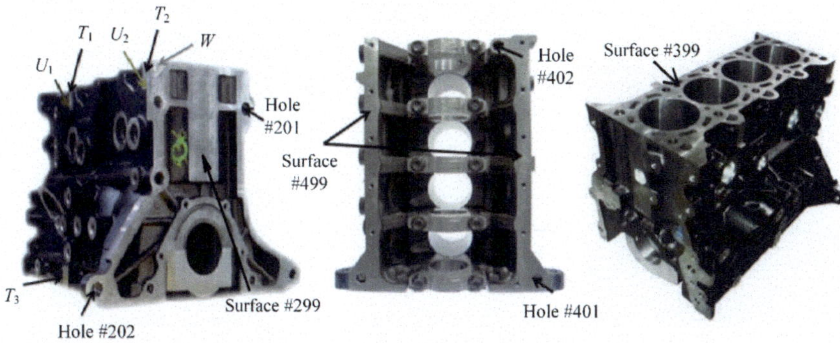

Fig. 6.29 The automotive engine cylinder block

[80], and its precision significantly impacts the engine's sealing performance [81, 82]. Consequently, this section concentrates on the machining of the top surface and selects three exemplary sequential stages from actual production. These stages capture the variation propagation inherent in MMPs and encompass various types of locating pairs, thereby effectively validating the proposed method for managing KCC variations.

Figure 6.30 provides a concise overview of the three stages. Comprehensive details, including the datum features and the nominal positions of the machining features relative to the reference CS, are presented in Table 6.10. Notably, OP10 employs the standard 3-2-1 locating scheme, with T_1, T_2, T_3, U_1, U_2, and W serving as the rough datum features of the workpiece. The deviation of feature #299 machined at OP10, along with the deviations of two holes drilled at another stage, constitute variation sources for the machining features at OP20. In a similar vein, the datum variations from OP20 accumulate deviations at OP30, which together form the System of Variations (SoV). Both OP20 and OP30 utilize the pin-hole locating scheme, encompassing a four-way pin and a two-way pin locating pair. Post OP30, the machined workpiece is transferred to inspection to assess the Key Product Characteristic (KPC) defined by surface #399. The dimensional tolerance for this KPC is specified as $\pm$ 0.05 mm, and a parallelism tolerance of 0.03 mm is required with respect to the GD&T datum features #499.

In this scenario, a total of 19 KCCs are taken into account for variation management ($r = 1, \ldots, 19$), which includes six locators in OP10 ($P_1 - P_6$), three locators and two pins in both OP20 ($P_7 - P_{11}$) and OP30 ($P_{12} - P_{16}$), and one cutting tool in each stage ($Q_1 - Q_3$). The goal of variation management for MMPs is to allocate optimal tolerances and replacement cycles for each KCC in order to attain a balance between quality and cost.

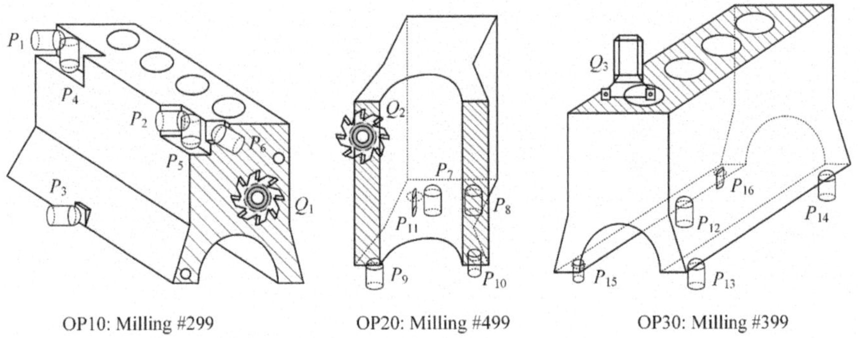

Fig. 6.30 Three stages of the MMP

6.3.3.2 Numerical Analysis

A numerical analysis was performed using the parameters detailed in Table 6.14 to refine the decision parameters **T** and **a**. Various multi-objective optimization techniques exist in the literature for addressing such models. The weighted method is frequently employed but it introduces subjectivity or bias when assigning weights to combine disparate objectives into a single scalar function [83]. Similarly, Archimedean and non-Archimedean goal programming methods grapple with establishing a hierarchical order of goals. The Multi-Criteria Goal Programming (MCGP) approach [84] circumvents these challenges and is utilized in this section. The rationale behind employing the Chebyshev distance metric is to achieve equilibrium. This method aims to balance the accomplishment of multiple objectives, contrasting with the lexicographic approach that prioritizes certain goals over others, or the weighted approach that selects decision variable values to minimize an achievement function [85]. Consequently, the MCGP approach is capable of yielding an optimal solution where equilibrium in the levels of goal satisfaction is essential [86].

For the specific procedural steps, the model is initially decomposed into two single-objective optimization problems, both subject to the same constraints. Each of these problems is addressed using the *fmincon* function in MATLAB, which employs the Sequential Quadratic Programming (SQP) method. The SQP method is an efficient algorithm for nonlinear programming, renowned for its rapid convergence. The solution process in SQP involves solving a series of quadratic sub-problems. The objective values obtained from the solutions to these two decomposed problems are summarized in Table 6.15.

The values of the two objectives are not directly comparable. Minimizing the variation management cost typically results in higher quality loss, while reducing quality loss often leads to an increase in cost. Since neither extreme is optimal, the MCGP method is invoked to find a compromised solution. This approach can be interpreted within the framework of fuzzy programming, where the fuzzy membership functions for the two objectives are defined [87].

Table 6.13 Process description and nominal locations of key features

Stage	Datum features	Process descriptions	${}^{0}\boldsymbol{\omega}_{0}^{R}$	${}^{0}\mathbf{t}_{0}^{R}$
OP10	$T_1, T_2, T_3, U_1, U_2, W$	Mill flank surface #299	[0, 0, 0]	[– 170.5, 138, 0]
OP20	#299, #201, #202	Mill bottom surface #499	[0, π/2, π/2]	[17.5, 248, – 30.5]
OP30	#499, #401, #402	Mill top surface #399	[0, – π/2, π/2]	[– 218.5, 248, – 30.5]

$$\mu_1 = \begin{cases} 1, & \Phi(\mathbf{T},\mathbf{a}) \leq \Phi(\mathbf{T},\mathbf{a})_1^* \\ \frac{\Phi(\mathbf{T},\mathbf{a})_2^* - \Phi(\mathbf{T},\mathbf{a})}{\Phi(\mathbf{T},\mathbf{a})_2^* - \Phi(\mathbf{T},\mathbf{a})_1^*}, & \Phi(\mathbf{T},\mathbf{a})_1^* \leq \Phi(\mathbf{T},\mathbf{a}) \leq \Phi(\mathbf{T},\mathbf{a})_2^* \\ 0, & \Phi(\mathbf{T},\mathbf{a}) \geq \Phi(\mathbf{T},\mathbf{a})_2^* \end{cases} \tag{6.95}$$

$$\mu_2 = \begin{cases} 1, & \Psi(\mathbf{T},\mathbf{a}) \leq \Psi(\mathbf{T},\mathbf{a})_2^* \\ \frac{\Psi(\mathbf{T},\mathbf{a}) - \Psi(\mathbf{T},\mathbf{a})_1^*}{\Psi(\mathbf{T},\mathbf{a})_2^* - \Psi(\mathbf{T},\mathbf{a})_1^*}, & \Psi(\mathbf{T},\mathbf{a})_2^* \leq \Psi(\mathbf{T},\mathbf{a}) \leq \Psi(\mathbf{T},\mathbf{a})_1^* \\ 0, & \Psi(\mathbf{T},\mathbf{a}) \geq \Psi(\mathbf{T},\mathbf{a})_1^* \end{cases} \tag{6.96}$$

The best distance from two objectives' worst values is the compromised solution which can be obtained by maximizing δ. It is formulated as

$$\max\ \delta \tag{6.97}$$

subjected to $\delta \leq \frac{\Phi(\mathbf{T},\mathbf{a})_2^* - \Phi(\mathbf{T},\mathbf{a})}{\Phi(\mathbf{T},\mathbf{a})_2^* - \Phi(\mathbf{T},\mathbf{a})_1^*}$, $\delta \leq \frac{\Psi(\mathbf{T},\mathbf{a}) - \Psi(\mathbf{T},\mathbf{a})_1^*}{\Psi(\mathbf{T},\mathbf{a})_2^* - \Psi(\mathbf{T},\mathbf{a})_1^*}$, and all constraints in original model.

The solution to the model represents the equilibrium point between quality and cost, specifically $\delta = 0.814$, with $\Phi(\mathbf{T},\mathbf{a})^* = 1.820$ and $\Psi(\mathbf{T},\mathbf{a})^* = 0.279$. The optimal tolerance and replacement cycle for each KCC are detailed in Table 6.13. Variation management strategies vary for different KCCs because each KCC impacts the quality loss to a different extent. For instance, the variation of a particular locator may have a minimal effect on the final workpiece quality; therefore, a more lenient tolerance and a longer replacement cycle are assigned. This approach can help to reduce costs while maintaining the quality at an acceptable level (Table 6.16).

6.3.3.3 Comparison and Discussion

In alignment with the three principal contributions of this section, the advantages of the proposed method are contrasted and discussed across three dimensions: (i) application scenarios, (ii) integration of geometric tolerances, and (iii) performance within the manufacturing system.

Regarding the comparison of application scenarios, this case encompasses two distinct types of locating schemes. However, previous research on KCC variation management for MMPs has been limited to addressing the conventional 3-2-1

Table 6.14 Parameters for variation management in the case study

Point-to-surface locating pair related (mm)		Pin-hole locating pair related (mm)		Cutting tool related (mm)		Cost function related ($)		Quality loss related (unit/mm^2)		Constrain related (mm)	
G_k	1.9×10^{-7}	μ_0	5×10^{-7}	E_k	0.0113	w_i	200	q	1	a	108
μ_Δ	8.7×10^{-8}	μ_1	1×10^{-6}	F_k	0.0019	c_{0i}^f	200			b	326.5
σ_Δ	2.0×10^{-7}	β	1×10^{-3}	μ_Δ	3.4×10^{-4}	c_j^c	50			$T_i^{\max}$	0.1
		μ_Δ	5.0×10^{-7}	σ_Δ	6.8×10^{-3}	c_{0j}^c	100				
		σ_Δ	5.0×10^{-5}								

Table 6.15 Solution of two single-objective problems

	min $\Phi(\mathbf{T}, \mathbf{a})$	min $\Psi(\mathbf{T}, \mathbf{a})$	|Difference|
Value of $\Phi(\mathbf{T}, \mathbf{a})$	$\Phi(\mathbf{T}, \mathbf{a})_1^* = 1.678$	$\Phi(\mathbf{T}, \mathbf{a})_2^* = 2.441$	0.763
Value of $\Psi(\mathbf{T}, \mathbf{a})$	$\Psi(\mathbf{T}, \mathbf{a})_1^* = 0.297$	$\Psi(\mathbf{T}, \mathbf{a})_2^* = 0.275$	0.022

Table 6.16 Optimal variation management strategy for each KCC

i	1	2	3	4	5	6	7	8	9	10
T_i	0.028	0.021	0.031	0.093	0.087	0.016	0.031	0.028	0.029	0.018
a_i^f	3.43	2.32	3.67	11.82	10.18	1.96	3.41	3.58	3.21	1.98
i	11	12	13	14	15	16	j	1	2	3
T_i	0.013	0.029	0.025	0.027	0.092	0.088	a_j^c	4.10	4.94	5.11
a_i^f	1.51	3.54	3.11	3.22	11.27	11.0				

[a] the unit of T is "mm", the unit of a_i^f is "$\times 10^4$ operations", and the unit of a_j^c is "$\times 10^3$ operations"

locating scheme, characterized by point-to-surface contact. This section extends the variation distribution to include a new type of KCC in a degradation state and incorporates it into the novel variation management model. This advancement overcomes the previous limitation and broadens the range of applicable scenarios.

To highlight the enhancement of the variation management method following the integration of geometric tolerances, 10,000 Monte Carlo simulations were conducted for two MMPs. These two MMPs are delineated by the output of variation management strategies: one, consistent with previous studies, disregards geometric tolerances, while the other implements the strategy derived from the proposed method. For each simulation, a random value for each KCC is generated based on its probability density function. To emulate the impact of un-modeled noise, a random variable following a normal distribution $N\left(0,\ (0.005/6)^2\right)$ is included in each simulation. The simulation outcomes reveal that the proposed method ensures that 99.21% of the workpieces meet the final quality specifications, whereas only 90.44% can be guaranteed if geometric tolerance is excluded from the variation management strategy. The discrepancy in the proportion of qualified workpieces stems from the fact that omitting geometric tolerance consideration results in looser tolerances when formulating optimization problems. Consequently, by accounting for the impact of geometric tolerance on MMPs, the variation management method increases the number of workpieces within specifications by 9.7%, underscoring the importance of integrating geometric tolerances. Additionally, Fig. 6.31 illustrates the probability density distribution of KPC deviation following the Monte Carlo simulations. It verifies that, for the case study, the KPC variability can be approximated as a normal distribution, even though some KCCs do not follow a normal distribution.

For a more comprehensive comparison of manufacturing system performance, two prevalent approaches for KCC variation management in the automotive industry are examined. The first approach relies solely on experience, uniformly setting the

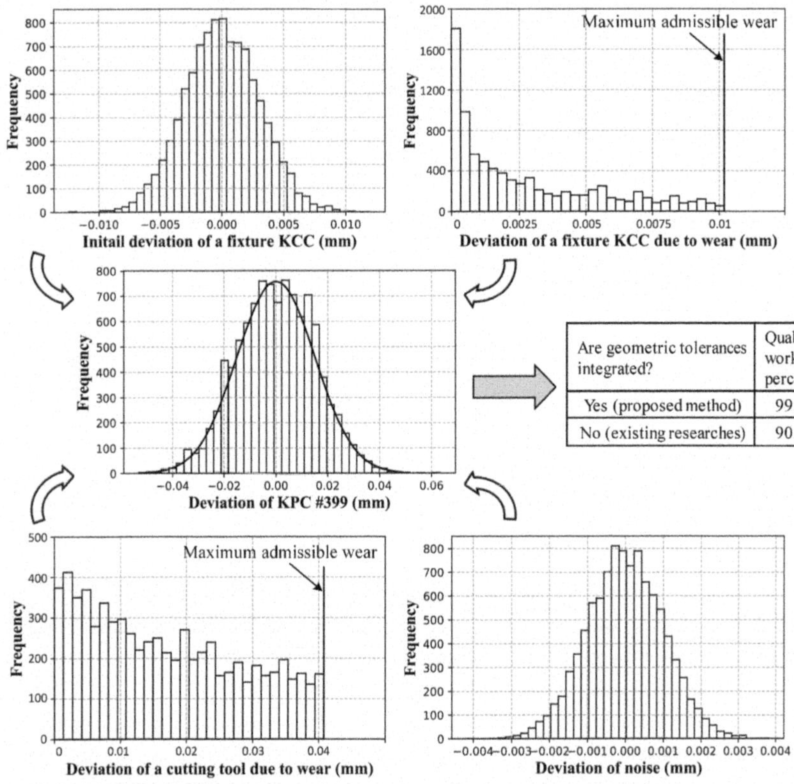

Are geometric tolerances integrated?	Qualified workpiece percentage
Yes (proposed method)	99.21%
No (existing researches)	90.44%

Fig. 6.31 The distribution of KCCs and KPC deviation after 10,000 simulations

tolerances of fixture elements to 0.1 mm. The second approach is the process-oriented tolerancing method, which assigns specific tolerances to each fixture element to ensure a high-quality product. In these conventional methods, the replacement cycle for fixture elements is arbitrarily set to 10,000 operations, and for cutting tools, it is set to 1000 operations, based on experience. The proposed method integrates both tolerance and maintenance planning for KCCs, and the comparison with these two traditional methods is presented in Table 6.17.

Compared to the engineering experience approach, the process-oriented tolerancing method optimizes tolerance configurations based on the impact of each KCC on product quality, resulting in a significantly lower long-term average quality loss without increasing the total cost. This highlights the importance of tolerance allocation for KCCs in the manufacturing system. The method proposed in this section extends process-oriented tolerancing by incorporating maintenance planning, creating a comprehensive variation management strategy. Although this method incurs a higher initial setup cost due to tighter tolerances, it achieves a similar long-term average KCC cost to the process-oriented tolerancing method because

Table 6.17 Comparison with conventional methods in manufacturing industry

Item	Formula	Engineering experience method	Process-oriented tolerancing method	The proposed method
KCC cost at the first setup ($)	$\sum_{i}^{p} \frac{w_i}{T_i} + \sum_{j}^{q} c_j^c$	32,000	30,628	113,955
Long-term average KCC cost ($/operation)	$\lim_{t\to\infty} \frac{\sum_{\tau=0}^{t} E\left(\sum_{i\in S_t} \frac{w_i}{T_i} + \sum_{j\in S_t} c_j^c\right)}{t}$	1.380	1.321	1.394
Long-term average variation management cost ($/operation)	$\lim_{t\to\infty} \frac{\sum_{\tau=0}^{t} E\left(\sum_{i\in S_t} \left(\frac{w_i}{T_i} + c_{0i}^f\right) + \sum_{j\in S_t} \left(c_j^c + c_{0j}^c\right)\right)}{t}$	2.783	2.724	1.820
Long-term average quality loss (unit/operation)	$\lim_{t\to\infty} \frac{\sum_{\tau=0}^{t} E(L(\mathbf{Y}(\tau)))}{t}$	0.554	0.281	0.279

[a] The KCC cost includes the tolerance cost of fixture element and the cost of cutting tool itself

of extended service life. More importantly, the proposed method offers a significant advantage in long-term average maintenance cost due to reduced maintenance frequency, ultimately leading to a 33.2% reduction in total variation management cost. In terms of quality, the process-oriented tolerancing method significantly enhances product quality and reduces the long-term average quality loss by 49.3%. The proposed method maintains a comparable level of quality loss while greatly reducing variation management costs. Therefore, in terms of balancing quality and cost, the method proposed in this section outperforms the two conventional methods widely adopted in the manufacturing industry.

The deviation measurement outcomes of KPCs from actual production indicate that the implementation of the proposed method has also enhanced the performance of the manufacturing system. Based on the process capacity calculations from the real production workshop, the process capacity index for the top surface machining has been elevated to over 1.33, with the defective rate decreasing from 0.27% to 0.11%. Given that the annual production volume of this engine cylinder block is 350,000 units, with each unit costing nearly 70 USD, the reduction in the defective rate alone translates to an annual cost saving of nearly 400,000 USD. This does not account for the additional cost savings achieved through the optimization of the maintenance strategy. Consequently, this section has thoroughly discussed the advantages of the proposed method, which can significantly enhance the manufacturing system's performance in both quality and cost aspects.

6.3.4 Conclusion

A novel variation management framework for KCCs in MMPs has been introduced, aiming to achieve a balance between quality and cost. This innovative approach integrates process-oriented tolerancing with maintenance planning. Within the state space model of MMPs, the variation distribution for each KCC is established during its degradation phase. The optimization model then allocates the most effective variation management strategy for each KCC, considering its influence on the manufacturing system. The framework addresses and overcomes the limitations found in prior research regarding application scenarios, comprehensive quality specifications, and the equilibrium between quality and cost. A case study on the MMP of an automotive engine cylinder block is presented to illustrate the potential of the proposed variation management methodology. Findings indicate that neglecting geometric tolerance constraints in variation management strategies can reduce the qualification rate of workpieces; the proposed method, however, ensures optimal manufacturing system performance with respect to quality and cost, covering various locating schemes.

The example of a three-stage automotive engine cylinder block machining process is utilized to demonstrate the proposed method. It is noteworthy that the overarching framework, the state space model for variation propagation, and the KCC degradation model are generic to various MMP types, indicating a strong capacity for scalability across different production scenarios. For future research, the incorporation of machining parameters is suggested. The optimization of these parameters can significantly enhance production efficiency and reduce costs. However, adjustments to machining parameters might accelerate KCC degradation, potentially diminishing product quality. Thus, the concurrent optimization with machining parameters presents an intriguing area for variation management. Additionally, adaptive preventive maintenance, which adjusts the maintenance schedule based on real-time process measurements, reflects the dynamic nature of maintenance. Consequently, the selection of maintenance strategies based on measurement costs and practicality merits further investigation.

References

1. Du S, Xi L, Ni J, Pan E, Liu CR (2008) Product lifecycle-oriented quality and productivity improvement based on stream of variation methodology. Comput Ind 59(2):180–192
2. Du S, Lv J, Xi L (2012) On-line classifying process mean shifts in multivariate control charts based on multiclass support vector machines. Int J Prod Res 50:6288–6310
3. Du S, Lv J (2013) Minimal Euclidean distance chart based on support vector regression for monitoring mean shifts of auto-correlated processes. Int J Prod Econ 141:377–387
4. Du S, Huang D, Lv J (2013) Recognition of concurrent control chart patterns using wavelet transform decomposition and multiclass support vector machines. Comput Ind Eng 66:683–695
5. Bai D, Yun H (1996) Optimal allocation of inspection effort in a serial multi-stage production system. Comput Ind Eng 30(3):387–396
6. Shetwan ALG, Vilentin IV, Tjahjono B (2011) Allocation of quality control stations in multistage manufacturing systems. Comput Ind Eng 60:473–484

7. Jin J, Shi J (1999) State space modeling of sheet metal assembly for dimensional control. J Manuf Sci Eng 121(4):7
8. Ding Y, Ceglarek D, Shi J (2002) Design evaluation of multi-station assembly processes by using state space approach. J Mech Des 124:408
9. Camelio JA, Hu SJ, Ceglarek D (2004) Impact of fixture design on sheet metal assembly variation. J Manuf Syst 23(3):182–193
10. Huang Q, Shi J, Yuan J (2003) Part dimensional error and its propagation modeling in multi-operational machining processes. J Manuf Sci Eng 125(2):255–262
11. Djurdjanovic D, Ni J (2001) Linear state space modeling of dimensional machining errors. Transactions-North American Manufacturing Research Institution of SME, pp 541–548
12. Huang Q, Shi J (2004) Variation transmission analysis and diagnosis of multi-operational machining processes. IIE Trans 36(9):807–815
13. Loose JP, Zhou S, Ceglarek D (2007) Kinematic analysis of dimensional variation propagation for multistage machining processes with general fixture layouts. IEEE Trans Autom Sci Eng 4(2):141–152
14. Loose JP, Zhou Q, Zhou S, Ceglarek D (2010) Integrating GD&T into dimensional variation models for multistage machining processes. Int J Prod Res 48(11):3129–3149
15. Cao Y, Subramaniam V, Chen R (2012) Performance evaluation and enhancement of multistage manufacturing systems with rework loops. Comput Ind Eng 62:161–176
16. Zhou S, Huang Q, Shi J (2003) State space modeling of dimensional variation propagation in multistage machining process using differential motion vectors. IEEE Trans Robot Autom 19(2):296–309
17. Abellan-Nebot JV, Liu J, Subirón FR, Shi J (2012) State space modeling of variation propagation in multistation machining processes considering machining-induced variations. J Manuf Sci Eng 134(2):021002–021002
18. Shi J (2006) Stream of variation modeling and analysis for multistage manufacturing processes. CRC Press, New York
19. Shi J, Zhou S (2009) Quality control and improvement for multistage systems: a survey. IIE Trans 41(9):744–753
20. Du S, Yao X, Huang D (2015) Engineering model-based Bayesian monitoring of ramp-up phase of multistage manufacturing process. Int J Prod Res 53(15):4594–4613
21. Craig JJ (2004) Introduction to robotics: mechanics and control. Prentice Hall
22. Cai W, Hu SJ, Yuan J (1997) A variational method of robust fixture configuration design for 3-d workpieces. J Manuf Sci Eng 119(4):593–602
23. Hu SJ (1997) Stream-of-variation theory for automotive body assembly. CIRP Ann Manuf Technol 46(1):1–6
24. Jin J, Shi J (1999) State space modeling of sheet metal assembly for dimensional control. J Manuf Sci Eng-Trans Asme 121(4):756–762
25. Mantripragada R, Whitney DE (1999) Modeling and controlling variation propagation in mechanical assemblies using state transition models. IEEE Trans Robot Autom 15(1):124–140
26. Ding Y, Ceglarek D, Shi J (2002) Design evaluation of multi-station assembly processes by using state space approach. J Mech Des 124(3):408–418
27. Djurdjanovic D, Ni J (2003) Dimensional errors of fixtures, locating and measurement datum features in the stream of variation modeling in machining. J Manuf Sci Eng-Trans Asme 125(4):716–730
28. Huang Q, Shi J (2003) Part dimensional error and its propagation modeling in multi-operational machining processes. J Manuf Sci Eng-Trans Asme 125(2):255–262
29. Abellan-Nebot JV, Liu J, Subirón FR, Shi J (2012) State space modeling of variation propagation in multistation machining processes considering machining-induced variations. J Manuf Sci Eng-Trans Asme 134(2):021002
30. Luan X, Zhang S, Li G (2018) Modified power prediction model based on infinitesimal cutting force during face milling process. Int J Precision Eng Manuf-Green Technol 5(1):71–80
31. Zhang H, Zhang Q, Ren Y, Shay T, Liu G (2018) Simulation and experiments on cutting forces and cutting temperature in high speed milling of 300 M steel under CMQL and dry conditions. Int J Precis Eng Manuf 19(8):1245–1251

32. Liu S, Lin M (2019) Thermal-mechanical coupling analysis and experimental study on CNC machine tool feed mechanism. Int J Precis Eng Manuf 20(6):993–1006
33. Yang F, Jin S, Li Z (2017) A modification of DMVs based state space model of variation propagation for multistage machining processes. Assem Autom 37(4):381–390
34. Qazani M, Pedrammehr S, Nategh M (2018) An investigation on the motion error of machine tools' hexapod table. Int J Precis Eng Manuf 19(4):463–471
35. Liu T, Li Z, Jin S, Chen W (2018) A variation analysis method for linkage mechanism with consideration of joint clearance and deformation. Int J Precis Eng Manuf 19(10):1495–1506
36. Du S, Yao X, Huang D, Wang M (2015) Three-dimensional variation propagation modeling for multistage turning process of rotary workpieces. Comput Ind Eng 82:41–53
37. Lee K, Lee J, Yang S (2018) Optimal on-machine measurement of position-independent geometric errors for rotary axes in five-axis machines with a universal head. Int J Precis Eng Manuf 19(4):545–551
38. Echerfaoui Y, El Ouafi A, Chebak A (2018) Experimental investigation of dynamic errors in coordinate measuring machines for high speed measurement. Int J Precis Eng Manuf 19(8):1115–1124
39. Ahn H, Kang H, Ghim Y, Yang H (2019) Touch probe tip compensation using a novel transformation algorithm for coordinate measurements of curved surfaces. Int J Precis Eng Manuf 20(2):193–199
40. Wang J, Wang Q, Li H (2019) The method of geometric error measurement of NC machine tool based on the principle of space vector's direction measurement. Int J Precis Eng Manuf 20(4):511–524
41. Wang K, Yin Y, Du S, Xi L, Xia T (2017) State space modeling of multi-scale variation propagation in machining process using matrix model. In: 2017 IEEE international conference on industrial engineering and engineering management, pp 770–774
42. Yang F, Jin S, Li Z (2016) A comprehensive study of linear variation propagation modeling methods for multistage machining processes. Int J Adv Manuf Technol 90(5–8):2139–2151
43. Zeng W, Rao Y (2019) Modeling of assembly deviation with considering the actual working conditions. Int J Precis Eng Manuf 20(5):791–803
44. Chen H, Jin S, Li Z, Lai X (2014) A comprehensive study of three dimensional tolerance analysis methods. Comput Aided Des 53(5):1–13
45. Desrochers A, Ghie W, Laperrière L (2003) Application of a unified Jacobian-Torsor model for tolerance analysis. J Comput Inf Sci Eng 3(1):2–14
46. Chen H, Jin S, Li Z, Lai X (2015) A modified method of the unified Jacobian-Torsor model for tolerance analysis and allocation. Int J Precis Eng Manuf 16(8):1789–1800
47. Wang H, Huang L, Yao C, Kou M, Wang W, Huang B et al (2015) Integrated analysis method of thin-walled turbine blade precise machining. Int J Precis Eng Manuf 16(5):1011–1019
48. Zuo X, Li B, Yang J, Jiang X (2013) Application of the Jacobian-torsor theory into error propagation analysis for machining processes. Int J Adv Manuf Technol 69(5–8):1557–1568
49. Kamali Nejad M, Vignat F, Desrochers A, Villeneuve F (2010) 3D Simulation of manufacturing defects for tolerance analysis. J Comput Inf Sci Eng 10(2):1–13
50. Desrochers A, Clement A (1994) A dimensioning and tolerancing assistance model for CAD/CAM systems. Int J Adv Manuf Technol 9(6):352–361
51. Teissandier D, Couetard Y, Gerard A (1999) A computer aided tolerancing model: proportioned assembly clearance volume. Comput Aided Des 31(13):805–817
52. Ding Y, Jin J, Ceglarek D, Shi J (2005) Process-oriented tolerancing for multi-station assembly systems. IIE Trans 37(6):493–508
53. Abellán-Nebot JV, Liu J (2013) Variation propagation modelling for multi-station machining processes with fixtures based on locating surfaces. Int J Prod Res 51(15):4667–4681
54. Wang K, Du S, Xi L (2020) Three-dimensional tolerance analysis modelling of variation propagation in multi-stage machining processes for general shape workpieces. Int J Precis Eng Manuf 21(1):31–44
55. Wang K, Li G, Du S, Xi L, Xia T (2020) State space modelling of variation propagation in multistage machining processes for variable stiffness structure workpieces. Int J Prod Res 1–20

56. Chen Y, Ding Y, Jin J, Ceglarek D (2006) Integration of process-oriented tolerancing and maintenance planning in design of multistation manufacturing processes. IEEE Trans Autom Sci Eng 3(4):440–453
57. Cai W (2008) A new tolerance modeling and analysis methodology through a two-step linearization with applications in automotive body assembly. J Manuf Syst 27(1):26–35
58. Andolfatto L, Thiebaut F, Lartigue C, Douilly M (2014) Quality- and cost-driven assembly technique selection and geometrical tolerance allocation for mechanical structure assembly. J Manuf Syst 33(1):103–115
59. Tsutsumi D, Gyulai D, Kovacs A, Tipary B, Ueno Y, Nonaka Y, Fujita K (2020) Joint optimization of product tolerance design, process plan, and production plan in high-precision multi-product assembly. J Manuf Syst 54:336–347
60. Jin S, Zheng C, Yu K, Lai X (2010) Tolerance design optimization on cost–quality trade-off using the Shapley value method. J Manuf Syst 29(4):142–150
61. Zheng C, Jin S, Lai X, Yu K (2011) Assembly tolerance allocation using a coalitional game method. Eng Optim 43(7):763–778
62. Awad MI, Hassan NM (2018) Joint decisions of machining process parameters setting and lot-size determination with environmental and quality cost consideration. J Manuf Syst 46:79–92
63. Huang Q, Shi J (2003) Simultaneous tolerance synthesis through variation propagation modeling of multistage manufacturing processes. In: 31st North American manufacturing research conference, pp 515–522
64. Liu J, Shi J, Hu SJ (2009) Quality-assured setup planning based on the stream-of-variation model for multi-stage machining processes. IIE Trans 41(4):323–334
65. Chen S, Wang H, Huang Q (2009) Multistage machining process design and optimization using error equivalence method. In: ASME International manufacturing science & engineering conference
66. Zhang X, Wang H, Chen S, Huang Q (2018) A novel two-stage optimization approach to machining process selection using error equivalence method. J Manuf Syst 49:36–45
67. Celen M, Djurdjanovic D (2020) Integrated maintenance and operations decision making with imperfect degradation state observations. J Manuf Syst 55:302–316
68. Alimian M, Ghezavati V, Tavakkoli-Moghaddam R (2020) New integration of preventive maintenance and production planning with cell formation and group scheduling for dynamic cellular manufacturing systems. J Manuf Syst 56:341–358
69. Hu J, Jiang Z, Liao H (2020) Joint optimization of job scheduling and maintenance planning for a two-machine flow shop considering job-dependent operating condition. J Manuf Syst 57:231–241
70. Abellán-Nebot JV, Liu J, Subirón FR (2013) Process-oriented tolerancing using the extended stream of variation model. Comput Ind 64(5):485–498
71. Ding Y, Ceglarek D, Shi J (2000) Modeling and diagnosis of multistage manufacturing processes: part I state space model. In: Japan-USA symposium of flexible automation
72. Bloch HP, Geitner FK (1985) Practical machinery management for process plants. Vol 3: machinery component maintenance and repair. Gulf Professional Publishing, Burlington, MA, USA/Oxford, UK
73. Archard JF (1953) Contact and rubbing of flat surfaces. J Appl Phys 24(8):981–988
74. Jin J, Chen Y (2001) Quality and reliability information integration for design evaluation of fixture system reliability. Qual Reliab Eng Int 17(5):355–372
75. ASM-Handbook (2003) ASM Metals Handbook, vol. 16: machining. ASM International
76. Wang P, Liang M (2005) Simultaneously solving process selection, machining parameter optimization, and tolerance design problems: a bi-criterion approach. J Manuf Sci Eng-Trans ASME 127(3):533–544
77. Pignatiello JJ (1993) Strategies for robust multiresponse quality engineering. IIE Trans 25(3):5–15
78. Prisco U, Giorleo G (2002) Overview of current CAT systems. Integr Comput Aided Eng 9(4):373–387

79. Roy U, Li B (1999) Representation and interpretation of geometric tolerances for polyhedral objects. II.: size, orientation and position tolerances. Comput Aided Des 31(4):273–285
80. Shao Y, Wang K, Du S, Xi L (2018) High definition metrology enabled three dimensional discontinuous surface filtering by extended tetrolet transform. J Manuf Syst 49:75–92
81. Shao Y, Yin Y, Du S, Xia T, Xi L (2018) Leakage monitoring in static sealing interface based on three dimensional surface topography indicator. J Manuf Sci Eng 140(10):101003
82. Shao Y, Yin Y, Du S, Xi L (2019) A surface connectivity-based approach for leakage channel prediction in static sealing interface. J Tribol 141(6):062201
83. Lin Y, Chang P, Yeng L, Huan S (2019) Bi-objective optimization for a multistate job-shop production network using NSGA-II and TOPSIS. J Manuf Syst 52:43–54
84. Zimmermann HJ (1978) Fuzzy programming and linear programming with several objective functions. Fuzzy Sets Syst 1(1):45–55
85. Jones D, Tamiz M (2010) Practical goal programming. Springer, New York
86. Wang P, Liang M (2005) An integrated approach to tolerance synthesis, process selection and machining parameter optimization problems. Int J Prod Res 43(11):2237–2262
87. Sakawa M, Kubota R (2000) Fuzzy programming for multiobjective job shop scheduling with fuzzy processing time and fuzzy duedate through genetic algorithms. Eur J Oper Res 120(2):393–407

The manufacturer's authorised representative in the EU is Springer
Nature Customer Service Centre GmbH, Europaplatz 3, 69115 Heidelberg,
Germany. If you have any concerns regarding our products, please
contact ProductSafety@springernature.com

Printed and bound by CPI Group (UK) Ltd, Croydon, CR0 4YY
01/10/2026
02225935-0014